GEOSTATISTICS BANFF 2004

Volume 1

Quantitative Geology and Geostatistics

VOLUME 14/1

The titles published in this series are listed at the end of this volume.

GEOSTATISTICS BANFF 2004

Volume 1

Edited by

OY LEUANGTHONG

*University of Alberta,
Edmonton, Canada*

and

CLAYTON V. DEUTSCH

*University of Alberta,
Edmonton, Canada*

A C.I.P. Catalogue record for this book is available from the Library of Congress.

ISBN-10 1-4020-3515-2 (HB) (indivisible set)
ISBN-13 978-1-4020-3515-9 (HB) (indivisible set)
ISBN-10 1-4020-3610-8 (e-book)
ISBN-13 978-1-4020-3610-1 (e-book)

Published by Springer,
P.O. Box 17, 3300 AA Dordrecht, The Netherlands.

www.springeronline.com

Printed on acid-free paper

All Rights Reserved
© 2005 Springer
No part of this work may be reproduced, stored in a retrieval system, or transmitted
in any form or by any means, electronic, mechanical, photocopying, microfilming, recording
or otherwise, without written permission from the Publisher, with the exception
of any material supplied specifically for the purpose of being entered
and executed on a computer system, for exclusive use by the purchaser of the work.

Printed in the Netherlands.

TABLE OF CONTENTS

VOLUME 1

Foreword ... xv

Acknowledgements ... xvi

List of Participants ... xvii

PLENARY

Accounting for Geological Boundaries in Geostatical Modeling of Multiple Rock Types
P. Larrondo and C.V. Deutsch ... 3

Data Integration using the Probability Perturbation Method
J. Caers ... 13

Spectral Component Geologic Modeling: A New Technology for Integrating Seismic Information at the Correct Scale
T. Yao, C. Calvert, G. Bishop, T. Jones, Y. Ma and L. Foreman ... 23

Joint Simulations, Optimal Drillhole Spacing and the Role of the Stockpile
A. Boucher, R. Dimitrakopoulos and J. A. Vargas-Guzmán ... 35

Theory of the Cloud Transform for Applications
O. Kolbjørnsen and P. Abrahamsen ... 45

Probability Field Simulation: A Retrospective
R. M. Srivastava and R. Froidevaux ... 55

Sequential Spatial Simulation Using Latin Hypercube Sampling
P. C. Kyriakidis ... 65

Field Scale Stochastic Modeling of Fracture Networks: Combining Pattern Statistics with Geomechanical Criteria for Fracture Growth
X. Liu and S. Srinivasan ... 75

Direct Geostatistical Simulation on Unstructured Grids
J. Manchuk, O. Leuangthong and C. V. Deutsch ... 85

Directional Metropolis: Hastings Updates for Posteriors with Nonlinear Likelihoods
H. Tjelmeland and J. Eidsvik ... 95

Detection of Local Anomalies in High Resolution Hyperspectral Imagery using Geostatistical Filtering and Local Spatial Statistics
P. Goovaerts ... 105

Covariance Models with Spectral Additive Components
D. Marcotte and M. Powojowski ... 115

A Statistical Technique for Modelling Non-stationary Spatial Processes
J. Stephenson, C. Holmes, K. Gallagher and A. Pintore ... 125

Conditioning Event-based Fluvial Models
M. J. Pyrcz and C. V. Deutsch ... 135

3D Geological Modelling and Uncertainty: The Potential-field Method
C. Aug, J.-P. Chilès, G. Courrioux and C. Lajaunie ... 145

Accounting for Non-stationarity and Interactions in Object Simulation for Reservoir Heterogeneity Characterization
D. Allard, R. Froidevaux and P. Biver ... 155

Estimating the Trace Length Distribution of Fractures from Line Sampling Data
C. Lantuéjoul, H. Beucher, J.P. Chilès, C. Lajaunie and
H. Wackernagel ... 165

On Some Controversial Issues of Geostatistical Simulation
L. Y. Hu and M. Le. Ravalec-Dupin ... 175

On the Automatic Inference and Modelling of a Set of Indicator Covariances and Cross-Covariances
E. Pardo-Igúzquiza and P. A. Dowd ... 185

On Simplifications of Cokriging
J. Rivoirard ... 195

Efficient Simulation Techniques for Uncertainty Quantification on Continuous Variables: A Process Preserving the Bounds, including Uncertainty on Data, Uncertainty on Average, and Local Uncertainty
P. Biver ... 205

Higher Order Models using Entropy, Markov Random Fields and Sequential Simulation
C. Daly ... 215

Beyond Covariance: The Advent of Multiple-Point Geostatistics
A. G. Journel ... 225

Non-Stationary Multiple-point Geostatistical Models
S. Strebelle and T. Zhang ... 235

A Workflow for Multiple-point Geostatistical Simulation
Y. Liu, A. Harding, R. Gilbert and A. Journel ... 245

A Multiple-scale, Pattern-based Approach to Sequential Simulation
G. B. Arpat and J. Caers ... 255

Sequential Conditional Simulation Using Classification of
Local Training Patterns
T. Zhang, P. Switzer and A. Journel ... 265

A Parallel Scheme for Multi-scale Data Integration
O. I. Tureyen and J. Caers ... 275

Stochastic Modeling of Natural Fractured Media: A Review
J. P. Chilès ... 285

Geostatistical Simulation of Fracture Networks
R. M. Srivastava, P. Frykman and M. Jensen ... 295

The Promises and Pitfalls of Direct Simulation
O. Leuangthong ... 305

Sample Optimization and Confidence Assessment of Marine Diamond Deposits using Cox Simulations
G. Brown, C. Lantuéjoul and C. Prins ... 315

Inverse Conditional Simulation of Relative Permeabilities
J. Gómez-Hernández and C. Guardiola-Albert ... 325

Quantifiable Mineral Resource Classification: A Logical Approach
C. Dohm ... 333

MINING

Geostatistics in Resource/Reserve Estimation: A Survey of the Canadian Mining Industry Practice
M. Dagbert ... 345

Integration of Conventional and Downhole Geophysical Data in Metalliferous Mines
M. Kay, R. Dimitrakopoulos and P. Fullagar ... 351

The Kriging Oxymoron: A Conditionally Unbiased and Accurate Predictor (2nd Edition)
E. Isaaks 363

Post Processing of SK Estimators and Simulations for Assessment of Recoverable Resources and Reserves for South African Gold Mines
D. G. Krige, W. Assibey-Bonsu and L. Tolmay 375

The Practice of Sequential Gaussian Simulation
M. Nowak and G. Verly 387

Spatial Characterization of Limestone and Marl Quality in a Quarry for Cement Manufacturing
J. A. Almeida, M. Rocha and A. Teixeira 399

A Non-linear GPR Tomographic Inversion Algorithm Based on Iterated Cokriging and Conditional Simulations
E. Gloaguen, D. Marcotte and M. Chouteau 409

Application of Conditional Simulation to Quantify Uncertainty and to Classify a Diamond Deflation Deposit
S. Duggan and R. Dimitrakopoulos 419

Geostatistical Simulation Techniques Applied to Kimberlite Orebodies and Risk Assessment of Sampling Strategies
J. Deraisme and D. Farrow 429

Modelling 3D Grade Distributions on the Tarkwa Paleoplacer Gold Deposit, Ghana, Africa
T. R. Fisher, K. Dagdelen and A. K. Turner 439

Conditional Simulation of Grade in a Multi-element Massive Sulphide Deposit
N. A. Schofield 449

The Ultimate Test – Using Production Reality. A Gold Case Study
P. Blackney, C. Standing and V. Snowden 457

Ore-Thickness and Nickel Grade Resource Confidence at the Koniambo Nickel Laterite (A Conditional Simulation Voyage of Discovery)
M. Murphy, H. Parker, A. Ross and M.-A. Audet 469

Mineral Resource Classification Through Conditional Simulation
T.M. Wawruch and J.F. Betzhold 479

Geostatistical Investigation of Elemental Enrichment in Hydrothermal Mineral Deposits
A. R. Samal, R. H. Fifarek, and R. R. Sengupta 491

Valuing a Mine as a Portfolio of European Call Options: The Effect of Geological Uncertainty and Implications for Strategic Planning
E. Henry, D. Marcotte, and M. Samis 501

Classification of Mining Reserves using Direct Sequential Simulation
A. Soares 511

Using Unfolding to Obtain Improved Estimates in the Murrin Murrin Nickel-cobalt Laterite Deposit in Western Australia
M. Murphy, L. Bloom, and U. Mueller 523

Measures of Uncertainty for Resource Classification
L. Eduardo de Souza, J.F.C.L. Costa, and J.C. Koppe 529

Incorporating Uncertainty in Coal Seam Depth Determination via Seismic Reflection and Geostatistics
V. C. Koppe, F. Gambin, J.F.C.L. Costa, J.C. Koppe,
G. Fallon, and N. Davies 537

Implementation Aspects of Sequential Simulation
S. Zanon and O. Leuangthong 543

VOLUME 2

PETROLEUM

Early Uncertainty Assessment: Application to a Hydrocarbon Reservoir Appraisal
G. Caumon and A. G. Journel 551

Reservoir Facies Modelling: New Advances in MPS
A. Harding, S. Strebelle, M. Levy, J. Thorne, D. Xie, S. Leigh, R. Preece
and R. Scamman 559

Fitting the Boolean Parameters in a Non-stationary Case
H. Beucher, M. B. García-Morales and F. Geffroy 569

Fine Scale Rock Properties: Towards the Spatial Modeling of Regionalized Probability Distribution Functions
M. García, D. Allard, D. Foulon and S. Delisle 579

A Combined Geostatistical and Source Model to Predict Superpermeability from Flowmeter Data: Application to the Ghawar Field
J. Voelker and J. Caers 591

Combining Methods for Subsurface Prediction
P. Abrahamsen 601

Process-based Reservoir Modelling in the Example of Meandering Channel
I. Cojan, O. Fouché, S. Lopéz and J. Rivoirard ... 611

Multiple Point Geostatistics: Optimal Template Selection and Implementation in Multi-Threaded Computational Environments
A. E. Barrera, J. Ni, and S. Srinivasan ... 621

Direct Assessment of Uncertainty using Stochastic Flow Simulation
J. Y. Leung and S. Srinivasan ... 631

Preservation of Multiple Point Structure when Conditioning by Kriging
W. Ren, L. B. Cunha and C. V. Deutsch ... 643

Stochastic Modeling of the Rhine-Meuse Delta using Mutiple-point Geostatistics
A. Maharaja ... 653

Stochastic Simulation of Undiscovered Petroleum Accumulations
Z. Chen, K. G. Osadetz and H. Gao ... 661

Prediction of Spatial Patterns of Saturation Time-lapse from Time-lapse Seismic
J. Wu, T. Mukerji and A. Journel ... 671

Statistical Scale-up: Concepts and Application to Reservoir Flow Simulation Practice
L. W. Lake, S. Srinivasan and A. John ... 681

Coupling Sequential-Self Calibration and Genetic Algorithms to Integrate Production Data in Geostatistical Reservoir Modeling
X. H. Wen, T. Yu, and S. Lee ... 691

High Resolution Geostatistics on Coarse Unstructured Flow Grids
G. Caumon, O. Grosse, and J. L. Mallet ... 703

Assessment of Uncertainty in Reservoir Production Forecasts Using Upscaled Flow Models
O.P. Lødøen, H. Omre, L.J. Durlofsky, and Y. Chen ... 713

Sensitivity of Oil Production to Petrophysical Heterogeneities
A. Skorstad, O. Kolbjørnsen, B. Fjellvoll, J. Howell, T. Manzocchi, and J.N. Carter ... 723

Scaling Relations and Sampling Volume for Seismic Data
P. Frykman, O.V. Vejbæk, and R. Rasmussen ... 731

Hidden Markov Chains for Identifying Geological Features from Seismic Data
J. Eidsvik, E. Gonzalez, and T. Mukerji ... 737

Evaluation of Stochastic Earth Model Workflows, Vertical Up-Scaling and Areal Up-scaling Using Data from the Eunice Monument South Unit (New Mexico) and the LL-652 Central Fault Block (Venezuela) Reservoirs
W.S. Meddaugh ... 743

Application of Design of Experiments to Expedite Probabilistic Assessment of Reservoir Hydrocarbon Volumes (OOIP)
W. S. Meddaugh, S.D. Griest, and S.J. Gross ... 751

Stochastic Reservoir Model for the First Eocene Reservoir, Wafra Field, Partitioned Neutral Zone (PNZ)
W.S. Meddaugh, D. Dull, S.D. Griest, P. Montgomery, and G. McNaboe ... 757

Multiple-point Statistics to Generate Pore Space Images
H. Okabe and M.J. Blunt ... 763

Local Updating of Reservoir Properties for Production Data Integration
L. Zhang, L.B. Cunha, and C.V. Deutsch ... 769

ENVIRONMENTAL

Comparison of Model Based Geostatistical Methods in Ecology: Application to Fin Whale Spatial Distribution in Northwestern Mediterranean Sea
P. Monestiez, L. Dubroca, E. Bonnin, J. P. Durbec and C. Guinet ... 777

Simulation-based Assessment of a Geostatistical Approach for Estimation and Mapping of the Risk of Cancer
P. Goovaerts ... 787

Air Quality Assessment using Stochastic Simulation and Neural Networks
A. Russo, C. Nunes, A. Bio, J. Pereira and A. Soares ... 797

Mapping Land Cover Changes with Landsat Imagery and Spatio-temporal Geostatistics
A. Boucher, K. Seto and A. Journel ... 809

Spherical Wavelets and their Application to Meteorological Data
H. S. Oh ... 819

Multivariate Geostatistical Mapping of Atmospheric Deposition in France
O. Jaquet, L. Croisé, E. Ulrich and P. Duplat ... 833

Geostatistical and Fourier Analysis Applied to Cross-hole Tomography Seismic Data
J. M. Carvalho and A. T. Cavalheiro ... 843

The Importance of De-clustering and Uncertainty in Climate Data: A Case Study of West African Sahel Rainfall
A. Chappell and M. Ekström ... 853

S. GeMS: The Stanford Geostatistical Modelling Software: A Tool for New Algorithms Development
N. Remy ... 865

Evaluating Techniques for Multivariate Classification of Non-collocated Spatial Data
S. A. McKenna ... 873

Travel Time Simulation of Radionuclides in a 200 m Deep Heterogeneous Clay Formation Locally Disturbed by Excavation
M. Huysmans, A. Berckmans and A. Dassargues ... 883

Geostatistics and Sequential Data Assimilation
H. Wackernagel and L. Bertino ... 893

Spatial Properties of Seasonal Rainfall in Southeast England
M. Ekström and A. Chappell ... 899

Geostatistic Indicators of Waterway Quality for Nutrients
C. Bernard-Michel and C. de Fouquet ... 907

Application of Geostatistical Simulation to Enhance Satellite Image Products
C. A. Hlavka and J. L. Dungan ... 913

Geostatistical Noise Filtering of Geophysical Images: Application to Unexploded Ordnance (UXO) Sites
H. Saito, T. C. Coburn, and S. A. McKenna ... 921

THEORY & SELECTED TOPICS

Modeling Skewness in Spatial Data Analyis without Data Transformation
P. Naveau and D. Allard ... 929

Gradual Deformation of Boolean Simulations
M. Le Ravalec-Dupin and L. Y. Hu ... 939

When can Shape and Scale Parameters of a 3D Variogram be Estimated?
P. Dahle, O. Kolbjørnsen and P. Abrahamsen ... 949

Comparison of Stochastic Simulation Algorithms in Mapping Spaces of Uncertainty of Non-linear Transfer Functions
S. E. Qureshi and R. Dimitrakopoulos ... 959

Integrating Multiple-point Statistics into Sequential Simulation Algorithms
J. M. Ortiz and X. Emery ... 969

Post-processing of Multiple-point Geostatistical Models to Improve Reproduction of Training Patterns
S. Strebelle and N. Remy ... 979

Improving the Efficiency of the Sequential Simulation Algorithm using Latin Hypercube Sampling
G. G. Pilger, J. F. C. L. Costa and J. C. Koppe ... 989

Exact Conditioning to Linear Constraints in Kriging and Simulation
J. Gómez-Hernández, R. Froidevaux and P. Biver ... 999

A Dimension-reduction Approach for Spectral Tempering Using Empirical Orthogonal Functions
A. Pintore and C. C. Holmes ... 1007

A New Model for Incorporating Spatial Association and Singularity in Interpolation of Exploratory Data
Q. Cheng ... 1017

Lognormal Kriging: Bias Adjustment and Kriging Variances
N. Cressie and M. Pavlíková ... 1027

Evaluating Information Redundancy Through the Tau Model
S. Krishnan, A. Boucher and A. Journel ... 1037

An Information Content Measure Using Multiple-point Statistics
Y. Liu ... 1047

Internal Consistency and Inference of Change-of-support Isofactorial Models
X. Emery and J.M. Ortiz ... 1057

History Matching Under Geological Control: Application to a North Sea Reservoir
B. T. Hoffman and J. Caers ... 1067

A Direct Sequential Simulation Approach to Streamline-Based History Matching
J. Caers, H. Gross and A. R. Kovscek ... 1077

Mapping Annual Nitrogen Dioxide Concentrations in Urban Areas
D. Gallois, C. de Fouquet, G. Le Loc'h, L. Malherbe and G. Cárdenas ... 1087

A Step by Step Guide to Bi-Gaussian Disjunctive Kriging
J. Ortiz, B. Oz, and C.V. Deutsch ... 1097

Assessing the Power of Zones of Abrupt Change Detection Test
E. Gabriel and D. Allard 1103

Experimental Study of Multiple-support, Multiple-point Dependence and its Modeling
S. Krishnan 1109

Validation of First-order Stochastic Theories on Reactive Solute Transport in Highly Stratified Aquifers
D. Fernàndez-Garcia and J.J. Gómez-Hernández 1117

Geostatistical and Fourier Analysis Approaches in Mapping an Archeological Site
A.T. Cavalheiro and J.M. Carvalho 1123

BATGAM© Geostatistical Software Based on GSLIB
B. Buxton, A. Pate, and M. Morara 1131

INDEX 1137

FOREWORD

The return of the congress to North America after 20 years of absence could not have been in a more ideal location. The beauty of Banff and the many offerings of the Rocky Mountains was the perfect background for a week of interesting and innovative discussions on the past, present and future of geostatistics.

The congress was well attended with approximately 200 delegates from 19 countries across six continents. There was a broad spectrum of students and seasoned geostatisticians who shared their knowledge in many areas of study including mining, petroleum, and environmental applications. You will find 119 papers in this two volume set. All papers were presented at the congress and have been peer-reviewed. They are grouped by the different sessions that were held in Banff and are in the order of presentation.

These papers provide a permanent record of different theoretical perspectives from the last four years. Not all of these ideas will stand the test of time and practice; however, their originality will endure. The practical applications in these proceedings provide nuggets of wisdom to those struggling to apply geostatistics in the best possible way. Students and practitioners will be digging through these papers for many years to come.

Oy Leuangthong
Clayton V. Deutsch

ACKNOWLEDGMENTS

We would like to thank the industry sponsors who contributed generously to the overall success and quality of the congress:

>De Beers Canada
>Earth Decision Sciences
>Maptek Chile Ltda.
>Mira Geoscience
>Nexen Inc.
>Petro-Canada
>Placer Dome Inc.
>Statios LLC
>Total

We would also like to thank all the chair people, reviewers, volunteers, contributors and delegates who helped to make the congress a success. Special thanks to R. Mohan Srivastava for his help with the artwork. We thank the International Geostatistics committee for their comments and suggestions:

>Winfred Assibey-Bonsu (South Africa)
>Jef Caers (USA)
>Clayton Deutsch (Canada)
>Roussos Dimitrakopoulos (Australia)
>Peter Dowd (Australia)
>J. Jaime Gómez-Hernández (Spain)
>Christian Lantuéjoul (France)

The staff and students of the Centre for Computational Geostatistics at the University of Alberta gave generously of their time to all aspects of organizing the congress. Amanda Potts (research administrator) provided essential administrative support. Sandra Correa, Paula Larrondo, Jason McLennan, Chad Neufeld, Weishan Ren, Stefan Zanon, and Linan Zhang (graduate students) provided invaluable assistance before and during the congress.

LIST OF PARTICIPANTS

Abrahamsen, Petter, Norwegian Computing Center, Po Box 114 Blindern, Gaustadalleen 23, OSLO, NO0314, NORWAY

Alapetite, Julien, Earth Decision Science, 3 Route De Grenoble, MOIRANS, FR38430, FRANCE

Almeida, Jose, CIGA, Faculdade Ciencias Tecnologia, Universidade Nova Lisboa, MONTE DA CAPARICA, PORTUGAL

Arpat, Burc, Stanford University, Department of Petroleum Engineering, STANFORD, CA 94305, USA

Assibey- Bonsu, Winfred, Gold Fields Mining Services Ltd, 24 St. Andrews Rd, ZA 2193, SOUTH AFRICA

Assis Carlos, Luis, Anglo American Brasil Ltda, 502- Setor Santa Genoveva, Av. Interlandia, GOIANIA 74672-360, BRAZIL

Aug, Christophe, Centre de Geostatistique, 35 Rue Saint- Honore, Ecole Des Mines De Paris, FONTAINEBLEAU F77305, FRANCE

Bankes, Paul, Teck Cominco Limited, 600-200 Burrard Street, VANCOUVER, BC V6C 3L9, CANADA

Barbour, Russell, EPH/Yale School of Medicine, 60 College St. Rm 600, P. O. Box 208034, NEW HAVEN, CT 06520, USA

Barrera, Alvaro, University of Texas at Austin, 4900 E Oltorf St Apt 538, AUSTIN, TX 78741, USA

Bellentani, Giuseppe, ENI E&P, Via Emilia 1-5 Pal., Uff.-5013 E, SAN DONATO MILANESE, 20097, ITALY

Berckmans, Arne, NIRAS, Kunstlaan 14, BRUSSSELS, 1210, BELGUIM

Bernard- Michel, Caroline, Ecole des Mines de Paris, 35 Rue Saint Honore FONTAINEBLEAU, F77305, FRANCE

Bertoli, Olivier, Quantitative Geoscience Pty Lt, Po Box 1304, FREMANTLE, 6959, AUSTRALIA

Beucher, Helene, Ecole des Mines de Paris, 35 Rue Saint Honore, Centre De Geostatistique, FONTAINEBLEAU, F77305, FRANCE

Biver, Pierre, TOTAL SA, C S T J F, Ba 3112, Avenue Larribau, PAU, 64000, FRANCE

Blackney, Paul, Snowden Mining In. Consultants, P O Box 77, West Parth, 87 Colin St, W A 6872, PERTH, 6005, AUSTRALIA

Blaha, Petr, Holcim Group Support Ltd., Holderbank, HOLDERBANK, 5113, CHILE

Boucher, Alexandre, Stanford University, 151 Calderon Ave, Apt 232, MOUNTAIN VIEW, CA 94041, USA

Bourgault, Gilles, Computer Modelling Group Ltd, 3512 33 Street N W, Office #200, CALGARY, AB T2L 2A6, CANADA

Brega, Fausto, ENI S.P.A- E& P Division, Via Emilia, 1, San Donato Milanese, MILAN, 20097, ITALY

Brown, Gavin, De Beers Consolidated Mines, PO Box 350, CAPE TOWN, 8000, SOUTH AFRICA

Brown, Steven, Nexen Inc, 801-7th Avenue S W, CALGARY, AB T2P 3P7. CANADA

Buecker, Christian, R W E Dea A G, Ueberseering 40, HAMBURG, 22297, GERMANY

Bush, David, De Beers, P Bag X01 Southdale, Cnr. Crownwood Rd & Diamond Dr, JOHANNESBURG, ZA2135, SOUTH AFRICA

Buxton, Bruce, Battelle Memorial Institute, 505 King Avenue, COLUMBUS, OH, 43201, USA

Caers, Jef, Stanford University, Petroleum Engineering, 367 Panama St, STANFORD, CA, 94305, USA

Carvalho, Jorge, Dep. Minas, Faculty Of Engineering, PORTO, 4200-465, PORTUGAL

Castro, Scarlet, Stanford University, 361 Green Earth Sciences Bldg., 367 Panama Street, STANFORD, CA, 94305, USA

Caumon, Guillaume, Stanford University, 367 Panama St., Petroleum Engineering Dept, STANFORD, CA, 94305, USA

Chappell, Adrian, University of Salford, Environment & Life Sciences, Peel Building, The Crescent, SALFORD, M5 4WT, UK

Chelak, Robert, Roxar Canada, 1200 815-8 Th Ave SW, CALGARY, AB, T2P 3P2, CANADA

Chen, Zhuoheng, Geological Survey of Canada, 3303-33rd Street N.W., CALGARY, AB, T2L 2A7, CANADA

Cheng, Qiuming, York University, 4700 Keele Street, NORTH YORK, ON, M3J 1P3, CANADA

Chiles, Jean- Paul, Ecole des Mines de Paris, 35 Rue Saint Honore, Centre De Geostatistique, FONTAINEBLEAU, FR77305, FRANCE

Coburn, Timothy, Abilene Christian University, Acu Box 29315, ABILENE, TX, 79699, USA

Cornah, Alastair, University of Exeter, Camborne School Of Mines, Trevenson Rd, CORNWALL, TR15 3SE, UK

Correa Montero, Sandra, University of Alberta, Department of Civil & Environmental. Engineering, 3-133 Markin/CNRL Natural Resources Engineering Facility, EDMONTON, AB, T6G 2W2, CANADA

Costa, Joao Felipe, UFRGS, Av. Osvaldo Aranha 99/504, PORTO ALEGRE, 90035190, BRAZIL

Costa, Marcelo, BRAZIL

Da Silva, Emidio, University A. Neto, Faculdade De Engenharis-d De, Eng. Minas Av. 21 De Janeiro, LUANDA, 1756, ANGOLA

Dagbert, Michel, Geostat Systems Int. Inc, 10 Blvd Seigneurie E, Suite 203, BLAINVILLE, QC, J7C 3V5, CANADA

Dahle, Pal, Norwegian Computing Center, P O Box 114, Blindern, Gaustadalleen 23, OSLO, NO0314, NORWAY

Daly, Colin, Roxar Ltd, Pinnacle House, 17-25 Hartfield Rd, WIMBLEDON, SW19 3SE, UK

De Visser, Jan, RSG Global

Della Rossa, Ernesto, ENI S.P.A, Eni E& P- Apsi, Via Emilia 1, San Donato Milanese, MILANO, 20097, ITALY

Deutsch, Clayton, University of Alberta, Department. of Civil & Environmental Engineering, 3-133 Markin/CNRL Natural Resources Engineering Facility, EDMONTON, AB, T6G 2W2, CANADA

Dimitrakopoulos, Roussos, University of Queensland, Wh Bryan Mining Geo. Res. Cen, Richards Building, BRISBANE, 4072, AUSTRALIA

Dohm, Christina, Anglo Operations Ltd., P O Box 61587, MARSHALLTOWN, 2107, JOHANNESBURG, 2000, SOUTH AFRICA

Dose, Thies, RWE Dea AG, Ueberseering 40, Hamburg, HAMBURG, D22297, GERMANY

Dowd, Peter, Faculty of Engineering, Computer and Mathematical Sciences, University of Adelaide, ADELAIDE, SA 5005, AUSTRALIA

Dube, Pascal, Cameco, 243 Beckett Green, SASKATOON, SK S7N 4W1, CANADA

Duggan, Sean, De Beers Consolidated Mines, 117 Hertzog Boulevard, CAPE TOWN, 8001, SOUTH AFRICA

Dungan, Jennifer, NASA Ames Research Center

Eidsvik, Jo, Statoil, Ark Ebbels V 10, Statoil Research Center, TRONDHEIM, NO7032, NORWAY

Emery, Xavier, University of Chile, Dept. Of Mining Engineering, Avenida Tupper 2069, SANTIAGO, 8320000, CHILE

Faechner, Ty, Assiniboine Community College, 1430 Victoria Ave East, BRANDON, MB, R7A 2A9, CANADA

Fisher, Thomas, Colorado School of Mines, 1211-6th Street, GOLDEN, CO, 80403, USA

Francois Bongarcon, Dominique, Agoratek International, 345 Stanford Center Pmb# 432, PALO ALTO, CA, 94304, USA

Froidevaux, Roland, FSS Consultants SA, 9, Rue Boissonnas, GENEVA, 1256, SWITZERLAND

Frykman, Peter, Geological Suvery of Denmark, G E U S, Oster Voldgade 10, COPENHAGEN, 1350, DENMARK

Gabriel, Edith, INRA, Domaine St Paul, Site Agropic, AVIGNON, 84914, FRANCE

Gallagher Jr., Joseph, ConocoPhillips, Reservoir Sciences, 269 Geoscience Bldg, BARTLESVILLE, OK, 74004, USA

Garcia, Michel, FSS International, 1956, Avenue Roger Salengro, CHAVILLE, FR92370, FRANCE

LIST OF PARTICIPANTS

Garner, David, Conoco Phillips Canada Ltd., P.o. Box 130, 401- 9th Ave S W, CALGARY, AB, T2P 2H7, CANADA

Glacken, Ian, Snowden Mining In. Consultants, P O Box 77 West Perth, 87 Colin St, WA 6872, PERTH, 6005, AUSTRALIA

Gloaguen, Erwan, Ecole Polytechnique, C. P. 6079 Succ. Centre-ville, MONTREAL, QC, H3C 3A7, CANADA

Gomez- Hernandez, Jaime, Univ. Politecnica de Valencia, Escuela De Ing. De Caminos, Camino De Vera S/n, VALENCIA, EP46022, SPAIN

Gonzalez, Eric, Maptek South America, 5 Norte 112, Vina Del Mar, VINA DEL MAR, 2520180, CHILE

Goovaerts, Pierre, BioMedware, 516 N State St, ANN ARBOR, MI, 48104, USA

Gray, James, Teck Cominco Ltd, 600-200 Burrard St, VANCOUVER, BC, V6C 3L9, CANADA

Grills, John Andrew, De Beers Consolidated Mines, 117 Hertzog Boulevard, PO Box 350, CAPE TOWN, 8000, SOUTH AFRICA

Gringarten, Emmanuel, Earth Decision Sciences, 11011 Richmond Ave, Suite 350, HOUSTON, TX, 77042, USA

Guenard, Cindy, Talisman Energy Inc., Suite 3400, 888-3rd Street S.W., CALGARY, AB, T2P 5C5, CANADA

Guibal, Daniel, SRK Consulting, P. O. Box 943 West Perth, 1064 Hay Street, WEST PERTH, 6005, AUSTRALIA

Harding, Andrew, Chevron Texaco, 6001 Bollinger Canyon Road, DANVILLE, CA, 94506, USA

Hauge, Ragnar, Norwegian Computing Center, P. O. Box 114 Blindern, OSLO, 0314, NORWAY

Hayes, Sean, Talisman Energy Inc., Suite 3400, 888-3rd Street S W, CALGARY, AB, T2P 5C5, CANADA

Henry, Emmanuel, AMEC/Ecole Polytechnique, De Montreal, Suite 700, 2020 Winston Park Dr, OAKVILLE, ON, L6H 6X7, CANADA

Hlavka, Christine, NASA, N A S A/ A M E S Res. Center, Mail Stop 242-4, MOFFETT FIELD, CA, 94035, USA

Hoffman, Todd, Stanford University, 367 Panama Street, Green Earth Sciences Bldg., STANFORD, CA 94305, USA

Hu, Lin Ying, I F P, 1 4 A. De Bois Preau, RUEIL- MALMAISON, 92870, FRANCE

Huysmans, Marijke, University of Leuven, Redingenstraat 16, LEUVEN, 3000, BELGIUM

Isaaks, Edward, Isaaks & Co., 1042 Wilmington Way, REDWOOD CITY, CA, 94062, USA

Jackson, Scott, Quantitative Geoscience Pty Lt, P O Box 1304, FREMANTLE, 6959, AUSTRALIA

Jaquet, Olivier, Colenco Power Engineering Ltd, Taefernstrasse 26, BADEN, 5405, CHILE

John, Abraham, University of Texas at Austin, Cpe 2.502, Department of Petroleum & Geosystems, AUSTIN, TX, 78712, USA

Journel, Andre, Stanford University, Petroleum Engineering Department., STANFORD, CA, 94305, USA

Jutras, Marc, Placer Dome Inc, P. O. Box 49330 Bentall Sa., Suite 1600-1055 Dunsmir St., VANCOUVER, BC, V7X 1P1, CANADA

Kashib, Tarun, EnCana Corporation, 150, 9th Avenue S.W, .P O Box 2850, CALGARY, AB, T2P 2S5, CANADA

Keech, Christopher, Placer Dome Inc, P O Box 49330 Bentall Station, 1600-1055 Dunsmuir St, VANCOUVER, BC, V7X 1P1, CANADA

Khan, Dan, 1601, 8708 - 106th Street, EDMONTON, AB, T6E 4J5, CANADA

Kolbjornsen, Odd, Norwegian Computing Center, P.O. Box 114 Blindern, OSLO, 0314 NORWAY

Koppe, Jair, Univ. of Rio Grande do Sul, Av Cavalhada 5205 Casa 32, PORTO ALEGRE, 91.751-831, BRASIL

Krige, Daniel G, Po Box 121, FLORIDA HILLS, 1716, SOUTH AFRICA

Krishnan, Sunderrajan, Stanford University, #353 Green Earth Sciences, Geological & Environmental Sci, STANFORD, CA, 94305, USA

Kyriakidis, Phaedon, University of California, Dept. of Geography, Ellison Hall 5710, SANTA BARABRA, CA, 93106, USA

LIST OF PARTICIPANTS

Larrondo, Paula, University of Alberta, Department of Civil & Environmental Engineering, 3-133 Markin/CNRL Natural Resources Engineering Facility, EDMONTON, AB, T6G 2W2, CANADA

Le Loc`h, Gaelle, Ecole des Mines de Paris, 35 Rue Saint Honore, Centre De Geostatistique, FONTAINEBLEAU, F77305, FRANCE

Le Ravalec, Mickaela, I F P, 1 4 A. Bois Preau, Rueil-malmaison., 92852, FRANCE

Leuangthong, Oy, University of Alberta, Department of Civil & Environmental Engineering, 3-133 Markin/CNRL Natural Resources Engineering Facility, EDMONTON, AB, T6G 2W2, CANADA

Lewis, Richard, Placer Dome Asia Pacific Ltd., G P O Box 465, Brisbane, 4001, 90 Alison Road, RANWICK, AU 2031, AUSTRALIA

Liu, Yuhong, ExxonMobil Upstream Research, P. O. Box 2189, HOUSTON, TX, 77252, USA

Lodoen, Ole Petter, Norwegian Univ. of Science & T, Dept. Of Mathematical Sciences, TRONDHEIM, N-7491, NORWAY

Maharaja, Amisha, Stanford University, 51 Dudley Lane, Apt 424, STANFORD, CA, 94305, USA

Marcotte, Denis, Ecole Polytechnique, C. P. 6079 Succ. Centre-ville, SAINT-BRUNO, QC, J3V 4J7, CANADA

Mattison, Blair, Petro- Canada, 150-6th Ave. SW, CALGARY, AB, T2P 3E3, CANADA

Maunula, Tim, Wardrop Engineering, 2042 Merchants Gate, OAKVILLE, ON, L6M 2Z8, CANADA

Mc Kenna, Sean, Sandia National Laboratories, P O Box 5800 M S 0735, ALBUQUERQUE, NM, 87185, USA

Mc Lennan, Jason, University of Alberta, Dept. Of Civil & Environmental. Engineering., 3-133 Markin/CNRL Natural Resources Engineering Facility, EDMONTON, AB, T6G 2W2, CANADA

Meddaugh, William, ChevronTexaco Energy Tech. Co., 4800 Fournace Place, P O Box 430, BELLAIRE, TX, 77401, USA

Merchan, Sergio, Encana, 421-7th Avenue SW, Calgary, AB, T2P 4K9, CANADA

Monestiez, Pascal, INRA, Domaine St Paul Site Agropic, Avignon, AVIGNON, 84914, FRANCE

Murphy, Mark, Snowden Mining Ind. Consultant, P O Box 77, West Perth,, 87 Colin St, WA, 6872, PERTH, 6005, AUSTRALIA

Myers, Donald, University of Arizona, Dept. Of Mathematics, 617 North Santa Rita, TUCSON, AZ, 85721, USA

Naveau, Philippe, University of Colorado, Applied Mathematics Dept., 526 U C B, BOULDER, CO, 80309, USA

Nel, Stefanus, De Beers Consolidated Mines, Private Bag 1, Southdale, Gauteng, ZA 2135, SOUTH AFRICA

Neufeld, Chad, University of Alberta, Department. Of Civil & Environment Engineering, 3-133 Markin/CNRL Natural Resources Engineering Facility, EDMONTON, AB, T6G 2W2, CANADA

Nicholas, Grant D., De Beers, The Dtc, Mendip Court, Bath Rd, South Horrington, WELLS, BA5 3DG, UK

Norrena, Karl, Nexen Canada Ltd, 801-7th Ave SW, CALGARY, AB, T2P 3P7, CANADA

Norris, Brett, Paramount Energy Trust, 500, 630-4th Ave SW, CALGARY, AB, T2P 0L9, CANADA

Nowak, Marek, Nowak Consultants Inc., 1307 Brunette Ave, COQUITLAM, BC, V3K 1G6, CANADA

Okabe, Hiroshi, Imperial College London, Japan Oil, Gas & Metals Nat.co, 1-2-2 Hamada, Mihama-ku, CHIBA- SHI, 261-0025, JAPAN

Ortiz Cabrera, Julian, Universidad de Chile, Av. Tupper 2069, SANTIAGO, 837-0451, CHILE

Osburn, William, St John River Water Management, 4049 Reid St., P O Box 1429, PALATKA, FL, 32178, USA

Parker, Harry, AMEC Inc, 19083 Santa Maria Ave, CASTRO VALLEY, CA, 94546, USA

Perron, Gervais, Mira Geoscience Ltd, 310 Victoria Avenue, Suite 309, WESTMOUNT, QC, H3Z 2M9, CANADA

Pilger, Gustavo, Federal Univ of Rio Grande do, Av. Osvaldo Aranha 99/504, PORTO ALEGRE, 90035-190, BRAZIL

Pintore, Alexandre, Imperial College of London, 11 Walton Well Road, OXFORD, OX2 6ED, UK

LIST OF PARTICIPANTS

Pontiggia, Marco, ENI E&P, Via Emilia 1-5 Pal., Uff.-5013 E, SAN DONATO MILANESE, 20097, ITALY

Porjesz, Robert, CGG, Casa Bote B, Casa 332, LECHERIA, VENEZULA

Potts, Amanda, University of Alberta, Department of Civil & Environmental Engineering, 3-133 Markin/CNRL Natural Resources Engineering Facility, EDMONTON, AB, T6G 2W2, CANADA

Prins, Chris, The DTC, Mendip Court, Bath Road, Minrad, Wells, SOMERSET, BA5 3DG, UK

Pyrcz, Michael, University of Alberta, Dept. Of Civil & Environmental Engineering, 3-133 Markin/CNRL Natural Resources Engineering Facility, EDMONTON, AB, T6G 2W2, CANADA

Remy, Nicolas, Standford University, G E S Department, Braun Hall, STANFORD, CA, 94305, USA

Ren, Weishan, University of Alberta, Dept. Of Civil & Environmental Engineering, 3-133 Markin/CNRL Natural Resources Engineering Facility, EDMONTON, AB, T6G 2W2, CANADA

Riddell, Marla, EnCana, #21, 18-20 Hillcrest Road, LONDON, W5 1HJ, UK

Rivoirard, Jacques, Ecole des Mines de Paris, Centre De Gostatistique, 35 Rue Saint-Honore, FONTAINEBLEAU, 77210, FRANCE

Russo, Ana, CMRP/ IST, Av. Rovisco Pais, 1, Lisboa, 1049-001. PORTUGAL

Saito, Hirotaka, Sandia National Lab., P O Box 5800 M S 0735, ALBUQUERQUE, NM, 87185, USA

Saldanha, Paulo, 4 Dom Thomas Murphy St., WY, BRAZIL

Samal, Abani, Southern Illinois University, 2000 Evergreen Terrace Dr. W., Apt 07, CARBONDALE, IL, 62901, USA

Savoie, Luc, Candian Natural Resources Ltd, 2500, 855-2nd St SW, CALGARY, AB, T2P 4J8, CANADA

Schirmer, Patrice, TOTAL, 2 Place Coupole, La Defense, PARIS, .092604, FRANCE

Schnetzler, Emmanuel, Statios, 1345 Rhode Island Street, SAN FRANCISCO, CA, 94107, USA

Schofield, Neil, Hellman & Schofield Pty. Ltd, P. O. Box 599, Beecroft, Nsw, 2119, Suite 6, 3 Trelawney St, EASTWOOD, 2122, AUSTRALIA

Scott, Anthony, Placer Dome, P O Box 49330 Bentall Station, Suite 1600-1055 Dunsmuir St., VANCOUVER, BC, V6G 3J3, CANADA

Seibel, Gordon, AngloGold, P O Box 191, VICTOR, CO, 80860, USA

Shi, Genbao, Landmark Graphics, Two Barton Skyway, 1601 S. Mopac Expressway, AUSTIN, TX, 78746, USA

Skorstad, Arnem Norwegian Computing Center, P O Box 114 Blindern, Gaustadalleen 23, OSLO, NO0314, NORWAY

Soares, Amilcar, CMRP/ IST, Av. Rovisco Pais, 1, LISBOA, 1049-001, PORTUGAL

Soulie, Michel, Ecole Polytechnique, P. O. Box 6079, Station Centre-ville, MONTREAL, QC, H3C 3A7, CANADA

Srinivasan, Sanjay, University of Texas at Austin, Petroleum & Geosystems Eng., 1 University Station C0300, AUSTIN, TX, 78712, USA

Srivastava, Mohan, FSS Canada, 42 Morton Road, TORONTO, ON, M4C 4N8, CANADA

Stavropoulos, Achilles, Canadian Natural Resources Ltd, 900, 311-6th Ave. SW, CALGARY, AB, T2P 3H2, CANADA

Stephenson, John, Imperial College, Dept. Of Earth Science & Eng, Royal School Of Mines, LONDON, SW7 2AZ, UK

Strebelle, Sebastien, ChevronTexaco ETC, 6001 Bollinger Canyon Rd,, Room D1200, SAN RAMON, CA, 94583, USA

Suzuki, Satomi, Stanford University, 367 Panama St., STANFORD, CA, 94305, USA

Syversveen, Anne Randi, Norwegian Computing Center, P O Box 411 Blindern, OSLO, 0314, NORWAY

Thurston, Malcolm, De Beers Canada, 65 Overlea Blvd, Suite 400, TORONTO, ON, M4H 1P1, CANADA

Tjelmeland, Haakon, Norwegian University, Of Science & Technology, Dept. Of Mathematical Sciences, TRONDHEIM, 7491, NORWAY

Tolmay, Leon, Gold Fields, 4 Cedar Avenue, WESTONARIA, 1779, SOUTH AFRICA

Toscano, Claudio, ENI S.P.A -E & P Division, Via Emilia 1, San Donato Milanese, MILAN, 1-20097, ITALY

LIST OF PARTICIPANTS

Tran, Thomas, Chevron Texaco, 1546 China Grade Loop, Room A7, BAKERSFIELD, CA, 93308, USA

Tureyen, Omer Inanc, Stanford University, 367 Panama St., 065 Green Earth Sciences Bldg., STANFORD, CA, 94305, USA

Vann, John, Quantitative Geoscience Pty Lt, P O Box 1304, FREMANTLE, 6959, AUSTRALIA

Vasquez, Christina, XU Power, 3010 State St. #309, DALLAS, TX, 75204, USA

Verly, Georges, Placer Dome, P. O. Box 49330, Bentall Station, VANCOUVER, BC, V7X 1P1, CANADA

Voelker, Joe, Stanford University, P O Box 12864, STANFORD, CA, 94309, USA

Wackernagel, Hans, Ecole des Mines De Paris, 35 Rue Saint Honore, Centre De Geostatistique, FONTAINEBLEAU, F-77305, FRANCE

Wagner, Jayne, De Beers, Private Bag X01, Southdale, 2135 Cornerstone Bldg., JOHANNESBURG, 2135, SOUTH AFRICA

Wain, Anthony, Talisman Energy Inc., 888-3rd Street SW, Suite 3400, CALGARY, AB, T2P 5C5, CANADA

Walls, Elizabeth, Petro-Canada, 208, 930 18th Ave SW, CALGARY, AB, T2T 0H1, CANADA

Watson, Michael, Husky Energy, Box 6525, Station D, 707 8th Ave. S W, CALGARY, AB, T2P 3G7, CANADA

Wawruch, Tomasz, Anglo American Chile, Mailbox 16178, Correo 9, 291 Pedro De Valdivia Ave, SANTIAGO, 6640594, CHILE

Wen, Xian-Huan, Chevron Texaco ETC, 6001 Bollinger Canyon Road, D2092, SAN RAMON, CA, 94583, USA

Wu, Jianbing, SCRF, Stanford University, Petroleum Engineering Dept, 367 Panama Street, STANFORD, CA, 94305, USA

Yao, Tingting, ExxonMobil., Upstream Research Company, P. O. Box 2189, S W 508, HOUSTON, TX, 77252, USA

Yarus, Jeffrey, Quantitative Geosciences, LLP, 2900 Wilcrest Suite 370, HOUSTON, TX, 77042, USA

Zanon, Stefan, University of Alberta, Dept. Of Civil & Environmental Engineering, 3-133 Markin/CNRL Natural Resources Engineering Facility, EDMONTON, AB, T6G 2W2, CANADA

Zhang, Linan, University of Alberta, Dept. Of Civil & Environmental Engineering, 3-133 Markin/CNRL Natural Resources Engineering Facility, EDMONTON, AB, T6G 2W2, CANADA

Zhang, Tuanfeng, Stanford University, 113 A, E. V, STANFORD, CA, 94305, USA

PLENARY

ACCOUNTING FOR GEOLOGICAL BOUNDARIES IN GEOSTATISTICAL MODELING OF MULTIPLE ROCK TYPES

PAULA LARRONDO and CLAYTON V. DEUTSCH
Department of Civil & Environmental Engineering, 220 CEB, University of Alberta, Canada, T6G 2G7

Abstract. Geostatistical simulation makes strong assumptions of stationarity in the mean and the variance over the domain of interest. Unfortunately, geological nature usually does not reflect this assumption and we are forced to subdivide our model area into stationary regions that have some common geological controls and similar statistical properties. This paper addresses the significant complexity introduced by boundaries. Boundaries are often soft, that is, samples near boundaries influence multiple rock types.

We propose a new technique that accounts for stationary variables within rock types and additional non-stationary factors near boundaries. The technique involves the following distinct phases: (i) identification of the rock types and boundary zones based on geological modeling and the timing of different geological events, (ii) optimization for the stationary statistical parameters of each rock type and the non-stationary mean, variance and covariance in the boundary zones, and (iii) estimation and simulation using non-stationary cokriging. The resulting technique can be thought of as non-stationary cokriging in presence of geological boundaries.

The theoretical framework and notation for this new technique is developed. Implementation details are discussed and resolved with a number of synthetic examples. A real case study demonstrates the utility of the technique for practical application.

1 Introduction

The most common geostatistical techniques, such as kriging and Gaussian/indicator simulation, are based on strong assumptions of stationarity of the estimation domains. In particular, they are based in a second order stationary hypothesis, that is, the mean, variance and covariance remain constant across the entire domain and they do not depend on the location of the support points but only in the distance between them.

Once estimation domains have been selected, the nature of the boundaries between them must be established. Domain boundaries are often referred to as either 'hard' or 'soft'. Hard boundaries are found when an abrupt change in the mean or variance occurs at the contact between two domains. Hard boundaries do not permit the interpolation or extrapolation across domains. Contacts where the variable changes transitionally across the boundary are referred as soft boundaries. Soft domain boundaries allow selected data from either side of a boundary to be used in the estimation of each domain.

It is rather common that soft boundaries are characterized by a non-stationary behavior of the variable of interest in the proximities of the boundary, that is, the mean, variance or covariance are no longer constant within a zone of influence of one rock type into the other, and their values depends on the location relative to the boundary. An example is the increased frequency of fractures towards a boundary between geological domains of structural nature. Faults or brittle zones are examples of this transition. The fractures may cause the average to increase close to the boundary. The increase in the presence of fractures will often lead to an increase in the variance closer to the boundary.

Although soft boundaries are found in several types of geological settings due to the transitional nature of the geological mechanisms, conventional estimation usually treats the boundaries between geological units as hard boundaries. This is primarily due to the limitations of current estimation and simulation procedures. We will show that non-stationary features in the vicinity of a boundary can be parameterized into a local model of coregionalization. With a legitimate spatial model, estimation of grades can be performed using a form of non-stationary cokriging. This proposal provides an appealing alternative when complex contacts between different rock types exist. We develop the methodology in the context of mining geostatistics, but it is widely applicable in many different settings.

2 Theoretical Background

The technique involves the identification of stationary variables within each rock type and additional non-stationary components near boundaries for the mean, variance and covariance. For a geological model with K rock types or estimation domains, there are a maximum of $K(K-1)/2$ boundary zones to be defined. Then, the continuous random function $Z(\mathbf{u})$ that represents the distribution of the property of interest can be decomposed into K stationary random variables $Z_k(\mathbf{u})$ $k=1,...,K$ and a maximum of $K(K-1)/2$ non-stationary boundary variables $Z_{kp}(\mathbf{u})$, with $k,p=1,...K$ and $Z_{kp}(\mathbf{u})=Z_{pk}(\mathbf{u})$ (Figure 1). By definition, the non-stationary variable will take values only for locations within the maximum distance of influence of rock type k into rock type p.

The maximum distance of influence orthogonal to the boundary of rock type k into rock type p is denoted $dmax_{kp}$. A boundary zone is defined by two distances: $dmax_{kp}$ and $dmax_{pk}$, since there is no requirement that the regions on each side of the boundary are symmetric, that is, $dmax_{kp} \neq dmax_{pk}$. The modeler using all geological information available and his expertise should establish these distances.

When more than two rock types converge at a boundary, two or more rock types may influence the boundary zone in the adjacent domain. In this case, precedence or ordering rules should determine the dominant boundary zone. Although the behavior of a property near a boundary could be explained by the overlapping of different geological controls, the task of identifying the individuals effects of each rock type and their interactions can be quite difficult. Geological properties are not usually additive and therefore the response of a combination of different rock types is complex. Only one non-stationary factor will be considered at each location. The modeler should put together the precedence rules based on the geology of the deposit. The relative timing of intrusion, deposition or mineralisation events, geochemistry response of the protolith to an alteration or mineralisation process could be used to resolve timing and important variables. If the geological data does not provide sufficient information to establish a geological order of events, a neutral arrangement can be chosen. In this case, the precedent rock type p at a location will be the one to which the distance to the boundary is the minimum over all surrounding rock types.

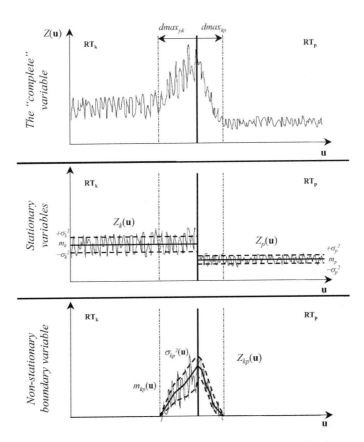

Figure 1: Decomposition of a one-dimensional random function Z(u) in two stationary variables $Z_k(u)$ and $Z_p(u)$, with constant mean and variance, and a non-stationary boundary variable $Z_{kp}(u)$, with a mean and variance that are functions of the distance to the boundary.

STATIONARY AND NON-STATIONARY STATISTICAL PARAMETERS

The mean function of the continuous random function $Z(\mathbf{u})$ for a specific rock type k will be the mean of the stationary variable $Z_k(\mathbf{u})$ plus the mean of any corresponding non-stationary variable $Z_{kp}(\mathbf{u})$. The stationary component of the mean (m_k) is independent of location and is a constant value. The non-stationary component of the mean (m_{kp}) is a function of the distance to the boundary, $d_{pk}(\mathbf{u})$ and takes values different than zero for locations within the boundary zone defined by rock types k and p. The mean of rock type k is:

$$E\{Z(\mathbf{u}_i)\} = \begin{cases} m_k & \text{, if } d_{pk}(\mathbf{u}_i) \geq dmax_{pk} \\ m_k + f(d_{pk}(\mathbf{u}_i)) & \text{, otherwise} \end{cases} \quad \text{where } \mathbf{u}_i \in RT_k$$

where p is the adjacent rock type that shares a boundary with rock type k and $f(\cdot)$ is an arbitrary function that describes the mean as a function of distance to the boundary.

Similarly, the variance of $Z(\mathbf{u})$ for rock type k will be the sum of a constant stationary variance (σ_k^2) due to $Z_k(\mathbf{u})$ and the independent non-stationary variance (σ_{kp}^2) due to $Z_{kp}(\mathbf{u})$. The variance of a random function $Z(\mathbf{u})$ in a rock type k is:

$$E\left\{\left(Z(\mathbf{u}_i) - E\{Z(\mathbf{u}_i)\}\right)^2\right\} = \begin{cases} \sigma_k^2 & \text{, if } d_{pk}(\mathbf{u}_i) \geq dmax_{pk} \\ \sigma_k^2 + g(d_{pk}(\mathbf{u}_i)) & \text{, otherwise} \end{cases} \quad \text{where } \mathbf{u}_i \in RT_k$$

where p is the adjacent rock type that shares a boundary with rock type k and $g(\cdot)$ is an arbitrary function that describes the variance as a function of distance to the boundary.

As with the mean and variance, the covariance structure between two rock types that share a local non-stationary boundary consists of a stationary and a non-stationary component.

$$Cov_Z(\mathbf{u}_i, \mathbf{v}_i) = E\left\{\left(Z(\mathbf{u}_i) - m(\mathbf{u}_i)\right) \cdot \left(Z(\mathbf{v}_i) - m(\mathbf{v}_i)\right)\right\} = Cov_Z^S(\mathbf{h}) + Cov_Z^{NS}(\mathbf{u}_i, \mathbf{v}_i)$$

where $\mathbf{h} = \mathbf{u}_i - \mathbf{v}_i$. Since $Z_k(\mathbf{u})$ and $Z_{kp}(\mathbf{u})$ are independent random variables, the cross terms are zero, therefore the covariance of $Z(\mathbf{u})$ is the sum of the stationary and non-stationary components. The combination of these components corresponds to a local linear model of coregionalization.

The stationary component of the covariance can be calculated and modeled from data pairs within the internal stationary portion of a rock type, that is $\mathbf{u}_i$ and $\mathbf{v}_i$ belong to rock type k, and do not belong to any boundary zone.

To obtain the non-stationary component of the covariance model we will assume that the shape of the spatial correlation of the non-stationary variable $Z_{kp}(\mathbf{u})$ $k,p=1,...,K$ is stationary and that it can be specified by the modeler. Due to the non-stationary nature of variable $Z(\mathbf{u})$ at the boundary zone, this relative stationary spatial model has to by scaled at each point by a non-stationary mean and variance. The relative standardized variogram model for the boundary zone is:

$$\hat{\gamma}_{kp}(\mathbf{u}_i, \mathbf{v}_i) = \frac{1}{2} \cdot E\left\{ \left[\frac{Z(\mathbf{u}_i) - m(\mathbf{u}_i)}{\sigma(\mathbf{u}_i)} - \frac{Z(\mathbf{v}_i) - m(\mathbf{v}_i)}{\sigma(\mathbf{v}_i)} \right]^2 \right\}$$

where $m(\mathbf{u}) = m_{kp}(\mathbf{u}) + m_k$ and $\sigma(\mathbf{u}) = \sigma_{kp}(\mathbf{u}) + \sigma_k$. Expanding and reordering the terms of the squared difference, and since $E\{Z(\mathbf{u}_i)^2\} = \sigma(\mathbf{u}_i)^2 + m(\mathbf{u}_i)$ and $E\{Z(\mathbf{u}_i)\} = m(\mathbf{u}_i)$, the previous expression becomes:

$$\hat{\gamma}_{kp}(\mathbf{u}_i, \mathbf{v}_i) = 1 - \frac{Cov_Z^{NS}(\mathbf{u}_i, \mathbf{v}_i)}{\sigma(\mathbf{u}_i) \cdot \sigma(\mathbf{v}_i)}$$

Reordering the terms and replacing the mean and variance by the sum of their stationary and non-stationary components, we obtain an expression for the non-stationary covariance model:

$$Cov_Z^{NS}(\mathbf{u}_i, \mathbf{v}_i) = E\{Z(\mathbf{u}_i) \cdot Z(\mathbf{v}_i)\} - (m_{kp}(\mathbf{v}_i) + m_k) \cdot (m_{kp}(\mathbf{u}_i) + m_k)$$
$$= (1 - \hat{\gamma}_{kp}(\mathbf{u}_i, \mathbf{v}_i)) \cdot (\sigma_{kp}(\mathbf{u}_i) + \sigma_k) \cdot (\sigma_{kp}(\mathbf{v}_i) + \sigma_k)$$

Currently we assume that the shape, anisotropies and nugget effect of the relative standardized variogram are inputs from the modeler; only the range must be established through an optimization algorithm.

3 Optimization of the Statistical Parameters

We need to find the optimum $f(d_{pk}(\mathbf{u}_i))$, $g(d_{pk}(\mathbf{u}_i))$ and $Cov_Z^{NS}(\mathbf{u}_i, \mathbf{v}_i)$ that fit the distribution of the random variable $Z(\mathbf{u})$ at the boundary zone given the stationary components of mean, variance and covariance, a set of precedence rules and the maximum distances of influence within the rock type model.

We will consider that the non-stationary components of the mean and variance follow a linear function of the distance to the boundary (d_{pk}). In this scenario, the optimization of the parameter m_{kp} and σ_{kp}^2 will be equivalent to optimizing estimates of the intercepts at zero distance to boundary: a_{kp} and b_{kp}, considering $a_{kp} = a_{pk}$ and $b_{kp} = b_{pk}$.

The mean m_{kp} is optimized given that m_k is known from the experimental average of data within rock type k, outside any boundary zone. The objective function is:

$$O_m = \sum_{k=1}^{K}\sum_{p=1}^{P}\sum_{i=1}^{N_{kp}}\left[z(\mathbf{u}_i)-(\hat{m}_k+m_{kp}(\mathbf{u}_i))\right]^2$$

where $z(\mathbf{u}_i)$ is the outcome value at every data location in the boundary zone, N_{kp} is the total number of data in zone $k\text{-}p$, $\hat{m}_k$ is the experimental average of all data in $\mathbf{RT}_k$ and outside any boundary zone, and $m_{kp}(\mathbf{u}_i)$ is the non-stationary mean at location $\mathbf{u}_i$ calculated as:

$$m_{kp}(\mathbf{u}_i) = \begin{cases} \dfrac{(dmax_{kp}-d_{kp}(\mathbf{u}_i))}{dmax_{kp}} \cdot a_{kp} & \text{for } 0 \leq d_{kp}(\mathbf{u}_i) \leq dmax_{kp} \\[2ex] \dfrac{(dmax_{pk}-d_{pk}(\mathbf{u}_i))}{dmax_{pk}} \cdot a_{kp} & \text{for } 0 \leq d_{pk}(\mathbf{u}_i) \leq dmax_{pk} \\[2ex] 0 & \text{for } d_{kp}(\mathbf{u}_i) \geq dmax_{kp} \text{ and } d_{pk}(\mathbf{u}_i) \geq dmax_{pk} \end{cases} \quad (1)$$

The optimization of the mean can be achieved by iteratively modified a_{kp} $\forall k,p$, in a random fashion while accepting all changes in a_{kp} that reduce the objective function. This is a simplified version of the simulated annealing formalism.

The optimum σ_{kp}^2, will be the one that minimizes the following objective function:

$$O_{\sigma^2} = \sum_{k=1}^{K}\sum_{p=1}^{P}\sum_{i=1}^{N_{kp}}\left[r(\mathbf{u}_i)^2 - (\hat{\sigma}_k^2 + \sigma_{kp}^2(\mathbf{u}_i))\right]^2$$

where $r(\mathbf{u}_i)$ is the residual value at every location in the boundary zone. $\hat{\sigma}_k^2$ is the experimental variance of all data within the stationary region of rock type k and $\sigma_{kp}^2(\mathbf{u}_i)$ is the non-stationary variance at location $\mathbf{u}_i$ calculated from a linear expression for the intercept b_{kp} similar to Equation 1.

Figure 2 shows the stationary and non-stationary mean and variance for a 1D synthetic example. The optimum intercepts a_{kp} and b_{kp} are in agreement with the reference.

To find the optimum covariance model we minimize the following objective function:

$$O_{Cov} = \sum_{i=1}^{N}\left[\hat{C}(z(\mathbf{u}_i),z(\mathbf{v}_i)) - C_{MOD}(z(\mathbf{u}_i),z(\mathbf{v}_i))\right]^2$$

where $\hat{C}$ denote the experimental covariance of the pair located at $\mathbf{u}_i$ and $\mathbf{v}_i$, which is just the multiplication of the two residual values: $r(\mathbf{u}_i) \cdot r(\mathbf{v}_i)$, and C_{MOD} the modeled boundary covariance, corresponding to the sum of the stationary and non-stationary component.

Finding the optimum covariance model of a boundary zone is equivalent to optimizing the range of the relative standardized variogram scaled by the non-stationary standard deviation. The range is iteratively modified by a random amount until the difference between the experimental and modeled covariance is minimized. For this 1D example, the optimum range of the non-stationary covariance structure (Figure 3) is 6.4 meters, acceptably similar to the 10 meters range of the variogram used to obtain the reference.

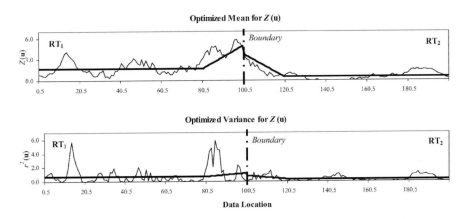

Figure 2: 1D example stationary and optimized non-stationary mean and variance.

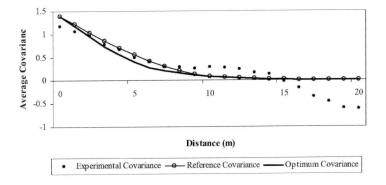

Figure 3: Optimum non-stationary covariance of 1D example (solid line), experimental covariance from pairs within the boundary zone (dots) and original covariance of the non-stationary component used to build the synthetic dataset (line/dots).

4 Estimation in presence of local non-stationary boundaries

The basic linear regression equation for non-stationary simple cokriging is:

$$z^*(\mathbf{u}) - m(\mathbf{u}) = \sum_{\alpha=1}^{n} \lambda_\alpha(\mathbf{u}) \cdot \left[z(\mathbf{u}_\alpha) - m(\mathbf{u}_\alpha)\right]$$

where $z^*(\mathbf{u})$ is the estimate at unsampled location $\mathbf{u}$, $m(\mathbf{u})$ is the stationary plus the non-stationary mean value at location $\mathbf{u}$, $\lambda_\alpha(\mathbf{u})$ is the weight assigned to datum $z(\mathbf{u}_\alpha)$, n is the number of close data to the location $\mathbf{u}$ being estimated, and $m(\mathbf{u}_\alpha)$ are the n stationary plus the non-stationary mean values at the data locations.

To find the optimal weights $\lambda_\alpha(\mathbf{u})$, $\alpha=1,\ldots,n$ the kriging system must be solved:

$$\sum_{\beta=1}^{n} \lambda_\beta(\mathbf{u}) \cdot Cov(\mathbf{u}_\alpha, \mathbf{u}_\beta) = Cov(\mathbf{u}, \mathbf{u}_\alpha) \quad \text{with } \alpha, \beta = 1, \ldots, n$$

where $\lambda_\alpha(\mathbf{u})$, $\alpha=1,\ldots,n$ are the simple kriging weights, $Cov(\mathbf{u}_\alpha, \mathbf{u}_\beta)$, $\alpha,\beta=1,\ldots,n$ correspond to the data-to-data covariances, and $Cov(\mathbf{u}, \mathbf{u}_\alpha)$, $\alpha=1,\ldots,n$ are the data-to-unknown location covariances. In the presence of local non-stationary boundaries, the terms of the data covariance matrix and the vector of data-to-estimate covariances are obtained combining the stationary and non-stationary covariance model components. If both locations are in the same rock type and both are in the same boundary zone, the covariance is the stationary plus the non-stationary covariances; otherwise, it is only the stationary component. If they are in different rock types and both samples are in the same boundary zone the covariance is the non-stationary component only. The covariance is zero in all other cases.

For the 1D example, the kriging estimates reproduce well the reference, using as conditioning data one of four grid nodes of the reference (Figure 4).

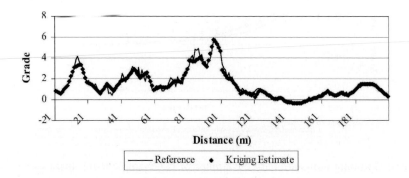

Figure 4: Reference versus kriging estimates, 1D example.

5 Application

The rock type model of a porphyry copper deposit in Northern Chile was used to create a reference image with simulated grades (Figure 5). This reference image was sampled in a 100x100 meters grid. The geological model has five rock types and six non-stationary soft boundaries.

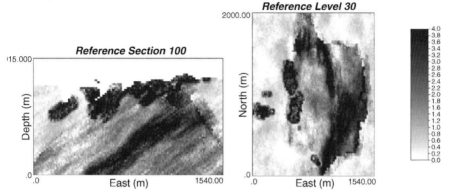

Figure 5: Section and level maps of the reference used for the 3D application.

The reference intercepts for the non-stationary mean and variance are well reproduced by the optimization subroutines for all boundary zones, as well as the optimum ranges compared to the range used in the transformed unconditional simulation.

The correlation between the estimates and the reference value is around 0.8 for each boundary zone. The reference stationary means of each rock type is reproduced almost exactly by the kriging estimation. The variance of the estimates is lower than the reference, which is expected since kriging has a smoothing effect. The non-stationary behavior of the mean is also very well reproduced by the proposed non-stationary kriging as shown in Figure 6. Although the variance of the estimates in the boundary zone is lower than the reference, as expected, the increasing trend toward the boundary is well reproduced.

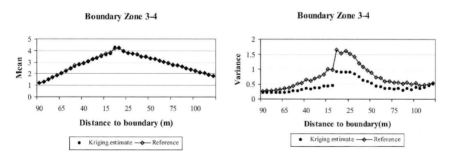

Figure 6: Mean and variance of the kriging estimates versus the reference image at the one of non-stationary boundary zone. Each point corresponds to the average/variance of all grid nodes within a 5 meters interval of the distance to the boundary.

Cross validation results show that the data are reliably estimated both in the stationary and the non-stationary regions. In particular, for all data within the non-stationary regions, if compared with ordinary kriging using a typical soft boundary approach, the proposed methodology shows a higher coefficient of correlation (Figure 7).

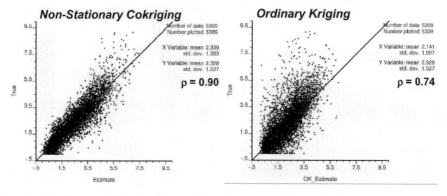

Figure 7: Cross validation comparison between the proposed methodology, non-stationary cokriging, and ordinary kriging with soft boundaries.

6 Conclusions

This new technique provides a theoretically robust methodology to handle non-stationary soft boundaries. The non-stationary features of the mean, variance and covariance are parameterized into a legitimate local model of coregionalization. Through this spatial model a non-stationary form of cokriging accounts for the changes in mean and variance at the vicinity of boundaries. The kriging estimates reproduce the non-stationary behavior of the conditioning data at the geological contacts, and it also reproduces the stationary means of each rock type in the model. A decrease in the global variance is due to the smoothing effect of kriging.

By construction, the kriging variance also has a non-stationary component. Since the kriging variance is the missing variability that is reintroduced in simulation, its implementation in the presence of local non-stationary boundaries will be delicate and is part of the future work.

Acknowledgements

We would like to acknowledge the industry sponsors of the Centre for Computational Geostatistics at the University of Alberta for supporting this research. We also thank Canada's National Sciences and Engineering Research Council (NSERC) for funding.

DATA INTEGRATION USING THE PROBABILITY PERTURBATION METHOD

JEF CAERS
Department of Petroleum Engineering,
Stanford University, Stanford, CA 94305-2220, USA

Abstract. A new method, termed probability perturbation, is developed for solving non-linear inverse problem under a prior model constraint. The method proposed takes a different route from the traditional Bayesian inverse models that rely on prior and likelihood distribution for stating, then sampling, from a posterior distribution. Instead, the probability perturbation method relies on so-called pre-posterior distributions, which state the distribution of the unknown parameter set given each individual data type (linear or non-linear). Sampling consists of perturbing the probability models used to generate the model realization, by which a chain of realizations is created that converge to match any type of data. The probability perturbations are such that the underlying spatial structure (prior model) of the stochastic algorithm is maintained through all perturbations. A simple example illustrates the approach.

1 Introduction

Conditioning stochastic simulations is of the utmost importance in many applications of geostatistics. Most of the current algorithms can condition to data that are linear or pseudo-linear (i.e. linearized using transformations) and of a single-point nature, by which it is understood that there is a linear relationship between data and the unknown taken one at the time. For example, the technique of sequential simulation, either under Gaussian or non-Gaussian assumptions, can be conditioned to hard data, (pseudo-linear) block average data or soft data, the latter through some form of (linear) co-kriging.

Many applications of geostatistics call for the inclusion of non-linear and multiple-point data. The relationship between data and unknown is provided through a complex multi-dimensional transfer function, also termed a forward model. This function often is modelled numerically through a partial differential equation (or its numerical implementation) such as in aquifer models, pollution models, ecological models and for models of flow in oil & gas reservoirs. Integrating this type of data into stochastic simulation calls for an iterative solution (trial-and-error) of an often ill-posed inverse problem. Sampling solutions such a Markov chain Monte Carlo within a Bayesian framework have been proposed (Mosegaard and Tarantola, 1995; Omre and Tjelmeland, 1997) but are often prohibitive in terms of CPU when the forward model is expensive to compute.

In this paper a new and practical approach within the context of Bayesian inverse modelling is presented. The method allows to condition stochastic simulations to virtually any type of non-linear data. The principle of this method is simple: by perturbing the probabilities models used to generate the model realization, a chain of realizations is created that converge to match any type of data. It is shown that the probability perturbations are such that the underlying spatial structure of the stochastic algorithm is maintained through all perturbations. The probability perturbations can be parameterized by a single parameter or by multiple parameters in order to provide enough flexibility to match large models with a possible large set of non-linear data.

2 Bayesian inverse modelling

Inverse modeling consists of finding a set of model parameters **m** given some data **d**. In the Earth Sciences the model parameters are often unknown material or rock properties located on a 3D grid, e.g. unknown soil type or unknown petrophysical properties in the subsurface. Most inverse problems are underdetermined, meaning that a joint distribution of model parameters is possible given the data. In this paper, we will divide the data into two sets: (1) $\mathbf{d}_1$, or "easy data" which have a simple linear or pseudo-linear relationship with the model parameters, and (2) $\mathbf{d}_2$ or "difficult data" which exhibit a multi-point, non-linear relationship with **m**. For the data $\mathbf{d}_1$, many fast and robust direct sampling methods exist for sampling the distribution of possible model realizations **m**. To condition to data $\mathbf{d}_2$, iterative sampling is required. The posterior distribution from which these samples are drawn is, in a Bayesian context, decomposed into a likelihood and prior distribution

$$f(\mathbf{m}|\mathbf{d}_1,\mathbf{d}_2) = \frac{f(\mathbf{d}_1,\mathbf{d}_2|\mathbf{m})f(\mathbf{m})}{f(\mathbf{d}_1,\mathbf{d}_2)} \simeq \frac{f(\mathbf{d}_1|\mathbf{m})f(\mathbf{d}_2|\mathbf{m})f(\mathbf{m})}{f(\mathbf{d}_1,\mathbf{d}_2)} \qquad (1)$$

where the likelihood $f(\mathbf{d}_1, \mathbf{d}_2|\mathbf{m})$ is further decomposed into $f(\mathbf{d}_1 | \mathbf{m})$ and $f(\mathbf{d}_2| \mathbf{m})$ under the assumption of conditional independence. This assumption makes inference of the likelihood feasible. The assumption of conditional independence is difficult to verify yet may have considerable consequence to the model definition (model for the posterior distribution).

The prior density $f(\mathbf{m})$ describes the dependency between the model parameters. In a spatial context such dependency refers to the spatial structure of **m**. The likelihood density $f(\mathbf{d}|\mathbf{m})$ models the stochastic relationship between the observed data and each particular model **m** retained. This likelihood would account for model and measurement errors. In the absence of any such errors, the data **d** and model **m** are related through a forward model g

$$\mathbf{d} = g(\mathbf{m})$$

Markov chain Monte Carlo methods encompass a set of iterative sampling techniques for drawing samples from this posterior distribution. Popular sampling methods are rejection sampling and the Metropolis sampler (Metropolis et al., 1953; Besag and

Green, 1993; Mosegaard and Tarantola, 1995; Omre and Tjelmeland, 1997). These samplers avoid specification of $f(\mathbf{d})$ and are iterative in nature in order to obtain a single sample $\mathbf{m}^{(\ell)}$ of $f(\mathbf{m}|\mathbf{d})$. Generating multiple (conditioned to $\mathbf{d}$) samples $\mathbf{m}^{(\ell)}$, $\ell=1,\ldots,L$ in this manner quantifies the uncertainty modeled in $f(\mathbf{m}|\mathbf{d})$.

While theoretically sounds, there are some important practical limitations to this approach. First, obtaining iterative samples are CPU demanding and may take many thousand of evaluations to converge. This is impractical when the forward model g takes a few hours to compute (e.g. flow simulations, solving elastic wave equations). Secondly, for reason of analytical convenience a Gaussian model is often adopted for either likelihood and/or prior distribution. A Gaussian model limits modelling realistic spatial structures on $\mathbf{m}$. Moreover, the assumption of conditional independence in Eq. (1) limits the proper modelling of the full dependence between data $\mathbf{d}_1$ and the data $\mathbf{d}_2$. In this paper we propose a method for dealing with both issues: (1) realistic non-Gaussian prior and (2) alternatives to the conditional independence hypothesis.

3 Methodology

3.1 SAMPLING THE PRIOR

To emphasize non-Gaussianity, the methodology will be developed for binary model parameters, although the method works equally well for multi-category and continuous variables. At each location of a 3D grid an unknown model parameter m_i is modelled through a binary indicator variable

$$I(\mathbf{u}_i) = \begin{cases} 1 & \text{if the "event" occurs at } \mathbf{u}_i \\ 0 & \text{else} \end{cases}$$

where "event" could represent any spatially distributed phenomenon. The model parameters are then given by the set of binary indicators

$$\mathbf{m} = \{I(\mathbf{u}_1), I(\mathbf{u}_2), \ldots, I(\mathbf{u}_N)\}$$

with joint (prior) distribution

$$f(\mathbf{m}) = \text{Prob}\{I(\mathbf{u}_1) = i(\mathbf{u}_1), I(\mathbf{u}_2) = i(\mathbf{u}_2), \ldots, I(\mathbf{u}_N) = i(\mathbf{u}_N)\}$$

In this paper we will use sequential simulation methods to sample from either prior or posterior distribution, by relying on the following decomposition of the joint distribution

$$f(\mathbf{m}) = \text{Prob}\{I(\mathbf{u}_1) = 1\} \times \text{Prob}\{I(\mathbf{u}_2) = 1 | i(\mathbf{u}_1)\} \times \ldots \times \text{Prob}\{I(\mathbf{u}_N) = 1 | i(\mathbf{u}_1), \ldots, i(\mathbf{u}_{N-1})\}$$

Sequential sampling from each of these conditional distribution amounts to sampling from a joint prior distribution. In actual field cases, prior information on $\mathbf{m}$ comes in the form of limited statistics (e.g. a spatial covariance). The type of multi-variate density f always needs to be assumed. In all sequential simulation approaches, except Gaussian simulation, the decision of distribution type is not made on the joint distribution, but on the conditional distributions. An example of such approach is direct sequential simulation (*dssim*, Journel, 1993), where the conditional distribution can be of any type, as long as they have mean and variance provided by a simple kriging system. Another example is *snesim* where the conditional distributions are derived from training images (Strebelle, 2002).

3.2 SAMPLING THE POSTERIOR

To sample from the posterior, a similar sequential decomposition approach is considered. For simplicity, the data $\mathbf{d}_1$ constitute direct observations (hard data) of the model parameters at a set of n spatial locations, but in general could constitute any linear data,

$$\mathbf{d}_1 = \{i(\mathbf{u}_\alpha), \alpha = 1, \ldots, n\}$$

The relationship between the non-linear data and model parameters is modelled through a forward model g

$$\mathbf{d}_2 = g(\mathbf{m}) = g(I(\mathbf{u}_1), I(\mathbf{u}_2), \ldots, I(\mathbf{u}_N))$$

The goal is to draw samples from the joint (posterior) distribution of the model parameters given the two data sets

$$f(\mathbf{m} \mid \mathbf{d}_1, \mathbf{d}_2) = \mathrm{Prob}\{I(\mathbf{u}_1) = i(\mathbf{u}_1), I(\mathbf{u}_2) = i(\mathbf{u}_2), \ldots, I(\mathbf{u}_N) = i(\mathbf{u}_N) \mid \{i(\mathbf{u}_\alpha), \alpha = 1, \ldots, n\}, \mathbf{d}_2\}$$

To make this practically feasible, the following decomposition is used:

$$\begin{aligned} f(\mathbf{m} \mid \mathbf{d}_1, \mathbf{d}_2) = &\mathrm{Prob}\{I(\mathbf{u}_1) = 1 \mid \{i(\mathbf{u}_\alpha), \alpha = 1, \ldots, n\}, \mathbf{d}_2\} \times \\ &\mathrm{Prob}\{I(\mathbf{u}_2) = 1 \mid i(\mathbf{u}_1), \{i(\mathbf{u}_\alpha), \alpha = 1, \ldots, n\}, \mathbf{d}_2\} \times \ldots \times \\ &\mathrm{Prob}\{I(\mathbf{u}_N) = 1 \mid i(\mathbf{u}_1), \ldots, i(\mathbf{u}_{N-1}), \{i(\mathbf{u}_\alpha), \alpha = 1, \ldots, n\}, \mathbf{d}_2\} \end{aligned} \quad (2)$$

Generating a sample of a (not explicitly stated) posterior distribution is equivalent to generating sequential samples from conditional distributions of the type

$$\begin{aligned} &\mathrm{Prob}\{I(\mathbf{u}_j) = 1 \mid i(\mathbf{u}_1), \ldots, i(\mathbf{u}_{j-1}), \{i(\mathbf{u}_\alpha), \alpha = 1, \ldots, n\}, \mathbf{d}_2\} = \mathrm{Prob}(A_j \mid B_j, C) \\ &\text{with } A_j = \{I(\mathbf{u}_j) = 1\}; \ B_j = \{i(\mathbf{u}_1), \ldots, i(\mathbf{u}_{j-1}), \{i(\mathbf{u}_\alpha), \alpha = 1, \ldots, n\}\}; \ C = \mathbf{d}_2 \end{aligned} \quad (3)$$

A simpler notation in terms of 'A' (unknown), 'B' (easy data) and 'C' (difficult data) has been used to make further development clear. To further specify the conditionals in Eq. (3), we propose a decomposition of $\mathrm{Prob}(A_j|B_j,C)$ into two pre-posteriors $\mathrm{Prob}(A_j|B_j)$ and $\mathrm{Prob}(A_j|C)$ using Journel's decomposition (or tau-model, Journel, 2002) of the type

$$\text{Prob}(A_j \mid \mathbf{B}_j, \mathbf{C}) = \frac{1}{1+x} \text{ with } x = b\left(\frac{c}{a}\right)^{\tau} \text{ where:} \quad (4)$$

$$b = \frac{1-\text{Prob}(A_j \mid \mathbf{B}_j)}{\text{Prob}(A_j \mid \mathbf{B}_j)}, \quad c = \frac{1-\text{Prob}(A_j \mid \mathbf{C})}{\text{Prob}(A_j \mid \mathbf{C})}, \quad a = \frac{1-\text{Prob}(A_j)}{\text{Prob}(A_j)}$$

Working with pre-posteriors will lead to an approach that is different from a classical Bayesian inversion which would involve the likelihoods Prob($\mathbf{B}|A_j$) and Prob($\mathbf{C}|A_j$). This difference will lead to a fundamentally different sampling method as well. Stating "pre-posteriors", instead of likelihoods, allows using (non-iterative) sequential simulation, instead of (iterative) McMC.

The τ-value in Eq. (4) allows modeling explicitly the full dependency between the **B**-data and **C**-data. The case when $\tau=1$ is equivalent to an assumption of standardized conditional independence. In the context of sequential simulation, the pre-posterior Prob($A_j|\mathbf{B}_j$) is simply the conditional distribution of the unknown A_j given any previously simulated nodes. The remaining pre-posterior Prob($A_j|\mathbf{C}$) cannot be directly estimated, instead, a new sampling technique termed probability perturbation is introduced.

3.3 PROBABILITY PERTURBATION

Using sequential simulation, a sample realization can be drawn from the prior model, conditioned to the data $\mathbf{d}_1$. If the pre-posterior Prob($A_j|\mathbf{C}$) were known, then including the data $\mathbf{d}_2$ could be achieved through Eq. (4) and a sequential simulation from the conditionals through Eq (2). Since this is not the case, the initial sample conditioned to $\mathbf{d}_1$, will be used as an initial guess for further matching the data $\mathbf{d}_2$ iteratively. To achieve this, the unknown pre-posterior Prob($A_j|\mathbf{C}$) is modelled using a single parameter model in the following equation:

$$\text{Prob}(A_j \mid \mathbf{C}) = \text{Prob}(I(\mathbf{u}_j) = 1 \mid \mathbf{C}) = (1-r_C) \times i_B^{(o)}(\mathbf{u}_j) + r_C \times P(A_j), \quad j=1,\ldots,N \quad (5)$$

where r_C is a parameter between [0,1], not dependent on $\mathbf{u}_j$. $\{i_B^{(0)}(\mathbf{u}_j), j=1,\ldots,N\}$ is an initial realization conditioned to the $\mathbf{d}_1$ data (B-data) only. Given Eq. (5), Prob($A_j|\mathbf{C}$) can be calculated for a given value of r_C and for a given initial realization constrained to the B-data. Next, the probability Prob($A_j|\mathbf{C}$) is combined with Prob($A_j|\mathbf{B}_j$) to form the conditionals Prob($A_j|\mathbf{B}_j,\mathbf{C}$) by which sequential simulation is possible, Eq. (2), and a new realization $\{i^{(1)}(\mathbf{u}_j), j=1,\ldots,N\}$ is generated. The new realization is dependent on the initial realization and the value of r_C. To get some more insight into the role of the value r_C, consider the examples in Figure 1. Each row shows in its first column an initial realization $\{i_B^{(0)}(\mathbf{u}_j), j=1,\ldots,N\}$ generated with different sequential simulation methods. The next columns contain realizations $\{i^{(1)}(\mathbf{u}_j), j=1,\ldots,N\}$ for various values of r_C and using a random seed s' different from the random seed used to generate the initial realization. The important message of Figure 1 is that regardless of the value of r_C the each realization honors the same spatial statistics as the initial model, i.e. the prior distribution is maintained.

In case $r_C=0$, then $\text{Prob}(A_j|\mathbf{C})= i_B^{(0)}(\mathbf{u}_j)$, hence per Eq (4) $\text{Prob}(A_j|\mathbf{B},\mathbf{C})= i_B^{(0)}(\mathbf{u}_j)$. In other words, the initial realization is re-created. In case $r_C=1$, then $\text{Prob}(A_j|\mathbf{C})=P(A_j)$, a simple calculation using Eq. (4) shows that in that case $\text{Prob}(A_j|\mathbf{B},\mathbf{C})= \text{Prob}(A_j|\mathbf{B}_j)$. Since the seed s' is different from the seed s, the realization $\{i^{(1)}(\mathbf{u}_j), j=1,\ldots,N\}$ is equiprobable with the initial realization $\{i_B^{(0)}(\mathbf{u}_j), j=1,\ldots,N\}$. In other words, $r_C=1$ entails a "maximum perturbation" within the prior model constraints.

A value r_C between $(0,1)$ will therefore generate a perturbation $\{i^{(1)}(\mathbf{u}_j,r_C), j=1,\ldots,N\}$ between the initial realization and another equiprobable realization both conditioned to the data $\mathbf{d}_1$ and each honoring the prior model statistics. An optimal value for r_C can be picked by selecting the perturbation for which the mismatch between the forward model simulation and actual data $\mathbf{d}_2$, namely

$$O(r_C) = \|g(i^{(1)}(\mathbf{u}_j,r_C)) - \mathbf{d}_2\| \qquad (6)$$

is minimal.

3.4 PROBABILITY PERTURBATION ALGORITHM

The probability perturbation of the initial realization is likely to reduce the objective function in Eq. (6), however, minimizing $O(r_C)$ would only achieve a local minimum since the perturbation takes place between just two equiprobable realizations. To further reduce the objective function, the perturbations are iterated in the following algorithm:

- choose random seed
- generate an initial realization $i^{(0)}(\mathbf{u}_j), \forall j$
- change random seed
- Until the data $\mathbf{d}_2$ are matched to some desired level
    - Minimize to get r_C^{opt}

    $$O(r_C) = \|g(i^{(1)}(\mathbf{u}_j,r_C)) - \mathbf{d}_2\|$$

    - Change random seed
    - Assign

    $$i^{(0)}(\mathbf{u}_j) \leftarrow i^{(1)}(\mathbf{u}_j,r_C^{opt}), \forall j$$

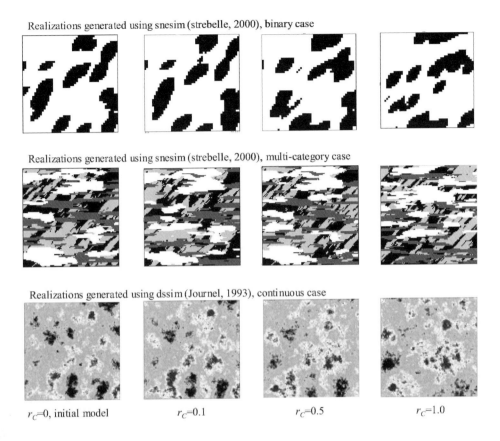

Figure 1: Left picture of each row is an initial guess realization, then followed by perturbation of this initial realization parameterized by a parameter r_C.

3.5 REGIONAL PROBABILITY PERTURBATION

The probability perturbation method generates a perturbation between an initial guess realization and another equiprobable realization that is parameterized using a single parameter. In a spatial context this induces a perturbation of each individual model parameter $i(\mathbf{u}_j)$ that is, *in probability*, the same for all $\mathbf{u}_j$. Parameterizing a perturbation using a single parameter may not effectively solve complex spatial inverse problem.

The above presented method does not restrict a higher order parameterization: the value of r_C can be made dependent on location

$$\text{Prob}(A_j \mid \mathbf{C}) = \text{Prob}(I(\mathbf{u}_j) = 1 \mid \mathbf{C}) = (1 - r_C(\mathbf{u}_j)) \times i_B^{(s)}(\mathbf{u}_j) + r_C(\mathbf{u}_j) \times P(A_j) \quad (7)$$

The use of (7) in the probability perturbation method now requires a multi-dimensional optimization on all $r_C(\mathbf{u}_j)$, $j=1,...,N$. To avoid a potentially difficult full multi-dimensional search for the best $r_C(\mathbf{u}_j)$, $j=1,...,N$, a region-wise parameterization of these parameters is proposed. Consider M regions in the domain of study, each region R_m, $m=1,..., M$ consists of a set of grid node locations,

$$R_m = \{\mathbf{u}_i^{(m)}, \mathbf{u}_j^{(m)},...\}$$

Which nodes belong to which region is a problem specific question. The number of regions M however is likely to be considerably less than the number of grid nodes N. The pre-posterior of Eq. (7) is rewritten using a region-wise parameterization as follows:

$$\text{Prob}(A_j^{(m)} | C) = \text{Prob}(I(\mathbf{u}_j^{(m)}) = 1 | C) = (1 - r_C^{(m)}) \times i_B^{(s)}(\mathbf{u}_j^{(m)}) + r_C^{(m)} P(A_j^{(m)}), \quad j = 1,...,N$$

where the parameter $r_C^{(m)}$ is the same for all grid nodes $\mathbf{u}_j^{(m)}$ of region R_m. An efficient strategy for jointly optimizing on all M $r_C^{(m)}$ parameters is discussed in Hoffman and Caers (2003).

4 Example

The aim of this paper is to present the inverse theory behind the probability perturbation method which has been extensively researched and applied to real cases in the context of inversion of flow data in oil reservoirs (Caers, 2003; Hoffman and Caers, 2004). We refer the reader to these paper for practical examples.

In this paper, a simple but rather revealing example is presented and illustrated in Figure 2. The model consists of a grid with three nodes, $\mathbf{u}_1$, $\mathbf{u}_2$ and $\mathbf{u}_3$. Each node can be either black, $I(\mathbf{u})=1$ or white, $I(\mathbf{u})=0$. The model $\mathbf{m}$ is therefore simply

$$\mathbf{m} = \{I(\mathbf{u}_1), I(\mathbf{u}_2), I(\mathbf{u}_3)\}$$

The spatial dependency of this simple 1D model is described by a 1D training image shown in Figure 2. One can extract, by scanning the training image with a 3 x 1 template, the prior distribution, $f(\mathbf{m})$, of the model parameters, as shown in Figure 2. To test the probability perturbation method we consider two data: the first datum is a point measurement (B-data, or "easy data") namely, $i(\mathbf{u}_2)=1$ (a black pixel in the middle), the second one is $I(\mathbf{u}_1)+I(\mathbf{u}_2)+I(\mathbf{u}_3)=2$ (C-data or "difficult data"). The problem posed is:

What is $\text{Prob}(I(\mathbf{u}_1) = 1 | i(\mathbf{u}_2) = 1, I(\mathbf{u}_1) + I(\mathbf{u}_2) + I(\mathbf{u}_3) = 2)$?
or in simple notation P(A|B,C) ?

Training image

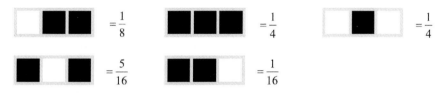

Figure 2: illustrative example: (top) 1D training image (bottom) derived from the training image are the prior probabilities of the model m

To get the answer we use four alternative techniques:

1. Get the true answer by elimination from the prior: $\text{Prob}(A|B,C) = \dfrac{\frac{1}{16}}{\frac{1}{8}+\frac{1}{16}} = \dfrac{1}{3}$

2. Using conditional independence (standardized)

$\text{Prob}(A) = \dfrac{4}{16}+\dfrac{1}{16}+\dfrac{5}{16} = \dfrac{5}{8} \Rightarrow a = \dfrac{3}{5}$; $\text{Prob}(A|B) = \dfrac{\frac{1}{4}+\frac{1}{16}}{\frac{1}{4}+\frac{1}{16}+\frac{1}{8}+\frac{1}{4}} = \dfrac{5}{11} \Rightarrow b = \dfrac{6}{5}$

$\text{Prob}(A|C) = \dfrac{\frac{5}{16}+\frac{1}{16}}{\frac{5}{16}+\frac{1}{16}+\frac{1}{8}} = \dfrac{3}{4} \Rightarrow c = \dfrac{1}{3}$

Applying Eq. (4) with $\tau=1$: $x = \dfrac{2}{3} \Rightarrow \text{Prob}(A|B,C) = \dfrac{3}{5}$

3. Using Monte Carlo simulation on the probability perturbation algorithm (with $\tau=1$ in Eq. (4), $\text{Prob}(A|B,C) = 0.35$

4. Using Monte Carlo simulation on the "gradual deformation of sequential simulation" (Hu et al., 2001) $\text{Prob}(A|B,C) = 0.27$

It is clear from comparing [1.] and [2.] that the conditional independence hypothesis is not valid for this case.

While the PPM relies on the same assumption of conditional independence the result is much closer to the true posterior probability. The reason for the latter observation can be explained by means of Eq (5). In this equation, the pre-posterior $\text{Prob}(I(\mathbf{u}_j)=1|C)$ is a function of the data C through the parameter r_C, *and*, a function of an initial realization

$\{i(\mathbf{u}_1),i(\mathbf{u}_2)=1,i(\mathbf{u}_3)\}$. This initial realization depends itself on the pre-posterior $\text{Prob}(I(\mathbf{u}_j)=1|\ B_j)$ with B_j depending on the random path taken. Hence, Eq. (5) forces an explicit dependency between the $\text{Prob}(I(\mathbf{u}_j)=1|B_j)$ and $\text{Prob}(I(\mathbf{u}_j)=1|C)$ *prior* to combining both into $\text{Prob}(I(\mathbf{u}_j)=1|\ B_j\ ,C)$ using a conditional independence hypothesis (Eq. (4) with $\tau=1$). At least from this simple example, one can conclude that the *sequential* decomposition of the posterior into pre-posteriors has robustified the estimate of the true posterior under the conditional independence hypothesis.

The same conclusion can be reached for the gradual deformation of sequential simulation. In gradual deformation of sequential simulation one perturbs gradually the random numbers used to draw from the various conditional distributions in Eq.(2), not the conditional distributions themselves as in the probability perturbation method. It appears that the gradual deformation of sequential simulation has an implicit model of dependency between the B and C data different from the probability perturbation, and more importantly different from the actual dependence.

The differences between the various methods are considerable. One can therefore conclude that future research should focus on understanding better the basic model assumptions, such as conditional independence, rather than focussing on developing precise samplers of models that are based on poorly understood assumptions. Such assumptions will have a first order effect on the ultimate space of uncertainty created.

References

Besag, J, and Green, P.J.,. Spatial statistics and Bayesian Computation, *Journal of the Royal Statistical Society*, B, v. 55, 1993, p. 3-23.

Caers, J., History matching under a training image-based geological model constraint. *SPE Journal*, SPE # 74716, 2003, p. 218-226

Hoffman, B.T. and Caers, J., Geostatistical history matching using the regional probability perturbation method. *In* SPE Annual Conference and Technical Exhibition, Denver, Oct. 5-8. SPE # 84409, 2003, 16pp. Society of Petroleum Engineers.

Hoffman, B.T. and Caers, J., History matching with the regional probability perturbation method – applications to a North Sea reservoir *In*: Proceedings to the ECMOR IX, Cannes, Aug 29 - Sept 2, 2004.

Hu, L.Y, Blanc, G. and Noetinger, B., Gradual deformation and iterative calibration of sequential stochastic simulations. *Mathematical Geology.*, v 33, 2001, p. 475-490.

Journel, A.G., Geostatistics: roadblocks and challenges. In Soares, A. ed., Geostatistics-Troia, v1: Kluwer Academic, Dordrecht, 1993, p. 213-224.

Journel, A.G., Combining knowledge from diverse data sources: an alternative to traditional data independence hypothesis. *Mathematical. Geology.*, v. 34, 2002, p. 573-596.

Metropolis, N., Rosenbluth, A.W., Rosenbluth, M.N., Teller, A.H. and Teller, E., Equation of state calculation by fast computing machines. *J. Chem. Phys.*, v. 21, 1953, p. 1087-1092.

Moosegard, K. and Tarantola, A., Monte Carlo sampling of solutions to inverse problems. *Journal of Geophyical. Research, B*, v. 100, 1995, p. 12431-12447.

Neal, R.M., Probabilistic inference using Markov chain Monte Carlo methods. Technical report CRG-TR-93-1, 1993 Department of Computer Science, University of Toronto.

Omre, H. and Tjelmeland, H. Petroleum Geostatistics. *In* Proceeding of the Fifth International Geostatistics Congress, ed. E.Y Baafi and N.A. Schofield, Wollongong Australia, v.1, 1997, p. 41-52. Kluwer Academic Publishers.

Strebelle, S., Conditional simulation of complex geological structures using multiple-point geostatistics. *Mathematical Geology*, v. 34, 2002, p. 1-22.

SPECTRAL COMPONENT GEOLOGIC MODELING: A NEW TECHNOLOGY FOR INTEGRATING SEISMIC INFORMATION AT THE CORRECT SCALE

TINGTING YAO, CRAIG CALVERT, GLEN BISHOP,
TOM JONES, YUAN MA, LINCOLN FOREMAN
ExxonMobil Upstream Research Company,
Houston, TX 77252, U.S.A.

Abstract. Spectral component geologic modelling (SCGM) is a new technology developed to properly account for both the scale and accuracy of any and all interpretations derived from seismic data in building geologic models. Seismic data can be integrated as spectral components which are volume- or map-based property interpretations representing a specific and measurable scale. The SCGM method starts with combining different spectral components together to build what is referred to as a "tentative geologic model", accounting for different scales and measurement accuracy of information in each component. The tentative geologic model will then be further constrained to honor the spatial continuity by substituting the amplitude spectrum of current tentative model with the desired amplitude spectrum from the target variogram model through spectral simulation. After that, the model will then be post processed to first honor the target histogram and then well data. In addition to honoring one single global variogram model, as do traditional geostatistical algorithms, SCGM has the capability to model local variations or trends in the continuity range and dominant azimuth direction of spatial continuity, by modifying the amplitude spectrum using spectral simulation.

1 Introduction

Geologic modeling has been widely used in reservoir management to characterize the rock-property heterogeneity that control pore-fluid storage and flow in a reservoir. For many reservoirs, particularly those in discovery through early production stages, well data may be sparse, and the well data alone are often insufficient to adequately constrain the assignment of reservoir properties between the wells in the geologic model. For such reservoirs, 3D seismic data have been increasingly used as an aid to assign these properties in the geologic model.

However, the utilization of seismic data for modeling reservoir properties faces some severe problems, possibly the most important being that of the difference in scale and accuracy between the seismic and the well data. The traditionally used geostatistical modeling methods integrate seismic data through kriging with local varying mean (Goovaerts, 1997), block cokriging (Behrens, Macleod, and Tran, 1996; Yao, 2000), or simulated annealing (Deutsch, Srinivasan, and Mo, 1996). These methods either wrongly treat the seismic data at the same scale as the geologic model, or make some strong assumption about the relationship between the coarse-scale seismic data and fine-scale well data (linear average as in block co-kriging). As a result, the geologic model may not fully exploit information contained in the seismic data and may not honor the input information.

The primary incentive for developing the SCGM method was to properly account for both the limitation of scale and accuracy of any and all interpretations derived from seismic data. Secondary incentives were to obtain a method that provides advanced capabilities for controlling rock-property continuity in the geologic model, and that can build or truly update a geologic model with new information quickly, based on spectral simulation (Calvert et. al., 2000, 2001, 2002).

2 Review of spectral simulation

Spectral simulation is gaining wider application in building geologic models due to the advantage of better honoring the spatial continuity of petrophysical properties, such as reservoir property and shale volume. The spatial continuity structure is characterized by a covariance/variogram model in the space domain and is represented by a density spectrum in the frequency domain. Distinct from sequential simulation methods, spectral simulation is a global method in the sense that a global density spectrum is calculated once from variogram model and the inverse Fourier transform is performed on the Fourier coefficient only once to generate a realization.

A spectral-simulation method, called Fourier Integral Method (FIM), has been proposed to generate geologic-model realizations that honor the spatial structure of a random field $z(u)$ in one-, two-, or three dimensions (Borgman, Taheri, and Hagan, 1984; Gutjahr, Kallay, and Wilson, 1987; Mckay, 1988; Pardo-Iguzquiza and Chica-Olmo, 1993). This method is performed in the frequency domain, as opposed to the usual sequential-Gaussian-simulation method performed in the space domain.

The spatial structure of a random field $z(u)$ is characterized by the covariance $C_z(h)$ or variogram $\gamma_z(h)$ in the space domain. In 1D, the covariance of $z(u)$ is defined as the convolution product (Bracewell, 1986):

$$C_z(h) = \int_{-\infty}^{+\infty} z(u) \cdot z(u+h) du = z * \check{z}, \text{ where } \check{z}(u) = z(-u) \tag{1}$$

The Fourier transform (FT) of the covariance into the density spectrum of $z(u)$ in the frequency domain exchanges convolution and multiplicative products:

$$s(\varpi) = FT(C_z) = FT(z) \cdot FT(\check{z}) = Z(\varpi) \cdot Z^*(\varpi) = |Z(\varpi)|^2 \tag{2}$$

where $Z(\varpi) = FT(z) = \int z(u) e^{-i\varpi u} du$, and $Z^*(\varpi)$ is the complex conjugate. The term $s(\varpi)$ is referred to as the density spectrum, $|Z(\varpi)|$ as the amplitude spectrum, $Z(\varpi) = |Z(\varpi)| e^{-i\phi(\varpi)}$ as the Fourier coefficient, and $\phi(\varpi)$ as the phase. The spectral-simulation method is based on the correspondence between the space property, $z(u)$, and the frequency counterparts, $s(\varpi)$ and $\varphi(\varpi)$, as illustrated in Figure 1. The implementation details can be referred to Yao (1998, 2002)

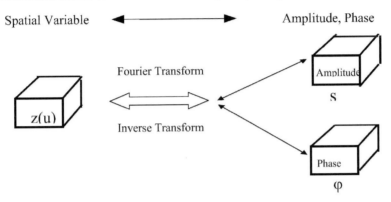

Figure 1. The correspondence between the space domain variable, z, and the frequency counterparts, $s(\varpi)$ and $\varphi(\varpi)$.

There are several advantages of spectral simulation over traditional geostatistical simulation. The spectral-simulation method is fast, particularly when based on the Fast Fourier Transform (Kar, 1994; Lam, 1995; Bruguera, 1996; Mckay, 1998). It is a global method in the sense that all of the amplitude-spectrum values over the whole field are used simultaneously to generate the simulated property. Therefore, the amplitude spectrum, or variogram model in the space domain, can be honored globally over the whole field instead of only within search neighborhoods as with the traditional sequential-Gaussian simulation method. Actually, the variogram model is honored over half field size, see Yao, 2002. A related advantage is that the separation of amplitude (spatial continuity) and phase (spatial location) allows updating of models if new information about either spatial continuity or location is obtained, or allows the conditioning of models to local information; see Calvert et al (2001, 2002). In addition, the advantage of spectral simulation in separating amplitude and phase information allows the spatial continuity to be modified to account for this traditionally unaccountable local information (Calvert et al., 2001, 2002).

3 Scale and accuracy of interpreted rock-property information: spectral component

All data that we use for geologic interpretation are limited in the scale of rock-property information that they contain, although we do not always appreciate the fact. For example, seismic data cannot directly be used to predict high-frequency variability in rock properties because the seismic data contain no information at high frequencies. If we attempt to use seismic data to estimate rock properties at scales that are outside of the data frequency band, interpretation errors can result. A spectral component is a volume- or map-based property interpretation representing only a specific and measurable scale of the property to be modeled, such as porosity. It could be derived from any data source or even from analogue information, representing all or a portion of the reservoir volume being modeled. The following data represent some of the spectral components (Figure 2):

- 3D volume from seismic amplitude calibration, which contains information only within the seismic frequency band, typically about 15-75 Hz.
- 2D map from seismic facies or geologic interpretation, such as average porosity map. This provides no information about the vertical variability in porosity values, hence contains no information at any frequency above zero Hz in vertical direction.
- 1D trend from well data, such as compaction trend of porosity observed, i.e., pososity generally decreasing with burial depth according to a fairly predictable function. This contains only low frequency information, e.g., 0-5 Hz, because slow vertical changes represent low-frequency vertical variability.

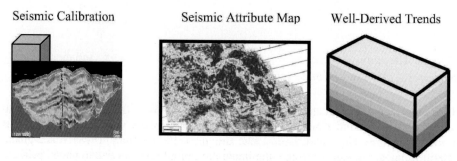

Figure 2. Examples of spectral components at different scales.

These spectral components are generated from different sources - some may be generated from data interpretation, whereas others may be generated from a concept or an analogue. Different frequency components might have different accuracy - those directly measured from well logs will be more accurate than others from qualitative interpretation. In addition, a spectral component often contains information that spans a bandwidth of frequencies. The measurement/interpretation accuracy could also change with different frequency component. The new SCGM method will account for the uncertainty or accuracy about each spectral component.

4 Overview of SCGM method

SCGM method involves constructing a geologic model by first mathematically combining different spectral components together. The combined volume is referred to as the "tentative geologic model", which might not represent all desired reservoir characteristics and needs to be further constrained to honor the target statistics such as variogram, histogram and well data. The constrained model can be further post processed to represent local variations in continuity trend and continuity azimuth, based on spectral simulation. The generalized SCGM process is given in Figure 3.

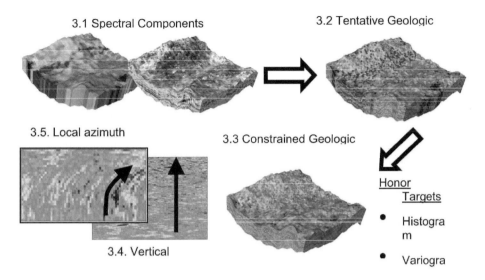

Figure 3. General process of SCGM method.

4.1 BUILDING THE TENTATIVE GEOLOGIC MODEL

The SCGM method starts with combining different spectral components together to build what is referred to as a "tentative geologic model", accounting for different scales of information in each component and the different measurement accuracy. The tentative geologic model represents the integration of all relevant data types to produce an *a priori* geologic model. The least-complex tentative geologic model that can be built is one in which all spectral components represent distinct or complementary frequency bands. Such a tentative geologic model can be constructed by simply summing the independent components. However, in reality, there will always be missing scales of information. For any tentative geologic model that is built, if a frequency-band of information is missing (e.g., high-frequency information), then this missing band of information must be simulated within SCGM and added to the tentative geologic model. It is also possible that information may be missing over a specific region within the model area; in this case the data are simulated and added to the model, but only in that specific region.

The process of building the tentative geologic model gets somewhat more complicated when the individual spectral components overlap in their frequency content. The spectral components may completely overlap in frequency content or they may partially overlap. To properly integrate spectral components that have overlapping spectra, the measurement or interpretation accuracy of each overlapping component must be known. The accuracy can be quantified by a value between 0.0 and 1.0. Those components having higher accuracy values will have relatively greater influence on the resulting tentative geologic model, but only over those frequencies that overlap. Measurement or interpretation accuracy can also vary spatially. For example, the spectral component in one location within the modeling area may be more accurate than in another due to the interpreter's diligence. In this case, the tentative geologic model is constructed through weighted averaging of the spectral components *by location* (i.e., different average weight at different locations). Figure 3.2 shows a simple tentative model generated by simply adding up different frequency components in Figure 3.1.

4.2 CONSTRAINING THE TENTATIVE GEOLOGIC MODELS

The tentative geologic model will not have all of the desired properties of geologic model, e.g., it likely will not honor the well data or the target variogram and histogram. The tentative geologic model is further constrained to honor these targets. The constraining process is sequential, in that the model is modified first to honor the variogram, second to honor the histogram, and finally to honor the well data.

- *Honoring the desired spatial continuity.* Spectral simulation is a perfect application to update the tentative geologic model in an attempt to honor the spatial continuity represented by the target variogram model. From the tentative geologic model, we calculate its amplitude and phase spectrum. We only keep part of the amplitude spectrum which we believe are reliable and substitute the other part with the target one (representing the target variogram model), and keep the phase spectrum to generate new Fourier coefficients. The inverse Fourier transform provides a model that honors the spatial continuity represented by the target amplitude spectrum, as well as the spatial distribution of high and low values observed in the tentative geologic model, see Figure 4.

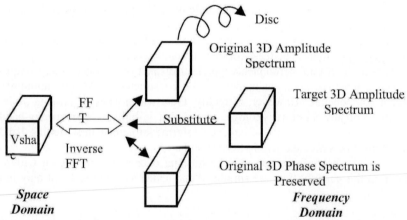

Figure 4. Schematic illustration of the process of spectral simulation, as implemented in SCGM.

- *Honoring the desired histogram.* Quartile transform is used to force the distribution of geologic model matches the target histogram. Each rock-property value in each cumulative distribution function (CDF) corresponds to a probability quintile. The p quintile of the cumulative distribution function (CDF) is transformed to the same p quintile of the CDF for the target histogram so that the CDF of the tentative geologic model matches the target CDF. The quintile transform does not change the relative rank of the data, hence also referred to as "rank-preserved transform".

- *Honoring the well data.* Following the steps of honoring the variogram model and the histogram, the rock-property values at the well locations are reset to match the actual well-data values. Resetting these values results in changes in the properties assigned to these cells. These changes [(actual well value) - (current model value)] are propagated to all tentative geologic model cells in the neighborhood of each well; the magnitudes of the changes are weighted as a function of inverse distance from the well.

Figure 3.3 shows a constraint model which honors the target variogram, histogram, and well data. Note that the sequential implementation of first honoring variogram, then honoring histogram, finally honoring well data might distort the target parameters honored first such as varigoram and histogram by later honoring other target parameters such as well data. Therefore, the ideal implementation would be iteratively repeat the sequential honoring process to ensure all the target parameters are honored in the same degree. However, many tests show that so long as the target parameters are consistent with each other, the first iteration does 90% of the job of constraining the model to the target. For speed purpose, we used only one iteration, but strongly suggest checking the model to make sure all the targets are met without much distortion.

The SCGM process described above *does* properly account for the scale of the input data, both in terms of the spectral component information (as represented in the phase spectrum) and the target variogram model (as represented in the target amplitude spectrum). As a result, SCGM can honor both the compositional information contained in the tentative geologic model and the target variogram model, without compromising either. Given the same input of variogram, histogram and well data, SCGM is proved to honor the input information better than the traditional geostatistical methods such as kriging with locally varying mean or collocated cokriging (Calvert et al., 2002).

In addition to honoring one single global variogram model, as do traditional geostatistical algorithms, SCGM has the capability to model local variations or trends in the range and dominant azimuth direction of spatial continuity, using spectral simulation.

4.3 CONDITIONING TO THE LOCAL SPATIAL CONTINUITY TREND

In a geologic model, the three-dimensional spatial continuity of a rock property is commonly controlled with geostatistical algorithms and a variogram that quantifies the spatial variability of the rock property as a function of both separation distance and direction. Geostatistical algorithms used in constructing geologic models assume stationarity in the geologic characteristics of the modeled region, i.e., they assume that a modeled rock property can be represented by a single set of statistical measures, which are often referred to as "global" measures. For example, a single, global variogram model would be used to represent the spatial continuity of the rock property over the entire modeled region.

However, we know that the geologic characteristics of the subsurface are non-stationary. For example, the global spatial continuity of bed thickness in a reservoir can be characterized by a global spherical variogram model with a range of 10 feet. However, we can also observe a trend of thicker beds at the bottom and thinner beds on the top: the thicker beds at the bottom might have a range of 20 feet and the thinner beds on the top might have a range of 5 feet. To account for the local continuity trend beyond the global variogram model, we use the local (variogram models with longest and shortest ranges) and global variogram models to calculate the spectral amplitude ratios of the local spectral amplitudes vs. the global one at different frequency bins. We interpolate the local amplitude ratios in between according to the trend. This will provide one local amplitude ratio for each frequency bin at each cell. Then, we decompose the model that honors the global spectrum into different frequency components (represent each frequency bin), and multiply each component by the corresponding local amplitude ratio. The summed result of all the multiplied spectral components displays the local continuity trend. The implementation details can be referred to Yao (2003).

Figure 3.4 gives examples of applying this method to impose a trend in vertical continuity on a geologic model of shale volume. This example applied only one continuity trend on the geologic model. If additional trends in spatial continuity are desired, then treat the geologic model created before as a new starting geologic model and apply the local trend using different global and local amplitude spectra that represent the new trends. This will generate a geologic model that honors multi-dimensional trends in spatial continuity.

Other algorithms available to account for non-stationary spatial continuity usually separate the whole modeling area into different sub-areas and use a different variogram model for each sub-area. Such a method addresses the non-stationarity of the large area, but at the cost of artifact boundaries between sub-areas. Using spectral simulation and manipulating the amplitude spectrum allows us to address explicitly the gradually changing continuity trend.

4.4 CONDITIONING TO THE LOCAL SPATIAL CONTINUITY AZIMUTHS

Rock property continuity within a reservoir often shows anisotropy, i.e., continuity is greater in one direction than in another. In addition, the local direction of greatest continuity might change from one location to another within the reservoir. Consider sediments deposited in a river channel. Paleo-hydrodynamics often control the distribution of the lithological and petrophysical properties within the channel. We know that the continuity of these properties is anisotropic, typically greatest along channel and less continuous across channel. We also know that sinuosity may cause the channel to locally vary in direction; therefore, the rock-property continuity will also locally vary in direction, as with a meandering pattern.

Methods used to introduce variable directions in rock-property continuity into the geologic model are generally based on methods published by Xu (1996). Continuity direction is varied according to an input grid of azimuths, which represent local variations in continuity direction. Often, this grid is based on interpreted seismic facies, i.e., the shape of the interpreted facies (e.g., sinuosity of a channel) is represented in the azimuth data. To conditioning to the locally varying continuity azimuth, we first identify the strings of connected nodes from the azimuth grids (Jones, et al., 2001). Then, we simulate each string to have maximum continuity along that string, using 1D spectral simulation. Finally, we put the simulated values back to the original nodes along the string. Therefore, the continuity along a path that bends according to the azimuth data as desired is reproduced (Craig, et al., 2002). The traditional pixel-based geostatistical methods require that this path be represented locally by a straight line. If those segments are small (i.e., the range of continuity is short), the curved line can be approximated well with straight-line segments. However, if the segments are long (i.e., the range of continuity is long), the curved line can not be approximated with straight-line segments. This limitation practically manifests itself as a trade-off between honoring the azimuth data and the target variogram range. The new SCGM method can simulate continuous rock properties along a bent path, therefore, it should produce better results in situations when long-range continuity is to be represented along a curved geologic feature, see Figure 3.5.

4.5 UPDATING AN EXISTING GEOLOGIC MODEL

The process of updating of any existing geologic model with any new information is very straightforward. For example, a new or alternative spectral component (e.g., from seismic-volume interpretation) became available after the original model was built. To incorporate this new information, we could update the original model by the following process:
- From the existing model, filter out and discard the information that is of the same scale (frequency band) as the new spectral component,
- Combine this filtered model with the new spectral component to create a tentative geologic model
- Constrain the new tentative geologic model to satisfy other targets.

A new variogram or histogram target model can also be incorporated efficiently to update the existing model.

5 Conclusions

SCGM is a new technology for integrating all the relevant data at their correct scale. It starts with combining different spectral components together to build what is referred to as a "tentative geologic model", accounting for different scales of information in each component and the different measurement accuracy. Then, the tentative geologic model is constrained to honor all the desired properties of geologic model such as honoring the spatial continuity, histogram and well data. Spectral simulation is applied to honor the spatial continuity globally as well as to gain speed advantage. In addition to honoring one single global variogram model, as do traditional geostatistical algorithms, SCGM has the capability to model local variations or trends in the range and dominant azimuth direction of spatial continuity, by modifying the amplitude spectrum using spectral simulation.

References

Behrens, R.A., Macleod, M.K., and Tran, T.T., 1996. Incorporating seismic attribute maps in 3-D reservoir models: SPE 36499, 31-36.
Borgman, L., Taheri, M., and Hagan, R., 1984. Three-dimentional frequency-domain simulations of geological variables, in Verly, G., Journel, A.G., and Marechal, A., eds., Geostatistics for natural resources characterization: NATO ASI Series, Reidel Publ., Dordrecht, The Netherlands, p. 517-541.
Bracewell, R., 1986. The Fourier transform and its application: McGraw Hill, Inc., Singapore, 474 p.
Bruguera, J., 1996. Implementation of the FFT butterfly with redundant arithmetic: IEEE Transaction on Circuits and Systems, Part II: Analog and Digital Signal Processing, v. 43, no. 10, p. 717-723.
Calvert, C.S., Bishop, G.W., Ma, Y.Z., Yao, T., Foreman, J.L., Sullivan, K.B., Dawson, D.C., and Jones, T.A., 2000. Method for constructing 3D geologic models by combining multiple frequency pass band, U.S. Patent Application Pending.
Calvert, C.S., Yao, T., Bishop, G.W., and Ma, Y.Z., 2001. Method for locally controlling spatial continuity in geologic models, U.S. Patent Application Pending.
Calvert, C.S., Jones, T.A., Bishop, G.W., Yao, T., Foreman, J.L., and Ma, Y.Z., 2002. Method for conditioning a random field to have directionally varying anisotropic continuity, U.S. Patent Application Pending.
Calvert, C.S., Bishop, G.W., Yao, T., Ma, Y., Forema, J.L., 2002. Spectral Component Geologic Modeling: an overview, ExxonMobil internal report.
Deutsch, C.V., Srinivasan, S., and Mo, Y., 1996. Geostatistical reservoir modeling accounting for precision and scale of seismic data: SPE 36497, 9-15.
Goovaert, P., 1997. Geostatistics for natural resources evaluation, Oxford Univ. Press.
Gutjahr, A., Kallay, P., and Wilson, J., 1987. Stochastic models for two-phase flow: A spectral-perturbation approach: Eos, Transaction, American Geophysical Union, v. 68, no. 44, p. 1266-1267.
Jones, T.A., Foreman, J.L., Yao, T., 2001. Method to honor channel directionality when building 3-D petrophysical models, U.S. Patent Application Pending.
Kar, D., 1994. On the prime factor decomposition algorithm for the discrete sine transform: IEEE Transactions on Signal Processing, v. 42, p. 3258-3260.
Lam, K.-M., 1995. Computing the inverse DFT with the in-place, in-order prime factor FFT algorithm: IEEE Transactions on Signal Processing, v. 43, p. 2193-2194.
Mckay, D., 1988. A fast Fourier transform method for generation of random fields: Unpubl. Master's thesis, New Mexico Institute of Mining and Technology, 92 p.
Pardo-Iguzquiza, E., and Chica-Olmo, M., 1993. The Fourier integral method: An efficient spectral method for simulation of random fields: Math. Geology, v. 25, no. 4, p. 177-217.
Xu, W., 1996. Conditional curvilinear stochastic simulation using pixel-based algorithms: Math. Geology, v.28, no. 7., p. 937-949.
Yao, T., 1998. Conditional spectral simulation with phase identification: Math. Geology, v. 30, no. 3, p. 285-308.
Yao, T., 1998. SPECSIM: A fortran-77 program for conditional spectral simulation in 3D: Computer & Geosciences, v. 24, no. 10, p. 911-921.
Yao, T., 1998. Automatic modeling of (cross) covariance tables using fast Fourier transform: Math. Geology, v. 30, no. 6., p. 589-615.
Yao, T., and Journel, A.G., 2000. Integrating seismic attribute maps and well logs for porosity modeling in a west Texas carbonate reservoir: addressing the scale and precision problem: Journal of Petroleum Science and Engineering, v. 28, p. 65-79.
Yao, T., 2002. Reproduction of mean, variance, and variogram model in spectral simulation: Math. Geology, accepted.
Yao, T., Calvert, C., Jones, T., Bishop, G., Ma, Y., Foreman, L., 2003. Spectral simulation and its advanced capability of conditioning to local continuity trends in geologic modeling: Journal of Petroleum Science and Engineering, submitted.

Autobiography:

Tingting Yao, BS from University of Petroleum in 1993, majoring in petroleum geology; PhD in Geostatistics from Stanford University in 1998. Her PhD research focuses on automatic covariance modeling to characterize the spatial continuity of earth science phenomena and conditional spectral simulation through phase identification to honor the spatially sampled conditioning data. She Joined Mobil Technology Company in July 1998 and transferred to ExxonMobil Upstream Research Company in February 2000. She has been working on improved methods for more accurate and efficient geologic modeling and reservoir characterization through integration of various data type.

JOINT SIMULATIONS, OPTIMAL DRILLHOLE SPACING AND THE ROLE OF THE STOCKPILE

A. BOUCHER, R. DIMITRAKOPOULOS and J.A. VARGAS-GUZMAN
WH Bryan Mining Geology Research Centre
The University of Queensland, Brisbane Qld 4072, Australia

Abstract. Infill and grade control drilling are a major cost in any mining operation. Reduction of drilling density can considerably enhance the profitability of an operation provided the cost from block misclassification is less than the savings in drilling. This paper presents a general simulation based approach to assess the performance of potential drilling schemes from the available deposit information. The approach integrates joint simulation of correlated variables with the computationally efficient minimum/maximum autocorrelation factors, multi-elements ore classification, and mine planning considerations. The latter employs key indicators such as profit per tonne mined and profit per tonne milled, as well as the potential use of a stockpile and its discounting. A case study at the Murrin Murrin nickel-cobalt deposit, Western Australia, is used to elucidate the proposed approach and to show the critical effect of planning decisions on drilling.

1 Introduction

Infill drilling is a critical information collection process in mining operations leading to substantial investment that can be in the order of millions of dollars. As a result, the ability to assess the performance of potential drilling schemes, prior to drilling is important. A reduction in drilling density could enhance the profitability of an operation, if misclassification cost does not exceed the saving in drilling. At the same time, additional information becomes counterproductive after the point of diminishing returns, i.e. the cost of additional information exceeds its benefit. Past work in geostatistically assessing additional drilling was based on estimation variances which largely reflect the geometry of drilling configurations (Goovaerts, 1997) without any consideration of the local grade variability and uncertainty (Ravenscroft, 1992), economic cost/benefit analysis, or a link to mine planning decisions.

A stochastic simulation framework can be used to realistically address the assessment of infill drilling patterns (e.g., Dimitrakopoulos, 2003). In the general case of multi-element deposits, joint simulation of pertinent correlated variables is used to produce realisations of an exhaustively known deposit. Such a realisation is treated as an "actual" deposit and is subsequently virtually drilled. This new drilling information can then be used to re-simulate the deposit leading to comparisons of block classifications and other indicators to the actual exhaustively known deposit. This way different drilling schemes can be compared to make informed selection of a drilling strategy.

Computationally efficient joint simulation methods are essential for generating realistic representation of "actual" deposit. Conventional co-simulation methods (e.g., Verly, 1993; Chiles and Delfiner, 2000) become inefficient when more than two variables are considered. Collocated cosimulation with a Markov-type coregionalisation (Almeida and Journel 1993) assumes a very specific coregionalisation and does not extend well beyond two attributes. Therefore, conditional simulation with the so-called minimum/maximum autocorrelation factors or MAF (Switzer and Green 1984; Desbarats and Dimitrakopoulos 2000) is advocated herein. MAF transforms attributes of interest to uncorrelated factors that are independently simulated by any simulation method and then reconstructed to realisations of the original variables reproducing their cross and auto-correlation.

In addition to the orebody geology, mine planning aspects, such as stockpiling, also affect the performance of the infill drilling patterns,. When low-grade ore blocks are stockpiled, the performance of an infill drilling scheme is a function of how, when, and if the stockpile would be processed in the future. The uncertainty linked to the stockpiling strategy can be factored into the selection of a drilling scheme by depreciating the stockpile value with a discount rate. If an ore block sent to the stockpile is considered lost, the stockpile can be regarded as waste. Alternatively, if the stockpile will be processed in coming years, a misclassified ore block in the stockpile makes no difference and no penalty is necessary. Discount rates enable the comparison of the different schemes by linking the two extreme scenarios.

In the next sections, the drilling optimisation method is outlined and is followed by a brief discussion of simulation with MAF and the definition of economic indicators for comparing drilling efficiency. Finally, the intricacies of the method are detailed in a case study at the Murrin Murrin nickel-cobalt deposit, Western Australia.

2 A method for infill drilling assessment and optimisation

The following method, also schematized in Figure 1, is suggested to assess and compare the performance of drilling patterns for a multi-element deposit:

Step 1: From the exploration drilling data available within the pit, jointly simulate a representation of the deposit using min/max autocorrelation factors for the attributes under study. This first realization is called the "actual" deposit.

Step 2: Sample the above actual deposit with the different infill drilling schemes of interest.

Step 3: For each drilling scheme, jointly simulate with MAF the elements of interest conditional to the data from the "actual" deposit in Step 2 above, to obtain several joint realisations. Re-block the realizations for the attributes simulated to produce models of mining blocks to be assessed.

Step 4: Do grade control and classify the blocks (e.g. from their average grades) for each sampling scheme (e.g., milling ore, stockpile and waste). Compare the classification to the "actual" classification using, as economic indicators, the profit per tonne mined and profit per tonne milled; calculated as discussed in a subsequent paragraph.

Step 5: Graph and assess the results with respect to the point of diminishing returns. Repeat to assess sensitivity of the results.

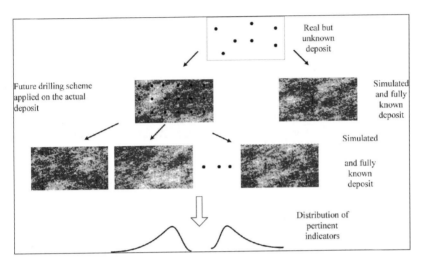

Figure 1 Schematic workflow for the proposed methodology

3 Multivariate simulation with minimum/maximum autocorrelation factors

The multivariate deposit is cosimulated by orthonalising the grade attributes into three factors deemed uncorrelated from each other. Each of these factor is then simulated independently, and the simulated grade is obtained by back-rotating the factors into the attribute space.

The minimum/maximum autocorrelation factors (MAF) is an orthogonalisation similar to the well-known principal components analysis (PCA). The advantage of the MAFs over the PCs is the extension of orthogonalisation to the non-zero lags. Principal components (David, 1988) are uncorrelated at all lags only if they are derived from a random field with an intrinsic coregionalisation. The MAFs uncorrelate a RF for all lags with a linear model of coregionalisation containing at most two structures (Switzer and Green 1984; Desbarats and Dimitrakopoulos 2000). Boucher (2003) and Dimitrakopoulos and Fonseca (2003) have successfully used the MAF method to jointly simulate grades in a mining environment.

Switzer and Green (1984), later reviewed in Desbarats and Dimitrakopoulos (2000), show how to obtain the factors through the eigenvectors of the matrix $2\Gamma_Z(h)\mathbf{B}^{-1}$, where

$$\mathbf{B} = \text{cov}[\ \mathbf{Z}(u), \mathbf{Z}(u)\ ]$$
$$2\Gamma_Z(h) = \text{cov}[\ \mathbf{Z}(u) - \mathbf{Z}(u+h), \mathbf{Z}(u) - \mathbf{Z}(u+h)\ ]$$

where $\mathbf{B}$ is the variance/covariance matrix of $\mathbf{Z}(u)$, a multiGaussian RF, and $\Gamma_Z(h)$ is the variogram matrix at lag h.

The matrix $\mathbf{A}$ of orthogonalisation coefficients is such that

$$2\Gamma_Z(h)\mathbf{B}^{-1} = \mathbf{A}^T \Lambda \mathbf{A} \qquad (1)$$

Refer to Desbarats and Dimitrakopoulos (2000) for an equivalent but computationally more efficient method to derive the coefficients $\mathbf{A}$ by performing two successive principal component decompositions.

Each orthogonal factor $\mathbf{Y}_i(u), i=1,...,p$ is obtained with the coefficients $\mathbf{a}_i$ constituting the ith row of $\mathbf{A}$.

$$\mathbf{Y}_i(u) = \mathbf{a}_i\, \mathbf{Z}(u),\ \ i=1,...,p \qquad (2)$$

The new vector RF $\mathbf{Y}(u)$, is then, by construction orthogonalised at lag 0 and at lag h.

$\mathbf{Z}(u)$ is simulated by independently simulating each factor $y_i^*(u), i=1,...,p$ and back-rotating them with the coefficient matrix:

$$\mathbf{z}^*(u) = \mathbf{A}^T \cdot \mathbf{y}^*(u) \qquad (3)$$

In practice, the attributes are first normal score transformed and then rotated into orthogonal factors. This prior transformation reduces the effect of skewed distributions but potential problems may occur if the rank and the Pearson coefficient of correlation of the original attribute differ too much.

4 Economic indicators

The key indicators suggested here are the profit per tonnes mined and profit per tonnes milled. These are defined here, without loss of generality, using the case of a deposit with two revenue generating elements and a third one that adversely affects production. Consider an example of a Ni laterite deposit producing nickel and cobalt where magnesium is a "penalty" element increasing processing costs.

Taking into account the quantity of the penalty element (M^{Mg}) and a penalty factor (f_k^{Mg}) the cost of classifying a block at location **u** in category k, $C_k^{Total}(\mathbf{u})$ is expressed as

$$C_k^{Total}(\mathbf{u}) = \left(C^{Drilling} + C_k^{Mining} + C_k^{Processing}\right) + M^{Mg}(\mathbf{u}) \cdot f_k^{Mg}$$

The revenue from a block is expressed as

$$R_k(\mathbf{u}) = \sum_{i=1}^{n} M_k^i(\mathbf{u}) r_k^i p^i$$

where M^i is the quantity of revenue generated by metal i, (e.g., M^{Ni} is the quantity of Nickel), r_k^i is the recovery of metal i when classified in category k and p^i is the price for attribute i.

The gross profit $F_k(\mathbf{u})$ generated by classifying a block at location **u** in group k is

$$F_k(\mathbf{u}) = R_k(\mathbf{u}) - C_k^{Total}(\mathbf{u}) \qquad (4)$$

The final classification of a block at location **u** is such that it maximizes the gross profit. A block will be classified in group k' such that

$$k' = \arg\max_{j} F_j(\mathbf{u})$$

The optimal drilling pattern is the one that would maximize the gross profit, such that the sum of all $F_{k'}(\mathbf{u}_j), j = 1,..., N$ is maximal. Excessive infill drilling would increase the cost and insufficient drilling would decrease the revenue.

Two indicators are used to assess the performance of the sampling scheme. The primary one is the profit per tonne mined, which is the sum of the profit generated by the N blocks inside the domain

$$P_{mined} = \frac{\sum_{j=1}^{N} F_{k'}(\mathbf{u}_j)}{N} \qquad (5)$$

which is an indication of the efficiency of the selection. The second indicator is the profit per tonne milled

$$P_{milled} = \frac{\sum_{j=1}^{N} F_{k'}(\mathbf{u}_j)}{\sum_{j=1}^{N} I_{milled}(\mathbf{u}_j)} \qquad (6)$$

where $I_{milled}(\mathbf{u}_j)$ takes the value one if the block located at $\mathbf{u}_j$ is sent to the mill, and takes zero otherwise. The profit per tonne milled indicates the quality of the ore being selected. The difference between the two can be seen with a simple example. Consider a case where some economics material has to be stockpiled to allow only very high grade to the mill. Being profitable, the misclassification of this stockpile material in the mill material will increase the revenue and, the tones mined being constant, the ratio profit per tones mined will also increase. In contrast, the profit per tone milled will decrease as the misclassified material generates less revenue than the high grade material.

The conditional distribution of profit per tonne mined P_{mined} (5) and per tonne milled P_{milled} (6) are computed for each of the N_D sampling scheme Ω_j from their respective conditional joint simulations, obtained from Eq. (1) to (3).

$$\Pr(P_{mined} < X \mid \mathbf{z}(u_i), \Omega_j, \; i=1,...,n) \quad j=1,...,N_D$$

$$\Pr(P_{milled} < X \mid \mathbf{z}(u_i), \Omega_j, \; i=1,...,n) \quad j=1,...,N_D$$

5 Application at the Murrin Murrin deposit

The Murrin-Murrin nickel-cobalt deposit is located in the Eastern Goldfields Province, Western Australia. It is hosted in weathered peridotites comprising of a ferrugious zone, which is predominantly waste, and two ore bearing horizons, a smectite unit with a transitional boundary to a magnesium enriched saprolite horizon (Jaine, 2003) The Murin Murin operation provides a 4 Mtpa supply to the processing plant that recovers nickel and cobalt. Given the magnesium content of the ore, the response of the mill feed to pressure-acid leaching and the cost of acid consumption is a metallurgical issue.

In the case study that follows, exploration drillholes within the saprolite zone are available from one of the open pits at Murrin Murin. Those holes, approximately gridded on a 50x50 metre spacing, give 263 one metre composites. Grade control infill drilling is typically performed on a 12.5 by 12.5 metre grid, with block size of 15x15 metre. A reduction in drilling would lead to direct saving in pre-mining costs whilst additional information could improve the quality of mill feed, thus reducing contaminant penalties and improving ore selection in addition to improving short term scheduling performance of the mine. The choice of a bench height is also looked at, two and three metre bench thicknesses are considered.

5.1 SIMULATING Ni, Co AND Mg WITH MAF

From the exploration drillholes, 4 actual deposits are simulated on point support: 2 with two metre bench and 2 with three metre bench. After normal score-transform, the nickel, cobalt and magnesium attributes are rotated into MAF space with expression (2). Each of the factors is then simulated independently one from the other. Then rotated back together into the Gaussian space and finally back-transformed into the original space. Figure 2 shows some joint realisations of nickel, cobalt and magnesium of one actual deposit. The cross-variograms, shown in Figure 3, are well-reproduced thus preserving the important spatial relationships between the attributes.

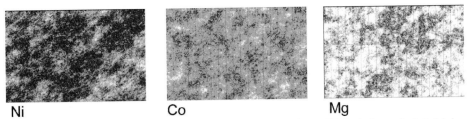

Figure 2 Joint realisations of nickel, cobalt and magnesium. Light is low, dark is high.

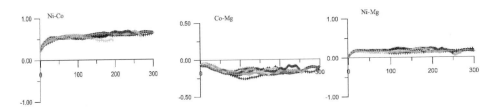

Figure 3 Reproduction of cross-correlation at all lags. The black crosses are the experimental variogram values.

5.2 SIMULATING THE DRILLING AND CLASSIFICATION PROCESS

The infill drilling information $\mathbf{z}(\mathbf{u}_\alpha) = \{z_{Ni}(\mathbf{u}_\alpha), z_{Co}(\mathbf{u}_\alpha), z_{Mg}(\mathbf{u}_\alpha)\}$, where $\mathbf{u}_\alpha$ are the sampling locations, is obtained by virtually drilling the actual deposits with a specific drilling scheme. Four ($N_D = 4$) regular sampling schemes, $\Omega_i, i = 1,...,4$, are considered:

Ω_1:	12m x 12m	(512 holes)
Ω_2:	18m x 12m	(320 holes)
Ω_3:	18m x 18m	(210 holes)
Ω_4:	25m x 25m	(210 holes)

For each of these sampling, 30 cosimulations $\mathbf{z}^*(\mathbf{u})$ are performed conditional to the prior $\mathbf{z}(\mathbf{u}_i)$ (exploration holes) and posterior $\mathbf{z}(\mathbf{u}_\alpha)$ (the virtual infill-sampling). There are 480 (4 actual deposits x 4 sampling schemes x 30 cosimulations) simulated deposits to which the economics indicators would be applied. All those point-support cosimulations are then upscaled into 338 (26x13) blocks of dimension 15x15 metre. The block selection for each sampling scheme is based on the E-type mimicking the actual selection process for a mine operation. Finally, the profit per tonne mined and profit per tonne milled are calculated with expressions (5) and (6).

Considering the material stockpiled as waste, Figure 4 shows the histograms of P_{mined} for all sampling configurations for the first actual on two and three metre bench heights. First, all the schemes are profitable, i.e. no loss occurred, but some are more profitable than the others. It is also noticeable that the mean decreases with a sparser drilling pattern while the variance (and the coefficient of variation) increases. The 12x12 scheme is the most advantageous when considering both the efficiency and the uncertainty.

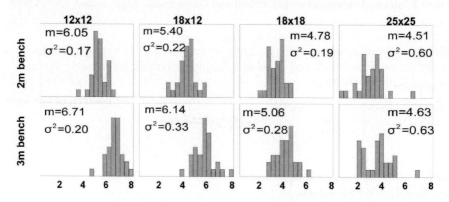

Figure 4 Histogram of profit per tones mined.

5.3 DISCOUNTING THE STOCKPILE AND EFFECTS ON DRILLING

In reality, the stockpile has a value that is neither ore nor waste. Stockpiling an economic block may be seen as 'money in the bank' without interest, thus inducing an opportunity cost. The cost of misclassifying an ore block as stockpile increases according to the stockpile strategy, i.e. when that block will be mined. From the banking analogy, the increase in cost depends on a discount rate and the number of years the material is to be stockpiled. The revenue generated from the stockpile, say to be mined in Δt years, is expressed in today's dollar (t =0)

$$F_{t=0} = (1+i)^{-\Delta t} F_{t=\Delta t} \qquad (7)$$

where i is the discount rate. A higher discount rate, i.e. a higher uncertainty about the stockpile strategy, will increase the misclassification cost.

The performance of the four infill drilling schemes is revisited by considering uncertainty in the stockpiled strategy modelled by discount rates. The cost of stockpiling marginal ore is investigated with six different discount rates on a period of 10 years. The profit generated by the stockpile is transformed into dollars in ten years time with a specific discount rate. With a discount rate of zero, suggesting no cost of opportunity, the stockpile is considered as ore. A high discount rate indicates that the stockpile lost all its value, thus it is considered as waste. All other discount rates are intermediate scenarios between these two "end-point" cases.

The profit per tonne mined is expressed in Australian dollar increment based on the currently used 12m x 12m sampling grid. The median profit per tonne mined for the four sampling scheme applied on actual #1 is shown in Figure 5. The upper left graph considers the stockpile as waste and the lower right graph as ore. The profit per tonne milled is shown in the same format in Figure 6.

For an increase of 8 cents per tonne in drilling cost between 12x12m and 25x25m, the profit per tonne mined improves up to $2 (35%) at a high cut-off (stockpile as waste) and of $0.50 at a lower cut-off (stockpile as waste). The 12x12m scheme is more profitable, even at a low cut-off (stockpile as ore). The 12x18m scheme does not decrease the profit too much and could also be appropriate for the deposit. The results seem insensitive to the bench height. For actual #1 the 2m is better than the 3m, the inverse is observed for the actual #2 therefore mining with 3m benches is appropriate.

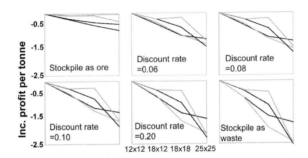

Figure 5 Median increment of profit per tonne mined. The median of the 12x12 metre scheme is set to zero. The black lines are for the actual deposits on two metre bench, the gray lines are for those on three metre bench.

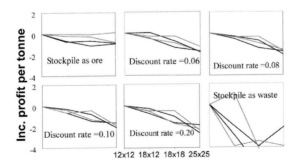

Figure 6 Median increment of profit per tonne milled. The median of the 12x12 metre scheme is set to zero. The black lines are for the actual deposit on two metre bench, the grey lines are for those on three metre bench.

6 Comments and Conclusions

This study shows that multivariate deposits can be efficiently simulated by first orthogonalising the attributes with the minimum/maximum autocorrelation factorisation. Once the simulated values are back-rotated, the cross-correlation at all lags between attributes is restored. The resulting realizations are then a better representation of the deposit and are therefore more appropriate for further processing.

The economic consequences of the drilling patterns on a multivariate deposit is then regarded on a large scale that takes into account some aspect of long-term planning, specifically with regards to the strategy and uncertainty related to the stockpiled material. The uncertainty of this material is translated into a discount rate, which indicates the risk of losing a profitable block when stockpiled. The performance of the drilling scheme can be assessed in a larger perspective than by the traditional misclassification parameters. This study demonstrates how the main mine planning decisions impact the lower level activities such as the spacing of the infill drilling.

Acknowledgment

The work presented herein was part of a research project funded by Anaconda Operations, Anglo Gold, BHP Billiton, Highlands Pacific, MIM Holdings (Xstrata), Pasminco, Rio Tinto and WMC Resources.

References

Almeida, A.S. and Journel, A.G. *Joint simulation of multiple-variables with a Markov-type coregionalization model*. Mathematical Geology vol 26, no. 5, 1994, p. 565-588.

Boucher, A. Conditional joint simulation of random fields on block support. Earth Sciences. Brisbane, University of Queensland: 2003

Chilès, J.-P. and Delfiner P., *Geostatistics modeling spatial uncertainty.*, 1999

Desbarats, A. J. and Dimitrakopoulos, R. *Geostatistical simulation of regionalized pore-size distributions using min/max autocorrelation factors*. Mathematical Geology vol 32, no. 8, 2000 p. 919-942.

Dimitrakopoulos, R., 2003. Orebody *Uncertainty, Risk Assessment and Profitability in Recoverable Reserves, Ore Selection, and Mine Planning: Simulation Models, Concepts And Applications for the Mining Industry*. BRC, Brisbane, 342p.

Dimitrakopoulos, R. and Fonseca, M.B., *Assessing risk in grade tonnage curves in a complex copper deposit, Northern Brazil, based on an efficient joint simulation of multiple correlated variables*. APCOM 2003, Cape Town

Dimitrakopoulos, R. and Luo, X., *Generalized sequential Gaussian simulation on group size v and screen effect approximations for large field simulations*. Mathematical Geology vol 36, no. 5, 2004, p. 567-591.

Goovaerts, P., *Geostatistics for Natural Resources Evaluation*, Oxford University Press, 1997.

Switzer, P. and Green A. A.., *Min/Max autocorellation factors for multivariate spatial imagery*. Stanford University, Department of Statistics. 1984.

Jaine, O J. *Mine planning at Murrin-Murrin - modelling to determine the optimum path*. in Proceedings of Twelfth International Symposium on Mine Planning and Equipment Selection (AusIMM: Melbourne), 2003

Ravenscroft, P.J., *Risk analysis for mine scheduling by conditional simulation*, Transaction of the institution of mining and metallurgy Section A, vol. 101, 1992, p. A104-A108

Verly, G. W., *Sequential Gaussian cosimulation; a simulation method integrating several types of information. In* Soares and Amilcar (Eds.) Geostatistics. Kluwer Academic Publishers, Dordrect, p. 543-554

THEORY OF THE CLOUD TRANSFORM FOR APPLICATIONS

ODD KOLBJØRNSEN and PETTER ABRAHAMSEN
Norwegian Computing Center, Oslo, Norway

Abstract. We present the multidimensional cloud transform and propose an estimator for the transform. The estimation procedure is based on scatter plot smoothing. The resulting transform does not introduce artificial discontinuities in the transformed data, which is a common problem for the traditional estimates. The method is compared to a traditional estimate in a synthetic example.

Key words: Non-Gaussian distribution, stochastic simulation, seismic conditioning

1 Introduction

Seismic data provide valuable information with high lateral resolution that improves reservoir models. Geophysical variables such as acoustic impedance, shear impedance and Poisson ratio are often available through out the reservoir as results of seismic inversions. The cloud transform, see Bashore et al. (1994), is a frequently used tool when incorporating one explanatory variable such as the acoustic impedance into the reservoir model. The multi dimensional cloud transform incorporates multiple explanatory variables in the transform. This is useful as elastic inversions that provide multiple geophysical variables now are quite common.

Traditional estimates of cloud transforms are constructed by introducing non-geological facies, e.g. impedance classes. This method introduces artificial discontinuities in the petrophysical simulations, and requires a large amount of well data in order to obtain a reliable result. When the explanatory data have multiple dimensions the traditional binning estimates will suffer due to lack of accuracy and precision of the estimates because the number of bins increases dramatically with the dimension.

In the current work we present the cloud transform using a probabilistic terminology and propose estimators for the cloud transform that is based on scatter plot smoothing. The major difference between the current approach and other scatter plot smoothing approaches, e.g. Xu and Journel (1995) and Deutsch (1996), is that we work with the cloud transform directly and do not consider the joint density. The resulting estimates yield continuous transforms. Asymptotic expressions for accuracy and precision are presented and discussed. Asymptotic convergence rates are obtained such that for a given target distribution the asymptotically ideal smoothing factor can be computed. The convergence rate of the estimator is the same as the convergence rate for the estimator of the density of the explanatory variables in the transform. Thus it converges faster than the kernel estimator of the joint density of response variable and explanatory variables.

A presentation of the cloud transform is given in section 2, the estimator and asymptotic properties are given in section 3, synthetic example with comparison of proposed estimators to the traditional estimate is given in section 4. At the end there is a discussion and concluding remarks in section 5 and 6 respectively.

In what follows the function f denotes a generic density, where the random variable(s) in question is implied by the argument(s) of f, e.g. $f(\boldsymbol{x})$ and $f(y)$ denotes the density of $\boldsymbol{X}$ and Y respectively. Bold letters are used to denote vectors, e.g. $\boldsymbol{x} \in \mathcal{R}^d$. The function F denotes a cumulative distribution function the random variable in question is again implied by the argument, e.g. $F(y) = \text{Prob}(Y < y)$. Consequently $f(y|\boldsymbol{x})$ and $F(y|\boldsymbol{x})$ denotes the conditional pdf and cdf for Y given $\boldsymbol{X} = \boldsymbol{x}$ respectively. The quantile function that corresponds to the cdf F is denoted F^{-1} such that by definition $x = F^{-1}(F(x))$.

2 The cloud transform

The cloud transform is a conditional inverse probability transform. Let X and Y denote the explanatory and response variable respectively. Typically X is the acoustic impedance and Y is the porosity. A stochastic simulation from $f(x, y)$ can be obtained by the following algorithm:

Algorithm 1:

i) Compute $F(x)$
ii) Sample $u_1^* \sim \text{Uniform}[0, 1]$
iii) Let $x^* = F^{-1}(u_1^*)$
iv) Compute $F(y|x^*)$
v) Sample $u_2^* \sim \text{Uniform}[0, 1]$ independent of u_1^*
vi) Let $y^* = F^{-1}(u_2^*|x^*)$
vii) Return (x^*, y^*).

The transform in step $iii)$ is an inverse probability transform. The transform in step $vi)$ is the cloud transform. The multi dimensional cloud transform denotes the case when the explanatory variable is multi dimensional, i.e. $y = F^{-1}(u|\boldsymbol{x})$. For example can the components of $\boldsymbol{x}$ be the acoustic and the Poisson ratio.

In a spatial setting the cloud transform is applied pointwise, i.e.

$$Y(\boldsymbol{s}) = F^{-1}(U(\boldsymbol{s})|\boldsymbol{x}(\boldsymbol{s})), \tag{1}$$

with $\boldsymbol{s}$ being the spatial reference, $Y(\boldsymbol{s})$ being the response field, $U(\boldsymbol{s})$ being a p-field and $\boldsymbol{x}(\boldsymbol{s})$ being the given explanatory field. A p-field has the property that the stationary distribution of a realisation of $U(\boldsymbol{s})$ is uniform on $[0, 1]$. A spatially correlated p-field can for example be obtained as $U(s) = \Phi(Z(s))$, with Φ being the standard normal cdf; and $Z(s)$ being a standard normal random field.

On a bounded domain the response field $Y(s)$ defined in expression (1) is almost surely continuous if $f(x, y)$ is a density, $U^*(s)$ is almost surely continuous and $x(s)$ is continuous almost everywhere.

The following algorithm use the cloud transform to reproduce the conditional distributions of $Y(s)$ given $x(s)$: *Algorithm 2:*

i) For all s in grid: compute $F(y|x(s))$
ii) Sample a p-field $u^*(s)$ independent of $x(s)$
iii) For all s in grid: let $y^*(s) = F^{-1}(u^*(s)|x(s))$
iv) Return $y^*(s)$.

In step *ii)* the term independent is used in terms of independent stationary distribution; i.e. all information regarding $Y(s)$ given by $x(s)$ is given through the transform.

The cloud transform can also be used to reproduce joint multivariate distributions, by sampling in a sequential manner. The first variable is sampled according to an inverse probability transform; the next variables are sampled using the cloud transform given the previously sampled variables.

One can also imagine combinations of these two uses, by first simulating porosity given geophysical variables and next simulate permeability given geophysical variables and porosity.

The cloud transform become storage intensive as the dimension of the explanatory variable increase. Its hard store a cloud transform with a reasonable resolution if the dimension of the explanatory variable exceed four. In place of storing the transform one may consider to estimate it each time it is needed. The time requirement in this approach is prohibitive. In addition the number of data needed for a reliable estimate increase rapidly with the dimension.

3 Estimation of multi dimensional cloud transform

In an applied setting the cloud transform is unknown, and must be estimated from data. The estimator proposed here is based on the theory of kernel density estimation as presented in Silverman (1986), main results are summarised below. Other methods of density estimation see e.g. Donoho et al. (1996) and more refined approaches to kernel smoothing see e.g. Sain and Scott (1996), can also be developed into the setting of the cloud transform.

3.1 DENSITY ESTIMATION

Let $X_1, X_2, ..., X_n$ be a given multivariate data set whose underlying density is to be estimated. The kernel density estimator of the joint density is then

$$\hat{f}(x) = \frac{1}{nh^d} \sum_{i=1}^{n} k_d\left(\frac{x - X_i}{h}\right), \qquad (2)$$

with h being the bandwidth; and the kernel $k_d : \mathcal{R}^d \to \mathcal{R}$ being a radially symmetric unimodal probability density function such that

$$\int_{\mathcal{R}^d} x_i x_j k_d(\boldsymbol{x}) d\boldsymbol{x} = \delta_{ij},$$

where δ_{ij} is one if $i = j$; zero otherwise. Define further the constant

$$\beta_d = \int_{\mathcal{R}^d} |k_d(\boldsymbol{x})|^2 d\boldsymbol{x},$$

that is specific for the kernel k_d.

A standard argument using Taylor expansions yields the asymptotic expression for bias,

$$\mathrm{E}\left\{\hat{f}(\boldsymbol{x})\right\} - f(\boldsymbol{x}) \doteq \frac{h^2}{2} \nabla^2 f(\boldsymbol{x}), \qquad (3)$$

with ∇^2 being the Laplace operator in $\mathcal{R}^d$. The asymptotic variance is

$$\mathrm{Var}\left\{\hat{f}(\boldsymbol{x})\right\} \doteq \frac{\beta_d}{nh^d} f(\boldsymbol{x}). \qquad (4)$$

Combining the two yields the mean squared error

$$\begin{aligned}\mathrm{MSE}\left\{\hat{f}(\boldsymbol{x})\right\} &= \left[\mathrm{E}\left\{\hat{f}(\boldsymbol{x})\right\} - f(\boldsymbol{x})\right]^2 + \mathrm{Var}\left\{\hat{f}(\boldsymbol{x})\right\} \\ &\frac{h^4}{4}|\nabla^2 f(\boldsymbol{x})|^2 + \frac{\beta_d}{nh^d} f(\boldsymbol{x}).\end{aligned} \qquad (5)$$

From this expression one obtain the optimal rate of convergence for the bandwidth being,

$$h_{\mathrm{opt}} \sim n^{-1/(d+4)} \qquad (6)$$

yielding the convergence rate of the mean squared error,

$$\mathrm{MSE}\left\{\hat{f}(\boldsymbol{x})\right\} \sim n^{-4/(d+4)}. \qquad (7)$$

The integrated mean square error (IMSE) is a common measure of error in density estimation and is used to identify a common bandwidth for all $\boldsymbol{x} \in R^d$. It is however not possible to find a universal bandwidth that is applicable of all densities since the MSE and IMSE depend on the target density, see expression (5).

Consider also estimation of the cumulative distribution in one dimension, $F(x)$, using the ordinary count estimator,

$$\hat{F}(x) = \frac{\sum_{i=1}^n I(X_i \leq x)}{n}, \qquad (8)$$

with $I(X_i \leq x)$ being one if its argument is true zero otherwise. This estimator is unbiased and has variance according to the estimator of the probability in a binomial distribution. The mean squared error is thus identical to the variance which is

$$\text{Var}\left\{\hat{F}(x)\right\} = \frac{F(x)\left[1-F(x)\right]}{n}. \tag{9}$$

The convergence rate for the MSE of the count estimator is hence of order n^{-1}. This should be compared with the convergence rate of the density estimator, see expression (7). In one dimension the convergence rate for the density is $n^{-1+1/5}$. The convergence rate of the cdf corresponds to $d=0$ in expression (7).

3.2 CLOUD TRANSFORM ESTIMATION

Let $(Y_1, \boldsymbol{X}_1), (Y_2, \boldsymbol{X}_2), ..., (Y_n, \boldsymbol{X}_n)$ be a multivariate dataset for which the cloud transform is estimated with Y and $\boldsymbol{X}$ being the response and explanatory variable respectively. The kernel estimator of the joint density is then,

$$\hat{f}(\boldsymbol{x}, y) = \frac{1}{n\,h^d h_y} \sum_{i=1}^{n} k_d\left(\frac{\boldsymbol{x}-\boldsymbol{X}_i}{h}\right) \cdot k_1\left(\frac{y-Y_i}{h_y}\right), \tag{10}$$

where the kernel is separated for $\boldsymbol{x}$ and y; and h_y is the bandwidth used for the response variable. The target for the estimation is the conditional cumulative distribution $F(y|\boldsymbol{x})$. When using the density estimate in expression (10) one can obtain the estimator of $F(y|\boldsymbol{x})$ as,

$$\hat{F}(y|\boldsymbol{x}) = \frac{\sum_{i=1}^{n} k_d\left(\frac{\boldsymbol{x}-\boldsymbol{X}_i}{h}\right) \cdot K_1\left(\frac{y-Y_i}{h_y}\right)}{\sum_{i=1}^{n} k_d\left(\frac{\boldsymbol{x}-\boldsymbol{X}_i}{h}\right)}, \tag{11}$$

with $K_1(y) = \int_{-\infty}^{y} k_1(t)\,dt$. The bias in the estimator in expression (11) has the complexity

$$\text{E}\{\hat{F}(y|\boldsymbol{x})\} - F(y|\boldsymbol{x}) \sim o(h^2 + h_y^2).$$

The asymptotic variance has the complexity

$$\text{Var}\left\{\hat{F}(y|\boldsymbol{x})\right\} \sim o\left(\frac{1}{nh^d}\right).$$

The bound for the asymptotic variance is independent of h_y. This is intuitively explained by the fact that $K_1(y/h_y)$ in expression (11) is bounded whereas $k_1(y/h_y)/h_y$ in expression (10) is unbounded when h_y approaches zero. The usual trade off between bias and variance is not needed in the direction of the response variable. Thus let $h_y = 0$ an introduce the unnormalised conditional cdf of Y given $\boldsymbol{X}$,

$$G(y; \boldsymbol{x}) = \int_{-\infty}^{y} f(t, \boldsymbol{x})dt = F(y|\boldsymbol{x})f(\boldsymbol{x}).$$

The asymptotic bias for the case of $h_y = 0$ is

$$\text{E}\{\hat{F}(y|\boldsymbol{x})\} - F(y|\boldsymbol{x}) \doteq \frac{h^2}{2f(\boldsymbol{x})} \left[\nabla_{\boldsymbol{x}}^2 G(y; \boldsymbol{x}) - F(y|\boldsymbol{x})\nabla_{\boldsymbol{x}}^2 f(\boldsymbol{x})\right], \tag{12}$$

and the asymptotic expression for the variance is

$$\operatorname{Var}\left\{\hat{F}(y|\boldsymbol{x})\right\} \doteq \frac{\beta_d}{nh^d f(\boldsymbol{x})} \left[F(y|\boldsymbol{x})\left(1 - F(y|\boldsymbol{x})\right)\right]. \tag{13}$$

It is interesting to compare this variance with the one obtained for estimating empirical cumulative distributions in 1D. The factor $\left[F(y|\boldsymbol{x})\left(1 - F(y|\boldsymbol{x})\right)\right]/[nh^d f(\boldsymbol{x})]$ can be interpreted as the binomial uncertainty given $[nh^d f(\boldsymbol{x})]$ data, see expression (9). The factor β_d is related to the kernel smoothing, see expression (4).

By combining the bias in expression (12) and the variance in expression (13) to the mean squared error one see that the optimal rate of convergence for the bandwidth is obtained by

$$h_{\text{opt}} \sim n^{-1/(d+4)},$$

yielding the convergence rate for the mean squared error to be

$$\operatorname{MSE}\{\hat{F}(y|\boldsymbol{x})\} \sim n^{-4/(d+4)}. \tag{14}$$

This is the same rate of convergence as obtained for density estimation, see expression (7), but in expression (14) the dimension d refers to the dimension of the explanatory variables.

Note in particular that an estimator of the cloud transform that is based on the optimal kernel density estimator will have the convergence rate $n^{1/(d+5)}$, which is suboptimal.

The factor $1/f(\boldsymbol{x})$ which occur both in expression (12) and (13) is large in the flanks of $f(\boldsymbol{x})$. This factor may be reduced by transforming the explanatory variable to be approximately uniform on $[0,1]^d$. In the case of a one dimensional explanatory variable, the rank transform is uniform. In higher dimensions it is possible to obtain approximate uniform distributions by sequentially estimating the conditional transforms in the same manner as for the cloud transform, but applying them to the explanatory variables. This transform reduce the variance in the estimate but unfortunately the bias is increased trough the factor $\nabla_{\boldsymbol{x}}^2 G(y;\boldsymbol{x})$. The advantage is that the transform remains stable at the flanks. When the cloud transform is used to model spatial phenomena the histogram based on well logs have a smaller support than the histogram of the full field, due to the number of samples. It is therefore importance to have a reliable estimate of the cloud transform also towards the flanks of the distribution of the explanatory variable.

Note that the kernel estimator of the cloud transform corresponds to a density estimator for the explanatory variables. If the stationary distribution of the explanatory variable is an exhaustive sampling of this distribution, the marginal histogram of the dependent variables is exactly reproduced in the full field.

3.3 SPECIAL CASES

It is interesting to investigate the proposed estimator for cases in which one can see the effect directly.

In the trivial case where there is no explanatory variable, the estimator is identical to the empirical cdf of the response variable, see expression (8).

If the response variable is independent of the explanatory variable, the estimator introduces local bias for the simulated response variable. However the stationary distribution of the response variable will still be reproduced. In this case it is obvious that an infinite bandwidth is optimal for the explanatory variable.

When the response variable is discrete, the estimator is identical to estimates obtained by kernel density estimation for each level of the response variable. The kernel density estimates for all classes have a common bandwidth. The probability of a class at a given value of the explanatory variable is proportional to the density estimate weighted with the number of data in this class.

If there is a functional relationship between the explanatory and the response variables, the estimator blurs this relation. This introduces artificial uncertainty in the predictions. The obvious choice in this case is to estimate this deterministic relation instead of introducing the cloud transform which is a stochastic transform.

4 Example

The properties of the estimators are investigated in a synthetic example where a relation between acoustic impedance and porosity is considered. In Figure 1 the scatter plot of the data that are used to estimate the transform is displayed together with the cloud transform. A vertical line in the cloud transform yields a cumulative distribution for the porosity increasing monotonically from zero at low porosity values to one for high porosity values. In the figure both extreme ends are coloured white in order to highlight the active region.

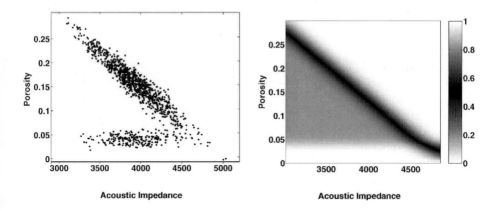

Figure 1. Data and original transform. On the left is the scatter plot of the 1200 well observations that are used to estimate the transform. On the right is the original cloud transform of the joint distribution of acoustic impedance and porosity. All cumulative distributions are zero for low porosity values and one for high values.

The binned estimator is compared with two estimators based on scatter plot smoothing. The first use the basic variable, i.e. acoustic impedance, the second use the rank transform of the basic variable as explanatory variable in the cloud transform. In Figure 2 the three estimated cloud transforms are displayed together with the original transform. Both estimates based on scatter plot smoothing are continuous whereas the binned estimate has clear discontinuities as the acoustic impedance crosses the boarder between bins. The binned estimate and the untransformed scatter plot smoother become unstable at the ends. This is a result of the high variance in the estimate in the extreme ends. The transformed scatter plot smoother is stable at the ends, but the bias is evident in the figure.

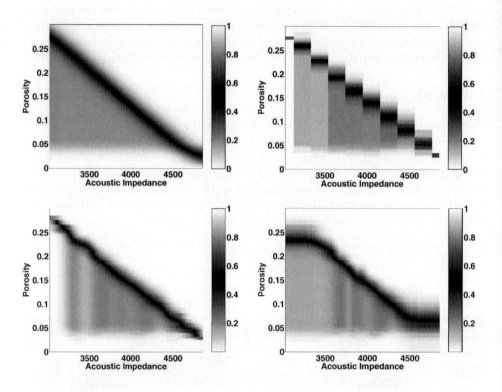

Figure 2. Original and estimated transforms. On the top left is the original cloud transform i.e. the target of estimation, top right is the binned transform, on bottom left is the estimate based on scatter plot smoothing, bottom right is the estimate based on a rank transform of the acoustic impedance. All three estimates are based on the 1200 well observations displayed in the scatter plot in Figure 1.

In order to compare statistical properties of the estimators of the cloud transform the root integrated mean square error,

$$\text{RIMSE}\left\{\hat{F}(y|x)\right\} = \left(\int_{\mathcal{R}} \text{MSE}\{F(y|x)\}\, dy\right)^{1/2},$$

is computed for each value of the acoustic impedance. The mean squared error is approximated by Monte Carlo integration using the following procedure. Generate 1000 independent data sets all consisting of 1200 data pairs. For each data set estimate the transform and compute the squared deviation between this and the true transform. The average of the 1000 squared deviations is the approximation to the mean squared error. The results for the three estimators are displayed in Figure 3, the density of the acoustic impedance, i.e. $f(x)$, is overlaid in the figure.

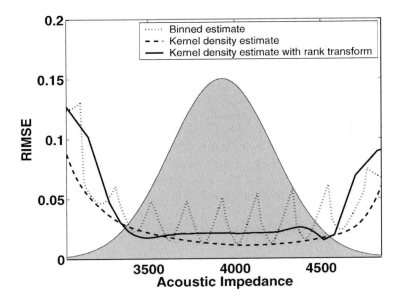

Figure 3. Pointwise root integrated mean square error. The root integrated mean square error for three estimators considered. Overlaid is the density of acoustic impedance. The range in the figure is about six times the standard deviation for acoustic impedance.

In terms of root integrated mean square error both estimates based on scatter plot smoothing outperform the binned estimate. The scatter plot smoother that use the basic variable is better than the one based on the rank transform.

5 Discussion

The cloud transform can be used to reproduce any multivariate distribution; however the spatial dependence is hidden in the p-field. A scatter plot such as the one in Figure 1 may indicate an underlying dichotomous random field. It is not obvious how to create a p-field with desired spatial properties, e.g. channels. However if the wells are dense compared with the correlation length of the random fields a p-field originating from a transformed Gaussian field may be satisfactory. An alternative approach is to build a facies model with the desired spatial properties and build separate cloud transforms for petrophysical modelling within each facies.

Scatter plots that are used for estimation of the cloud transform usually come from well observations. This will result in data that are correlated and not independent which is assumed in the calculations above. This will most likely have the effect that the variance of the estimator is larger than given in expression (13) above. The scatter plot may also come from rock physics simulations. In which case it is likely that the data are independent and the results are strictly valid.

6 Concluding remarks

We have proposed two estimators for the cloud transform. Both estimators are based on scatter plot smoothing and result in continuous estimates. The optimal bandwidth of the estimator has the same convergence rate as for the density estimation in the space of explanatory variables. There is however no need to smooth in the direction of the response variable as this introduces additional bias.

In a test example both proposed estimators are found to perform better than the traditional binning estimate in terms of root integrated mean squared error, also the estimators are more appealing visually as they preserve continuity in the estimate. The estimator based on the basic explanatory variable is the best of the two estimators in terms of root integrated mean squared error. This estimator does however have large variance at the ends of the interval resulting in an unstable estimate. This can be unfortunate, see discussion in end of section 3.2.

When choosing estimator for the cloud transform one should not only consider its theoretical properties, but also how the resulting estimate will be applied. The authors prefer a slightly higher bias in order to preserve good properties, i.e. smoothness, of the estimated transform at the flanks of the distribution of the explanatory variable.

Acknowledgements

The authors acknowledge Alistair MacDonald of Roxar ASA. The Norwegian research council has in part provided the funding for this work.

References

Bashore, W M.; Araktingi U.G.; Levy M; Schweller U.G. *Importance of a Geological Framework for Reservoir modelling and subsequent Fluid -Flow Predictions* AAPG Computer application in geology., No.3; Jeffery M. Yarus and Richard L Chambers eds.; 1994.

Deutsch C.V. *Constrained Smoothing of Histogram and Scatter plots with Simulated Annealing.* Technometrics, Vol. 38, No. 3,pp. 266-274,1996.

Donoho D.L.; Johnstone I.M.; Kerkyacharian G. and Picard D. *Density Estimation by Wavelet Thresholding* The Annals of Statistics, Vol. 24, No. 2, pp. 508-539, 1996.

Sain S. R.; Scott D. W. *On Locally Adaptive Density Estimation* Journal of the American Statistical Association, Vol. 91, No. 436, pp. 1525-1534, 1996.

Silverman B.W. *Density Estimation for Statistics and Data Analysis.*Chapman & Hall 1986.

Xu, W. and Journel, A.G. *Histogram and Scattergram Smoothing Using Convex Quadratic Programming* Mathematical Geology.Vol 27 , No.1; pp. 83-103 ; 1995.

PROBABILITY FIELD SIMULATION: A RETROSPECTIVE

R. MOHAN SRIVASTAVA[1] and ROLAND FROIDEVAUX[2]
[1] *FSS Canada Consultants, Toronto, Canada*
[2] *FSS Consultants SA, Geneva, Switzerland*

Abstract.
The practical advantages and theoretical disadvantages of P-field simulation are reviewed in the light of more than a decade of application and research since it was first introduced. A case study example highlights the enduring attractions of the algorithm: its flexibility and speed.

1 Introduction

When first introduced, probability field simulation was well-suited to certain types of problems that were not well handled by other simulation algorithms available. In particular, it adapted well to the situation where *a priori* local distributions were available. As it rapidly gained practical acceptance, largely because of its speed, "P-field" simulation was also dismissed by some as a procedure lacking a proper theoretical foundation — more of a clever algorithmic trick than a properly conceived approach to stochastic spatial simulation.

In the past decade, the advantages and shortcomings of the procedure have been illuminated through continued widespread application and theoretical research. This paper begins with an overview of the theoretical background and the usual practical implementation of P-field simulation. It then discusses theoretical concerns and assesses their practical implications. A mining case study example illustrates two enduring strengths of P-field simulation: flexibility and speed.

2 Overview and implementation

Let $F[\mathbf{u}; z]$ denote the cumulative distribution function (cdf) at location $\mathbf{u}$ of an attribute Z. Any simulated value, z_{sim}, represents a specific quantile of this local cdf: the z-value at which $F[\mathbf{u}; z]$ reaches a probability $p(\mathbf{u})$:

$$z_{\text{sim}} = F^{-1}[\mathbf{u}; p(\mathbf{u})] \qquad (1)$$

The p values are not spatially independent; this would preclude reproduction of almost any desired spatial autocorrelation in Z. Instead, the p values must be regarded as a realization of a random function $P(\mathbf{u})$, and simulated with an appropriate pattern of spatial continuity.

P-field simulation therefore proceeds as follows:
1. Generate a non-conditional realization of $P(\mathbf{u})$, i.e. a grid of spatially autocorrelated values that are uniformly distributed between 0 and 1.[1]
2. Use $P(\mathbf{u})$ to sample the local cdf $F[\mathbf{u}; z]$.

This procedure ensures that two of the common goals of conditional simulation are met: conditioning data are honored, as is the target global distribution. The global distribution of $Z(\mathbf{u})$ is honored because the local cdfs are sampled using $U[0, 1]$ values. As long as local cdfs correctly model local distributions of uncertainty, sampling these with $U[0, 1]$ values will preserve the global distribution. Conditioning data are honored because local cdfs collapse to a spike at data locations. Regardless of the probability value used to sample these zero-width distributions, the simulated value will match the conditioning data value.

The third common goal of conditional simulation, the reproduction of the variogram of $Z(\mathbf{u})$, is not exactly guaranteed. Since the $p(\mathbf{u})$ values are spatially autocorrelated, the $z_{\text{sim}}(\mathbf{u})$ will also be spatially autocorrelated, but the precise nature of the autocorrelation of the z_{sim} values is not directly controlled. The resulting variogram of the z_{sim} values will not necessarily reflect the intended target Z variogram model. As discussed later, it will often be very close to the desired target but there are situations in which, despite having *some* spatial continuity, the z_{sim} values do not have exactly the desired pattern of spatial continuity.

3 Theoretical considerations

The first P-field papers (Srivastava, 1992; Froidevaux 1993) focused on algorithmic details; little theoretical justification was provided and the acceptance of the procedure was due to its practical success. Theoretical investigations soon followed, however, and links between the P-field approach and other conditional simulation methods were eventually elucidated (e.g. Journel and Ying, 2001).

Although Journel (1995) proved that, in the absence of conditioning data, P-field simulation correctly reproduces univariate and bivariate properties of Z, Pyrcz and Deutsch (2001) pointed out that: i) if a stationary covariance model is used for P, the covariance of Z is not stationary and is biased in the vicinity of conditioning data and, ii) conditioning data usually appear as local extremes in the realizations.

3.1 INFERRING THE LOCAL CDFS

P-field simulation does not concern itself with the determination of the local cdfs; it considers them to have already been established. The origin of the local cdfs

---

[1] This is usually done by generating non-conditional Gaussian values, $Y(\mathbf{u})$, and then using the inverse of the cumulative Gaussian distribution to transform the Y values to P values: $P(\mathbf{u}) = G^{-1}[Y(\mathbf{u})]$. The generation of spatially autocorrelated Gaussian values can be done extremely rapidly using fast Fourier transform (FFT) algorithms or by efficient moving averages. In applications where the local cdfs are Gaussian, the transformation from Y values to P values can be skipped and the Y values are simply linearly transformed to a Gaussian distribution with the proper mean and variance.

does play a role in the theoretical analysis of the spatial structure of P. It is useful, therefore, to elucidate some common cases for the determination of local cdfs and to discuss how these impinge on the variogram model for P:

Case 1: Local cdfs not locally data-conditioned but are identified instead with the prior marginal distribution of Z: $\text{Prob}\{Z(\mathbf{u}) < z\} = F(z)$.

Case 2: Local cdfs are not locally data-conditioned but are identified instead with non-stationary prior distributions of Z: $\text{Prob}\{Z(\mathbf{u}) < z\} = F(\mathbf{u}; z)$.

Case 3: Local cdfs are estimated from existing sample data using an appropriate geostatistical technique: $\text{Prob}\{Z(\mathbf{u}) < z\} = F[\mathbf{u}; z|(n)]$.

The single most important issue here is conditioning to sample data. This will have a direct impact on the inference of the variogram model.

3.2 P-FIELD VARIOGRAM MODEL: STATIONARY OR NOT?

P-field simulation usually uses a stationary variogram model for P. This normal practice follows from the original suggestion of Froidevaux (1993): that the P variogram be modelled from the experimental variogram of the uniform transform of the available data. As Pyrcz and Deutsch (2001) pointed out, howeer, if the Z values are assumed to be second-order stationary, then the use of a stationary variogram model for P is inconsistent. If P is defined as

$$P(\mathbf{u}) = F[\mathbf{u}; Z(\mathbf{u})] \qquad (2)$$

using data-conditioned local cdfs, then second-order stationarity of Z entails lack of second-order stationarity for P. Cassiraga (1999) has shown that the range of autocorrelation of P is linked to the spatial density of the conditioning data.

To date, theoretical analysis of P-field simulation has proceeded from the assumption that the Z values have second-order stationarity, and that the P values are defined using Equation 2 above. One could, however, take a different approach: assume that the P values have second-order stationarity and that the Z values are defined using Equation 1. P and Z play complementary roles in Equations 1 and 2, and the results of Cassiraga (1999) can be extended to demonstrate that if we choose a random function model in which the P values are second-order stationary, then the consequence is that the Z values cannot be; or, as also pointed out by Pyrcz and Deutsch (2001), the stationarity of the P-field covariance makes the covariance structure of Z dependent on the nearby conditioning data.

So with two alternate random function models — one that is better researched and that chooses second-order stationarity for Z; the other that chooses second-order stationarity for P and whose theory has barely been explored — the question arises: which one is more appropriate? Though the tradition of geostatistics has been to choose second-order stationarity as a Z property, it is worth considering the pro's and con's of bestowing this property on P instead.

It is clear from the construction of the P-field that the P values are, globally, first-order stationary; if they are not, then the local cdfs do not properly quantify the local probability distribution of Z. It is equally clear, from practice, that in most interesting earth science applications, first-order stationarity of Z is a questionable choice. Geostatistics has adopted the good practice of using local search neighborhoods so that the dependence on stationarity becomes local; but the practical success of local customization of estimation and simulation parameters is consistent with the view that an assumption of global first-order stationarity is rarely appropriate for Z.

Moving from the consideration of first-order stationarity to second-order stationarity, if the Z values are not first-order stationary, why does it make sense to assume that they are, globally, second-order stationary? Might it not be better to assign the property of second-order stationarity to a random variable, P, that is known, by construction, to be globally first-order stationary?

The technical literature on P-field simulation has elucidated the fact that P and Z cannot both be second-order stationary. Research remains to be done on the theoretical consequences of the user's choice on which of the two complementary random variables this property will be assigned to.

3.3 LOCAL EXTREMES AT DATA LOCATIONS

When hard data are used to locally condition cdfs, a sample at $\mathbf{u}$ will typically have a very strong influence on the cdf at an adjacent location, $\mathbf{u}'$. If the local cdf $F[\mathbf{u}'; z|(n)]$ has been estimated geostatistically, then its mean will tend to be very close to the adjacent data value, $z(\mathbf{u})$, and its variance will be small. Given this situation, if the p values in the vicinity of $\mathbf{u}$ are significantly less than 0.5, then $z(\mathbf{u})$ will be a local maximum in the realization. Conversely, if the nearby p values are larger than 0.5, then $z(\mathbf{u})$ will be a local minimum. The conditioning data, $z(\mathbf{u})$, will not be noticeable as a locally extreme artifact in the realization only if the nearby probability field values are around 0.5.

4 Discussion

4.1 DECOUPLING CDF ESTIMATION FROM SAMPLING

The practical advantage of P-field simulation stems from the decoupling of the sampling of cdfs from their estimation. As with other geostatistical simulation procedures, the local cdfs in a P-field approach can be established through some form of kriging; they can also be derived directly from secondary information. In many petroleum applications, for example, geophysics or petrophysics can provide constraints on rock properties such as porosity and permeability, and on structural properties such as thickness and depth to top of reservoir. In such situations, local cdfs can be built directly from geophysical data and no kriging is required; all that remains is the appropriate sampling of the geophysically-derived local cdfs.

In studies that involve resource estimation, the decoupling of cdf estimation from cdf sampling has another benefit: it is easier to ensure that simulated outcomes do not imply outlandish or aberrant resource estimates. Though the concept that simulations fluctuate around the expected value is well understood theoretically, in practice it can be hard to ensure that the average of many realizations is suitably close to an already-calculated resource estimate. There are many situations, especially in mining applications, where conditional simulations are being considered (for grade control, for example, or for blending studies) and where a well accepted and carefully developed resource block model already exists. The use of a P-field approach that incorporates previously accepted and trusted local cdfs avoids the embarassment and confusion that results when simulated outcomes depart, significantly on average, from the "best estimate" of the global resource.

4.2 HONORING THE VARIOGRAMS

As noted above, the definition and inference of the P variogram is theoretically troublesome if the Z values are assumed to be second-order stationary. The practical impact of this issue is, however, usually minor. With the real goal being reproduction of the Z variogram; the P variogram is an intermediate stepping-stone. Even if the P variogram is theoretically ill-defined, the user can still adopt a variogram model based on analysis of the uniform transform of Z and can adjust this model if the resulting variogram of the z_{sim} values is unacceptable.

Luster (1985) discussed departures between target variogram models and experimental variograms of realizations. He noted that, by virtue of being conditioned by hard data, realizations have a pattern of spatial continuity whose mid- and long-range structure is controlled not by the variogram model but rather by available data. In practice, the critical aspect of the P variogram model is, therefore, its behavior at short distances (up to the nominal spacing of data). With the short-scale characteristics of the P variogram model well chosen, especially directional anisotropy and relative nugget effect, the results of P-field simulation are usually well within the fluctuations normally tolerated in conditional simulation studies.

Compared to sequential methods, P-field simulation is more successful at creating realizations with very low nugget effects and strong short-scale continuity (such as those typical of thickness or top of structure in petroleum applications). The realizations from sequential methods often have too much short-scale variability[2] and need to be post-processed to remove such artifacts (e.g. Tran, 1994).

4.3 LOCAL EXTREMES AT DATA LOCATIONS

To solve the problem of local extremes at data locations, Goovaerts (2002) proposed the use of a conditional probability field with fixed probability values of 0.5 at data locations. This entails a preferential sampling of the central part of the cdfs in the immediate vicinity of conditioning data. Although this method removes the artifacts, it does so at the expense of execution speed, which is one of the most attractive features of P-field simulation. Moreover, the justification and the consequences of forcing an arbitrary fixed p value still remain to be explored.

---

[2] A consequence of unstable kriging weights caused by strong screen effects in the end-stages of sequential simulation on a dense regular grid.

A practical application for which local extremes at data locations are clearly undesirable is flow and transport modelling. If wells or bore holes coincide with local minima and maxima in the permeability field, attempts to predict flow and transport may be seriously biased. In this sense, the P-field artifact of local extremes is similar to the "striping" or "banding" artifact often seen in realizations from the turning bands method and in realizations from sequential methods that do not randomize the sequential path. While such artifacts may not have any practical impact in certain types of studies, they may be serious flaws in others.

5 Case study: uncertainty on a mineralized envelope

Studies of mineral resource estimates typically incorporate a "mineralized envelope", an outer bounding limit beyond which grades are not estimated. Many case studies have demonstrated that the failure to adequately constrain the domain within which grades are estimated can lead to very unrealistic block models that overstate the tonnage of mineralized material, with peripheral grades being overestimated and grades in the heart of the deposit being underestimated.

Though some kind of mineralized envelope is necessary, the traditional approach, unfortunately, is to treat this boundary as deterministic. The limits of mineralization identified in drill holes typically serve as control points for a 3D solid or "wireframe". With the mineralized envelope thus frozen, the impact of the uncertainty of this envelope on resource estimates is very difficult to quantify. Even when simulation is used to study grade fluctuations within the envelope, the additional uncertainty due to the wireframe definition itself is rarely addressed.

Figure 1 shows an example of a simple wireframe constructed from exploration holes drilled on a 100m grid. These holes identify a deposit that lies in a shear zone between two faults, with a sharp hangingwall contact that can usually easily be correlated from hole to hole across the deposit. The footwall contact, which is more diffuse, does not appear to be a structural contact and is not clearly associated with any particular geological characteristic. In places, the footwall coincides with feldspathic alteration; in other places, it occurs where the intensity of shearing drops.

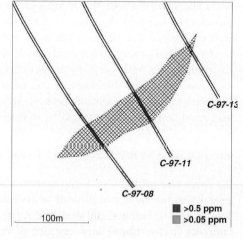

Figure 1. Drill holes and interpretation of mineralized envelope from initial drilling.

All three holes on the north-facing section shown in Figure 1 intersect anomalous mineralization which, for this project, was defined as more than five consecutive meters of drill core with total precious metals (TPM) in excess of 0.5 g/t. Small changes in the grade threshold or the length of interval have little impact on the definition of the hangingwall contact but have a more appreciable impact on the footwall.

The wireframe developed from initial drilling is necessarily simplistic, little more than a schematic cartoon that approximates the deposit's heart. In a second drilling campaign, in-fill holes were drilled from a development drift to penetrate the deposit from the west. Figure 2 shows the new holes with their mineralized intercepts, along with the old drill hole data and the mineralized envelope from Figure 1. The original wireframe provided good predictions of down-hole depth to the hangingwall, but its predictions of depth to the footwall are less precise.

Figure 3 shows the interpreted mineralized envelope, updated to honor all data currently available. With more closely spaced data, the shape of the wireframe has become slightly more complex. Though this new interpretation is an improvement, it is still far from perfect. If even more closely spaced holes were available, new short-scale complexities would be discovered in the shape of the mineralized zone. Rather than using the outline in Figure 3 as a single deterministic boundary for purposes of resource estimation, we would like to run several resource estimates, each one with a different but plausible version of the mineralized envelope, to study the impact on resources of uncertainty on the shape of the deposit. In conjunction with conditional simulations of TPM grades within the mineralized envelope, this will help determine whether or not additional definition drilling is required.

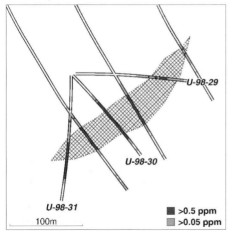

Figure 2. Drill holes after underground development drilling, with old interpretation of mineralized envelope.

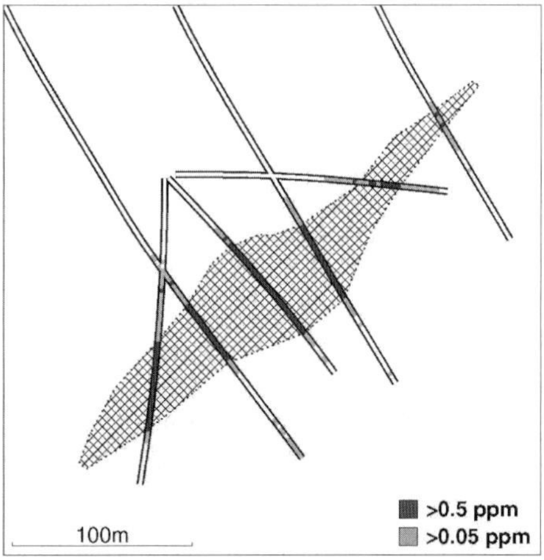

Figure 3. Drill holes after underground development drilling, with new interpretation of mineralized envelope.

P-field simulation has been used to produce alternate versions of the mineralized envelope, each one of which honors the drill hole data from the first two years. The attribute being simulated is $\Delta D(\mathbf{u})$, the deviation[3] of the true (but largely unknown) mineralized envelope from the current working interpretation shown in Figure 3. Close to existing drill holes, the current working interpretation is reliable and $\Delta D(\mathbf{u})$ is close to 0. As we move farther away from existing holes, the deviations between the actual and predicted surfaces will tend to become larger.

---

[3] Measured orthogonal to the wireframed surface, with the sign determining whether the deviation is outward (+) or inward (−).

For the 62 holes drilled in the second year, and which intersected the mineralized zone, Figure 4a shows the differences between actual depth to the hangingwall and predicted depths as a function of distance from a hole drilled in the first year. This plot, which shows us actual historical values of $\Delta D(\mathbf{u})$, can be used to calibrate possible future fluctuations. The dashed line in Figure 4a shows a model of $\pm$ one standard deviation of $\Delta D(\mathbf{u})$ as a function of distance from an existing drill hole; the dotted line shows $\pm$ two standard deviations.

Figure 4b shows the corresponding data and models for depth to the footwall. As noted earlier, the lack of a clear geological distinction at the footwall makes the wireframe a less reliable predictor of the footwall location than of the hangingwall location — $\Delta D(\mathbf{u})$ values are generally larger in magnitude on the footwall side.

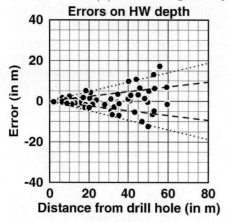
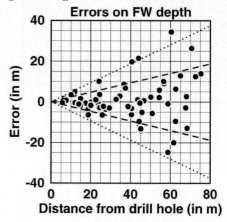

Figure 4a. $\Delta D(\mathbf{u})$ versus distance from nearest drill hole for hangingwall.

Figure 4b. $\Delta D(\mathbf{u})$ versus distance from nearest drill hole for footwall.

Using the dashed lines in Figure 4a and 4b, local cdfs of $\Delta D(\mathbf{u})$ can be constructed at every point on the mineralized surface. The distance from each point on the surface to the nearest drill hole intercept is calculated. Reading up from the x-axis on Figure 4 to the dashed line and across to the y-axis gives the standard deviation of $\Delta D(\mathbf{u})$ at that location; the mean is assumed to be zero and the shape of the cdf is assumed to be Gaussian. Figure 5 shows the median and $\pm 2\sigma$ bands of the local cdfs for the section shown in Figure 3.

With the local cdfs now established, all that remains is to sample them using a P-field with an appropriate range of

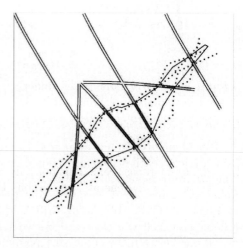

Figure 5. Local cdfs shown as a $\pm 2\sigma$ band around the median.

spatial autocorrelation. Figure 6 shows variograms of the uniform transform of the 62 $\Delta D(\mathbf{u})$ values from the second drilling campaign, along with their variogram models. Using these variograms, two 2D fields of spatially correlated probability values were created, one for use on the hangingwall and one for use on the footwall; these autocorrelated p values were then used to sample the local cdfs. Two of the resulting 100 realizations are shown in Figure 7.

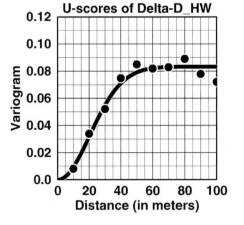
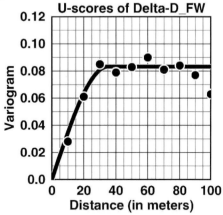

Figure 6a. Variogram for $\Delta D(\mathbf{u})$ on the hangingwall contact.

Figure 6b. Variogram for $\Delta D(\mathbf{u})$ on the footwall contact.

This example highlights the fact that local cdfs need not be estimated by kriging. In this case, they are established instead by a straightforward calibration based on historical data. This example also illustrates that the theoretical complexities of the P-field's statistical properties need not be an impediment to practical application. The uniform scores provide an experimental variogram that is easily modelled and that, when used to create unconditional P-fields, leads to geologically plausible results that greatly assist the assessment of project uncertainty and risk.

6 Conclusions

Even with exponential advances in computational speed, and the availability of many newer simulation algorithms, P-field simulation will likely remain one of the most often used geostatistical simulation procedures. Whenever local cdfs are already available and do not need to be generated using kriging, the P-field approach will be attractive since it decouples the issue of estimating cdfs from the task of sampling them. This not only reduces computational overhead, it also allows the user to generate realizations that fluctuate around a predetermined "base case", an advantage in many resource-based studies where a best estimate of global resources has already been established.

For studies that call for rapid generation of large numbers of conditional realizations, P-field will be attractive for its computational speed. Even as computer power has made it possible to run simulations hundreds of times faster than ten years ago, the appetite for larger simulations and for more realizations has kept

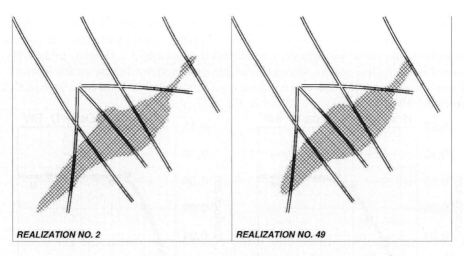

Figure 7. Two realization of the mineralized wireframe

pace. When a few realizations containing millions of grid nodes were once satisfactory, it is now not uncommon to generate hundreds of realizations containing tens of millions of grid nodes.

At the same time that P-field simulation will continue to be a good choice for many common applications, it is clear that there are many other common applications for which it is not a good choice. In particular, its tendency to create local extremes at conditioning locations makes it undesirable whenever downstream use of the realizations involves post-processing that is influenced by such artifacts. Fluid flow and contaminant transport studies are two examples of applications in which permeability extremes at wells or boreholes are clearly undesirable.

References

Cassiraga, E.F., 1999, *Incorporation de información blanda para la cuantificación de la incertitumbre: aplicación a la hidrogeologia*, Ph.D. Thesis, Universidad Politecnica de Valencia, Spain.

Froidevaux, R., 1993, "Probability Field Simulation", *Geostatistics Troia 1992*, A. Soares (ed.), Volume 1, Kluwer Academic Publishers, Dodrecht, Holland, pp. 73–84.

Goovaerts, P., "Geostatistical modelling of spatial uncertainty using p-field simulation with conditional probability fields", *International Journal of Geographical Information Sciences*, vol. 16, no. 2, pp. 167–178.

Journel, A.G., 1995, "Probability fields: another look and a proof", *SCRF Annual Report 8*, Stanford Center for Reservoir Forecasting, Stanford University, U.S.A.

Journel, A.G., and Ying, Z., 2001, "The theoretical links between sequential Gaussian simulation, Gaussian truncated simulation and probability field simulation, *Mathematical Geology*, vol. 33, no. 1, pp. 31–40.

Luster, G. R., 1985, *Raw Materials for Portland Cement: Applications of Conditional Simulation of Coregionalization*, Ph.D. Thesis, Stanford University, U.S.A.

Pyrcz, M., and Deutsch, C.V., 2001, "Two artifacts of probability field simulation", *Mathematical Geology*, vol. 33, no. 7, pp. 775–800.

Srivastava, R.M., 1992, "Reservoir characterization with probability field simulation", *SPE Paper No. 24753*, SPE Annual Conference and Exhibition, Washington, DC, U.S.A.

Tran, T., 1994, "Improving variogram reproduction on dense simulation grids", *Computers and Geosciences*, vol. 20, pp.1161–1168.

SEQUENTIAL SPATIAL SIMULATION USING LATIN HYPERCUBE SAMPLING

PHAEDON C. KYRIAKIDIS
Department of Geography, University of California Santa Barbara, Ellison Hall 5710, Santa Barbara, CA, 93106-4060, U.S.A.

Abstract. An efficient method is proposed for generating realizations from an arbitrary multivariate distribution using sequential simulation and Latin hypercube sampling. In a spatial context, this efficiency entails a reduction of sampling variability in statistics of spatially distributed model outputs when the inputs are realizations of random field models. The proposed method yields an unbiased reproduction of a target semivariogram, even for a small number of realizations, and consequently can be used for enhanced uncertainty and sensitivity analysis in complex spatially distributed models. In addition, the method is simple enough to be incorporated in virtually any geostatistical software for sequential simulation.

1 Introduction

Monte Carlo simulation is routinely used for uncertainty and sensitivity analysis of model outputs in a wide spectrum of scientific disciplines (Morgan and Henrion, 1990). Any realistic uncertainty analysis, however, calls for the availability of a representative distribution of such outputs, and can become extremely expensive in terms of both time and computer resources in the case of complex models and simple random (SR) sampling. This problem is far more pronounced for spatially distributed models, due to the large number of correlated (regionalized) variables comprising each input parameter map to such models, e.g., 3D rasters of hydraulic conductivity used for simulation of flow and transport in porous media.

An intelligent alternative to SR sampling is Latin hypercube (LH) sampling, a special case of stratified random sampling, which yields a more representative distribution of model outputs (in terms of smaller sampling variability of their statistics) for the same number of input simulated realizations. Analytical results demonstrating the efficiency of LH over SR sampling from univariate distributions are given in the (now classic) paper of McKay et al. (1979). A more recent comprehensive review of LH sampling for uncertainty and sensitivity analysis in complex systems can be found in Helton and Davis (2003).

LH sampling from a multivariate distribution, i.e., the task of inducing correlation in LH samples, is an important research theme in risk analysis and reliability engineering (Haas, 1999), which becomes critical in a spatial context for ensuring unbiased outputs of complex spatially distributed models. This paper makes a novel contribution to the literature of spatial uncertainty analysis, by proposing a simple and efficient method for sequential LH sampling from random field models.

2 Latin hypercube sampling

Consider a set of K independent continuous RVs $\{Y_k, k = 1, \ldots, K\}$, with $F_{Y_k}(y_k) = Prob\{Y_k \leq y_k)$ denoting the cumulative distribution function (CDF) of the k-th RV Y_k. Simple random (SR) sampling of N realizations from RV Y_k proceeds by first generating a $(N \times 1)$ vector $\mathbf{u}_k = [u_k^{(n)}, n = 1, \ldots, N]'$ of uniform random numbers in $[0, 1]$, which are treated as simulated probability values, and then transforming $\mathbf{u}_k$ into a $(N \times 1)$ vector $\mathbf{y}_k = [y_k^{(n)}, n = 1, \ldots, N]$ of simulated realizations as: $\mathbf{y}_k = F_{Y_k}^{-1}(\mathbf{u}_k)$, using the inverse CDF $F_{Y_k}^{-1}$ of RV Y_k.

Latin hypercube (LH) sampling of N realizations from the k-th RV Y_k calls for generating, independent of vector $\mathbf{u}_k$, a $(N \times 1)$ vector $\mathbf{p}_k = [p_k^{(n)}, n = 1, \ldots, N]'$ of random permutations of N integers $\{1, 2, \ldots, N\}$. A $(N \times 1)$ vector $\mathbf{z}_k = [z_k^{(n)}, n = 1, \ldots, N]'$ of stratified realizations is then obtained as (McKay et al., 1979):

$$\mathbf{z}_k = F_{Y_k}^{-1}\left(\frac{\mathbf{p}_k - \mathbf{u}_k}{N}\right) \tag{1}$$

where the argument $(\mathbf{p}_k - \mathbf{u}_k)/N$ of the inverse CDF $F_{Y_k}^{-1}$ ensures that the simulated probability values for the k-th RV Y_k are stratified, i.e., fall in N different probability strata. The monotonic transformation of the simulated probabilities incurred by the inverse CDF $F_{Y_k}^{-1}$ does not ruin stratification, which entails that each entry of vector $\mathbf{z}_k$ (each simulated value) falls within a different stratum in the original variable space, no matter the distributional form of $F_{Y_k}(y_k)$. The independence of vectors $\mathbf{p}_k$ and $\mathbf{u}_k$ ensures that there is a uniform probability $1/N$ for a simulated value within a particular stratum, i.e., there is no systematic placement of simulated values at the edges of strata. Variations of the above basic LH sampling procedure to further control sampling variability include variance reduction techniques, such as antithetic and control variates, as well as correlated sampling (Ang and Tang, 1984; Switzer, 2000).

A naive application of the above LH sampling procedure to correlated RVs fails to induce any correlation in the simulated values, simply because vectors $\mathbf{p}_k$ and $\mathbf{u}_k$ for the k-th RV Y_k are generated independent of other such vectors for other RVs. From these two sources that contribute to lack of correlation, the most important one is the vector $\mathbf{p}_k$ of random permutations because it dictates the strata within which the entries of $\mathbf{u}_k$ are distributed. To date, the most widely used method for generating LH samples from correlated RVs with a given rank correlation coefficient is the distribution-free method of Iman and Conover (1982). This method, however, is prohibitive for a large number $K > 10,000$ of RVs (typically the case in a spatial setting) because it calls for the Cholesky decomposition of an extremely large $(K \times K)$ variance-covariance matrix.

Stein (1987) proposed a (now also widely used) post-processing method for transforming a SR sample from K correlated RVs into a LH sample. Stein's method is independent of the simulation algorithm used to generate the original SR sample, and can be applied in principle to a large number K of RVs. Let $\mathbf{Y} = [\mathbf{y}_k, k = 1, \ldots, K]$ denote a $(N \times K)$ matrix containing a SR sample of size N from the K-variate CDF of the above K RVs; the k-th column $\mathbf{y}_k$ of this matrix corresponds to outcomes of the k-th RV Y_k. Matrix $\mathbf{Y}$ can be generated, for example, by simulation via the Cholesky decomposition of the covariance matrix, or via sequential simulation (Johnson, 1987). The SR sample $\mathbf{y}_k$ for the k-th RV Y_k is then transformed into a LH sample $\mathbf{z}_k$ for that RV, as:

$$\mathbf{z}_k = F_{Y_k}^{-1}\left(\frac{\mathbf{r}_k - \mathbf{u}_k}{N}\right) \qquad (2)$$

where $\mathbf{r}_k = [r_k^{(n)}, n = 1, \ldots, N]'$ denotes a $(N \times 1)$ vector containing the ranks of the entries of $\mathbf{y}_k$: the lowest $y_k^{(n)}$ simulated value for the k-th RV Y_k is assigned a rank of one, the second lowest a rank of two, and the highest a rank of N.

Stein's method is similar to the LH sampling method of Equation (1), with the sole, but extremely important, difference that the array $\mathbf{p}_k$ of random permutations in that equation is now replaced by the array $\mathbf{r}_k$ of ranks of $\mathbf{y}_k$. This substitution entails that the LH sample comprising the $(N \times K)$ matrix $\mathbf{Z} = [\mathbf{z}_k, k = 1, \ldots, K]$ is (column-wise) correlated, since it inherits correlation that is present in the SR sample $\mathbf{Y}$ via the corresponding $(N \times K)$ matrix $\mathbf{R} = [\mathbf{r}_k, k = 1, \ldots, K]$ of its ranks. In addition, the entries of any column of matrix $\mathbf{Z}$ are stratified, as opposed to the entries of any column of matrix $\mathbf{Y}$.

Figure 1 gives an example of a SR sample (A) and a LH sample generated using Stein's method (B), both of size $N = 10$, from two standard Gaussian RVs Y_1 and Y_2 with correlation coefficient $\rho_{12} = 0.7$. It can be easily appreciated that, for the LH sampling case, realizations for both RVs are marginally stratified, i.e., when viewed from either the abscissa or the ordinate, each stratum (delineated by vertical or horizontal solid lines, respectively) contains a single simulated value.

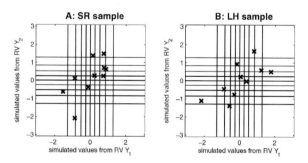

Figure 1. Examples of a SR sample (A), and a LH sample generated using Stein's method (B), both of size $N = 10$, from two correlated standard Gaussian RVs Y_1 and Y_2 with $\rho_{12} = 0.7$; solid lines delineate strata of equal probability.

Stein's method, however, underestimates the target correlation between any two RVs, because it does not fully account for the correlation in the original SR sample. More precisely, the sole vehicle for inducing correlation in the LH sample $\mathbf{z}_k$ for RV Y_k is the rank vector $\mathbf{r}_k$ of the original SR sample $\mathbf{y}_k$ for that RV; see Equation (2). The vector $\mathbf{u}_k$ of uniform random numbers in that equation is generated independent from any other such vector $\mathbf{u}_{k'}$ for any other RV $Y_{k'}$. For small sample sizes (small N) the displacement in the probability axis of the original uniform random vector that generated $\mathbf{y}_k$ (brought by the new vector $\mathbf{u}_k$) can be large; this affects the reproduction of a target correlation by the LH sample.

The above underestimation of a target correlation was also corroborated empirically in a spatial setting by Pebesma and Heuvelink (1999), who applied Stein's post-processing method to transform a SR sample generated via sequential Gaussian simulation to a LH sample. Their results showed that simulated realizations exhibited small-scale variability larger than that dictated by the target semivariogram model. This bias was also shown to be higher for small sample sizes, which unfortunately is precisely the reason for employing LH sampling in the first place.

In what follows, Stein's method is adopted not as a post-processing step, but as an integral part of sequential simulation for generating a LH sample from a multivariate distribution. To the author's knowledge, the proposed LH sampling method constitutes a novel contribution to the literature of importance sampling.

3 Sequential Latin hypercube sampling

Let $F_{Y_1,\ldots,Y_K}(y_1,\ldots,y_K|\mathbf{d}) = Prob\{Y_1 \leq y_1,\ldots,Y_K \leq y_K|\mathbf{d}\}$ denote the K-variate conditional CDF (CCDF) of K RVs $\{Y_k, k = 1,\ldots,K\}$, given a $(O \times 1)$ vector $\mathbf{d} = [d_o, o = 1,\ldots,O]'$ with known realizations (sample observations) of O RVs $\{Y_o, o = 1,\ldots,O\}$. Conditional stochastic simulation amounts to generating N alternative realizations from the multivariate CCDF $F_{Y_1,\ldots,Y_K}(y_1,\ldots,y_K|\mathbf{d})$, whereas unconditional simulation corresponds to absence of sample observations, in which case the data vector $\mathbf{d}$ is simply dropped from the notation.

The multiplication rule of probability allows one to decompose the above K-variate CCDF into a sequence of K univariate CCDFs as:

$$F_{Y_K,\ldots,Y_1}(y_K,\ldots,y_1|\mathbf{d}) = F_{Y_K}(y_K|y_{K-1},\ldots,y_2,y_1,\mathbf{d})\cdots F_{Y_2}(y_2|y_1,\mathbf{d})F_{Y_1}(y_1|\mathbf{d}) \quad (3)$$

which entails that the n-th SR sample from the above multivariate CCDF can be generated sequentially by first simulating a value $y_1^{(n)}$ from CCDF $F_{Y_1}(y_1|\mathbf{d})$, then a simulated value $y_2^{(n)}$ from CCDF $F_{Y_2}(y_2|y_1^{(n)},\mathbf{d})$, and so forth.

It is important to note that all the above univariate CCDFs, apart from the first one $F_{Y_1}(y_1|\mathbf{d})$, change from one realization to another, because the previously simulated values used as conditioning data are different for each realization. The CCDF of the k-th RV Y_k for the n-th realization should thus be denoted as: $F_{Y_k}^{(n)}(y_k|\mathbf{y}_{k-1}^{(n)},\mathbf{d})$, where $\mathbf{y}_{k-1}^{(n)} = [y_l^{(n)}, l = 1,\ldots,k-1]$ is the $(1 \times k - 1)$ vector of simulated values generated prior to $y_k^{(n)}$. In expected value (over a large number N of realizations), however, the CCDF for any RV Y_k tends towards its CCDF given only the data vector $\mathbf{d}$, i.e., $E\{F_{Y_k}(y_k|\mathbf{Y}_{k-1},\mathbf{d})\} \simeq F_{Y_k}(y_k|\mathbf{d})$, where $\mathbf{Y}_{k-1} = [\mathbf{y}_l, l = 1,\ldots, k-1]$ is the $(N \times k - 1)$ matrix of all simulated values for all RVs considered before Y_k in all N realizations.

The proposed LH sampling method from an arbitrary multivariate distribution capitalizes on the above decomposition, and amounts to embedding Stein's method into sequential simulation, which now proceeds in the following steps:

1. Establish a sequence for considering all K RVs. As long as all simulated values generated from any RV in this sequence are used as conditioning information (in addition to the data vector $\mathbf{d}$) for simulation from subsequent RVs, the order of the sequence is irrelevant: the resulting realizations constitute a genuine sample from the multivariate CCDF of Equation (3).
2. For the k-th RV Y_k in the above sequence:
   a) establish all N CCDFs $\{F_{Y_k}^{(n)}(y_k|\mathbf{Z}_{k-1},\mathbf{d}), n = 1,\ldots, N\}$, each corresponding to a particular realization n; $\mathbf{Z}_{k-1}$ is a $(N \times k - 1)$ matrix with the entire LH sample generated in all N realizations before considering RV Y_k.
   b) generate a $(N \times 1)$ vector $\mathbf{y}_k$ with a SR sample from RV Y_k; the n-th entry $y_k^{(n)}$ of vector $\mathbf{y}_k$ is drawn from the n-th CCDF $F_{Y_k}^{(n)}(y_k|\mathbf{z}_{k-1}^{(n)},\mathbf{d})$, where $\mathbf{z}_{k-1}^{(n)}$ is a $(1 \times k - 1)$ vector with the n-th LH sample generated from all RVs considered before Y_k, i.e., $\mathbf{z}_{k-1}^{(n)}$ is the n-th row of matrix $\mathbf{Z}_{k-1}$.
   c) transform the SR sample $\mathbf{y}_k$ into a LH sample $\mathbf{z}_k$, as:

$$\mathbf{z}_k = F_{Y_k|\mathbf{d}}^{-1}\left(\frac{\mathbf{r}_k - \mathbf{u}_k}{N}\right) \quad (4)$$

   where $F_{Y_k|\mathbf{d}}^{-1}$ denotes the inverse CCDF of RV Y_k given only the data vector $\mathbf{d}$, $\mathbf{r}_k$ is the rank transform of $\mathbf{y}_k$, and $\mathbf{u}_k$ is a vector of uniform random numbers in $[0, 1]$ (independent of $\mathbf{y}_k$).
   d) augment the LH sample matrix $\mathbf{Z}_{k-1}$ of step 2a to obtain the current LH sample matrix $\mathbf{Z}_k = [\mathbf{Z}_{k-1}\ \mathbf{z}_k]$ of size $(N \times k)$.
3. Consider the next RV Y_{k+1} in the sequence established in step 1, and repeat step 2 for generating LH samples from all remaining RVs $\{Y_l, l = k+1,\ldots, K\}$.

In the proposed approach, the LH sampling method of Stein is used as a post-processing tool (step 2c) *after* drawing a SR sample $\mathbf{y}_k$ from the N univariate CCDFs of RV Y_k (step 2b). But, unlike Stein's method, the LH sample $\mathbf{z}_k$ for RV Y_k is generated *before* proceeding to the simulation of the next SR sample $\mathbf{y}_{k+1}$ from the subsequent RV Y_{k+1} (step 3). Most importantly, that LH sample $\mathbf{z}_k$ is also considered as conditioning information for simulation from all subsequent RVs $\{Y_l, l = k+1,\ldots, K\}$ (step 2d), which leads to the reproduction of a target (conditional) correlation per the theory of sequential simulation (Journel, 1994).

In principle, any linear or non-linear regression scheme can be used to determine the CCDF of any RV Y_k (step 2a); the proposed LH sampling method, however, is independent of the particular scheme adopted for this CCDF determination. When the multivariate CCDF of Equation (3) is Gaussian, the CCDF of any RV Y_k (step 2a) is univariate Gaussian, and thus fully characterized by its conditional mean and variance which can be derived via generalized linear regression (Kriging); this is also the building block of sequential Gaussian simulation in a spatial context (Deutsch and Journel, 1998). LH samples from non-Gaussian RVs with specified pairwise rank correlations can also be generated by first simulating correlated deviates from K Gaussian RVs, and then transforming these deviates to correlated realizations of the original RVs using the inverse marginal or conditional CDF of each RV (Iman and Conover, 1982).

Since an unbiased (in expected value) reproduction of a target correlation is only ensured in sequential simulation under SR sampling, a hybrid approach between LH and SR sampling (still in a sequential mode) is also investigated in this paper. More precisely, this second proposal amounts to transforming the LH sample $\mathbf{z}_k$ for RV Y_k (step 2c above) to a new LH sample $\mathbf{x}_k$ that is as close as possible to the corresponding SR sample $\mathbf{y}_k$ for that RV, under the constrain that the elements of this new sample $\mathbf{x}_k$ remain in the strata used in the LH sampling procedure. In other words, the elements of the original LH sample $\mathbf{z}_k$ are "displaced" *within their strata* towards the corresponding elements of the SR sample $\mathbf{y}_k$ with the same rank. In the remainder of this paper, SRS denotes simple random sampling, LHSS denotes the LH sampling method of Stein, LHSP1 denotes the first proposal for LH sampling outlined in the flowchart given above, and LHSP2 denotes this second proposal for hybrid LH sampling.

Figure 2 gives the sampling distributions of correlation coefficients calculated from 10000 sets of LH samples, each of size $N = 10$, generated from two standard Gaussian RVs Y_1 and Y_2 with correlation $\rho_{12} = 0.8$ using the four sampling methods considered in this work. In this case, no data vector $\mathbf{d}$ is considered (unconditional simulation), and the CCDFs of RV Y_2 given realizations of RV Y_1 (step 2a) are determined via simple Kriging. The unbiased reproduction of the target correlation from SRS and LHSP2 (Figures 2A and D) is evident. Both LHSS and LHSP1 exhibit a bias in the reproduction of the target correlation. For LHSS (Figure 2B) that bias is -6%, whereas for LHSP1 (Figure 2C) it is reduced to -2%. Note that any bias decreases for larger sample sizes (larger N values).

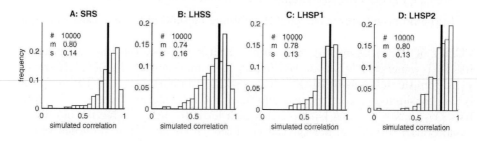

Figure 2. Sampling distributions of correlation coefficients calculated from 10000 sets of simulated pairs (each set of size $N = 10$) generated from two correlated standard Gaussian RVs Y_1 and Y_2 via: SRS (A), LHSS (B), LHSP1 (C), and LHSP2 (D); solid lines indicate the target correlation coefficient $\rho_{12} = 0.8$.

4 Spatial Latin hypercube sampling

In a spatial setting, when all K RVs pertain to a single spatial attribute Y (univariate case), the k-th RV $Y_k = Y(\mathbf{s}_k)$ is defined at a location with coordinate vector $\mathbf{s}_k$. The o-th entry $d(\mathbf{s}_o)$ of the data vector $\mathbf{d}$ denotes the sample attribute value at the o-th observation site with coordinate vector $\mathbf{s}_o$. The objective is then to generate simulated realizations (typically up to 3D) from the multivariate distribution of Equation (3), conditional or not on the data vector $\mathbf{d}$. Different sequential spatial simulation methods can be distinguished according to how each univariate CCDF is determined at each location (step 2a). Variants of Kriging are typically used for building such local CCDFs (Deutsch and Journel, 1998; Chilès and Delfiner, 1999), or more recently multi-point statistics when training images are available (Strebelle, 2000).

Sequential spatial simulation typically proceeds on a random path (different from one realization to another) for visiting each simulation location. This avoids the creation of artifact patterns in the realizations, when not all previously simulated values are used as conditioning information at any location along this path (Deutsch and Journel, 1998). In the proposed approach, that path can also be random, but it must be the *same* for all realizations (step 1); in any other case, a LH sample can only be obtained *after* sequential simulation, using the original method of Stein (1987) with its shortcomings for small N. In the examples of SR and LH sampling of this paper, a *single* random path is considered, and *all* previously simulated values are used as conditioning data at any simulation grid node to eliminate the impact of different search strategies on sampling variability.

The reproduction of target statistics from the four sampling methods considered was initially investigated using a single sample of $N = 10$ realizations generated via unconditional sequential Gaussian simulation at $K = 300$ nodes of a regular unit-spaced 1D grid. The stationary statistics included a zero mean, a unit variance, and a spherical semivariogram model of range 30 distance units. Figure 3 gives the reproduction of the marginal mean and variance at each grid node. Evidently, all LH sampling methods lead to a significant reduction in sampling variability with respect to SR sampling (compare Figure 3A with Figures 3B-D).

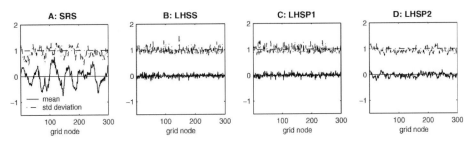

Figure 3. Mean and standard deviation of $N = 10$ simulated values at $K = 300$ unit-spaced nodes of a regular 1D grid, generated using unconditional sequential Gaussian simulation with: SRS (A), LHSS (B), LHSP1 (C), and LHSP2 (D). Horizontal solid and dashed-dotted lines indicate the target mean (0) and standard deviation (1), respectively.

Figure 4 gives the semivariogram reproduction of the $N = 10$ realizations, i.e., of the single sample whose marginal statistics are shown in Figure 3. Stein's LH sampling method (Figure 4B) leads to a higher mean simulated semivariogram than the target model at small lag distances; this critical underestimation of spatial correlation is significantly reduced by the original proposal LHSP1 (Figure 4C), and virtually eliminated by the hybrid proposal LHSP2 (Figure 4D). As expected, SR sampling leads to an unbiased mean simulated semivariogram, especially at the critical small lag distances (Figure 4A). Note also the somewhat smaller variance of the simulated semivariogram for LHSS and LHSP1 than for SR sampling (for this particular sample).

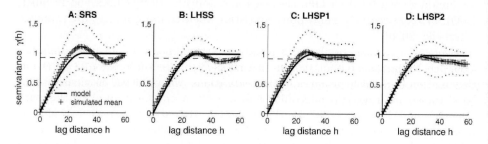

Figure 4. Semivariogram reproduction from a single sample of $N = 10$ simulated realizations at $K = 300$ unit-spaced nodes of a regular 1D grid, generated using unconditional sequential Gaussian simulation with: SRS (A), LHSS (B), LHSP1 (C), and LHSP2 (D). Solid lines indicate the target, unit-sill, spherical semivariogram model with range 30 distance units; crosses indicate the mean simulated semivariance at each lag; dotted lines delineate intervals of one standard deviation from either side of the mean semivariance; dashed horizontal lines indicate the average semivariogram value $\bar{\gamma}(V, V) = 0.972$, where V denotes the line segment support of 300 units.

To further assess the efficiency of spatial LH sampling, 1000 independent sets of $N = 10$ realizations were generated at $K = 100$ unit-spaced nodes of a regular 1D grid, using unconditional sequential Gaussian simulation and the four sampling methods considered in this work. The semivariogram model adopted was a unit-sill spherical semivariogram of 30 distance units; a stationary zero mean was also assumed. The statistic under consideration is the proportion of simulated values above threshold $G^{-1}(0.75) = 0.6745$, when these values are arranged in groups of three or more contiguous ("connected") nodes; here G^{-1} denotes the inverse Gaussian CDF. This latter connectivity consideration allows evaluating any bias incurred by a poor semivariogram reproduction: a larger than expected nugget effect, for example, will lead to a smaller than expected number of connected groups containing at least three nodes. The reference proportion 0.2219 of such connected nodes was established from a large SR sample of size $N = 1000$.

Figure 5 gives the sampling distribution of the simulated mean proportions for the four sampling methods considered in this experiment. The significant reduction in sampling variability incurred by the LH sampling methods with respect to SR sampling is evident (compare Figure 5A to Figures 5B-D). Such a reduction, however, comes at the expense of a bias in the case of Stein's method (Figure 5B); that bias is almost absent from the results of the proposed methods (Figures 5C-D).

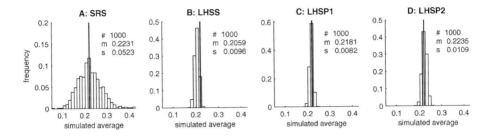

Figure 5. Sampling distributions of the mean proportion of "connected" simulated values above threshold 0.6745, for SRS (A), LHSS (B), LHSP1 (C), and LHSP2 (D); see text for details. Solid lines indicate the target proportion of 0.2219, calculated from a large SR sample of size $N = 1000$.

5 Discussion and conclusions

A novel method for LH sampling from random field models in a sequential mode has been presented in this paper. The original proposal consists of transforming a SR sample to a LH sample at each step of sequential simulation using Stein's method. A further improvement consists of additional "displacements" of the elements of the LH sample for a particular variable, *within their respective strata*, towards the corresponding elements of the SR sample with the same rank. It has been demonstrated that both proposals significantly reduce sampling variability in resulting marginal statistics, and thus make better use of the same number of realizations than SR sampling. The main advantage of these proposals over the comparable method of Stein is their better (less biased) reproduction of a target semivariogram at small lag distances, even from few realizations, a critical requirement in a spatial context to ensure unbiasedness of model outputs.

It should be noted that LH sampling leads to a smaller sampling variability in statistics of model outputs, when these models are monotonic in their inputs; such a reduction is also larger for linear models (McKay et al., 1979; Stein, 1987). It is therefore important that target marginal statistics be correctly estimated. If deemed necessary, uncertainty in these statistics should be incorporated in a formal Bayesian framework, rather than via ergodic fluctuations of SR sampling.

The proposed LH sampling method is not limited to Gaussian random field models, continuous variables, point support values, or two-point statistics, because it is independent of the algorithm used to determine the local CCDFs in sequential simulation. Any practical approximation in the implementation of sequential simulation, such as the consideration of a limited number of previously simulated values at each location, is nevertheless shared by the proposed LH sampling method. The impact of this latter approximation is typically alleviated via cascaded simulation on nested grids of increasing resolution (Deutsch and Journel, 1998). Moreover, in many cases, e.g., for simulation from auto-regressive processes, sequential simulation is the natural way to generate realizations from such processes. When the number of simulation locations is very large, and such locations do not lie on a regular grid, sequential simulation is perhaps the only feasible algorithm, due to precisely its practical implementation approximations.

Concluding, the proposed sequential method for spatial LH sampling can be readily used for enhanced uncertainty and sensitivity analysis, as well as subsequent risk assessment, in situations where complex spatially distributed models are involved. In addition, the method is simple enough to be incorporated in virtually any geostatistical software for sequential simulation, and can handle a very large number of simulation locations.

Acknowledgements

Funding for this research was partially provided by the National Science Foundation under Award BCS #0322349.

References

Ang, A.H-S., and Tang, W.H. (1984): *Probability Concepts in Engineering Planning and Design – Volume II: Decision, Risk, and Reliability*, John Wiley & Sons.

Chilès, J.P., and Delfiner, P. (1999): *Geostatistics: Modeling Spatial Uncertainty*, John Wiley & Sons.

Deutsch, C.V., and Journel, A.G. (1998): *GSLIB: Geostatistical Software Library and User's Guide, 2nd Edition*, Oxford University Press.

Haas, C.N. (1999): On modeling correlated random variables in risk assessment, *Risk Analysis*, 19(6), 1205-1214.

Helton, J.C., and Davis, F.J. (2003): Latin hypercube sampling and the propagation of uncertainty in analyses of complex systems, *Reliability Engineering & System Safety*, 81(1), 23-69.

Iman, R.L., and Conover, W.J. (1982): A distribution-free approach to inducing rank correlation among input variables, *Communications in Statistics, Part B–Simulation and Computation*, 11(3), 311-334.

Johnson, M.E. (1987): *Multivariate Statistical Simulation*, John Wiley & Sons.

Journel, A.G. (1994): Modeling spatial uncertainty: Some conceptual thoughts, *in: Geostatistics for the Next Century*, R. Dimitrakopoulos (Ed.), p. 30-43, Kluwer Academic Publishers.

McKay, M.D., Beckman, R.J., and Conover, W.J. (1979): A comparison of three methods for selecting values of input variables in the analysis of output from a computer code, *Technometrics*, 21(2), 239-245.

Morgan, M.G., and Henrion, M. (1990): *Uncertainty: A Guide to Dealing with Uncertainty in Quantitative Risk and Policy Analysis*, Cambridge University Press.

Pebesma, E.J., and Heuvelink, G.B.M. (1999): Latin hypercube sampling of Gaussian random fields, *Technometrics*, 41(4), 303-312.

Stein, M. (1987): Large sample properties of simulations using Latin hypercube sampling, *Technometrics*, 29(2), 143-151.

Strebelle, S. (2000): Conditioning simulation of complex geological structures using multi-point statistics, *Mathematical Geology*, 34(1), 1-21.

Switzer, P. (2000): Multiple simulation of spatial fields, *in: Proceedings of the 4th International Symposium on Spatial Accuracy Assessment in Natural Resources and Environmental Sciences*, G.B.M. Heuvelink, and M.J.P.M. Lemmens (Eds.), p. 629-635, Coronet Books Inc.

FIELD SCALE STOCHASTIC MODELING OF FRACTURE NETWORKS -
Combining pattern statistics with geomechanical criteria for fracture growth

XIAOHUAN LIU and SANJAY SRINIVASAN
Department of Petroleum & Geosystems Engineering
University of Texas at Austin, United States, 78712-0228

Abstract: According to recent estimates, the U.S. domestic potential for fractured oil reservoirs is on the order of tens of billion of barrels. Better technology for characterizing fracture flow paths, especially in deep, non-conventional plays and in carbonate rocks is a key to producing hydrocarbons economically from these reservoirs. The paper presents an approach for stochastic, field-scale modelling of fracture networks consistent with patterns observed on logs, the physical basis for fracture propagation and field-specific observations.

1. Introduction

Two aspects of research are presented. A stochastic simulation approach that utilizes fracture pattern information retrieved from analog models is presented first. Pattern characteristics are inferred from outcrop images using multipoint statistics and subsequently applied, after affine transformations to simulate fracture patterns in the target reservoir. A unique, stochastic fracture growth-based simulation algorithm is presented for imposing the multipoint fracture pattern characteristics on the simulation models.

Fracture patterns observed in outcrops or in subsurface reservoirs can be explained in terms of the structural geology of the reservoir and spatial variations in mechanical properties of rocks. Fracture growth model based on geomechanics can be used to perform physics-based numerical simulation of fracture patterns. However, geomechanical models can only generate fracture patterns up to a length scale of 1 kilometer and the uncertainty in fracture characteristics due to uncertainty in the stress field cannot be quantified. The paper presents a multipoint-based approach to characterize fracture patterns inferred from geomechanical models and these statistics are merged with the pattern statistics inferred from analogs such as outcrop or logs information in order to generate field-scale reservoir models. A Bayesian approach for incorporating uncertainty in reservoir stress field is also presented. The probability of the stress field conditional on the observed pattern in well logs is calibrated using the geomechanical model and later inverted to yield the probability of a fracture pattern given the uncertainty in the reservoir stress levels. Finally fractures are propagated in the reservoir by applying the multiple-point simulation approach constrained to the previously derived probabilities.

2. Geomechanical fracture classification and stochastic simulation

A natural fracture is a planar discontinuity in reservoir rock due to deformation or physical diagenesis[1]. Natural fracture patterns are frequently interpreted on the basis of laboratory-derived fracture patterns corresponding to models of paleo-stress fields and strain distribution in the reservoir at the time of fracture[2]. Sterns and Friedman[3] proposed a genetic classification of fracture systems based on stress/strain conditions in laboratory samples and features observed in outcrops and sub-surface settings. Based on their work, it can be concluded that complex stress and strain distributions in reservoir rocks can result in complex fracture patterns. Fracture patterns corresponding to different geological systems have key characteristics that can be used to classify and index fracture networks observed in outcrops and subsurface samples. Multiple point statistical measures can be used for identifying and classifying fracture patterns corresponding to different fracture systems[4].

Since stress boundary conditions strongly control the fracture pattern development subsurface at the time of fracturing, a geomechanics-based approach, where a physical understanding of the fracturing process is combined with measurements of mechanical properties of rock, is physically realistic to predict fracture network characteristics. This process-oriented approach can also provide a theoretical basis for deciding what types of fracture attribute distributions are physically reasonable, and how attributes such as length, spacing and aperture are inter-related. Additional geological information, such as the strain, pore pressure and diagenetic[5] history of the reservoir can provide further constraint on fracture network predictions.

In most cases, data available to model the fractured reservoir are sparse and information such as seismic maps and production response are related imprecisely to the fracture pattern characteristics, a probabilistic approach to fracture characterization is necessary. In the object-based modeling approaches, fractures are represented as objects defined by their centroid, shape, size and orientation. In "Random Disk" models[6], fractures are represented as two-dimensional convex circular disks located randomly in space. Although object-based models are easy to implement, their application is limited due to the assumed independence of the model parameters such as radii, orientation etc. A viable alternative is to employ pixel-based algorithms. Well established geostatistical algorithms such as sequential indicator simulation (sisim)[7] ensure reproduction of the two-point indicator variogram and can be used to classify nodes within the reservoir into fractures or matrix. However, models constrained only to two-point statistics are generally noisy and consequently inadequate for capturing clean-cut shapes such as fractures.

Although stochastic fracture models can be constructed that might be representative of analog fracture reservoir to some degree, it is still difficult to integrate geomechanical information such as stress boundary conditions in those models. A promising conditional multiple point simulation approach with integration of geomechanical data is developed as part of this research.

3. Multiple point approach to fracture growth simulation

3.1 Simulation algorithm

In the case of traditional two-point statistics based algorithms the cumulative conditional distribution function (CCDF) depicting the local uncertainty in attribute value is calculated on the basis of two-point correlation between pairs of data and between each data and simulation node. In multiple point statistics based algorithms[8, 9], this required conditional probability distribution is derived based on the entire data configuration on a spatial template, including the multiple-point interactions among the data and between the data and the unknown. Supposing there are n neighboring data events $A_\alpha, \alpha = 1,....., n$. An additional variable $t(n) = 1$ is assigned if all the elementary data events occur simultaneously. The conditional probability is[9]:

$$\text{Prob}\{A_0 = 1 | t(n) = 1\} = E\{A_0 = 1 | t(n) = 1\} \qquad (1)$$

A_0 is the unknown data at the unsampled location. Using Bayes' Theorem, the conditional probability in expression (1) can be written as:

$$\text{Prob}\{A_0 = 1 | t(n) = 1\} = \frac{\text{Prob}\{A_0 = 1, t(n) = 1\}}{\text{Prob}\{t(n) = 1\}} \qquad (2)$$

This implies that in order to derive the multiple-point conditional probability expression (1), we need to know the joint probability of observing the spatial pattern $A_0 = 1$ and $t(n) = 1$ as well as the prior probability of the occurrence of the template pattern $t(n) = 1$. Given an analog fracture model e.g. based on outcrop exposures, the required probabilities can be retrieved from that model. Defining a spatial template and translating that template over the analog model, the joint frequency of events such that $A_0 = 1$ and $t(n) = 1$ as well as the prior probability of events $t(n) = 1$ can be retrieved.

The fracture simulation approach adopted in this research exhibits a distinct departure from the current state-of-the-art multiple-point statistics based approaches in that the simulation event A_0 is itself considered to be a multiple-point event, obtained constrained to $t(n) = \cup A_\alpha$, a multiple-point event of arbitrary complexity. In contrast, in the traditional multiple-point simulation approaches, the simulation event A_0 is generally treated as single point event. As a consequence of this subtle and yet significant departure from other traditional methods, fractures are grown from each seed location based on the probability of the multiple point simulation events A_0 inferred from analog fracture models. The seed fracture locations are selected based on areal proportion maps that may be derived from seismic maps or other physical criteria such as surface curvature maps. Such a growth algorithm has the advantage that it is computationally efficient and permits integration of other physical criteria for fracture growth that might be controlled by variations in mechanical properties of the rock.

The simulation commences from an empty grid. The well locations with recorded fracture data are visited sequentially. The data configuration on a 27-point template

(Figure 1) surrounding the fracture location is examined. The conditioning data includes original well data as well as well nodes that have already simulated to be fractures. The analog fracture model is then scanned for the occurrence of that data configuration. Thus, if for example as in Figure 1, at the current stage of simulation, there are 23 points surrounding the central node that have been previously simulated to be fractures, then the analog model is scanned for the occurrence of that 24-point (23+1 central node) data configuration. This yields the probability $\text{Prob}\{t(n)=1\}$ corresponding to that data configuration. The simulation event A_0 can then be one of the following:

- None of the remaining three points on the template is a fracture
- One of the remaining three locations is a fracture. That location could be any one of the remaining nodes
- Two of the remaining three locations are fractures. There are three possible combinations.
- All three of the remaining three locations are also fractures.

The probability associated with all such multiple-point data events A_0 are retrieved by scanning the analog model. This is the joint probability $\text{Prob}\{A_0 = 1, t(n) = 1\}$ corresponding to each data event A_0. The conditional probability: $\text{Prob}\{A_0 = 1, | t(n) = 1\}$ is then derived as the ratio of the joint probability and the prior probability. A random value is drawn from the conditional probability distribution and this yields the set of nodes corresponding to the outcome A_0^* that are marked as fractures for the next step of the simulation algorithm.

3.2 Results discussion

As an example implementation of the simulation algorithm, Figure 2 is the training image of fracture distribution obtained as an unconditional realization of an object-based model. Since in most cases fracture patterns observed on a outcrop are on a 2-D plane, the analog model in Figure 2 as well as the subsequent multiple point simulation algorithm was implemented in 2-D. The spatial template used for retrieving the conditional probability distributions is shown in Figure 3. Figure 4 is the result of the simulation approach described above. It is easily to observe that the simulated model is consistent with the training image. For instance, there are three different fracture orientations (N-S, NE-SW, NW-SE) in training image (Figure 2) that can also be observed in simulation image; and both images exclude horizontal orientation fracture. It can be thus concluded that this multiple point statistical simulation approach can reproduce fracture patterns observed in training model/analog models.

4. Geomechanical Basis for Fracture growth

4.1 Principle of fracture growth

It is known that the observed strength of rocks in laboratory experiments is significantly lower than calculated theoretically values. This discrepancy is due to the remarkable strength reduction in rocks caused by stress concentrations at crack tips and subsequent propagation of pre-existing small flaws within rocks as well as other solid materials. The geomechanical modeling approaches discussed in research are based on the presence of pre-existing cracks known as Griffith cracks[5] within rocks. Stress concentrations and propagation will occur along cracks at an orientation consistent with the applied load. All modes (tensile, in-plane shear, anti-plane shear) of crack propagation with respect to different sizes, shapes and orientation of rocks and under various boundary conditions can be predicted by geomechanical analysis. Just as rocks have a critical tensile stress capacity[5], they also have a critical stress intensity factor K_c.

The crucial criterion to propagate a crack through the rock is that the stress intensity at the crack tips be at least equal to the critical stress intensity. In long-term loading systems such as in petroleum reservoirs, classic fracture mechanics may fail to accurately predict the crack growth especially in the presence of high temperature and chemical reactivity. Crack propagation can thus occur at a stress intensity value K less than the critical intensity. This has been observed in experiments using many materials including rocks and minerals and is referred to as subcritical crack growth.

4.2 Description of Geomechanical models

In order to model simultaneous propagation of fractures, a computer program developed by Olson[10] that is based on the conceptual formulation of joint growth[11] was utilized. This methodology utilizes a failure criterion and a propagation velocity model[12] given by:

$$v = A \left(\frac{K_I}{K_{IC}} \right)^n \qquad (3)$$

where K_{IC} is the critical fracture toughness and n is subcritical index. The fractures in this methodology are represented by series of equal-length boundary elements. Fracture pattern development is strongly influenced by the mechanical interactions of fractures through the fracture growth history. Based on the mechanical interaction behavior of nearby cracks and effects of other geological information, a fracture length model for larger opening mode fractures propagating through a material with randomly distributed, parallel flaws can be developed. The model requires input geological information such as reservoir thickness, subcritical index, size of stress field, stress boundary conditions and rock properties etc. The boundary element code assumes vertical fractures that are layer bound.

The fracture patterns shown in Figure 5 are generated using the geomechanical model and correspond to variations in the strain value $\left(10^{-3} m, 10^{-4} m, 10^{-5} m \right)$. The sub-critical index value is held constant at $\left(60\, Mpa.\sqrt{m} \right)$ and so is the bed thickness (10m). It is evident that the fracture patterns can be quite different corresponding to different geological conditions. With an increase in stress displacement, the number of fractures in the system increases and the pattern complexity also increases. Similar numerical experiments can be performed by varying the sub-critical index and bed thickness values. While physically realistic fracture patterns can be generated using the geomechanical model, currently the volume of investigation using such models is restricted to small areas of the stress field adjacent to flaws. The cost of simulation will increase significantly if the model is extended to a reservoir scale.

5. Incorporating information from geomechanical model

5.1 Simulation Approach

A key issue that remains to be addressed is the integration of pattern information from analog models together with information from geomechanical models so as to develop a stochastic model for the spatial distribution of subsurface fractures that is physically realistic as well as permits assessment of uncertainty. Fracture pattern information can normally be obtained from well logs or outcrop. However, since only indirect inference of the stress field is possible using borehole image and well core data, there is uncertainty in the predicted stress conditions and that has to be quantified. This uncertainty in reservoir stress values adds to the uncertainty in pattern information inferred on the basis of geomechanical simulations and has to be rigorously accounted for in the multiple-point geostatistical simulation technique.

The uncertainty in reservoir stress condition corresponding to an observed fracture pattern in well logs can be calibrated by applying Bayes' Theorem. Supposing T_{obs} is the fracture pattern observed in a borehole image. Using the geomechanical model and assuming a range of boundary stress values, fracture patterns corresponding to each boundary stress value can be simulated. Corresponding to each stress value B_i, a suite of fracture models can be generated by randomly locating the initial flaw locations. Other geomechanical parameters such as sub-critical index and layer thickness are measured independently and are assumed to be reliably known. These parameters are held constant during the geomechanical simulations. The probability of the fracture pattern T_{obs} in the K models corresponding to a particular boundary stress value B_i can be retrieved. The procedure is repeated for the N boundary stress values $B_i, i = 1,..,N$. At the end of this step, the conditional probability $\text{Prob}\{T_{obs} | B_i, i=1,..,N\}$ is obtained.

The likelihood of boundary stress value given an observed fracture pattern - Prob $\{B_i | T_{obs}\}$ can be calculated using Bayes' Theorem:

$$\text{Prob}\{B_i | T_{obs}\} = \frac{\text{Prob}\{T_{obs} | B_i\} \cdot \text{Prob}\{B_i\}}{\text{Prob}\{T_{obs}\}} \tag{4}$$

The $\text{Prob}\{T_{obs} | B_i, i=1,..,N\}$ have been calibrated using the procedure outlined earlier. Prob $\{B_i\}$ is the prior probability corresponding to the stress value B_i. In the absence of any expert information, we can assume each stress value to have the same prior probability i.e. $\text{Prob}\{B_i\} = \frac{1}{N}$. The probability $\text{Prob}\{T_{obs}\}$ is obtained concurrently with $\text{Prob}\{T_{obs} | B_i, i=1,..,N\}$ and is equal to:

$$\text{Prob}\{T_{obs}\} = \sum_{i=1}^{N} \text{Prob}\{T_{obs} \mid B_i\} \cdot \text{Prob}\{B_i\} \quad (5)$$

i.e. it is the probability of observing the pattern T_{obs} over the entire suite of $N \cdot K$ geomechanical fracture models.

The application of Expression (4) yields the updated distribution for the boundary stress values. This updated probability distribution is denoted as $\text{Prob}^*\{B_i\}$. In the stochastic simulation phase, at any step corresponding to a template partially filled with conditioning data (original data plus previously simulated values), the K images corresponding to a particular boundary stress value B_i are scanned for obtaining the probability of fracture patterns in the remaining empty nodes of the spatial template. This yields the probability $\text{Prob}\{A_o \mid t(n), B_i\}$ where A_0 implies the simulation data event, $t(n)$ is the partially filled fracture pattern. This probability is multiplied by the updated probability $\text{Prob}^*\{B_i\}$ to obtain the posterior probability corresponding to the simulation data event $\text{Prob}^*\{A_o \mid t(n), B_i\}$. By repeating this for all boundary stress values, the complete posterior CCDF characterizing the remainder uncertainty in fracture pattern can be constructed. The fracture pattern is propagated by sampling randomly from this posterior CCDF.

Fracture patterns simulated in this fashion rigorously incorporate the uncertainty in fracture pattern characteristics due to the lack of complete knowledge about the underlying physical process for fracture propagation. In addition, the models also incorporate the uncertainty in boundary stress values. Since the calibration process commences from the fracture patterns observed in image logs, the outlined approach is a viable technique for incorporating well log information into stochastic models for the fractured reservoir.

5.2 Discussion

Figure 6 is a fracture pattern observed on a log image. Figure 7 shows the fracture pattern corresponding to a stress displacement value of $8 \cdot 10^{-4}$ m and corresponding to two different initial distributions of Griffith cracks. Seven different stress displacement values were assumed and six different fracture patterns corresponding to each stress value were generated by varying the initial flaw locations randomly. As discussed earlier, the prior distribution of the stress values is assumed to be uniform (maximum uncertainty). Figure 8 shows the updated probability distribution of stress values based on the observed fracture pattern depicted in Figure 7. The posterior distribution indicates that the likelihood of the reservoir stress value being of the order $1 \cdot 10^{-4}$ m is higher. Better discrimination of the stress displacement value is possible if the pattern T_{obs} retrieved from well logs is more specific. In this case a generic pattern was retained for demonstration purposes.

Figure 9 is the final fracture pattern incorporating the uncertainty in reservoir stress conditions and variations in fracture pattern characteristics observed in the geomechanical models. We can observe that the simulation model has combined the fracture characteristics observed in the suite of geomechanical models such as the NW-SE orientation fractures, the occasional horizontal fractures observed in the some geomechanics models that have no obvious vertical fractures. Some other geomechanical models exhibit short vertical fractures that are also represented in the final simulation model. Another important characteristic of the geomechanics model is that some fractures propagate and terminate against previously existing fractures. This is physically plausible since pre-exiting fractures may reduce the stress at the tip of the daughter fractures, thereby causing the fracture propagation to stall. These characteristics can also be observed with some short fractures terminating against other fractures in simulation model.

The accuracy and robustness of the simulated fracture model is dependent upon the characteristics of the fracture pattern interpreted from image logs. If that pattern is highly specific, the resolution of the stress conditions will be more specific and consequently, only the dominant fracture patterns corresponding to that stress value will be manifested in the final simulation image. Nevertheless, it is possible to generate realistic fracture patterns using the proposed methodology to synthesize information from geomechanical model and well logs.

6. Conclusions

The research focused on developing a methodology for generating physically realistic models of fracture systems in reservoirs. The methodology hinges on the availability of training models of analogous fracture systems. When modeling a target reservoir, the multiple point statistical measures characterizing the patterns observed in the analog can be imposed on the model using a growth-based stochastic simulation technique proposed in this research.

Fracture initiation and growth are affected by a variety of physical geomechanical factors such as the regional stress field, spatial variations of rock properties, or bed thickness. The final model of the reservoir has to integrate the information obtained from geomechanical models and from analog outcrops; in order to yield more physically realistic representation of fracture systems. Furthermore, since important parameters such as the reservoir stress conditions can be only indirectly inferred, the uncertainty in stress field should be quantified and incorporated into stochastic models of the reservoir. That uncertainty can be rigorously quantified using the Bayesian procedure outlined in this paper. The Bayesian procedure is used to update a prior model for uncertainty in reservoir stress field into a posterior model based on the observed image log pattern. This updated probability of reservoir stress values is used to guide the selection of fracture growth patterns during the stochastic simulation phase of the model. Preliminary results obtained using the proposed procedures appear promising.

References

Nelson. Ronald A., 1985, "*Geologic Analysis of naturally fractured Reservoir*", Gulf Publishing Company, Houston, Texas.

Handin, J. and Hager, R.V., 1957, "Experimental determination of sedimentary rocks under confining pressure: Test at room temperature in dry samples", AAPG Bulletin, Vol. 41. pp 1-50.

Stearns, D.W. and Friedman, M., 1972, "*Reservoirs in Fractured Rock*", AAPG Memoir 16, pp 82-100.

Liu, X., Srinivasan, S. and Wong, D.W., 2002, "Geological characterization of naturally fractured reservoirs using multiple point geostatistics", SPE 75246, SPE/DOE Symposium on Improved Oil Recovery.

Atkinson, B.K., and Meredith, P.G., 1987, "The theory of subcritical crack growth with applications to minerals and rocks", In: Fracture mechanics of Rock (edited by Atkinson, B.K.). Academic Press London, 111-166.

Baecher, G.B., Einstein, H.H. and Lanney, N.A, 1977, "Statistical Descriptions of Rock Properties and Sampling", Proc. Of the 18th U.S. Symposium on Rock Mechanics, pp. 5C1.1-5C1.8.

Deutsch, C.V. and Journel, A.G., 1992, *GSLIB; Geostatistical Software Library and User's Guide*, Oxford University Press, New York, N.Y.

Guardiano, F. and Srivastava, R.M., 1992, " Multivariate Geostatistics: Beyond bivariate Moments", in Geostatistics Troia 92, A. Soares editor, Kluwer Acdemic Publisher, Vol. 1, pp. 133-144.

Journel, A.G., 1992, "Geostatistics, Roadblocks and Challenges", in Geostatistics Troia 92. A. Soares editor, Kluwer Academic Publisher, Vol. 1, pp. 213-224.

Olson, J. E., Holder, J. and Rijken, M.C., 2002, "Quantifying the fracture mechanics properties of rock for fractured reservoir characterization", SPE/ISRM 78207.

Segall, P., 1984, "Formation and Growth of Extensional Fracture Sets", Geological Society of America Bulletin 95, 454-462.

Atkinson, B.K., 1984, "*Subcritical Crack Growth in Geological Materials*", Journal of Geophysical Research 89, 4077-4114.

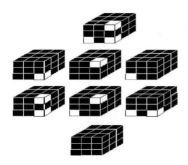

Figure 1: A 27-point 3D spatial template with 24 nodes identified.

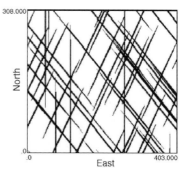

Figure 2: Training image generated object-based simulation.

Figure 3. conditional probability distribution from analog model

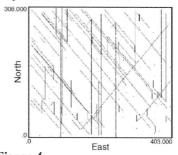

Figure 4. fracture growth based on training model

 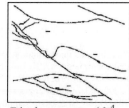 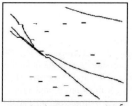

a. Displacement – 10^{-3}m b. Displacement – 10^{-4}m c. Displacement – 10^{-5}m

Figure 5. Fracture patterns corresponding to different strain value given constant subcritical index and bed thickness

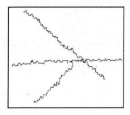

Figure 6. Analog image corresponding to displacement 10^{-4}m.

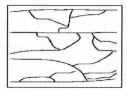 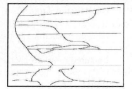

Figure 7. Patterns generated by geomechanics model with the displacement 8×10^{-4}m

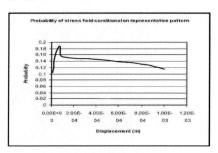

Figure 8. Probability of stress field conditional on observed pattern

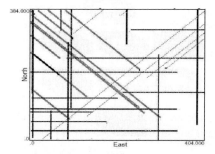

Figure 9. Final simulation image with uncertainty integration in stress field

DIRECT GEOSTATISTICAL SIMULATION ON UNSTRUCTURED GRIDS

JOHN MANCHUK, OY LEUANGTHONG and CLAYTON V. DEUTSCH
Department of Civil & Environmental Engineering, 220 CEB,
University of Alberta, Canada, T6G 2G7

Abstract. Unstructured grids are commonly used in reservoir modeling and are being increasingly considered in complex mining engineering applications. Block kriging of the attributes can be easily implemented; however, this implicitly assumes linear averaging, which is not the case after Gaussian transformation or with variables such as permeability. Direct simulation has been proposed as a solution; however, there are a number of important implementation considerations. This paper addresses the following considerations: (1) search for nearby relevant block and point data, (2) stabilization of the kriging equations and weights in presence of complex screening, (3) correction of the homoscedastic kriging variance to account for realistic proportional effect, (4) determination of valid conditional distribution shapes, (5) accounting for geological controls including stratigraphic surfaces and mixture of multiple facies within an unstructured grid block, and (6) accounting for directional permeability that does not average linearly. Direct simulation on unstructured grids is made practical by addressing these six considerations.

1 Introduction

Unstructured grids are used to model the complex geology and geometry of reservoirs and to provide better accuracy to important development areas. For example, tartan grids are used to provide a high cell density near wells and low cell density in less influential areas (Tran, 1995).

Sequential Gaussian simulation (SGS) (Isaaks, 1990) has become the most extensively used algorithm for continuous variable simulation; however, it is impractical when considering multiscale data, particularly when the data do not average linearly. Direct sequential simulation (DSS) (Xu and Journel, 1994) is an attractive alternative due to the increasing popularity of unstructured grids and the need to integrate multiscale data.

One advantage of DSS is that a wide variety of volume supports can be integrated. This requires that kriging is based on mean covariance/variogram values. There are various ways in which mean covariance calculations can be made more efficient (Pyrcz and Deutsch, 2002). While computational efficiency in this regard is important, an efficient search for nearby relevant data is just as important for practical implementation. The popular method when dealing with regular grids is the super block search strategy, a variation of which could be applied to unstructured grids; however, it may be advantageous to consider different search tree algorithms that may be more efficient.

The effects of screening remain an issue in the case of multiscale data. Proper filtering of data prior to kriging may be required to avoid anomalously high weights that may lead to extreme estimates. Some filtering techniques such as the octant search, iterative kriging, and the template technique have been used to mitigate screening.

The use of simple kriging (SK) results in an estimation variance that is independent of the data values; this independence is referred to as homoscedasticity. Unfortunately, real data may exhibit a heteroscedastic feature known as the proportional effect, wherein the local mean and variance are often quadratically related (Journel and Huijbreghts, 1978). This heteroscedasticity must be accounted for. An advantage of SK is that covariance reproduction only requires that the mean and variance of this distribution be defined by the SK mean and variance (Journel, 1994). A method of determining the local distribution shapes has been developed and will be revisited (Pyrcz and Deutsch, 2002; Deutsch et al, 2001; Oz et al, 2001).

The advantage of unstructured grids in capturing more complex geology also entails further complications related to geological controls such as stratigraphic surfaces and a mixture of multiple facies that may be represented within any particular block. An unstructured grid may not conform to the stratigraphic setting, which introduces problems relating to selecting relevant data for kriging and estimating grid blocks that contain multiple subsequence layers.

Further, the use of average variogram/covariance values in SK (inside DSS) for multiscale data has an implicit assumption of linear averaging of the model variables. This poses a problem when the variable of interest does not average linearly. Permeability is a classic example of such a variable. Accounting for the appropriate type of averaging is integral to the correct implementation of DSS.

This paper addresses these six important issues and proposes some novel approaches for resolution.

2 Search for Nearby Relevant Block and Point Data

When considering unstructured grids, data may consist of original data at a small scale, regularly gridded soft data, and grid blocks of varying sizes. There are several methods that can be used to deal with this array of data: A brute force method involving a matrix of distances n_{GB} by n_{GB} in size, where n_{GB} is the number of grid blocks; A super block search strategy (Deutsch and Journel, 1998); or the use of search trees. The brute force method is only applicable to small problems as larger data sets would be impractical for conventional computer memory availability. A super block strategy could be used; however, implementing certain types of search trees will be more efficient.

One type of search trees common in the computer graphics and computer gaming industry are quadtrees and octrees, (Figure 1) (Frisken and Perry, 2002). When considering graphics visualization, these search trees are used to quickly determine which polygons are in view such that only those polygons are drawn (this reduces memory requirements). Searching for nearest neighboring data is a similar task.

Quadtrees and octrees organize data in such a way that point location, region location, and nearest neighbor operations can be done easily. Frisken and Perry (2002) introduce a binary indexing system for quadtrees that allows for efficient execution of the above operations. This system can easily be applied to octrees for three dimensional data as well.

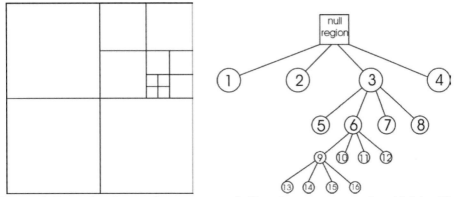

Figure 1: Example of a quadtree structure (left) and tree-representation (right). The quadtree is more refined in areas with higher data densities.

Implementing quadtrees or octrees to organize spatial data for simulation purposes allows for efficient acquisition of nearby data for each node to be simulated. The search for nearest neighboring leaf nodes is not dependant on the type of grid and tree traversal for finding and inserting points is a simple process. Having these characteristics along with low memory requirements makes search trees excellent for unstructured grid problems.

3 Stabilization of the Kriging Equations and Weights in the Presence of Complex Screening

Screening can cause extreme positive and negative weights that lead to erroneous estimates and estimation variances. One method of reducing the occurrence of extreme weights is to remove data from the kriging matrix: this iterative kriging technique will remove data until the absolute value of all the weights are below a specified maximum. Iterative kriging works; however, data that may be highly influential in estimating a location could be removed from the kriging matrix resulting in a less accurate result. Another method of reducing screening is the template technique which involves rejecting any data that are shadowed by a closer data (Figure 2). A downfall to the template technique is its high demand on computation time.

A new method of filtering data used in estimating a location is the sector search method, which is somewhat similar to the template technique. The sector search method uses input dip and azimuth tolerances to create sectors in which only the nearest data is selected for kriging (Figure 2).

The sector search subroutine works fast in two dimensions as the sectors are all pre-constructed and then translated to locations of interest; however, in three dimensions, the sectors are built as points are encountered making the process more time consuming.

Even though the sector search method removes many screened data, there may be unreasonable screening still present. For example, consider two points in adjacent sectors, the point closer to the location being estimated will screen the effect of the second point. Using larger sectors will keep screening to a minimum.

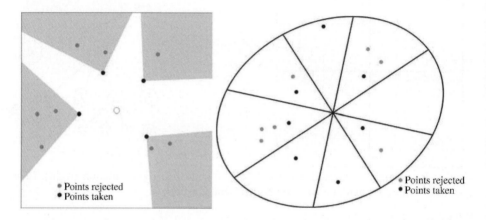

Figure 2: The template technique (left) and the sector search method (right) to reduce screening.

4 Correction of the homoscedastic kriging variance to account for realistic proportional effect

Data in original units are often heteroscedastic. High valued areas are more variable. This heteroscedastic behavior is commonly referred to as the proportional effect (Journel and Huijbregts, 1978). DSS relies on covariance reproduction through local distributions whose mean and variance are defined by SK (Xu and Journel, 1994). For data following the congenial Gaussian distribution this assumption is correct; however for data exhibiting heteroscedastic features, it is an unsuitable assumption due to the homoscedasticity of the kriging variance. The kriging variance must be adjusted such that the proportional effect is reproduced.

4.1 DSS USING LOGNORMAL DATA

To see the effects of directly simulating data that exhibit the proportional effect, a study was performed using a lognormal distribution. This distribution was chosen for a number of reasons: (1) although real data are not necessarily lognormal, most data exhibit a strong asymmetry similar to that characterized by the lognormal distribution, and (2) there is a clear mathematical link between the lognormal and the more common

Gaussian distribution that permits tractability of the results. Further, an equation describing the proportional effect of lognormal data exists (Journel and Huijbregts, 1978). Knowing these relations, the kriging variance can be calibrated to honor the heteroscedasticity inherent in lognormal data.

An exhaustive lognormal data set was generated by transforming an unconditional Gaussian model (Figure 3). The mean and variance of the lognormal data were arbitrarily chosen to be 100 and 10000, respectively. A set of 625 samples was drawn from the model and used for numerical experimentation.

4.1.1 Options of Simulation Explored

Three options were identified for evaluation:
- Option 1   Perform SGS
- Option 2   Perform DSS without correcting the kriging variance
- Option 3   Perform DSS and correct the kriging variance to honor the proportional effect

With lognormal data, an equation exists for correcting the variance using the mean or estimate:

$$\sigma_{Z,C}^2 = [z^*(\mathbf{u})]^2 (e^{\beta_G^2 \cdot \sigma_Y^2} - 1)$$

Where $\sigma_{Z,C}^2$ is the corrected variance, σ_Y^2 is the local variance in normal space, and β_G^2 is the global variance of $ln(Z)$. By determining a relation between the estimation variance in Gaussian space and that in lognormal space, the value of σ_Y^2 could be determined without having to perform kriging twice.

For each option, 100 realizations were generated and the E-type mean and variance was calculated (Figure 4). Reproduction of the global statistics and the variogram were verified. Figure 4 also shows similar results between DSS with a correction and SGS; however, with DSS and no correction as in Option 2, the variance is clearly homoscedastic. A more visual comparison of the three options is available in Figure 5 where the spatial distribution of the mean and standard deviation show reproduction of the proportional effect in a single realization.

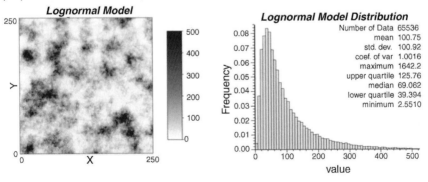

Figure 3: The lognormal model used for direct simulation experimentation.

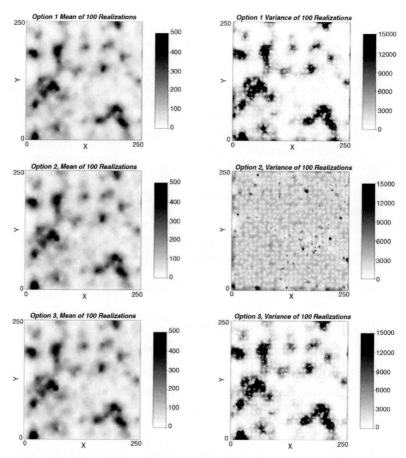

Figure 4: The mean and variance taken over 100 realizations for all three simulation approaches: Option 1 is the straightforward SGS, Option 2 refers to DSS, and Option 3 refers to DSS with variance correction to account for heteroscedasticity.

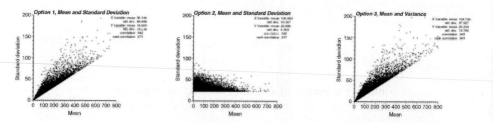

Figure 5: Local mean versus standard deviation at every estimated location for options 1 (left), 2 (middle), and 3 (right). Options 1 and 3 show the proportional effect and compare nicely. Option 2 shows a homoscedastic variance.

By performing simulation using lognormal data, it is possible to introduce a solution for dealing with the proportional effect. The lognormal distribution is particularly useful because the proportional effect is one of its prominent features and is analytically accessible.

We expect that the proportional effect could be fit from real data instead of using either the Gaussian model of no proportional effect of the lognormal model of a quadratic proportional effect.

5 Determination of valid conditional distribution shapes

A method to determine the shape of the local distributions in original units from the SK mean and variance is needed such that the global distribution is reproduced.

Figure 6 shows a numerical integration approach proposed by Oz et. al. (2001). If a specific probability p of a non-standard normal distribution with mean m and standard deviation σ is known, the corresponding direct space quantile can be calculated as:

$$Z(\mathbf{u}) = F^{-1}[G_{\{0,1\}}[G^{-1}_{\{m,\sigma\}}(p)]]$$

where $G_{\{0,1\}}$ is the cumulative distribution function (cdf) of a standard Gaussian distribution, $G_{\{m,\sigma\}}$ is the cdf of a non-standard Gaussian distribution with mean m and standard deviation σ, and F is the cdf of the representative data distribution. $Z(\mathbf{u})$ is the p quantile of the local distribution of uncertainty (Pyrcz and Deutsch, 2002).

By creating a series of Y-space distributions from a list of means and variances and repeating the above procedure for a range of quantiles, a set of Z-space distributions can be generated. The mean and variance of each Z-space distribution can be calculated and used as reference values. Upon kriging at a particular location, the resulting mean and variance can be used to look up the corresponding local distribution in original units, from which a simulated value can be drawn.

6 Accounting for geological controls including stratigraphic surfaces and mixture of multiple facies within an unstructured grid block

Some geological settings are characterized by a series of genetically related strata. The geology may consist of a sequence stratigraphic framework; the bounding surfaces between the layers correspond to a specific geologic time that separates two different periods of deposition or a period of erosion followed by deposition (Deutsch, 2002). This presents some potential issues related to the structure such as: the grid does not line up with the stratigraphic surfaces, grid blocks may contain multiple facies and subsequences (Figure 7), and searching for relevant data to estimate unknown locations.

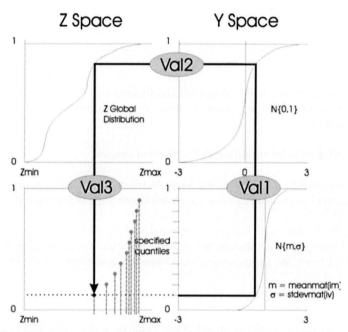

Figure 6: The graphical representation of the transformations applied to calculate the local distributions of uncertainty with a shape such that the global distribution is reproduced. The illustrated transformation is repeated for a sufficient number of quantiles to describe the local distribution.

A possible method of dealing with data within various subsequences is to flag the data by subsequence and only use data within genetically related strata. When simulating blocks that cross multiple subsequences, flagging and simulating its value poses a problem. One idea is to discretize the block into smaller "blocks", flag the smaller components and estimate them to obtain a value or multiple values and structure within a grid block. Since blocks may cross into multiple subsequences as well as contain multiple facies, a method to determine that portion of a grid block relevant to estimation is required. An idea of the subsequence structure within grid blocks being estimated as well as those being used for conditioning data is critical (Figure 8).

Upon estimating grid blocks, the proportion of facies within each block can be determined overall, but it may be better to retain the facies proportions within each subsequence in a block.

7 Accounting for directional permeability that does not average linearly

Because data exist in vastly different scales such as small core-based permeability and large scale production data, problems arise due to the scale difference and non linear averaging of permeability. By implementing a power law transform, permeability values will approximately average linearly, and can then be used in a direct simulation approach (Zanon et al, 2002).

The general formulae for power law averaging is

$$K_{eff} = \left[\frac{1}{v}\int_v k(\mathbf{u})^\omega d\mathbf{u}\right]^{\frac{1}{\omega}}$$

where v is the volume over which the average is calculated, $k(\mathbf{u})$ is the permeability at location $\mathbf{u}$ in v, and ω is an averaging exponent.

Since DSS utilizes kriging as an estimator, the model variables must average linearly with scale. By using a power law transformation prior to kriging, the problems generated by multiscale data can be avoided and transformed variables will average linearly with scale. A Gaussian transform would undo the benefit of the power-law transform; it is important to perform kriging and simulation in the correct units.

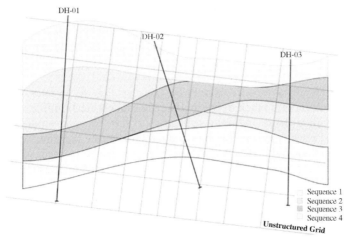

Figure 7: Stratigraphic surfaces and superimposed unstructured grid. Three hypothetical drill holes/wells are also shown.

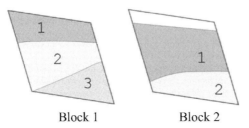

Figure 8: Unstructured grid block crossing multiple subsequence layers. If block 1 is being estimated using block 2, only data within block 2 and subsequence 1 should be used to estimate data within block 1 subsequence 1.

A concern of implementing power law transformation, especially when dealing with unstructured grids, is that ω may not be constant over every volume support. An unstructured grid may involve many different volume support sizes to be estimated and when the scale difference is large, ω may change. Other concerns that affect the value of ω are arbitrarily chosen boundary conditions and if the formation approaches the percolation threshold (Kirkpatrick, 1973).

8 Conclusions

Unstructured grids are practically relevant for realistic reservoir modeling. The distinction of simulating in the units of the original data provides significant benefits such accounting for multiscale data and permitting different local distributional shapes. In practice, implementation of DSS has been limited. Even something as seemingly straightforward as searching for data is complicated by the multiscale nature of the problem. In these instances, quadtrees or octrees may be particularly efficient. Screening may also lead to destabilization of the kriging matrix, thus a preferential filtering of the data through a sector search may be appropriate. Multiscale issues are further complicated by the very nature of the model variable, whether these variables average linearly or whether pre-processing transform such as the power law transform is required.

Unstructured grids allow for increasingly complex geology to be integrated; however, this presents issues in grid block definition and facies identification if the blocks are too large and/or if they cross multiple sequence or sub-sequence stratigraphic layers. Despite all these issues, perhaps the most important advance presented in this paper is the correction applied to the SK variance to account for the heteroscedastic nature that is often inherent to real data. The lognormal case was used to illustrate a corrective approach to effectively reproduce heteroscedasticity.

References

Deutsch, C.V., *Kriging with Strings of Data*, Mathematical Geology, Vol. 26, No. 5, 1994.
Deutsch, C.V., *Geostatistical Reservoir Modeling*, Oxford University Press Inc., 2002.
Deutsch, C.V. and Journel, A.G., *GSLIB Geostatistical Software Library and User's Guide*, Second Edition, Oxford University Press, 1998.
Deutsch, C.V. Tran, T. and Xie, Y., *A preliminary report on: An approach to ensure histogram reproduction in direct sequential simulation,* Technical report, Center for Computational Geostatistics, University of Alberta, Edmonton, AB, March 2001.
Frisken, S.F., and Perry, R.N., *Simple and Efficient Traversal Methods for Quadtrees and Octrees*, Mitsubishi Electrical Research Laboratories, 2002.
Journel, A.G. and Huijbregts, Ch.J., Mining Geostatistics, Academic Press Limited, 1978.
Kirkpatrick, S., *Percolation and Conduction*, Reviews of Modern Physics, 1973, Vol. 45, No. 4, pp 574-588
Leuangthong, O. and Deutsch, C.V., *Modeling Multivariate Multiscale Data*, Center for Computational Geostatistics, University of Alberta, Report 4: 2002.
Oz, B., Deutsch, C.V., Tran, T., and Xie, Y., *A fortan 90 program for direct sequential simulation with histogram reproduction*, Computers & Geosciences, page submitted 2001.
Tran, T.T., *Stochastic Simulation of Permeability Fields and Their Scale-up for Flow Modeling*, PhD Thesis, Stanford University, 1995.
Xu, W. and Journel, A.G., *DSSIM: A General Sequential Simulation Algorithm,* Stanford Center for Reservoir Forecasting, May 1994.
Zanon, S. Nguyen, H. and Deutsch, C.V., Power Law Averaging Revisited, Center for Computational Geostatistics, University of Alberta, Report 4, 2002.

DIRECTIONAL METROPOLIS: HASTINGS UPDATES FOR POSTERIORS WITH NONLINEAR LIKELIHOODS

HÅKON TJELMELAND
Department of Mathematical Sciences, NTNU, 7491 Trondheim, Norway

JO EIDSVIK
Statoil Research Center, 7005 Trondheim, Norway

Abstract. In this paper we consider spatial problems modeled by a Gaussian random field prior and a nonlinear likelihood linking the hidden variables to the data. We define a directional block Metropolis–Hastings algorithm to explore the posterior. The method is applied to seismic data from the North Sea. Based on our results we believe it is important to assess the actual posterior in order to understand possible shortcomings of linear approximations.

1 Introduction

Several applications in the earth sciences are preferably formulated by an underlying hidden variable which is indirectly observed via noisy measurements. Examples include seismic data, production data and well data in petroleum exploration: In seismic data the amplitudes are nonlinearly connected to the elastic parameters of the subsurface, see e.g. Sheriff and Geldart (1995). Production data contain the history of produced oil and gas, which is a complex functional of the permeability properties in the reservoir, see e.g. Hegstad and Omre (2001). Well data of radioactivity counts need to be transformed into more useful information, such as clay content in the rocks, see e.g. Bassiouni (1994). The Bayesian framework is a natural approach to infer the hidden variable; this entails a prior model for the variables of interest and a likelihood function tying these variables to observations.

In this paper we consider Gaussian priors for the underlying spatial variable, and nonlinear likelihood models. When using a nonlinear likelihood, the posterior is not analytically available. However, the posterior can be explored by Markov chain Monte Carlo sampling (see e.g. Robert and Casella (1999)), with the Metropolis–Hastings (MH) algorithm as a special case. We describe a directional MH algorithm in this paper, see Eidsvik and Tjelmeland (2003). We use this algorithm to update blocks of the spatial variable at each MH iteration. We show results of our modeling procedures for seismic data from a North Sea petroleum reservoir.

2 Methods

2.1 PRIOR AND LIKELIHOOD ASSUMPTIONS

The variable of interest is denoted $x = \{x_{ij} \in \mathcal{R}; i = 1, \ldots, n; j = 1, \ldots, m\}$, a spatial random field in two dimensions represented on a grid of size $n \times m$. We denote its probability density by $\pi(x) = \pi(x|\theta) = N(x; \mu(\theta), \Sigma(\theta))$, where $N(x; \mu, \Sigma)$ denotes a Gaussian density evaluated in x, with fixed mean μ and covariance matrix Σ. For generality we condition on hyperparameters θ, but in this study we treat θ as fixed parameters. The generalization to a vector variable, $x_{ij} \in \mathcal{R}^d$, is straightforward. For the application in Section 3 we have $x_{ij} \in \mathcal{R}^3$. A three dimensional grid, x_{ijk}, is of course also possible.

We assume the spatial variable x to be stationary and let the field be defined on a torus, see e.g. Cressie (1991). As explained below, this has important computational advantages, but the torus assumption also implies that one should not trust the results close to the boundary of the grid. Thus, one should let the grid cover a somewhat larger area than what is of interest.

The likelihood model for the data $z = \{z_{ij} \in \mathcal{R}; i = 1, \ldots, n; j = 1, \ldots, m\}$, given the underlying variable x, is represented by the conditional density $\pi(z|x) = N(z; g(x), S)$, where $g(x)$ is a nonlinear function. Hence, the conditional expectation of the data has a nonlinear conditioning to the underlying field. We assume that the likelihood noise is stationary with covariance matrix S. It is again straightforward to extend this model to vector variables at each location, $z_{ij} \in \mathcal{R}^d$, or three dimensional grids. For the application in Section 3 we have $z_{ij} \in \mathcal{R}^2$. We assume that a linearized version of the likelihood is available, and denote this by the conditional density $\pi_{x_0}^{lin}(z|x) = N(z; G_{x_0}x, S)$, where x_0 is the value of x used in the linearization.

The posterior of the hidden variable x conditional on the data is given by

$$\pi(x|z) \propto \pi(x)\pi(z|x), \qquad (1)$$

an analytically intractable posterior. The linearized alternative;

$$\pi_{x_0}^{lin}(x|z) \propto \pi(x)\pi_{x_0}^{lin}(z|x), \qquad (2)$$

for fixed x_0, can be written in a closed form, and is possible to evaluate and sample from directly. But note that in general this becomes computationally expensive in high dimensions. With our torus assumption discussed above, the covariance matrices involved become circular and the linearized posterior can then be evaluated and sampled from effectively in the Fourier domain (Cressie (1991), Buland, Kolbjørnsen, and Omre (2003)). The actual nonlinear posterior can also be evaluated, up to a normalizing constant, in the Fourier domain, by treating the prior and likelihood terms in equation (1) separately.

2.2 METROPOLIS–HASTINGS BLOCK UPDATING

A MH algorithm is an iterative sampling method for simulating a Markov chain that converges to a desired posterior distribution, see e.g. Robert and Casella (1999). Each iteration of the standard MH algorithm consists of two steps: (i) Propose a new value for the underlying variable, (ii) Accept the new value with a certain probability, else keep the value from the previous iteration.

We describe a MH algorithm which updates blocks of the random field at each iteration. Let $X^i = x$ denote the variable after the i-th iteration of the MH algorithm. For the $(i+1)$-th iteration we draw a block of fixed size $k \times l$ at random, where $k < n$, $l < m$. Since the grid is on a torus, there are no edge problems when generating this block. We denote the block by A, a defined boundary zone of the block by B, and the set of nodes outside the block and boundary by C, see Figure 1. Further, we split the variable into these blocks; $x = (x_A, x_B, x_C)$ as the parts in

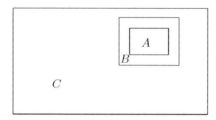

Figure 1. Block updating. The full size of the grid is $n \times m$. We illustrate the block A of gridsize $k \times l$, a boundary zone B, and the other parts of the field C.

the block, the boundary, and outside the block and boundary zone, respectively. Correspondingly, we denote the data vector by $z = (z_A, z_B, z_C)$. To propose a new value in the block we define a proposal density for the part in A, and denote the proposal on this block by y_A. The rest of x remains unchanged, and hence the proposed value is $y = (y_A, x_B, x_C)$. One proposal density in the block is the linearized posterior for y_A, conditional only on values in the boundary zone and data in A and B. We denote this by $\pi_x^{lin}(y_A|x_B, z_A, z_B) = N(y_A; m, T)$, where the mean m and covariance T can be calculated directly (Cressie, 1991). Note that the linearization is done at the current value $X^i = x$. The final step in the MH iteration is to accept or reject the proposed value y, and we thus obtain the next state X^{i+1}. Note that the results of the MH algorithm are the same no matter which block size we choose. The CPU time, on the other hand, will vary with the block size. In the application below we have chosen a block size similar to the range of the spatial correlation.

2.3 DIRECTIONAL METROPOLIS–HASTINGS

Directional MH algorithms are a special class of MH algorithms. Each iteration consists of the following steps; (i1) Generate an auxiliary random variable which defines a direction, (i2) Draw a proposed value on the line specified by the current state and the auxiliary direction, (ii) Accept the proposed value with a certain probability, else keep the variable from the previous iteration.

We present a directional updating scheme where the proposal step is done effectively in one dimension, see Eidsvik and Tjelmeland (2003). We outline our method following the block sampling setting in Section 2.2. Denote again the variable at the i-th iteration by $X^i = x = (x_A, x_B, x_C)$, and the data by $z = (z_A, z_B, z_C)$. We generate an auxiliary direction (step i1) as follows; First, draw w_A from $\pi_x^{lin}(\cdot|x_B, z_A, z_B)$. Next, define the auxiliary direction as $u = \pm \frac{w_A - x_A}{|w_A - x_A|}$, where we use $+$ or $-$ so that the first component of u is positive, see the discussion in Eidsvik and Tjelmeland (2003). Since this density for w_A is Gaussian, it is possible to calculate the density for the auxiliary unit direction vector u (Pukkila and Rao, 1988). We denote this density by $g(u|x_A, x_B, z_A, z_B)$.

At the last part of the proposal step (i2) we draw a one dimensional value t from some density $q(t|u, x, z)$, and set $y_A = x_A + tu$ as the proposed value for the block, and $y = (y_A, x_B, x_C)$ as the proposal for the entire field. This proposal is accepted, i.e. $X^{i+1} = y$, with probability

$$r(y|x) = \min\left\{1, \frac{\pi(y)}{\pi(x)} \cdot \frac{\pi(z|y)}{\pi(x|y)} \cdot \frac{g(u|y_A, x_B, z_A, z_B)}{g(u|x_A, x_B, z_A, z_B)} \cdot \frac{q(-t|u, y, z)}{q(t|u, x, z)}\right\}, \quad (3)$$

else we have $X^{i+1} = x$.

In particular, if we choose the one dimensional density as

$$q^*(t|u, x, z) \propto \pi(z|x_A + tu, x_B, x_C)\pi(x_A + tu, x_B, x_C)g(u|x_A + tu, x_B, z_A, z_B), \quad (4)$$

the acceptance probability in equation (3) is equal to unity (Eidsvik and Tjelmeland, 2003), and hence the proposed variable is always accepted. Two elements are important when considering the unit acceptance rate proposal in equation (4); (i) An acceptance rate of one is not necessarily advantageous in MH algorithms. It is advantageous to obtain fast mixing, i.e. to have a small autocorrelation between successive variables. This is best achieved with large moves at each iteration of the MH algorithm. (ii) It is not possible to sample directly from the q^* density. To obtain a sample we have to fit an approximation to q^* in some way, either by a parametric curve fit, or by numerical smoothing of a coarse grid approximation.

In this paper we use an alternative density which does not give unit acceptance rate. The adjusted density is

$$\tilde{q}(t|u, x, z) \propto (1 + |t|)^\lambda q^*(t|u, x, z), \quad \lambda > 0, \quad (5)$$

where the $(1+|t|)^\lambda$ term encorages t values away from $t = 0$. This makes it possible to have larger moves in the MH algorithm since $t = 0$ corresponds to $y_A = x_A$. We fit an approximation by first calculating $\tilde{q}$ on a coarse grid and then use linear interpolation in the log scale between the evaluated grid points. The approximation that we obtain with this approach is our proposal denoted q.

In Figure 2 we show the proposal with acceptance one, q^*, the adjusted proposal

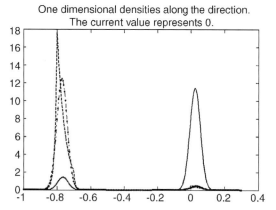

Figure 2. Sketch of the density q^* (solid), the adjusted density $\tilde{q}$ (dash-dots), and the fitted density q (dashed). This is a typical proposal density for t in our application in Section 3. The approximation q has an exponential form in each interval of length 0.1 on a grid (for example between -0.8 and -0.7).

$\tilde{q}$ and its fitted proposal q. This particular plot is obtained in our application in Section 3. We tuned λ in equation (5) so that the acceptance rate was about 0.5 in our application. This seems to be a reasonable acceptance rate considering the asymptotically optimal acceptance rates for random walk MH and Langevin MH algorithms at 0.25 and 0.5, respectively, see e.g. Robert and Casella (1999). We obtain this by setting $\lambda = 10$. Note that the proposal density for t has a bimodal shape. One mode is usually close to $t = 0$, while the other mode is quite far from 0. The mixing of the MH algorithm improves if we reach the mode away from 0 more often. This can be established by $\tilde{q}$ or q as shown in Figure 2. In the case of a linear likelihood the two modes illustrated in Figure 2 are always of equal size. A nonlinear likelihood causes them to have unequal mass, and most commonly the mode near $t = 0$ contains most of the probability mass.

3 Example

3.1 SEISMIC AMPLITUDE VERSUS OFFSET DATA

Seismic amplitude versus offset (AVO) analysis is commonly used to assess the underlying lithologies (rocks and saturations) in a petroleum reservoir, see e.g. Sheriff and Geldart (1995), and Mavko, Mukerji, and Dvorkin (1998). The reflection amplitude of seismic data changes as a function of incidence angle and as a function of the elastic properties which are indicative of lithologies.

We analyze reflection data at two incidence angles in the Glitne field, North Sea. The Glitne field is an oil-producing turbidite reservoir with heterogeneous sand

and shale facies, see Avseth et al (2001) and Eidsvik et al (2004). The domain of our interest is 2.5×2.5 km^2, and it is split into a grid of size 100×100, with each grid cell covering 25×25 m^2. The area covers what was interpreted as the lobe of the turbidite structure in Avseth et al (2001). Figure 3 shows the reflection amplitude along the grid at reflection angles zero (left) and thirty (right).

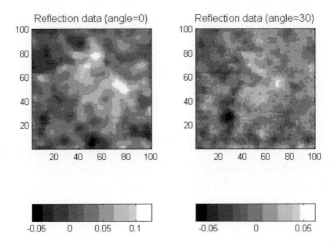

Figure 3. Reflection data at the 2D interface. Incidence angle 0 degrees (left) and 30 degrees (right).

The reflection amplitudes are hard to analyze directly because they are a result of the contrast in elastic properties in the cap rock covering the reservoir and the properties in the reservoir zone. We next use a statistical model for automatic analysis of the elastic reservoir properties.

3.2 STATISTICAL MODEL FOR SEISMIC DATA

The seismic data for the two reflection angles are denoted $z = (z^0, z^1)$, where z^0 refers to the zero offset reflection and z^1 to 30 degrees incidence angle (both are plotted in Figure 3). The statistical model that we use is closely connected to the one in Buland, Kolbjørnsen and Omre (2003).

The variables of interest are the pressure and shear wave velocities, and the density in the reservoir zone. We denote these by $x = (\alpha, \beta, \rho) = \{x_{ij} = (\alpha_{ij}, \beta_{ij}, \rho_{ij}); i = 1, \ldots, n; j = 1, \ldots, m\}$, where α is the logarithm of the pressure wave velocity, β is the logarithm of the shear wave velocity, and ρ is the logarithm of the density. The velocities and density are the elastic properties of the rocks which in some sense capture the rock mineral and saturation (Mavko, Mukerji, and Dvorkin, 1998). The units for the exponential equivalents of α, β and ρ are m/s for velocities and kg/m^3 for density. Let $(\alpha_0, \beta_0, \rho_0)$ be the logarithm of pressure and shear wave velocities, and the logarithm of density for the cap rock. These cap rock properties are treated as fixed values in this study, and are equal for all locations in the grid.

We assign a Gaussian prior density to the reservoir variables of interest, i.e. $\pi(x) = N(x; \mu, \Sigma)$, where μ now becomes a $3mn$ vector with the mean of the three log elastic properties. These prior mean values are fixed, and set to $\mu_\alpha = E(\alpha_{ij}) = 7.86$, $\mu_\beta = E(\beta_{ij}) = 7.09$, $\mu_\rho = E(\rho_{ij}) = 7.67$, for all (i,j). These mean values are assessed from well logs in Glitne, see Avseth et al (2001). The prior covariance matrix, Σ, is a $3mn \times 3mn$ matrix defined by a Kronecker product, giving a 3×3 block covariance matrix at the diagonal, and 3×3 matrices with this covariance function and a spatial correlation function on the off-diagonal, see Buland, Kolbjørnsen, and Omre (2003). The diagonal covariance matrix describing the marginal variability at each location is defined by $Std(\alpha_{ij}) = 0.06$, $Std(\beta_{ij}) = 0.11$, $Std(\rho_{ij}) = 0.02$, and correlations $Corr(\alpha_{ij}, \beta_{ij}) = 0.6$, $Corr(\alpha_{ij}, \rho_{ij}) = 0.1$, $Corr(\beta_{ij}, \rho_{ij}) = -0.1$. These parameters capture the variability expected from previous studies, see Buland, Kolbjørnsen, and Omre (2003). The spatial correlation function is the same for all three reservoir variables and is an isotropic exponential correlation function with range $250m$ (10 grid nodes).

The likelihood function is nonlinear, and is defined by approximations to the Zoeppritz equations, see e.g. Sheriff and Geldart (1995). The density for the seismic AVO data, given the underlying reservoir properties, is $\pi(z|x) = N(z; g(x), S)$, where the nonlinear function goes only locationwise, i.e. at grid node (i, j) the expectation term in the likelihood is a function of the variables at this gridnode only. For each location (i, j) and angle $\gamma = 0, 30$ we have

$$g_{ij,\gamma}(x) = g_{ij,\gamma}(\alpha_{ij}, \beta_{ij}, \rho_{ij}) = a_0(\alpha_{ij} - \alpha_0) + a_{1,ij}(\beta_{ij} - \beta_0) + a_{2,ij}(\rho_{ij} - \rho_0), \quad (6)$$

where

$$a_0 = \frac{1}{2}[1 + sin^2(\gamma)], \quad a_{1,ij} = -4\xi_{ij} sin^2(\gamma), \quad (7)$$

$$a_{2,ij} = \frac{1}{2}[1 - 4\xi_{ij} sin^2(\gamma)], \quad \xi_{ij} = \frac{\exp(2\beta_{ij}) + \exp(2\beta_0)}{\exp(2\alpha_{ij}) + \exp(2\alpha_0)}.$$

The noise covariance matrix of the likelihood, S, is a $2mn \times 2mn$ matrix defined from a Kronecker product. This covariance matrix has a block diagonal 2×2 matrix on the diagonal, and off-diagonal elements defined from an exponential correlation structure with range $250m$. The diagonal noise covariance matrix is defined by $Std(\gamma = 0) = 0.015$, $Std(\gamma = 30) = 0.012$, $Corr(\gamma = 0, \gamma = 30) = 0.7$. This likelihood noise model is specified using the parameters in Buland, Kolbjørnsen, and Omre (2003).

A linear likelihood model can be defined by fixing the ratio ξ_{ij} in equation (7). For a constant linearization point we have $\pi_\mu^{lin}(z|x) = N(z; G_\mu x, S)$. A similar linearization is used in Buland, Kolbjørnsen, and Omre (2003), and with this linearization they assess the analytically available posterior directly on a $3D$ dataset. For a linearized proposal density on the block A we have $\pi_x^{lin}(\cdot|x_B, z_A, z_B)$, where we use a block size of 9×9, and a boundary zone of width one grid node. The quality of the linearization varies across the lateral domain - it is better with dense sampling in time. It is important to remember that the choice of linearization is irrelevant for the results, it only influences the CPU time of the sampling algorithm.

3.3 RESULTS

The posterior is sampled using the block directional MH algorithm discussed above. We denote 144 updates as one iteration, i.e. on average each grid node is (proposed) updated about once in each iteration. Figure 4 shows trace plots for

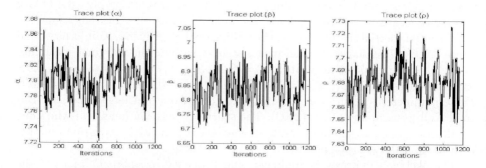

Figure 4. Trace plots of the three variables at one location in the grid. Log of pressure wave velocity α (left), log of shear wave velocity β (middle), and log of density (right).

the log elastic properties at one location in the grid. The traceplots are explained to some extent by the bimodal proposal density q (see Figure 2). In Figure 4 the variables move short distances at some iterations, while moves are large at other iterations, reflecting the bimodal density for the proposal.

In Figure 5 we show the estimates of the marginal mean and standard deviation for all three variables as images. Near grid coordinate (North,East) = (60, 80) (see Figure 5, top left image) both pressure and shear wave velocities are large. In the same area the reflection data (Figure 2) are large at both angles. Going south from gridnode (60, 80) (see Figure 5, top left image) the pressure wave velocity decreases, and so does the shear wave velocity, but to a smaller degree. In Figure 2 the reflection data become smaller in this area. These two regions comprise the lobe of the turbidite structure, see Avseth et al (2001). In Eidsvik et al (2004) these two regions were estimated to be water and oil saturated sands, respectively. Without moving on to classifying the velocity and density values, we merely note that pressure wave velocity is larger in water than oil saturated sands (Mavko, Mukerji, and Dvorkin, 1998). Our estimated velocities are hence in accordance with the results in Eidsvik et al (2004). The western part of the domain were predicted to contain mostly shales (a low velocity rock) in Eidsvik et al (2004).

The prior standard deviations for the three variables are $(0.06, 0.11, 0.02)$. In Figure 5 (right) the mean standard deviations in the posterior are $(0.033, 0.065, 0.02)$. This indicates that there is information about α and β in the AVO data (standard deviation decreases by a factor two), but not much about ρ.

Note that the standard deviations in Figure 5 (right) varies quite a lot across the field (a factor of two). The standard deviation for β is smaller where the velocities large. For a linear model [Buland, Kolbjørnsen, and Omre (2003)] the

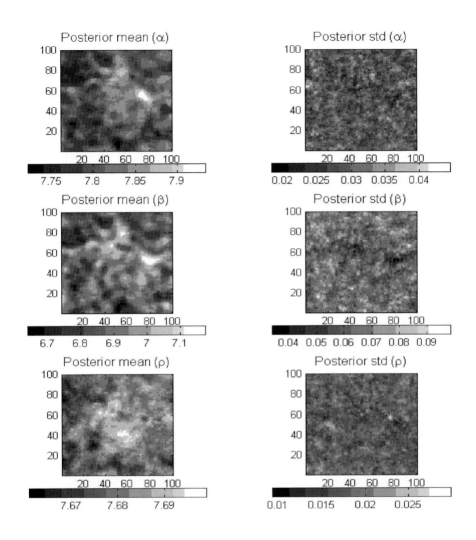

Figure 5. Mean and standard deviation of the three variables at each location in the grid. Logarithm of pressure wave velocity α (top). Logarithm of shear wave velocity β (middle), and logarithm of density (bottom).

standard deviations are constant across the field. The expected values also differ somewhat between a linear model and our nonlinear model; for example $E(\frac{\beta}{\alpha})$ is shifted significant between the two approaches. These differences suggest that the linearized Gaussian posterior in Buland, Kolbjørnsen, and Omre (2003) might introduce a bias in the estimation of the elastic parameters. One might want to correct for the possible bias or variance effects of a linear model, now that this effect is quantified by nonlinear sampling.

4 Closing Remarks

In this paper we consider Bayesian models with a Gaussian random field prior and nonlinear likelihood functions. Such models are common in the earth sciences, but are usually simplified (linearized) to make the posterior analytically available. We propose a directional block Metropolis–Hastings sampler for exploring the original nonlinear posterior. When we apply our methods to a seismic dataset from the North Sea, we recognize some differences between our results and the ones obtained by a linearized model. These differences indicate that it is useful to check the validity of a simplified likelihood model by sampling the full nonlinear models.

One of the current challenges with the Glitne field is uncertainty in the thickness of the turbidite structure, associated with the noise in seismic data due to overburden effects. A natural extension is hence to study the full 3D seismic data. An extension of our statistical methods is to assign priors to the hyperparameters in the statistical model, and hence include the variability of these parameters into the final results.

References

Avseth, P., Mukerji, T., Jørstad, A., Mavko, G., and Veggeland, T., *Seismic reservoir mapping from 3-D AVO in a North Sea turbidite system*, Geophysics, vol. 66, no. 4, 2001, p. 1157-1173

Bassiouni, Z., *Theory, measurement, and interpretation of well logs*, Society of Petroleum Engineers, 1994

Buland, A., Kolbjørnsen, O., and Omre, H., *Rapid spatially coupled AVO inversion in the Fourier domain*, Geophysics, vol. 68, no. 3, 2003, p. 824-835

Cressie, N.A.C., *Statistics for spatial data*, Wiley, 1991

Eidsvik, J., and Tjelmeland, H., *On directional Metropolis–Hastings algorithms*, Submitted for publication, Technical Report, Department of Mathematical Sciences, Norwegian University of Science and technology, http://www.math.ntnu.no/preprint/statistics/2003/S6-2003.pdf

Eidsvik, J., Avseth, P., Omre, H., Mukerji, T., and Mavko, G., *Stochastic reservoir characterization using pre-stack seismic data*, Geophysics, v. 69, no. 4, 2004, p. 978-993

Hegstad, B. K., and Omre, H., *Uncertainty in Production Forecasts based on well observations, seismic data and production history*, Society of Petroleum Engineering Journal, December 2001, p. 409-424

Mavko, G., Mukerji, T., and Dvorkin, J., *The Rock Physics Handbook*, Cambridge, 1998

Pukkila, T.M., and Rao, C.R., *Pattern recognition based on scale invariant functions*, Information Sciences, v. 45, 1988, p. 379-389

Robert, C.P., and Casella, G., *Monte Carlo Statistical Methods*, Springer, 1999

Sheriff, R.E., and Geldart, L.P., *Exploration seismology*, Cambridge, 1995

DETECTION OF LOCAL ANOMALIES IN HIGH RESOLUTION HYPERSPECTRAL IMAGERY USING GEOSTATISTICAL FILTERING AND LOCAL SPATIAL STATISTICS

PIERRE GOOVAERTS
BioMedware, Inc. 516 North State Street, Ann Arbor, MI 48104

Abstract. This paper describes a methodology to detect patches of disturbed soils in high resolution hyperspectral imagery, which involves successively a multivariate statistical analysis (principal component analysis, PCA) of all spectral bands, a geostatistical filtering of regional background in the first principal components using factorial kriging, and finally the computation of a local indicator of spatial autocorrelation to detect local clusters of high or low reflectance values as well as anomalies. The approach is illustrated using one meter resolution data collected in Yellowstone National Park. Ground validation data demonstrate the ability of the filtering procedure to reduce the proportion of false alarms, and its robustness under low signal to noise ratios. By leveraging both spectral and spatial information, the technique requires little or no input from the user, and hence can be readily automated.

1 Introduction

Spatial data are periodically collected and processed to monitor, analyze and interpret developments in our changing environment. Remote sensing is a modern way of data collecting and has seen an enormous growth since launching of modern satellites and development of airborne sensors. In particular, the recent availability of high spatial resolution hyperspectral (HSRH) imagery offers a great potential to significantly enhance environmental mapping and our ability to model spatial systems (Aspinall *et al.*, 2002; Marcus, 2002). Following Jacquez *et al.* (2002), HSRH images refer to images with resolutions of less than 5 meters and including data collected over 64 or more bands of electromagnetic radiation for each pixel.

High spatial resolution imagery contains a remarkable quantity of information that could be used to analyze spatial breaks (boundaries), areas of similarity (clusters), and spatial autocorrelation (associations) across the landscape. This paper addresses the specific issue of soil disturbance detection, which could indicate the presence of land mines or recent movements of troop and heavy equipment. A challenge presented by soil detection is to retain the measurement of fine-scale features (i.e. mineral soil changes, organic content changes, vegetation disturbance related changes, aspect changes) while still covering proportionally large spatial areas. An additional difficulty is that no ground data might be available for the calibration of spectral signatures, and little might be known about the size of patches of disturbed soils to be detected. Precise and accurate soil disturbance identification typically requires: (1) identification of a potential target (soil disturbance) of interest, (2) removal of confusion (the environmental setting), and (3) target (soil disturbance) confirmation. These different steps should be automated as much as possible to allow for the fast processing of multiple images, while false positives should be reduced to a manageable level.

A major challenge facing the use of HSRH data is the development of new, spatially explicit tools that exploit both the spectral and spatial dimensions of the data. Semivariograms allow one to detect multiple scales of spatial variability, and the spectral values can then be decomposed into the corresponding spatial components using factorial kriging (Goovaerts, 1997; Wackernagel, 1998). This technique has first been used in geochemical exploration to distinguish large isolated values (pointwise anomalies) from groupwise anomalies that consist of two or more neighboring values just above the chemical detection limit (Sandjivy, 1984). Ma and Royer (1988) have applied the same technique to image restoration, filtering and lineament enhancement, while Wen and Sinding-Larsen (1997) have analyzed sonar images. More recently, Van Meirvenne and Goovaerts (2002) applied factorial kriging to the filtering of multiple SAR images, strengthening relationships with land characteristics, such as topography and land use. None of these studies has however tackled the issue of automatic analysis and processing of large series of correlated spectral bands.

This paper describes a new approach that combines geostatistical filtering with local cluster analysis used in health sciences for the detection of clusters and outliers in cancer mortality rates (Jacquez and Greiling, 2003). The methodology is applied to HSRH imagery collected in Yellowstone National Park, and performances are assessed using ground data. Sensitivity analysis is conducted to investigate the impact of spectral resolution, signal to noise ratio, and kernel detection size on classification accuracy.

2 Geostatistical Methodology

Consider the problem of detecting, across an image, single or aggregated pixels that are significantly different from the surrounding ones. The information available consists of K variables (i.e. original spectral values or combinations of those) recorded at each of the N nodes of the image, $\{z_k(\boldsymbol{u}_i), i=1,...,N; k=1,...,K\}$. The proposed approach proceeds in two steps:
1. The regional variability (i.e. spatial background) of the image is filtered in order to highlight local anomalies, which are values that depart from the surrounding mean.
2. At each location across the filtered image the value of a detection kernel, whose size corresponds to the expected size of a patch of disturbed soil, is compared to neighborhood values and flagged as anomaly if its value is significantly higher or lower than surrounding pixel values.

2.1 GEOSTATISTICAL FILTERING

The first step involves removing from each image (i.e. original spectral bands or principal components) the low-frequency component or regional variability. For the k-th image, the low-frequency component, denoted m_k, is estimated at each location $\mathbf{u}$ as a linear combination of the n surrounding pixel values:

$$m_k(\mathbf{u}) = \sum_{i=1}^{n} \lambda_{ik} z_k(\mathbf{u}_i) \quad \text{with} \quad \sum_{i=1}^{n} \lambda_{ik} = 1 \qquad (1)$$

where λ_{ik} is the weight assigned to the *i*-th observation in the filtering window of size *n*. The main feature of this filtering technique is that the weights λ_{ik} are tailored to the spatial pattern of correlation displayed by each image and assessed using the semivariogram. These weights are computed as solution of the following system of linear equations (kriging of the local mean):

$$\sum_{j=1}^{n} \lambda_{jk} \gamma_k(\mathbf{u}_i - \mathbf{u}_j) + \mu(\mathbf{u}) = 0 \quad i = 1,\ldots,n$$
$$\sum_{j=1}^{n} \lambda_{jk} = 1 \qquad (2)$$

where $\gamma_k(\mathbf{u}_i - \mathbf{u}_j)$ is the semivariogram of the *k*-th image for the separation vector between $\mathbf{u}_i$ and $\mathbf{u}_j$, and $\mu(\mathbf{u})$ is a Lagrange multiplier that results from minimizing the estimation variance subject to the unbiasedness constraint on the estimator.

2.2 DETECTION OF ANOMALIES USING THE LISA STATISTIC

The second step amounts at scanning each filtered image, looking for local values that are significantly lower or higher than the surrounding values and might indicate the presence of disturbed soils. This procedure requires the definition of:
1. Detection kernel, whose size corresponds to the expected size of a patch of disturbed soil,
2. LISA neighborhood including the pixels surrounding the detection kernel,
3. Target area which is the area to be analyzed.

An example of these three parameters is provided in Figure 1.

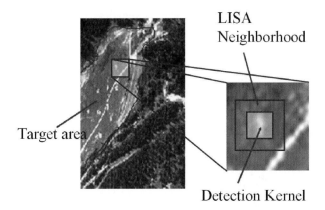

Figure 1. Illustration of key parameters used in the detection procedure.

The detection of local anomalies is based on the local Moran's I, which is the most commonly used LISA (Local Indicator of Spatial Autocorrelation) statistic (Anselin, 1995). It is computed for each pixel of coordinates ***u*** as:

$$\text{LISA}_k(\mathbf{u}) = \bar{r}_k(\mathbf{u}) \left[\frac{1}{J} \sum_{i=1}^{J} r_k(\mathbf{u}_i) \right] \quad (3)$$

where $\bar{r}_k(\mathbf{u})$ is the average value of the residuals, $r_k(u)=z_k(u)-m_k(u)$, over the detection kernel centered on pixel of coordinates ***u***, and J is the number of pixels in the LISA neighborhood (e.g. $J=12$ and kernel comprises 4 pixels for the example of Figure 1). Since the residuals have zero mean, the LISA statistic takes negative values if the kernel average is much lower (or higher) than the surrounding values. In other words the kernel average is below the global zero mean while the neighborhood average is above the global zero mean, or conversely, which indicates the presence of anomalies. Clusters of low or high values will lead to positive values of the LISA statistic (e.g. both kernel and neighborhood averages are jointly above zero or below zero).

In addition to the sign of the LISA statistic, its magnitude informs on the extent to which kernel and neighborhood values differ. To test whether this difference is significant or not, a Monte Carlo simulation is conducted, which consists in sampling randomly the target area and computing the corresponding simulated neighborhood averages. This operation is repeated many times (e.g. 1,000 draws) and these simulated values are multiplied by the detection kernel average $\bar{r}_k(\mathbf{u})$ to produce a set of 1,000 simulated values of the LISA statistic at ***u***. This set represents a numerical approximation of the probability distribution of the LISA statistic at ***u***, under the assumption of spatial independence. The observed LISA statistic, $\text{LISA}_k(u)$, can then be compared to the probability distribution, allowing the computation of the probability that this observed value could be exceeded (so-called *p*-value):

$$p_k(\mathbf{u}) = \text{Prob}\{L > \text{LISA}_k(\mathbf{u}) \mid \text{randomization}\} \quad (4)$$

Large *p*-values thus indicate large negative LISA statistic, corresponding to small values surrounded by high values or the reverse (anomalies or presence of negative local autocorrelation). Conversely, small *p*-values correspond to large positive LISA statistic which indicates clusters of high or low values (positive autocorrelation).

The last step is to combine the K *p*-values computed for the set of K images. Two novel statistics were developed to summarize for each node ***u*** the information provided by the K bands and to detect target pixels:

1. Average *p*-value over the subset of K' bands that display negative LISA statistic:

$$S_1(\mathbf{u}) = \frac{1}{K'} \sum_{k=1}^{K} i(\mathbf{u};k) p_k(\mathbf{u}) \quad \text{and} \quad K' = \sum_{k=1}^{K} i(\mathbf{u};k) \quad (5)$$

with $i(u;k) = 1$ if $\text{LISA}_k(u) < 0$, and zero otherwise. Large S_1 values indicate local anomalies (i.e. sample LISA statistic in the left tail of the distribution).

2. Average absolute deviation of *p*-values from 0.5 through the *K* bands:

$$S_2(\mathbf{u}) = \frac{1}{K} \sum_{k=1}^{K} | p_k(\mathbf{u}) - 0.5 | \tag{6}$$

Large S_2 values indicate either clusters or anomalies (i.e. sample LISA in either tails of the distribution).

The detection procedure requires applying a threshold to the maps of statistics S_1 or S_2 and classifying as disturbed soils all pixels exceeding this probability threshold. Instead of selecting a single threshold arbitrarily, it is better to select a series of thresholds and see how the proportion of pixels correctly or incorrectly classified as disturbed soils evolves. This information can then be summarized in the so-called Receiver Operating Characteristics (ROC) curve that plots the probability of false alarm versus the probability of detection.

3 Case study

The new methodology was tested on a vegetation plot located in the northern boundary area of Yellowstone National Park. The objective is to detect 4 blue tarps of 4m² area in the image (131×69 pixels). These four targets mainly correspond to the white pixels in the image of Figure 2 (left), and are denoted by the black squares in the right image. These data were collected using the Probe-1 sensor, a 128-band hyperspectral system operated by Earth Search Systems, Inc. To obtain 1 m resolution data, this sensor was mounted on a helicopter flying approximately 600 m above the ground. Following atmospheric correction, the images were degraded in order to investigate the robustness of the approach with respect to spatial resolution and signal to noise ratio. The data were first spectrally resampled to 2-5 times lower resolutions, by simply selecting one out of every 2 to 5 bands. Noise was added to simulate 50:1 signal-to-noise ratio (SNR) and 100:1 SNR, according to: $R_{sn}(\lambda) = R_s(\lambda)[1+\{N(0,1)/SNR(\lambda)\}]$, where $R_{sn}(\lambda)$ is the simulated, noisy spectrum, $R_s(\lambda)$ is the spectrum that has been spectrally resampled, N(0,1) is a Gaussian random number with a zero mean and unit variance, and SNR(λ) is the simulated signal-to-noise ratio.

Blue tarps

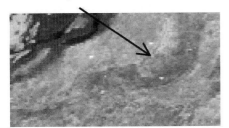
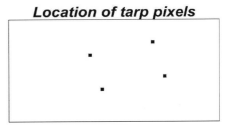

Figure 2. Probe 1, color-infrared image of the experimental site, and location of 16 tarp pixels (black) that are interpreted as disturbed soils to be detected.

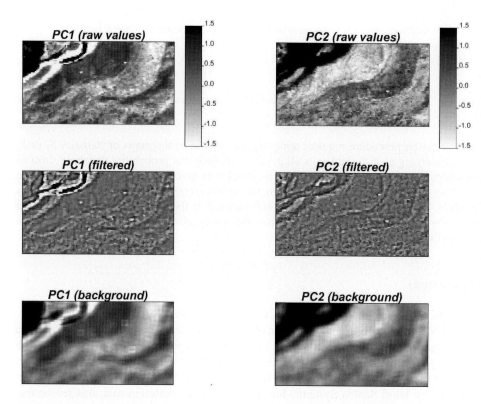

Figure 3. Maps of the first two principal components for the HSRH image, and the results of the geostatistical filtering of the regional background.

The analysis was first performed on the first 84 principal components (PC) of the data. Each image of principal components was decomposed into maps of local means and residuals or filtered values, using a 5×5 window centered on the pixel being filtered (i.e. $n=25$ in equation (1)). Figure 3 shows an example for the first 2 PCs. The original PC values are decomposed into the background values $m(u)$ and the residuals or filtered values $r(u)=z(u)-m(u)$. These images illustrate how the removal of regional variability, which might represent different soil or vegetation types, highlights the location of target pixels which appear as white in the filtered images. The information provided by either filtered or non filtered sets of 84 PCs was summarized using the statistic S_1 or S_2, see Figure 4. Dark pixels, corresponding to high values, indicate the presence of local anomalies for S_1 and clusters or anomalies for S_2. This figure clearly illustrates the benefit of the geostatistical filtering and use of statistic S_2, which reduces greatly the number of background pixels being wrongly detected as clusters or local anomalies and increases the similarity with the actual map of tarp pixels displayed in Figure 2.

The final step is to compute the ROC curves from the maps of statistic S_1 or S_2. A series of thresholds (probability of detection) are selected, and for each of them the pixels classified as disturbed soils are compared to ground data in order to compute the proportion of misclassified pixels (probability of false alarms). These two sets of probabilities are then plotted to generate the ROC curve. Figure 5 shows an example of such curves for detection using either statistic S_1 or S_2 computed from filtered or non-filtered images. The main conclusions are:

1. The filtering and use of statistic S_2 (black solid curve) allows the detection of all tarp pixels for a probability of false alarms not exceeding 0.20.
2. Detection of 60% of tarp pixels can be done with small probability of false alarm (vertical part of the ROC curve) and these pixels correspond to high purity in term of tarp content. Pixels that contain a mixture of tarp and other materials (i.e. bare soil, grass) are much more difficult to detect and generate an increase in the proportion of false alarms which can be fairly dramatic if no filtering is performed and only anomalies are searched (i.e. use of statistic S_1).

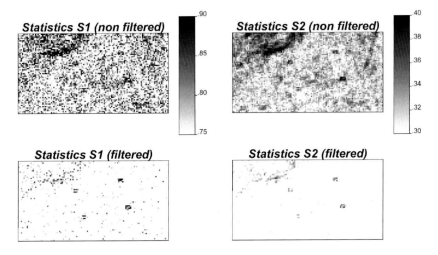

Figure 4. Maps of statistics S_1 and S_2 computed from the first 84 principal components before and after (bottom maps) filtering of the regional background.

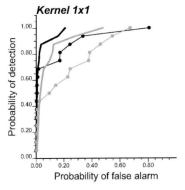

Figure 5. Receiver Operating Characteristics (ROC) curves obtained for the statistics S_1 (thin dotted line) and S_2 (solid line). Black curves are obtained from the filtered values, while the gray curves refer to original values (without geostatistical filtering).

Extensive sensitivity analysis has been conducted to assess the performance of the methodology under several conditions, such as:

1. Selection of subsets of PCs based on the strength of spatial correlation.
2. Choice of detection kernels of various sizes.
3. Decrease in signal to noise ratio and spectral resolution.

Instead of summarizing the information provided by the first 84 PCs, statistics S_1 and S_2 were computed for each PC separately and their average for both tarp and background pixels are plotted versus the rank/order of the PC in Figure 6 (top graph). Differences between tarp (black) and background (gray) pixels tend to attenuate as the order of the component increases and the spatial correlation of the image decreases (thick black curve). Subsets of PCs were thus retained based on a spatial correlation threshold of 0.5 or 0.25, plus the set of the first 25 PCs. The ROC curves indicate an increase in the proportion of false alarms when using fewer PCs. All ROC curves computed hereafter will be based on the first 25 PCs, thereby providing a balance between shorter CPU time (16.0 seconds versus 54.5 for 84 PCs on a Pentium 3.20 GHz) and slightly more false alarms.

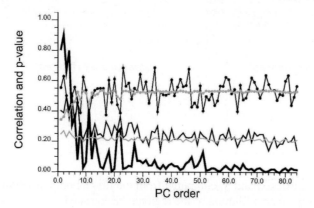

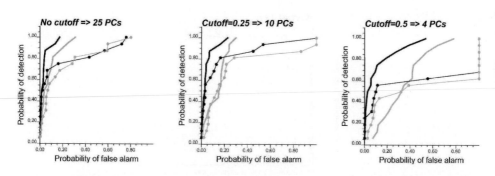

Figure 6. Plot of spatial correlation (lag=1 pixel) and value of statistics S_1 (thin dotted line) and S_2 (solid line) for either tarp pixels (black) or background pixels (gray), versus the order of the principal component (top graph). Bottom graphs show the ROC curves obtained for the first 25 PCs and two subsets based on the level of spatial correlation.

All results presented so far were obtained using a detection kernel of one pixel, which does not require any prior information regarding the size of the object to be detected. The benefit of tailoring the detection kernel to the size of the object was investigated by performing the classification and computing the ROC curves for three types of kernel: 1×1, 2×1 and 2×2. Figure 7 (top row) shows that that the use of kernels 2×1 and 2×2 improves detection performances of statistic S_1, while more false alarms occur when using statistic S_2. Indeed, statistic S_1 searches for local anomalies of size equal to the kernel, while S_2 detects both clusters and anomalies. The impact of the signal to noise (SN) ratio was investigated by adding a given proportion of noise to reflectance values before performing PCA. Figure 7 (middle row) shows the ROC curves obtained for increasing levels of noise (SN=100:1 to SN=50:1). As intuitively expected, noisy signals tend to blur the detection of anomalies, leading to a larger proportion of false alarms, although statistic S_2 on filtered signal is very robust.

The last test consisted in investigating how a decrease in spectral resolution would affect the quality of the detection. Figure 7 (bottom row) shows the ROC curves obtained for the original signal with 84 PCs, and then for one half (WV2, 42 PCs) and one third (WV3, 28 PCs) of the number of principal components. As for the signal to noise ratio, ROC curves indicate poorer performances when using the degraded image.

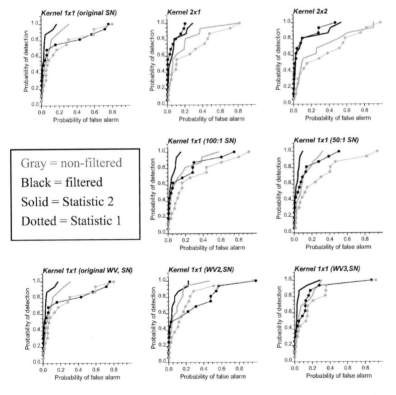

Figure 7. Receiver Operating Characteristics (ROC) curves obtained for three types of detection kernel, two signal to noise (SN) ratios, and three spectral resolutions (WV).

4 Conclusions

This paper presented and demonstrated the efficacy of a geostatistical approach to detecting disturbed soils in high spatial resolution hyperspectral imagery. The technique uses PCA to reduce dimensionality of the imagery, employs geostatistical filtering to remove regional background and enhance local signal, and applies a Local Indicator of Spatial Autocorrelation to identify patches of disturbed soils. In all scenarios, fewer false alarms were obtained when using the filtered signal and statistic S_2 to summarize information across bands. Image degradation through addition of noise or reduction of spectral resolution tends to blur the detection of anomalies, leading to more false alarms, in particular for the identification of the few mixed pixels.

In this paper the methodology was used to detect regular patches on a simple landscape. Similar results were obtained when applying the approach to more complex landscapes with multiple targets of various sizes and shapes (results not shown). Because it employs geostatistical filtering, the method is robust under low signal to noise ratios. By leveraging both spectral and spatial information, the technique requires little or no input from the user, and hence can be readily automated. Following our results a Pentium 3.20 GHz would allow the processing of a 1000×1000 scene including 25 bands within 18 minutes. Future research will investigate the benefit of processing directly the spectral bands instead of their principal components.

Acknowledgements

This work was supported by NAVAIR SBIR Phase I N02-172, and data were kindly provided by Andrew Marcus from the University of Oregon.

References

Anselin, L., Local indicators of spatial association-LISA, *Geographical Analysis*, vol. 27, 1995, p. 93-115.
Aspinall, R.J., Marcus, W.A., and Boardman, J.W., Considerations in collecting, processing, and analysing high spatial resolution hyperspectral data for environmental investigations, *Journal of Geographical Systems*, vol. 4, 2002, p. 15-29.
Goovaerts, P., *Geostatistics for Natural Resources Evaluation*, Oxford University Press, 1997.
Jacquez, G.M. and Greiling, D.A., Local clustering in breast, lung and colorectal cancer in Long Island, New York, *International Journal of Health Geography*, vol. 2, no. 3, 2003.
Jacquez, G.M., Marcus, W.A., Aspinall, R.J. and Greiling, D.A., Exposure assessment using high spatial resolution hyperspectral (HSRH) imagery, *Journal of Geographical Systems*, vol. 4, 2002, p. 15-29.
Ma, Y.Z., and Royer, J.J., Local geostatistical filtering : application to remote sensing, *Sciences de la Terre, Serie Informatique*, vol. 27, 1988, p. 17-36.
Marcus, W.A., Mapping of stream microhabitats with high spatial resolution hyperspectral imagery, *Journal of Geographical Systems*, vol. 4, 2002, p. 113-126.
Sandjivy, L., The factorial kriging analysis of regionalized data. Its application to geochemical prospecting. In *Geostatistics for Natural Resources Characterization*, edited by G. Verly, M. David, A.G. Journel, and A. Marechal, Dordrecht: Reidel, 1984, p. 559-571.
Van Meirvenne, M. and Goovaerts, P., Accounting for spatial dependence in the processing of multitemporal SAR images using factorial kriging, *International Journal of Remote Sensing*, vol. 23, 2002, p. 371-387.
Wackernagel, H., *Multivariate Geostatistics*, Springer-Verlag, 1998.
Wen, R. and Sinding-Larsen, R., Image filtering by factorial kriging—sensitivity analysis and application to Gloria side-scan sonar images, *Mathematical Geology*, vol. 29, no. 5, 1997, p. 433-468.

COVARIANCE MODELS WITH SPECTRAL ADDITIVE COMPONENTS

DENIS MARCOTTE and MIRO POWOJOWSKI
Département des génies civil, géologique et des mines,
École Polytechnique, C.P. 6079, Succ. Centre-ville, Montréal, QC, H3C 3A7

Abstract. We present a new model defining a whole class of variogram models: the spectral additive model (SAM). The model is obtained by linear combination of simple spectral components. The SAM parameters can be estimated linearly and without bias. The handling of mean drift is straightforward. In the spatial domain, the SAM possesses an analytic expression, a clear advantage over similar approaches based on covariance spectra obtained by FFT. The SAM is flexible as it can approximate any classical model, isotropic or anisotropic, to the desired degree of precision. A forward inclusion selection procedure enables avoiding over-parameterization of the model. This is especially useful in the general anisotropic case. Simulations illustrate the performance of the SAM for covariance function fitting.

1 Introduction

The choice of a suitable variogram or covariance model is an important step in any geostatistical study. This step remains largely handcrafted and resists automation. Current practice normally involves computing experimental variogram(s); a step involving its legion of more or less arbitrary decisions like the choice of directions, the angular tolerance, and the distance bins to adopt. Decisions on the characteristics of the model follow: stationary or non-stationary, isotropic or anisotropic, type of anisotropy, type of model or of combination of models to use. Finally the model parameters are adjusted, either manually or sometimes with the help of automatic fitting programs and cross-validation procedures (Marcotte, 1995). Most of the classical models being non-linear functions of distance, automatic fitting itself can be difficult to realize as many local optimums could exist. This partly explains why the parameters are still often obtained by visual fit.

Although a host of models are available (Chilès and Delfiner, 1999), one may question why data should necessarily comply with any of these models. There is a definite need to introduce greater flexibility and ease of estimation, especially if one considers implementation of geostatistical algorithms in wide general use software packages like GIS and statistical packages. The need for automation and flexibility is certainly present in the univariate stationary or non-stationary cases but it is even more compelling in the muzltivariate case.

We present a new class of models that are flexible and suitable for automatic estimation. This class of models is obtained by linear combinations of spectral components, thus the name spectral additive model (SAM). Because the model is linear, its parameters can be estimated by standard regression (or robust or weighted regression if preferred). There is no need to compute a variogram for the estimation because the fitting can be done easily considering each data pair available. The focus of this study is on the univariate case, stationary or non-stationary, isotropic or anisotropic.

The proposed approach bears resemblance to the approach suggested by Yao and Journel (1998). However, it is fundamentally different in many aspects. Here, the model defined is continuous and an analytical expression for the covariance function exists. In Tiao and Journel (1998) the model is discrete and numerically defined only at the lags used in the variogram computation.

2 Theory

2.1 A CLASS OF FLEXIBLE ISOTROPIC MODELS

Stationary (or homogeneous) random functions with absolutely integrable covariance function $C(h)$ possesses the spectral density $c(\omega)$. Together, they form a Fourier transform pair (Christakos, 1992, Yaglom, 1987). That is,

$$C(h) \xrightarrow{\mathfrak{I}} c(\omega)$$
$$C(h) \xleftarrow{\mathfrak{I}^{-1}} c(\omega) \qquad (1)$$

with $c(\omega) > 0$ for all frequencies ω and $c(\omega)$ symmetric.

The basic idea in our approach is to replace the continuous function $c(\omega)$ by a summation of piecewise continuous functions $c_i(\omega)$, that we call spectral components:

$$c_i(\omega) = f_i(|\omega|) \, 1_{r_{i-1} \leq |\omega| < r_i} \qquad \omega \in \mathfrak{R}^d \qquad (2)$$

with $0 = r_0 < r_1 < ... r_n \leq +\infty$ being an increasing finite sequence of positive numbers, and $1_{r_{i-1} \leq |\omega| < r_i}$ being an indicator function. The bins $r_0, r_1, ... r_n$ are selected so as to provide good coverage of the positive frequencies. Although there is freedom in the choice of $f_i(\omega)$, a simple and convenient choice is $f_i(\omega) = 1$ for all intervals except possibly for the last semi-infinite interval.

Now consider a linear combination of the spectral components:

$$c(\omega) = \sum_{i=1}^{n} a_i c_i(\omega) \qquad (3)$$

Its Fourier transform is:

$$C(h) = \sum_{i=1}^{n} a_i \mathfrak{I}^{-1}(c_i(\omega)) = \sum_{i=1}^{n} a_i C_i(h) \qquad (4)$$

The following results hold true:

R1. Any linear combination with coefficients $a_i \geq 0$ is the spectral density of an admissible covariance;

R2. Any admissible isotropic covariance having a spectral density can be approximated to an arbitrary degree of accuracy by model (4) (Powojowski, 2000). The accuracy of the approximation increases with the number of spectral components used to discretize the spectral density.

The isotropic covariance corresponding to the choice $f_i(\omega) = 1$ in Equation 2 is given by:

$$C_i(h) = h^{-d/2}(r_i^{d/2} J_{d/2}(r_i h) - r_{i-1}^{d/2} J_{d/2}(r_{i-1} h)) \qquad (5)$$

where $J_{d/2}$ is the order d/2 Bessel function of the first kind, where d is the dimension of the space. Note that for the limit where h->0:

$$C_i(0^+) = (r_i^{d/2} - r_{i-1}^{d/2})/\Gamma(1 + d/2) \qquad (6)$$

Figure 1 shows the isotropic covariance for a few selected spectral components after normalisation by $C_i(0^+)$ (Equation (6)) to ensure a unit sill.

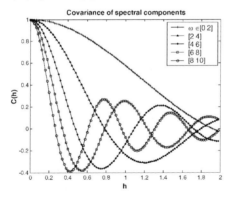

Figure 1. Equation 5 and normalized to a unit sill.

2.2 CONTROLLING MODEL BEHAVIOUR AT THE ORIGIN

In some cases, it is convenient to choose the last interval semi-open to infinity. This interval controls the behaviour of the covariance function at the origin. For example, the following choice ensures linear behavior at the origin:

$$c_n(\omega) = g(|\omega|) 1_{r_n < |\omega|} \qquad \omega \in \Re^d \qquad (7)$$

where $g(\omega)$ is the spectral density of the isotropic exponential covariance. When $d=2$, $g(|\omega|) = \frac{1}{a}\left(|\omega|^2 + \frac{1}{a^2}\right)^{-3/2}$, with a the range. Setting $f_n(\omega) = 0$ gives a differentiable covariance model.

2.3 APPROXIMATION OF CLASSICAL PARAMETRIC COVARIANCE MODELS BY SPECTRAL COMPONENTS : THE ISOTROPIC CASE

Figure 2 shows some 2D-isotropic parametric covariance models and their close approximation by a model with 4 (a and d) or 5 (b and c) spectral components. All spectral components are piecewise constants. However, in cases b) and c) an exponential spectral component is added to reproduce the model linear behavior. Clearly, the spectral additive model can match any isotropic classical model with few spectral components, thus demonstrating its great flexibility.

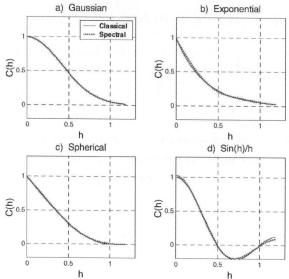

Figure 2. Examples of close approximation of various isotropic covariances by additive spectral models.

2.4 ESTIMATION OF PARAMETERS

A definite advantage of the spectral additive model is the possibility of estimating its parameters linearly. We assume, for generality, a model with an unknown mean that can include a trend.

The drift model is:

$$Z(x) = X\beta + \eta(x) \qquad (8)$$

where $\eta(x)$ is second order stationary with zero mean and real covariance K_Z and X is the $n \times p$ regression matrix used to model the drift.

The residuals Y_i, i=1...n, are obtained by ordinary least-squares estimation of β using the n available data:

$$Y = \left(I - X(X'X)^{-1}X'\right)Z = PZ \qquad (9)$$

where Y and Z are vectors of size nx1, I is the identitiy matrix of order n and P is the $n \times n$ projection matrix.

SPECTRAL ADDITIVE MODELS

The product of residuals is YY'. Its expectation is:

$$E[YY'] = PK_Z P \tag{10}$$

The covariances for the n data points computed from SAM are:

$$K_a = \sum_{j=1}^{q} a_j K_j \tag{11}$$

Each matrix K_j is $n \times n$. It represents the covariances associated with the "j^{th}" spectral component. There are q such spectral components.

The covariances for the residuals obtained with SAM are:

$$PK_a P = \sum_{j=1}^{q} a_j PK_j P = \sum_{j=1}^{q} a_j U_j \tag{12}$$

Estimators of a_j minimize the norm between the product of residuals YY' and the covariances computed with Equation 12:

$$min \left\| YY' - \sum_{j=1}^{q} \hat{a}_j U_j \right\|_2 \tag{13}$$

The least squares estimator is:

$$\hat{a} = \left[trace(U_i U_j) \right]^{-1} \left[Y' U_i Y \right] \text{ with } \hat{a}_i \geq 0, i = 1...q \tag{14}$$

where $\hat{a}$ is the $q \times 1$ vector of coefficients for the q spectral components. The notation $\left[trace(U_i U_j) \right]$ denotes a $q \times q$ matrix whose (i,j)- th entry is $trace(U_i U_j)$.

Note that a nugget effect can be included as an additional component in Equations 11 to 14.

2.5 CONSTRAINTS ON THE a_i COEFFICIENTS

For the SAM to be admissible, the coefficients a_i must be non-negative. Enforcing these constraints directly in the estimation procedure complicates the computation. One way to circumvent this problem is to forward select the spectral components one at a time. At each step, the spectral component providing the best-fit improvement to the product of residuals and, at the same time, providing a set of positive coefficients, is selected. The procedure is stopped when no further significant improvement is possible or when all the candidate spectral components provide inadmissible models (i.e. at least one coefficient becomes negative).

2.6 A DIFFERENT NORM

The norm used in Equation 13 gives the same weight to all products of residuals. It is well known that the experimental variogram or covariance function is more reliable at short distances than at large distances due to smaller fluctuations. Also, the covariance at short distances has more influence on geostatistical operators like kriging or simulations. Thus, it could be interesting to favour the covariance fit at short distances by considering instead the modified norm:

$$\min \left\| YY'*V - \sum_{j=1}^{q} \hat{a}_j U_j * V \right\|_2 \tag{15}$$

where $U*V$ is the Hadamard matrix product (i.e. element by element multiplication) and V is a $n \times n$ weighting matrix with elements defined by a positive non-increasing function of the distance separating the data.

With this weighting, the parameter estimates are now given by:

$$\hat{a} = [trace((U_i * V)U_j)]^{-1}[Y'(U_i * V)Y] \text{ with } \hat{a}_i \geq 0,\ i=1...q \tag{16}$$

2.7 SELECTION OF SPECTRAL BINS

Shannon's (1949) sampling theorem for data on a regular grid indicates the highest frequency component that can be estimated reliably from the data. This frequency, the Nyquist frequency, is given by:

$$f_{Nyq} = \frac{1}{2\Delta h} \tag{17}$$

where Δh is the grid step.

For irregularly spaced data, the Nyquist frequency is not defined. We compute a pseudo-Nyquist frequency using Equation 17 by substituting the grid spacing Δh by the average distance for the 30 nearest data pairs.

At frequencies higher than the Nyquist frequency, an exponential spectral component can be added to the set of piecewise constant spectral components to impose linear behaviour of the covariance function at the origin. Shannon's sampling theorem indicates this decision is essentially model-based and cannot be derived from the data. This may sound paradoxical as the first points of the experimental variogram show less fluctuation and are often well estimated. Nevertheless, the (non) differentiability of the process, that is the linear or parabolic behavior at the origin remains a modelling decision. In practical terms, the fact that the first variogram points define a straight line does not guarantee it extrapolates linearly to the intercept.

Having defined the highest frequency available, equal bins are used to define the various spectral components. A more elaborate spectral binning strategy is described in Powojowski (2000).

2.8 PROPERTIES OF THE ESTIMATOR

Powojowski (2000) studied in detail the properties of the estimator. He showed that when the true covariance function has the form of Equation 11, the estimator $\hat{a}$ is unbiased. Otherwise $\hat{a}$ is still a meaningful estimator in the sense that it is the closest to YY' for the chosen norm. It was shown that, with a known mean, the estimator is convergent under in-fill sampling - expanding domain conditions. However, for a compact domain the estimator has a residual variance even in the case of in-fill sampling. With an estimated mean, convergence is not ensured except if the weights in V (Equation 16) decay exponentially with distance.

2.9 A SPECTRAL ANISOTROPIC MODEL

The general methodology described above for isotropic models extends readily to general anisotropic models. The idea is to bin the 2D or 3D frequency domain. To illustrate, we consider only the 2D case. The covariance is an even function: $c(\omega_1, \omega_2) = c(-\omega_1, -\omega_2)$. Thus only half the frequency plane need be considered.

In 2D, to the spectral component:

$$c_{ij}(\omega_x, \omega_y) = 1 \begin{cases} \omega_{x,i-1} < \omega_x < \omega_{x,i} & -\omega_{x,i-1} > -\omega_x > -\omega_{x,i} \\ \omega_{y,j-1} < \omega_y < \omega_{y,j} & \text{and} \quad -\omega_{y,j-1} > -\omega_y > -\omega_{y,j} \end{cases} \quad (18)$$

$$\quad 0 \quad \text{elsewhere}$$

corresponds the anisotropic covariance:

$$C_{ij}(h_x, h_y) = \frac{\cos(\omega_{x,i}h_x + \omega_{y,j-1}h_y) + \cos(\omega_{x,i-1}h_x + \omega_{y,j}h_y)}{(\pi h_x h_y)} \dots$$
$$\frac{-\cos(\omega_{x,i}h_x + \omega_{y,j}h_y) - \cos(\omega_{x,i-1}h_x + \omega_{y,j-1}h_y)}{(\pi h_x h_y)} \quad (19)$$

The following limits for Equation 19 exist:

$$h_x \to 0; C(0^+, h_y) = \frac{(\omega_{x,j} - \omega_{x,j-1})(\sin(\omega_{y,i}h_y) - \sin(\omega_{y,i-1}h_y))}{\pi h_y}$$

$$h_y \to 0; C(h_x, 0^+) = \frac{(\omega_{y,j} - \omega_{y,j-1})(\sin(\omega_{x,i}h_x) - \sin(\omega_{x,i-1}h_x))}{\pi h_x} \quad (20)$$

$$h_x, h_y \to 0; C(0^+, 0^+) = \frac{(\omega_{x,i} - \omega_{x,i-1})(\omega_{y,j} - \omega_{y,j-1})}{\pi}$$

Figure 3 shows the good fit obtained for a few classical models presenting geometric anisotropy. Note that the SAM can accommodate also any kind of zonal anisotropy.

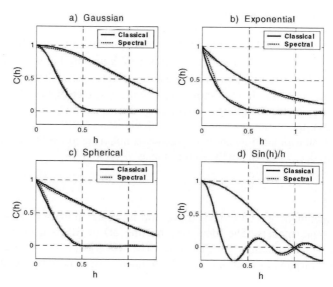

Figure 3. Examples of close approximation of various anisotropic covariances by additive spectral models.

3 Simulated examples

Two hundred data are simulated in a 100m x 100m square. The simulated model is spherical isotropic with a range of 30m, no nugget and a sill of 10. A global linear drift is added with m(x,y)=0.2*x+0.1*y where x and y are the spatial coordinates. Figure 4 shows the experimental covariance, the model covariance and the expected covariance of residuals obtained by fitting an isotropic SAM with drifts of order 0, 1 and 2. Note how the SAM retrieves very well the main characteristics of the simulation for drift orders 1 and 2. On the other hand, when adopting a zero order drift, the model is forced to include strong small frequency components (large range) to account for the drift not included in the model. Although the expected covariances match well the experimental covariances, the theoretical covariances are well above the experimental ones, a clear indication that the variance of the process can not be defined and therefore that the process is non-stationary.

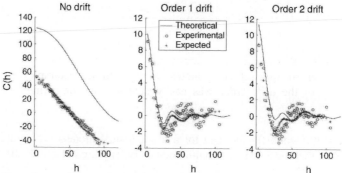

Figure 4. Experimental covariances, expected values of product of residuals and theoretical covariance for the simulated example. Isotropic case.

A similar example is simulated with geometric anisotropy ($a_x=50$, $a_y=25$). Figure 5 shows the results obtained with an anisotropic spectral model when specifying order 0 and order 1 drifts. When the right drift order is selected (i.e. order 1), the model correctly identifies the anisotropy present.

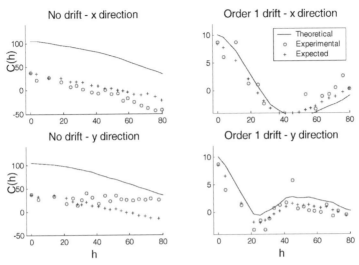

Figure 5. Experimental covariances, expected values of product of residuals and theoretical covariance for the simulated example. Anisotropic case.

5 Discussion and conclusion

The SAM approach reduces covariance model identification to a problem of linear regression with a simple positivity constraint on the regression coefficients. The control variable is the product of residuals. The regressors are the expectation of this product computed from each spectral component (Equation 12). This approach has interesting advantages. First, the class of models so defined is larger than for the usual parametric models. Second, as the parameters can be estimated linearly, it lends itself to automation. Third, all the tools of standard regression can be exploited: statistical tests, identification of outliers, ridge or robust procedures, etc. Fourth, possibly most importantly, it enables estimating the parameters of the model in the presence of mean drift as the effect of the drift is accounted explicitly when computing expected values of product of residuals. Finally, component selection procedures like forward inclusion or backward elimination, or a combination of both, can be used to limit the number of parameters and avoid possible difficulties due to colinearities between the regressors when the number of spectral components is high. In the examples presented, an underestimation of the drift order was easily detected by comparing the theoretical model to the expected value for the product of residuals.

Generalization of the approach to the multivariate case is possible. With "p" variables, "p" real coefficients (one for each spectral density) and $p(p-1)/2$ complex coefficients (one for each cross-spectral density) will have to be estimated for each spectral bin. The resulting complex coefficient matrix of size pxp must be Hermitian positive semi-definite for the model to be admissible (Wackernagel, 1995).

Acknowledgements

This research was financed by a NSERC grant. We thank Maria Annechione for editing the manuscript. Helpful suggestions from two anonymous reviewers are also acknowledged.

References

Chilès, J. P., and Delfiner, P., 1999, Geostatistics: Modeling spatial uncertainty: Wiley, New York, 695 p.
Christakos, G. 1992. Random Field Models in Earth Sciences. Academic Press, San Diego.
Marcotte, D., 1995. Generalized cross-validation for covariance model selection and parameter estimation. Mathematical Geology, v. 27, no. 6, 749-762.
Powojowski, M., 2000. Sur la modélisation et l'estimation de la fonction de covariances d'un processus aléatoire. Ph. D. thesis, University of Montreal, 195p.
Shannon, C.E., 1949. Communication in the presence of noise. Proc. Institute of Radio Engineers, v. 37, no. 1, 10-21.
Wackernagel, H., 1995. Multivariate Geostatistics. Springer, Berlin.
Yao, T. and Journel, A. G., 1998. Automatic modeling of (cross) covariance tables using fast Fourier transform. Mathematical Geology, v. 30, no. 6, 589-615.
Yaglom, A.M., 1987. Correlation theory of stationary random functions. Springer, New York.

A STATISTICAL TECHNIQUE FOR MODELLING NON-STATIONARY SPATIAL PROCESSES

JOHN STEPHENSON[1], CHRIS HOLMES[2], KERRY GALLAGHER[1] and ALEXANDRE PINTORE[2]
[1] Dept. Earth Science and Engineering, Imperial College, London. [2] Dept. of Mathematics, Oxford University

Abstract. A deficiency of kriging is the implicit assumption of second-order stationarity. We present a generalisation to kriging by spatially evolving the spectral density function of a stationary kriging model in the frequency domain. The resulting non-stationary covariance functions are of the same form as the evolved stationary model, and provide an interpretable view of the local effects underlying the process. The method employs a Bayesian formulation with Markov Chain Monte Carlo(MCMC) sampling, and is demonstrated using a 1D Doppler function, and 2D precipitation data from Scotland.

1 Introduction

The standard approach to spatial statistics assumes that the spatial dependence between two points is a function only of separation vector. These procedures fall under the generic label kriging, which are fully described in (Cressie, 1993). Such stationary models however, are unable to take account of localised effects such as geological (e.g. topography, river systems) or political (e.g. state governments conformance to air pollution measures in the US) boundaries, or rapid spatial variations. Although problematic, to date there are few generic non-stationary procedures.

One is the deformation approach of (Sampson and Guttorp, 1992), extended recently to a Bayesian framework in (Damian, Sampson, and Guttorp, 2001) and (Schmidt and O'Hagan, 2003). The more recent kernel-based methods of (Higdon, Swall, and Kern, 1999) and the spectral extension in (Fuentes, 2002) have been shown to be powerful and can be applied when only one observation is available at each site. Other approaches include orthogonal expansion (Nychka and Saltzman, 1998) and the localised moving window approach (Haas, 1990; Haas, 1995). Earlier work is summarised in (Guttorp and Sampson, 1994).

In this paper, we describe and extend our recent generalisation of kriging, involving the spatial evolution of the spectral density function of a stationary process, by manipulation in the frequency domain (Pintore and Holmes, 2003).

The new method we describe has a variety of attractive aspects, including an interpretable view of the non-stationary process, the definition of a global and analytical covariance structure (thereby making predictions at new locations trivial) and the ability to use the powerful framework developed within kriging directly.

2 Framework for non-stationary covariance functions

Here we use the standard stationary model and evolve a new class of non-stationary process. The emphasis lies in creating new covariance structures that are both non-stationary and interpretable. The proofs for the validity of these theorems can be found in (Pintore and Holmes, 2003).

2.1 STATIONARY GAUSSIAN PROCESSES

In the case of spatial interpolation, we use a stochastic model over the spatial variable s, defined over the p dimensional region $\mathbb{R}^p$. We adopt the standard, stationary approach to spatial statistics and consider our n irregularly sampled data $\mathbf{y}$ to be realisations of a Gaussian process, $Z(s) \sim \mathcal{N}_n(0, \sigma^2 \Sigma)$, where $\mathcal{N}_n$ is an n dimensional Gaussian distribution with covariance function Σ, scaled by σ^2. Subsequently we parameterise the covariance function as $\Sigma(s,t) = C(s,t) + \epsilon I_n$, with $C(s,t)$ representing the correlation between two spatial points s and t, ϵ a white noise effect (commonly known as the nugget), and I_n the n dimensional identity matrix. Common forms of $C(s,t)$ include the Gaussian, exponential and Matern stationary correlation functions.

2.2 EVOLUTION IN THE FREQUENCY DOMAIN

The new covariance functions are evolved by modifying stationary covariance functions in the frequency domain. For example, the Gaussian covariance function $C(s,t) = \exp(-\alpha \|s-t\|_p^2)$, has a spectral density function given by $f(\omega) = (4\pi\alpha)^{p/2} exp(\omega'\omega/4\alpha)$, where α is a global smoothing parameter commonly called the range and ω' represents the transpose. Non-stationarity is induced through a localised latent power process $\eta(s)$ acting on the stationary spectrum at location s, hence

$$f_{NS}^s(\omega) = h(s) \left[f(\omega)\right]^{\eta(s)} \qquad (1)$$

with the subscript NS now referring to the non-stationary versions of our process, and $h(s)$ a bounded function chosen to ensure constant power in the process. This is in effect saying that when $\eta(s) < 1$, greater emphasis is placed on lower frequencies, producing a smoother process and vice-versa. When $\eta(s) = 1$, we return to the original stationary covariance function. These effects are illustrated in figure 1(a).

We return to the spatial domain via the inverse Fourier transform, producing the non-stationary covariance function C_{NS}. For our example Gaussian function, the final non-stationary covariance function is given by

$$C_{NS}(s,t) = D_{s,t} \exp\left[-\beta_{s,t} \|s-t\|_p^2\right] \qquad (2)$$

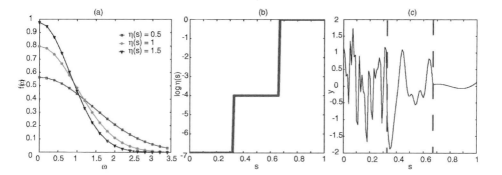

Figure 1. (a) Effect of latent process $\eta(s)$ on the spectral density of a Gaussian covariance function with $\alpha = 0.5$. (b) $\log \eta(s)$ parameterised as a step function. (c) A realisation from a Gaussian covariance function with the latent process defined in (b). The step changeovers are indicated by a dashed line in figure (c).

With

$$\beta_{s,t} = 2\alpha/[\eta(s) + \eta(t)], \qquad (3)$$

$$D_{s,t} = 2^{p/2} \frac{[\eta(s)\eta(t)]^{p/4}}{[\eta(s) + \eta(t)]^{p/2}} \qquad (4)$$

and is valid for $\eta(s) > 0$. The proof for the validity of this method with respect to Bochner's theorem (see (Levy, 1965)), and the valid choices of $\eta(s)$ are discussed for the Gaussian and Matern covariance functions in (Pintore and Holmes, 2003).

To further illustrate the effect of $\eta(s)$, and possible realisations from such models, we simulate data taken from a Gaussian non-stationary covariance function with $\alpha = 0.5$, and with $\eta(s)$ modelled as a step function (see figures 1(b) and 1(c)). Notice how the realisation **y** has high frequency content for $\log \eta(s)$ very negative, and is smooth (low frequency) for $\log \eta(s) \to 0$.

A key point to note is the similarity in form between the stationary and non-stationary covariance functions. This allows us to interpret the latent function $\eta(s)$ directly in terms of the changes in the underlying process.

3 A Bayesian approach

We now present a Bayesian extension to the method, using proper priors. For the moment, we assume a known constant mean of 0 across s to demonstrate the effect of $\eta(s)$ (N.B. The formulation presented is easily extendable to the general case where $Z(s) \sim \mathcal{N}_n(\mathbf{B}\beta, \sigma^2 \Sigma)$, with **B** representing a matrix of basis functions, and β a scaling vector included to account for deterministic trends in the data). References concerning Bayesian kriging are (Handcock and Stein, 1993) (using reference priors) and (Le and Zidek, 1992).

3.1 LIKELIHOOD, PRIORS AND POSTERIOR DISTRIBUTIONS

For mean equal to 0, the likelihood of the data $\mathbf{y}$ is expressed as

$$p(\mathbf{y}|\theta, \epsilon, \sigma^2) = (2\pi\sigma^2)^{-n/2}|\Sigma|^{-1/2} \exp\left(-\frac{\mathbf{y}'\Sigma^{-1}\mathbf{y}}{2\sigma^2}\right) \qquad (5)$$

where Σ is the covariance matrix, parameterised by the nugget effect ϵ and the parameters that define the covariance matrix θ. θ will contain the stationary (α) and the non-stationary (the parameterisation of $\eta(s)$) covariance parameters.

In order to facilitate calculation of the marginals later, we give σ^2 an inverse gamma prior density,

$$p(\sigma^2) \propto (\sigma^2)^{-(a+1)} \exp\left(\frac{-b}{\sigma^2}\right) \qquad (6)$$

with the two hyperparameters a and b set to 0.1, providing a wide, non-informative distribution.

For the covariance parameters $p(\theta, \epsilon)$, we assume independent uniform priors, expressing our lack of knowledge about the underlying system, and provide a fully flexible process. As we have assumed independence from σ^2, any other choice of informative prior is equally valid.

These definitions lead to the full posterior, given up to proportionality by Bayes theorem as

$$p(\theta, \epsilon, \sigma^2|\mathbf{y}) \propto (\sigma^2)^{-(n/2+a+1)}|\Sigma|^{-1/2} \exp\left(-\frac{\mathbf{y}'\Sigma^{-1}\mathbf{y} + 2b}{2\sigma^2}\right) p(\theta, \epsilon) \qquad (7)$$

3.2 POSTERIOR PREDICTIVE DENSITY

Our goal is to make a prediction of y_0 at a new position in $\mathbb{R}^p$ by integrating the posterior predictive density $p(y_0|\mathbf{y})$. As the integral is intractable, we solve it by sampling from the posterior distribution (equation 7) using MCMC, which for N samples gives the summation,

$$p(y_0|\mathbf{y}) \approx \frac{1}{N} \sum_{i=1}^{N} p(y_0|\theta_i, \epsilon_i, \sigma_i^2) \qquad (8)$$

where the first term is the conditional predictive distribution. This density is a shifted t distribution ((Le and Zidek, 1992)) with an expectation (for our simplified case) of $E\left[p(y_0|\theta_i, \epsilon_i, \sigma_i^2)\right] = c_i'\Sigma_i^{-1}\mathbf{y}$, where c_i is the vector of covariates between our data $\mathbf{y}$ and the new data point y_0. Then

$$E\left[p(y_0|\mathbf{y})\right] \approx \frac{1}{N} \sum_{i=1}^{N} c_i'\Sigma_i^{-1}\mathbf{y} \qquad (9)$$

in effect generating our average by the summation of the simple kriging predictions for each of the N drawn models i.

3.3 MARKOV CHAIN MONTE CARLO

In order to sample from the posterior distribution, we sample from a Markov chain. The most desirable method is to draw all of the parameters via Gibbs sampling, requiring us to know the full conditional posterior distributions. This however is only possible for the scaling parameter σ^2. After dropping constants in the posterior (equation 7), we have an inverse gamma distribution so that

$$\sigma^2 | \mathbf{y}, \theta, \epsilon, \sigma^2 \sim IG(n/2 + a, [\mathbf{y}'\Sigma^{-1}\mathbf{y} + 2b]/2) \qquad (10)$$

which we can sample from directly. This is not true of the remaining parameters which are tied up in Σ^{-1}, so we use Metropolis-Hastings sampling. To ensure better acceptance rates, we first marginalise σ^2, to give

$$p(\theta, \epsilon | y) = \int (p(\theta, \epsilon, \sigma^2 | y) d\sigma^2 \propto p(\theta, \epsilon) |\Sigma|^{-1/2} \left(\frac{\mathbf{y}'\Sigma^{-1}\mathbf{y} + 2b}{2} \right)^{-(n/2+a)} \qquad (11)$$

We use Gaussian proposal densities for ϵ and all members of θ, with variances chosen in each case to allow effective traversal of the model space.

4 Results

The two applications chosen to present the properties of our non-stationary method in a Bayesian setting, are a 1D synthetic Doppler function, and a 2D precipitation data set taken from UK Meteorological Office data over Scotland.

4.1 SYNTHETIC DOPPLER FUNCTION

We consider first the Doppler function examined in (Donoho and Johnstone, 1995) and (Pintore and Holmes, 2003), given as

$$f(s) = [s(1-s)]^{1/2} \sin[(2\pi)(1+r)/(s+r)] \quad s \in [0,1] \qquad (12)$$

with $r = 0.05$. The sample data $\mathbf{y}$ comprises 128 function evaluations positioned randomly within $s \in [0,1]$ (scaled to have a variance of 7), with added Gaussian white noise (with variance of 1). The function was then predicted at a further 500 points, uniformly sampled in the range $[0,1]$, using the stationary and non-stationary Gaussian covariance functions (equation 2). The accuracy as measured against the true function was then compared. See figure 2(a).

4.1.1 Latent process formulation
To parameterise the latent process $\eta(s)$ in the non-stationary covariance function, we follow the suggestion in (Pintore and Holmes, 2003) and use a regression spline with 5 nodes added to a linear function such that

$$\log \eta(s) = \gamma_0 + s\gamma_1 + \sum_{i=1}^{5} \phi_i \|s - u_i\| \qquad (13)$$

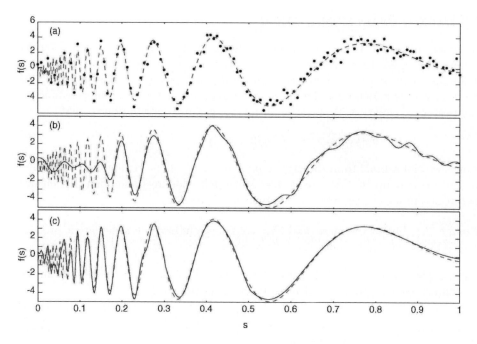

Figure 2. (a) Noisy Doppler function data set, plotted with the dashed predictive data set function. (b) Deterministic fit of the stationary Gaussian covariance function using REML. (c) Posterior predictive fit of the non-stationary Gaussian covariance function (from 4000 samples).

with u_i representing spline knot points, s, and $\{\gamma_0, \gamma_1, \phi\}$ a set of scaling coefficients. In this case, we choose to fix the knot set $\mathbf{u}$, using the kmeans algorithm, and vary the model using only the range, nugget and η scaling coefficients. The covariance parameter vector, θ, now comprises $\{\alpha, \epsilon, \gamma_0, \gamma_1, \phi\}$, which are all sampled using the Metropolis Hastings algorithm. From a run of 5000 iterations, the first 1000 were discarded as 'burn-in', and the remaining used to find the mean posterior prediction.

The stationary model was fitted using the technique of restricted maximum likelihood (REML) (Cressie, 1993), and optimised using a deterministic search (Nelder Mead algorithm).

4.1.2 Prediction Comparison

A comparison of the predictive powers of both samples can be found in figures 2(b) and 2(c). It is evident that the non-stationary method has performed far better than in the stationary case, which has been forced to comprimise between the very smooth data as $s \to 1$, and the higher frequencies as $s \to 0$.

In figure 3 we give the mean and 95% confidence intervals (an immediate advantage of using MCMC sampling) over the posterior latent process $\eta(s)$ for the 4000 samples. The figure is interpreted as placing higher emphasis on lower frequencies where $s < 0.24$ (where $\log \eta(s) < 0$), whilst providing an increasingly smooth covariance as $s \to 1$. This is further explained via figure 1(a).

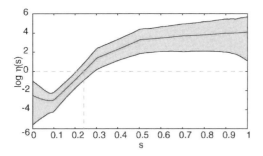

Figure 3. Log of the mean and 95% confidence intervals for the latent process $\eta(s)$. The stationary case corresponds to $\log \eta(s) = 0$.

4.2 PRECIPITATION IN SCOTLAND

For a real data example, we consider the UK Met Office 'Land Surface Observation Stations Data' held at the British Atmospheric Data Centre (BADC, 2004). The analysis of precipitation is important for agricultural as well as environmental reasons (eg. pollutant dispersal following the Chernobyl disaster). The purpose of this analysis was to demonstrate the interpretability of the latent process, rather than a comparison of stationary to non-stationary covariance functions.

Specifically, we extracted daily rainfall measurements from 1997, for the months of January, February and March, and analysed the measurements for 481 land stations in Scotland. The daily measurements within these three months were then averaged, to give three distinct data sets of average daily rainfall, (millimetres per day). A scatter plot of the January data is shown in figure 4.

4.2.1 *Latent process formulation*

To parameterise $\eta(\mathbf{s})$ (where $\mathbf{s}$ is now a 2 dimensional vector $[s_1, s_2]$ corresponding to longitude and latitude respectively), we choose to use the thin-plate spline with the form

$$\log \eta(\mathbf{s}) = \gamma_0 + \gamma_1 s_1 + \gamma_2 s_2 + \sum_{i=1}^{k} \phi_i \|\mathbf{s} - \mathbf{u}_i\|_p^2 \log \|\mathbf{s} - \mathbf{u}_i\|_p^2 \qquad (14)$$

where $\mathbf{u}_i$ is a set of k knot points, and $\{\gamma_0, \gamma_1, \gamma_2\}$ the scaling coefficients. In this case, we take $k = 20$ and fix the positions of the knot points using the kmeans clustering algorithm. Again choosing the Gaussian non-stationary covariance function, we perform MCMC over the spline scaling coefficients and the stationary parameters ϵ and α, again discarding the first 1000 iterations as 'burn-in', and averaging over the remaining 4000 samples.

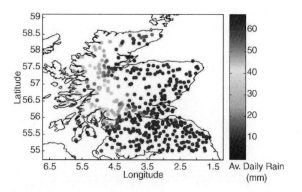

Figure 4. Positions and values of the average daily rainfall for Scotland during January 1997

4.2.2 Interpretation of $\eta(s)$ in January

We first look at the non-stationarity revealed in the January data set. A contour plot of the mean of $\eta(\mathbf{s})$, (see figure 5(a)) reveals a strong trend, moving from low values of $\eta(\mathbf{s})$ in the west, to high values in the east. Thus on the west coast, where rain is far more prevalent and protection from other landmasses is minimal, there is much greater variation from station to station. This is demonstrated by the greater emphasis placed on higher frequencies by $\eta(s)$, giving a local covariance function with a small range of influence. This contrasts with the smoother region in the east, where the level of rainfall is reduced, sheltered as it is by the topography in the Highlands.

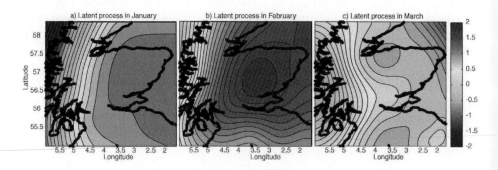

Figure 5. Contour plots of the latent process for (a) January, (b) February and (c) March during 1997.

To illustrate the significance of this west-east trend, we take a projection in the longitudinal direction and plot the 95% credible intervals (figure 6(a)). This demonstrates the small variation in the north-south direction, and the relatively tight credible intervals, reinforcing the notion of an west-east trend in stationarity.

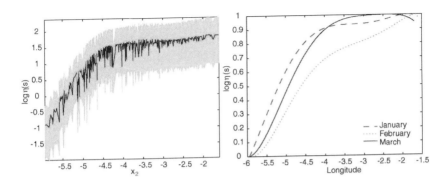

Figure 6. (a) log of the mean of $\eta(s)$ projected onto the longitudinal direction, plotted with 95% credible intervals. (b) Comparison of scaled longitudinal projections of $\eta(s)$ for the January, February and March data sets.

4.2.3 Comparison of $\eta(s)$ for three different months

To further demonstrate the importance of this east west trend, we carried out the same analysis on the two subsequent months and compared the posterior values of $\eta(s)$. These data sets are compared directly in figure 5, comprising a contour plot for each of the three months and again show strong west-east trends.

As the absolute values of $\eta(s)$ in these figures are influenced by the effect of the stationary covariance parameter α, we compare the longitudinal projections of $\log \eta(s)$ by first fitting a polynomial and then scaling the values to the range [0, 1]. This is shown in figure 6(b), and demonstrates concisely the consistency of the recognised trend.

The consistency of this result indicates that there is an underlying process causing a consistent non-stationarity in the data. Suggestions as to the cause of this observation are geographical effects such as coastal regions, shielding by topography and the consistent direction of weather patterns. Significantly, this demonstrates the ability of the method to reveal an accurate measure of the non-stationarity from only one realisation.

5 Discussion

In summary, we have presented a method for inducing non-stationarity in standard covariance functions, by use of latent power process $\eta(s)$ in the frequency domain. Such a treatment yields covariance functions that have the same analytical form of the base stationary function, and offers a direct and interpretable view of any non-stationary processes. A Bayesian methodology has been used and demonstrated using MCMC providing access to the full uncertainties.

The two applications have revealed an increased prediction accuracy when compared to standard stationary techniques, and demonstrated the ability to extract the underlying non-stationarity from a single realisation.

Now that the Bayesian context has been established, future work will involve using reversible jump MCMC when inferring $\eta(s)$ (Green, 1995), (providing the ability to change the position and number of knot points), as well as applying the method to spatio-temporal processes by treating η as a function of space and time.

Acknowledgements

We are grateful to the British Atmospheric Data Centre for providing access to the Met Office Land Surface Observation Stations Data, and to the UK Natural Environment Research Council for funding this work.

References

BADC, *Met Office - Land Surface Observation Stations Data*, http://badc.nerc.ac.uk/data/surface/, 2004.

Cressie, N. A. C., *Statistics for spatial data*, Wiley series in probability and mathematical statistics. Applied probability and statistics., 1993.

Damian, D., Sampson, P. D. and Guttorp, P., *Bayesian estimation of semi-parametric non-stationary spatial covariance structures*, Environmetrics, vol. 12, no. 2, 2001, p. 161-178.

Donoho, D. L. and Johnstone, I. M., *Adapting to unknown smoothness via wavelet shrinkage*, Journal of the American Statistical Association, vol. 90, no. 432, 1995, p. 1200-1224.

Fuentes, M., *Spectral methods for nonstationary spatial processes*, Biometrika, vol. 89, no. 1, 2002, p. 197-210.

Green, P. J., *Reversible jump Markov chain Monte Carlo computation and Bayesian model determination*, Biometrika, vol. 82, no. 4, 1995, p. 711-732.

Guttorp, P. and Sampson, P. D., *Methods for estimating heterogeneous spatial covariance functions with environmental applications*, in G.P. Patil and C.R. Rao (eds), Handbook of Statistics XII: Environmental Statistics. Elsevier, 1994, p. 663-90.

Haas, T. C., *Lognormal and Moving Window Methods of Estimating Acid Deposition*, Journal of the American Statistical Association, vol. 85, no. 412, 1990, p. 950-963.

Haas, T. C., *Local prediction of a spatio-temporal process with an application to wet sulfate deposition*, Journal of the American Statistical Association, vol. 90, no. 432, 1995, p. 1189-1199.

Handcock, M. S. and Stein, M. L., *A Bayesian-Analysis of Kriging*, Technometrics, vol. 35, no. 4, 1993, p. 403-410.

Higdon, D., Swall, J. and Kern, J., *Non-stationary spatial modeling*, in Bernado, J.M. et al. (eds), *Bayesian statistics 6 : proceedings of the Sixth Valencia International Meeting, June 6-10, 1998*. Oxford University Press, 1999.

Le, N. D. and Zidek, J. V., *Interpolation with Uncertain Spatial Covariances - a Bayesian Alternative to Kriging*, Journal of Multivariate Analysis, vol. 43, no. 2, 1992, p. 351-374.

Levy, P., *Processus stochastiques et mouvement brownien*. Gauthier-Villars, 1965.

Nychka, D. and Saltzman, N., *Design of Air Quality Monitoring Networks*, Case Studies in Environmental Statistics, no. 132, 1998, p. 51-76.

Pintore, A. and Holmes, C. C., *Constructing localized non-stationary covariance functions through the frequency domain*, Technical Report. Imperial College, London, 2003.

Sampson, P. D. and Guttorp, P., *Nonparametric-Estimation of Nonstationary Spatial Covariance Structure*, Journal of the American Statistical Association, vol. 87, no. 417, 1992, p. 108-119.

Schmidt, A. M. and O'Hagan, A., *Bayesian inference for non-stationary spatial covariance structure via spatial deformations*, Journal of the Royal Statistical Society Series B-Statistical Methodology, vol. 65, no. 3, 2003, p. 743-758.

CONDITIONING EVENT-BASED FLUVIAL MODELS

MICHAEL J. PYRCZ and CLAYTON V. DEUTSCH
*Department of Civil & Environmental Engineering, 220 CEB,
University of Alberta, Canada, T6G 2G7*

Abstract. A fluvial depositional unit is characterized by a central axis, denoted as a streamline. A set of streamlines can be used to describe a stratigraphic interval. This event-based (denoted as event-based to avoid confusion with streamline-based flow simulation) approach may be applied to construct stochastic fluvial models for a variety of reservoir types, fluvial styles and systems tracts. Prior models are calculated based on all available soft information and then updated efficiently to honor hard well data.

1 Introduction

Interest in North Sea fluvial reservoirs led to the development of object-based models for fluvial facies and geometries (see Deutsch and Wang, 1996 for a review of development). For these models conditioning is often problematic. These difficulties in conditioning spurred research in direct object modeling. Visuer et al. (1998) and Shmaryan and Deutsch (1999) published methods to simulate fluvial object-based models that directly honor well data. These algorithms segment the well data into unique channel and nonchannel facies and then fit channels through the segments. The channel center line is parameterized as a random function of departure along a vector and the geometry is based on a set of sections fit along the center line.

Yet, these techniques are only well suited to paleo valley (PV) reservoir types. The PV reservoir type geologic model is based on ribbon sandbodies from typically low net-to-gross systems with primary reservoir quality encountered in sinuous to straight channels and secondary reservoir rock based on levees and crevasse splays embedded in overbank fines (Galloway and Hobday, 1996; Miall, 1996).

More complicated channel belt (CB) fluvial reservoir types are common. Important examples include the McMurray Formation (Mossop and Flach, 1983, Thomas et al., 1987) and Daqing Oil Field, China (Jin et al., 1985, Thomas et al., 1987). These reservoirs include complicated architectural element configurations developed during meander migration punctuated by avulsion events. The application of the bank retreat model for realistic channel meander migration has been proposed by Howard (1992), applied by Sun et al. (1996) and Lopez et al. (2001)

to construct realistic models of CB type fluvial reservoirs. These methods lack flexibility in conditioning.

A event-based paradigm is introduced with (**1**) improved flexibility to reproduce a variety of fluvial reservoir styles with realistic channel morphologies, avulsion and meander migration and (**2**) a new efficient approach to condition to well data and areal reservoir quality trends. Fortran algorithms are available that apply this techniques, ALLUVSIM is an unconditional algorithm for the construction of training images and the ALLUVSIMCOND algorithm includes streamline updating for well conditioning. Greater detail on this work and the associated code is available in Pyrcz (2004).

This work was inspired by the developments of Sun et al. (1996) and Lopez et al. (2001), but it was conducted independent of Cojan and Lopez (2003) and Cojan et al. (2004). The reader is referred to these recent papers for additional insights into the construction of geostatistical fluvial models.

2 Event-based Stochastic Fluvial Model

The basic building block of this model is the *streamline*. A streamline represents the central axis of a flow event and backbone for architectural elements (Wietzerbin and Mallet, 1993). This concept is general and may represent confined or unconfined, fluvial or debris flows.

Genetically related streamlines may be grouped into *streamline associations*. Streamline associations are interrelated by process. For example, a streamline association may represent a channel fill architectural elements within a braided stream or lateral accretion architectural elements within point bar. Fluvial architectural elements are attached to streamlines and architectural element interrelationships are characterized by streamline associations. This is a logical technique for constructing fluvial models since all architectural elements are related to "flow events".

2.1 3-D STREAMLINES

The direct application of a cubic spline function to represent the plan view projection of a fluvial flow event is severely limited. As a function, a spline represented as $f^s(x)$ may only have a single value for any value x. In graphical terms, a function may not curve back on itself. This precludes the direct use of a spline function to characterize high sinuosity channel streamlines.

A streamline is modeled as a set of cubic splines. Each spline models the coordinates (x, y and z) with respect to distance along the spline (s). The advantages of this technique are: (**1**) continuous interpolation of streamline location in Cartesian coordinates at any location along the streamline, (**2**) relatively few parameters required to describe complicated curvilinear paths, (**3**) manipulation of splines is much more computationally efficient than modifying geometries and (**4**) other properties such as architectural element geometric parameters and longitudinal trends may be stored as continuous functions along the streamline. These issues are discussed in further detail below.

EVENT-BASED SIMULATION

The control nodes of a 3-D spline may be freely translated. The only requirement is that the second derivatives of the spline location parameters is recalculated after modification. This operation is very fast. The calculation of complicated geometries generally requires a high level of computational intensity or simplification. In the event-based models the geometric construction is postponed to the end of the algorithm. This results in very fast calculation and manipulation of complicated geometric morphologies and associations represented as 3-D splines.

Any properties may be attached to the 3-D spline and interpolated along the length of the spline. In the fluvial event-based model, the channel width, local curvature, relative thalweg location and local azimuth are included in the 3-D spline. Other information including architectural element type and additional property trends may be included. These properties are calculated at the control nodes and then splines are fit as with the location parameters.

2.2 STREAMLINE ASSOCIATIONS WITHIN EVENT-BASED MODELS

A streamline association is a grouping of interrelated 3-D splines. Streamline associations are characterized by their internal structure and interrelationship or stacking patterns. The internal structure is the relation of streamlines within the streamline association. The external structure is the interrelationship between streamline associations. Streamline associations may be tailored to reproduced features observed in each fluvial reservoir style.

A variety of stacking patterns may exist in the fluvial depositional setting. Compensation is common in dispersive sedimentary environments such as proximal alluvial fans, vertical stacking with little migration is common in anastomosing reaches and nested channel belts often form in incised valleys. These patterns include important information with regard to the heterogeneity of a reservoir and should be included in fluvial models.

2.3 STREAMLINE OPERATIONS

A suite of streamline operations is presented that allow for event-based models to be constructed by the creation and modification of streamlines. These operations include (**1**) initialization, (**2**) avulsion, (**3**) aggradation and (**4**) migration.

The streamline *initialization operator* is applied to generate an initial streamline or to represent channel avulsion proximal of the model area. The disturbed dampened harmonic model developed by Ferguson (1976) is applied.

The *avulsion operator* creates a copy of a specific channel streamline, selects a location along the streamline, generates a new downstream channel segment with same streamline sinuosity and the same geometric parameter distributions. The geometric parameters (e.g. channel width) of the new streamline are corrected so that the properties are continuous at the avulsion location. The initial azimuth is specified as the azimuth of the tangent at the avulsion location. There is no constraint to prevent the avulsed streamline from crossing the original streamline distal of the avulsion location.

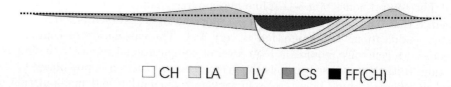

Figure 1. An illustration of the fluvial architectural elements applied in the event-based model.

Aggradation is represented by a incremental increase in the elevation of a streamline. The current implementation is to add a specified constant value to the elevation, z, parameter for all control nodes.

The streamline *migration operator* is based on the bank retreat model. The application of the bank retreat model for realistic channel meander migration has been proposed by Howard (1992), applied to construct fluvial models by Sun et al. (1996) and extended to construct meandering fluvial models that approximately honor global proportions, vertical and horizontal trends by Lopez et al. (2001).

Key implementation differences from the original work from Sun et al. (1996) include (**1**) standardization of migration steps, (**2**) integration of 3-D splines for location and properties, (**3**) application of various architectural elements. The meander migration along the streamline is standardized such that the maximum migration matches a user specified value. This removes the significance of hydraulic parameters such as friction coefficient, scour factor and average flow rate, since only the relative near bank velocity along the streamline is significant. Hydraulic parameters are replaced by the maximum spacing of accretion surfaces, which may be more accessible in practice.

2.4 FLUVIAL ARCHITECTURAL ELEMENTS

The available architectural elements include (**1**) channel fill (CH), (**2**) lateral accretion (LA), (**3**) levee (LV), (**4**) crevasse splay (CS), (**5**) abandoned channel fill (FF(CH)) and (**6**) overbank fines (FF) (see illustration in Figure 1). The geometries and associated parameters are discussed for each element in detail in Pyrcz (2004).

2.5 EVENT SCHEDULE

The event-based approach is able to reproduce a wide variety of reservoir styles with limited parametrization. This algorithm may reproduce braided, avulsing, meandering channels and may reproduce geometries and interrelationships of a variety of fluvial reservoir types. The algorithm is supplied with areal and vertical trends, distributions of geometric parameters, probabilities of events and architectural elements.

EVENT-BASED SIMULATION

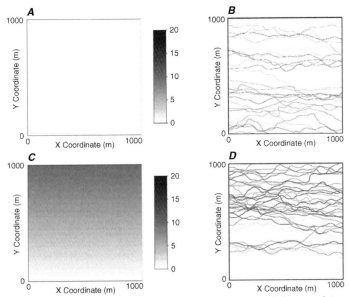

Figure 2. Example areal trends in channel density and the resulting streamlines. A and B - no areal trend supplied and C and D - a linear trend increasing in the y positive direction. Note areal trend is a relative measure without units.

2.6 AREAL CHANNEL DENSITY TRENDS

Analogue, well test and seismic information may indicate areal trends in reservoir quality. Although seismic vertical resolution is often greater than the reservoir thickness, seismic attributes calibrated to well data may indicate a relative measure of local reservoir quality. Well tests may provide areal information on the distribution of reservoir quality and may significantly constrain model uncertainty. Analogue information such as reservoir type may indicate a confined PV type or a more extensive and uniform SH type reservoirs. If the net facies are associated with CH, LV and CS elements then this areal trend information may be integrated by preferentially placing streamlines in areal locations with high reservoir quality.

The technique for honoring areal trends is to (**1**) construct a suite of candidate streamlines with the desired morphology, (**2**) superimpose each candidate streamline on the areal trend model and calculated average relative quality along the streamline and (**3**) for each streamline initialization drawn from this distribution of candidate streamlines (without replacement) weighted by the average quality index. This technique is efficient since the construction of hundreds or thousands of streamlines is computationally fast. This technique is demonstrated in Figure 2.

2.7 VERTICAL CHANNEL DENSITY TRENDS AND AGGRADATION SCHEDULE

Well data and analogue information may provide information on vertical trends in reservoir quality. Well logs calibrated by core are valuable sources of vertical trend information. Often, identification of systems tract and fluvial style will provide analogue information concerning potential vertical trends.

These trends may be honored by constraining the aggradation schedule. The current implementation is to apply the trend within a user defined number of constant elevation levels. Streamlines and associated architectural elements are generated at the lowest level until the NTG indicated by the vertical trend is reached for the model subset from the base of the model, to the elevation of the first level. Then the aggradation operator is applied to aggrade to the next level and the process is repeated through all user defined levels. For the highest level, the model is complete when the global NTG ratio is reached.

3 Conditional Event-based Simulation

There are a variety of available methods that may be applied to condition complicated geologic models; (**1**) dynamically constrain model parameters during model construction to improve data match (Lopez et al., 2001), (**2**) posteriori correction with kriging for conditioning (Ren et al.,2004), (**3**) pseudo-reverse modeling (Tetzlaff, 1990), (**4**) apply as a training image for multiple-point geostatistics (Strebelle, 2002) and (**5**) direct fitting of geometries to data (Shmaryan et. al., 1999 and Visuer et. al., 1998). Each of these techniques has limitations either in efficiency, robustness or the ability to retain complicated geometries and interrelationships.

An event-based model consists of associations of streamlines with associated geometric parameters and identified architectural elements. A prior model of streamline associations may be updated to reproduce well observations. The proposed procedure is: (**1**) construct the prior event-based model conditioned by all available soft information, (**2**) interpret well data and identify CH' element intervals (where CH' elements are channel fill elements without differentiation of CH, LA and FF(CH) elements), (**3**) update streamline associations to honor identified CH' element intervals and (**4**) correct for unwarranted CH' intercepts. This technique entails the manipulation of large-scale elements to honor small scale data; therefore, it is only suitable for settings with sparse conditioning data. Settings with dense data may be intractable.

3.1 INTERPRETED WELL DATA

The hard data from wells is applied to identified CH' element intervals. CH' elements are typically identified by erosional bases and normal grading. CH' element fills often occur in multistory and multilateral configurations. CH' elements often erode into previously deposited CH' elements to form amalgamated elements (Collinson, 1996, Miall, 1996).

The geologic interpretation of well data is performed prior to the updating step. The input data includes the areal location for each vertical well and a list of CH' element intervals with base and original top (prior to erosion). The geologic interpretation is often uncertain, especially with amalgamated CH' elements. Alternate geologic interpretations may be applied to account for this uncertainty.

3.2 UPDATING STREAMLINE ASSOCIATIONS TO HONOR WELL DATA

The model is updated by modifying the position of streamline associations to honor CH' element intercepts observed in well data. For each CH' element interval the following steps are performed. (**1**) The horizontal position is corrected such that the CH' element intercept thickness is within tolerance of the CH' element interval thickness. (**2**) Then the vertical location is corrected such that the CH' element intercept top matches the top of the CH' element interval. Entire streamline associations are corrected to preserve the relationships between streamlines within a streamline association. For example, if a streamline association includes a set of streamlines related by meander migration, the entire set of streamlines representing a point bar is shifted. If individual streamlines were modified independently this may change the nature of the streamline association.

The CH' element intervals are sequentially corrected. If there is no previous conditioning then streamline associations are translated (see A in Figure 3). If there is previous conditioning a smooth correction method is applied to the streamline association (see B in Figure 3). A step vector is constructed oriented from the nearest location on a streamline within the streamline association to the location of the well interval. The scale of the step of the sense is determined by an iterative procedure described below.

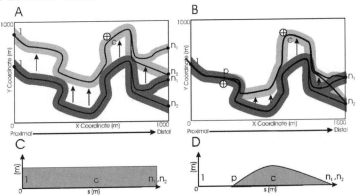

Figure 3. An illustration of methods for updating streamline associations with well data. For this example, there are two streamlines in the streamline association representing an avulsion event that are corrected to honor conditioning data (c). A - the case with no previous conditioning. B - the case with previous conditioning. C and D - the transverse correction with respect to location along the streamline.

3.3 ITERATIVE PROCEDURE FOR UPDATING STREAMLINE ASSOCIATIONS

Modifications of streamline associations has an impact on CH' element geometry. It would be difficult to directly calculate the precise translation of a streamline to result in the correct interval thickness at a well location. A simple iterative method is applied to correct the well intercept thickness. The thickness of the CH' element from a streamline association is calculated at the vertical well location. The error is calculated, if the thickness is less than indicated by the conditioning then the streamline association is shifted towards the well location. If the thickness is greater than indicated by the conditioning then the streamline association is shifted away from the well location. The procedure is repeated for all identified CH' element intercepts.

3.4 CORRECTION FOR UNWARRANTED WELL INTERCEPTS

The correction for unwarranted CH' element intercepts applies a robust iterative technique. For each unwarranted CH' element intercept the associated streamline association is checked for conditioning. If the streamline association is not anchored to conditioning data then the streamline association may be translated in the direction transverse to the primary flow direction. If the streamline association is anchored to conditioning data then a smooth modifications is applied.

The streamline association is modified until the thickness of the unwarranted CH' element intercept reaches zero. For each iteration the step size of the modification is increased and the direction is reversed. This method is robust since it does not become trapped with complicated streamline associations. This methodology is illustrated in Figure 4 with a complicated setting.

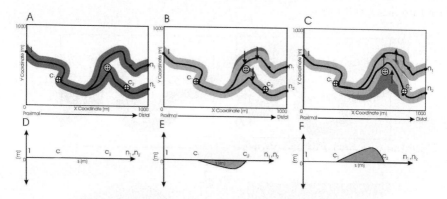

Figure 4. An illustration of the method for correcting streamline associations to remove unwarranted well intercepts. The two streamlines are related by avulsion in the streamline association and there are two previously conditioned locations (C_1 and C_2). A and D - the initial streamline association prior to correction. B and E - the first smooth modification (Oliver, 2002). C and F - the second iteration.

3.5 EXAMPLE CONDITIONAL EVENT-BASED MODELS

The ALLUVSIMCOND algorithm was applied to construct a conditional model. The streamlines include braided low to high sinuosity morphology. A single well is included with two CH' element intervals identified. Cross sections and streamline plan sections of the prior and updated models are shown in Figure 5. The morphology of the streamlines is preserved while the well intercepts are honored.

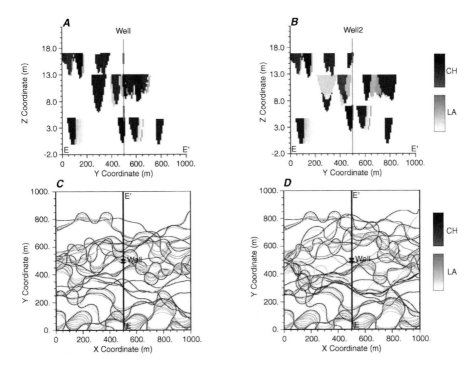

Figure 5. An example conditional event-based model from ALLUVSIMCOND. A and B - cross section of prior and updated model and C - D plan section of prior and updated model streamlines with cross section indicated.

4 Conclusions and Future Work

The event-based approach is a flexible and efficient tool for the construction of stochastic fluvial models. The building block approach allows for the modeling of a variety of fluvial reservoir styles, including the complicated architectures of CB type fluvial reservoirs. Event-based models may be constructed based on all available soft geologic information and then updated to honor hard well data.

Future implementation will address well observations of other architectural elements and the applications of the the event-based approach to a variety of depositional settings, such as deepwater (Pyrcz, 2004).

Acknowledgements

We would like to acknowledge the industry sponsors of the Centre for Computational Geostatistics at the University of Alberta and the National Science and Engineering Council of Canada for supporting this research.

References

Cojan, I. and Lopez, S., *Process-based Stochastic Modeling of Meandering Channelized Reservoirs*, AAPG Annual Meeting, Dallas, 2003, May

Cojan, I. and Fouche, O and Lopez, S., *Process-based Reservoir Modelling in the Example Meandering Channel*, Seventh International Geostatistics Congress, Banff, 2004 September

Collinson, J. D., *Alluvial Sediments*, in H. G. Reading, editor, *Sedimentary Environments: Processes, Facies and Stratigraphy*, Balckwell Science, 1996.

Deutsch, C. V. and Tran, T. T., *FLUVSIM: A Program for Object-Based Stochastic Modeling of Fluvial Depositional Systems*, Computers & Geosciences, vol. 28, 2002.

Ferguson, R. I., *Disturbed Periodic Model for River Meanders*, Earth Surface Processes, vol. 1, p. 337-347.

Galloway, W. E. and Hobday, D. K., *Terrigenous Clastic Depositional Systems*, Springer, 1997.

Howard, A. D., *Modeling Channel Migration and Floodplain Sedimentation in Meandering Streams*, in P. A. Carling and G. E. Petts, editor, *Lowland Floodplain Rivers*, John Wiley and Sons, 1992.

Jin, Y., Liu D. and Luo C., *Development of Daqing Oil Field by Waterflooding*, Journel Petroleum Technology, February, 1985, p. 269-274.

Lopez, S, Galli, A. and Cojan, I., *Fluvial Meandering Channelized Reservoirs: a Stochastic and Process-base Approach*, 2001 Annual Conference of the IAMG, International Association of Mathematical Geologists, Cancun, Mexico, 2001.

Miall, A. D., *The Geology of Fluvial Deposits*, Springer, 1996.

Mossop, G. D. and Flach, P. D., *Deep Channel Sedimentation in the Lower Cretaceous McMurray Formation*, Sedimentology, vol. 30, 1983, p. 493-509.

Oliver, D. S., *Conditioning Channel Meanders to Well Observations*, Mathematical Geology, vol. 34, 2002, p. 185-201.

Pyrcz, M. J. *Integration of Geologic Information into Geostatistical Models*, Ph. D. Thesis, University of Alberta, Edmonton, 2004.

Ren, W., Cunha, L. and Deutsch, C. V., *Preservation of Multiple Point Structure when Conditioning by Kriging*, Sixth International Geostatistics Congress, Banff, AB, 2004

Shmaryan, L. and Deutsch, C. V., *Object-Based Modeling of Fluvial/Deepwater Reservoirs with Fast Data Conditioning: Methodology and Case Studies*, SPE Annual Technical Conference and Exhibition, Society of Petroleum Engineers, Houston, TX, 1999.

Strebelle, S., *Conditional Simulation of Complex Geological Structures Using Multiple-Point Statistics*, Mathematical Geology, vol. 32, no. 9, 2002, p. 2937-2954.

Sun T., Meakin, P. and Josang, T., *A Simulation Model for Meandering Rivers*, Water Resources Research, vol. 32, no. 9, 1996, p. 2937-2954.

Tetzlaff, D. M., *Limits to the Predictive Ability of Dynamic Models the Simulation Clastic Sedimentation*, in T. A. Cross, editor, *Quantitative Dynamic Stratigraphy*, Prentice-Hall, 1990.

Thomas, R. G., Smith D. G., Wood, J. M., Visser, J., Calverley-Range, A. and Koster, E. H., *Inclined Heterolithic Stratification - Terminology, Description, Interpretation and Significance*, Sedimentary Geology, vol. 53, 1987, p. 123-179.

Viseur, S., Shtuka, A. and Mallet J. L., *New Fast, Stochastic, Boolean Simulation of Fluvial Deposits*, SPE Annual Technical Conference and Exhibition, Society of Petroleum Engineers, New Orleans, LA, 1998.

Wietzerbin, L. J. and Mallet J. L., *Parameterization of Complex 3D Heterogeneities: A New CAD Approach*, SPE Annual Technical Conference and Exhibition, Society of Petroleum Engineers, Houston, TX, 1993.

3D GEOLOGICAL MODELLING AND UNCERTAINTY: THE POTENTIAL-FIELD METHOD

CHRISTOPHE AUG [1], JEAN-PAUL CHILÈS [1],
GABRIEL COURRIOUX [2] and CHRISTIAN LAJAUNIE [1]
[1] Centre de Géostatistique, Ecole des Mines de Paris
35, rue Saint-Honoré
77305 Fontainebleau Cedex, France
[2] Bureau de Recherches Géologiques et Minières
3, avenue C. Guillemin – BP 6009
45062 Orléans Cedex 2, France

Abstract. The potential-field method (Lajaunie *et al.*, 1997) is used to create geological surfaces by interpolating from points on interfaces, orientation and fault data by universal cokriging. Due to the difficulty of directly inferring the covariance of the potential field, it is identified from the orientation data, which can be considered as derivatives of the potential. This makes it possible to associate sensible cokriging standard deviations to the potential-field estimates and to translate them into uncertainties in the 3D model.

1 Introduction

During the last ten years, 3D geological modelling has become a priority in several domains such as reservoir characterization or civil engineering. In geological mapping too, 3D digital pictures are created to model and visualize the subsurface and the relations between layers, faults, intrusive bodies, etc. While completing its 1:50 000 geological map programme for the entire French territory, B.R.G.M. (the French geological survey) started a research project for defining three-dimensional maps which could clearly represent the subsurface and underground geology. A new tool, the "Editeur Géologique", has been developed to face this particularly tough issue. It is based on the construction of implicit surfaces using the potential-field method.

2 Reminders on the potential-field method

2.1 PRINCIPLES

The problem is to model the geometry of geological layers using drill-hole data, digital geological maps, structural data, interpreted cross-sections, etc.
The method is based on the interpolation of a scalar field considered as a potential field. In this approach, a surface is designed as a particular isovalue of the field in 3D space. In all the following equations, $\mathbf{x}=(x,y,z)$ is a point in the three-dimensional space R^3. The potential is assumed to be a realization of a differentiable random function Z.

We first consider one interface or several sub-parallel interfaces (iso-surfaces related to the same potential field) and we will see later how to manage several fields.

2.2 DATA

The 3D model is obtained by integrating data originating from different sources.
The first kind of data is a set of points belonging to the interfaces to be modelled. They come from digitized contours on the geological map and from intersections with boreholes. The other type of data is structural data (orientation of surfaces).
For the interpolation of the potential field, these data are coded as follows:
- if we have a set J of n points on an interface, we use n-1 linearly independent increments of potential, all equal to zero; these increments are of the form:

$$Z(\mathbf{x}_j) - Z(\mathbf{x}_{j'}) = 0$$

$$e.g., \quad Z(\mathbf{x}_j) - Z(\mathbf{x}_{j-1}) = 0 \quad j = 2,...,n$$

If several interfaces are modelled with the same potential field, the data set J is the union of the elementary data sets relative to the various interfaces.
- orientation data are considered as gradients of the potential, namely polarized unit vectors, normal to the structural planes:

$$\frac{\partial Z}{\partial x}(\mathbf{x}_i) = G_i^x, \quad \frac{\partial Z}{\partial y}(\mathbf{x}_i) = G_i^y, \quad \frac{\partial Z}{\partial z}(\mathbf{x}_i) = G_i^z$$

2.3 SOLUTION

Determining a geological interface is an interpolation problem which can be solved by determining the potential at any point in the space and by drawing the iso-potential surface corresponding to the interface. The potential field is defined up to an arbitrary constant, because we only work with increments. Indeed we will interpolate the potential at $\mathbf{x}$ in comparison with the potential at some reference point $\mathbf{x}_0$. These increments of potential are estimated as:

$$[Z(\mathbf{x}) - Z(\mathbf{x}_0)]^* = \sum_i \left(\lambda_i G_i^x + \mu_i G_i^y + \nu_i G_i^z \right) + \sum_j \lambda_j \left[Z(\mathbf{x}_j) - Z(\mathbf{x}_{j-1}) \right] \quad (1)$$

The last term is equal to zero, but we introduce it here, because the weights λ_i, μ_i, ν_i are different from weights based on the gradient data alone.
The weights are the solution of a universal cokriging system of the form:

$$\begin{pmatrix} C_G & {}^tC_{GI} & {}^tU_G & {}^tF_G \\ C_{GI} & C_I & {}^tU_I & {}^tF_I \\ U_G & U_I & 0 & 0 \\ F_G & F_I & 0 & 0 \end{pmatrix} \begin{pmatrix} A \\ B \\ C \\ D \end{pmatrix} = \begin{pmatrix} C_G^0 \\ C_I^0 \\ U^0 \\ F^0 \end{pmatrix}$$

C_G and C_I are the covariance matrices for gradient and potential data respectively, and C_{GI} is their cross-covariance matrix.
U_G and U_I contain drift functions and F_G and F_I contain fault functions.
A, B, C, D are the solution of this linear system.
C^0_G is the covariance vector between the estimated increment and the gradient data and C^0_I is the covariance vector between the estimated increment and increment data.
U^0 and F^0 contain drift and fault functions at the estimated point.

Once the system has been solved, the iso-potential surface corresponding to the interface can be drawn. We can then visualize the 3D cube or cross-sections through it (Figure 1).

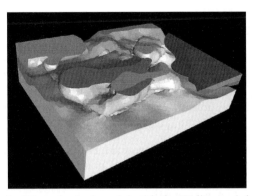

Figure 1. Example of a 3D geological model with the "Editeur géologique"

3 Variograms of orientation data illustrated by the Limousin dataset

The Limousin dataset, approximately a 70x70 km square, located in Centre France, is represented in Figure 2. Data sample a surface which is the top of a set of metamorphic rocks called "lower gneiss unit" (LGU). These data were all taken on the topographic surface.

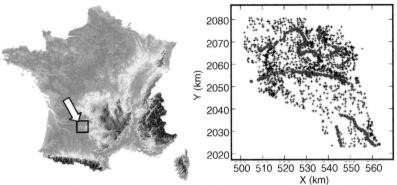

Figure 2. Base map of Limousin dataset. Black crosses: 1485 orientation data. Red discs: 133 interface data (digitized from the geological map).

3.1 ORIENTATION AND GRADIENT

Orientation data are vectors orthogonal to structural planes (e.g. foliation for metamorphic rocks, or stratification for sedimentary rocks) which are assumed to be parallel to the surfaces defined by the potential field. These data are sampled on the interface and also within the geological formations. They are considered as representing the gradient of the potential field. Since no intensity is usually attached to those gradient data, then vectors are arbitrarily considered as unit vectors. In practice, we work only with the three components of that vector. Let us mention that, in an orthonormal coordinate system, the mean of the components of a random unit vector is null and its variance is equal to one third.

3.2 COVARIANCE OF POTENTIAL AND GRADIENT

All the increments of potential are null, variograms of them are then useless. A first implementation of the method used a covariance given *a priori* by the user. But gradient data are algebraically linked with potential data. Therefore in this work, we use the only non-null data, namely gradient data, to infer the covariance model. Let $r = \sqrt{h_x^2 + h_y^2 + h_z^2}$ be the distance between two points and ${}'\mathbf{h} = (h_x, h_y, h_z)$ the vector joining these two points. Now, let K_P denote the covariance of Z and K_G^x, K_G^y, K_G^z the covariances of the three components of the gradient of Z. In order for Z to be differentiable, K_P must be twice differentiable (Chilès and Delfiner, 1999). Using the definition of differentiation, we can write the covariance of G^x, for instance:

$$K_{G^x}(\mathbf{h}) = -\frac{\partial^2 K_P(\mathbf{h})}{\partial h_x^2} \quad (2)$$

In the case of an isotropic covariance, $K_P(\mathbf{h}) = C(r)$ with C twice differentiable for $r \geq 0$ and we have:

$$K_G^x(\mathbf{h}) = -\left(\frac{C''(r)}{r^2} h_x^2 + C'(r)\left[\frac{1}{r} - \frac{h_x^2}{r^3}\right]\right) \quad (3)$$

More general formulas are available for anisotropic covariances.
The model parameters will be determined only with the sample variograms of the gradient data.
The cubic model with range a and sill C_0, chosen as basic model for K_p, is defined in the isotropic case by:

$$K_P(r) = C_0\left(1 - 7\left(\frac{r}{a}\right)^2 + \frac{35}{4}\left(\frac{r}{a}\right)^3 - \frac{7}{2}\left(\frac{r}{a}\right)^5 + \frac{3}{4}\left(\frac{r}{a}\right)^7\right) \quad \text{for } 0 \leq r \leq a \quad (4)$$

$K_P(r) = 0 \quad \text{for } r \geq a$

It is well adapted for geological contouring, because at the scale considered, geological surfaces are smooth and the cubic model has the necessary regularity at the origin.
Even if we assume the isotropy of K_P, K_G is necessarily anisotropic (Chauvet et al., 1976). If we consider the partial derivative G^x for example, the extreme cases are the direction of the derivative, namely x, and the direction orthogonal to it, here y, and in term of variogram we have:

$$\gamma_{G^x}(h_y) = \frac{14 C_0}{a^2}\left[\frac{15}{8}\frac{h_y}{a} - \frac{5}{4}\left(\frac{h_y}{a}\right)^3 + \frac{3}{8}\left(\frac{h_y}{a}\right)^5\right] \quad \text{for } h_y \leq a \quad (5)$$

$$\gamma_{G^x}(h_x) = \frac{28 C_0}{a^2}\left[\frac{15}{8}\frac{h_x}{a} - \frac{5}{4}\left(\frac{h_x}{a}\right)^3 + \frac{3}{8}\left(\frac{h_x}{a}\right)^5\right] - \frac{7 C_0}{a^2}\left[5\left(\frac{h_x}{a}\right)^3 - 3\left(\frac{h_x}{a}\right)^5\right] \quad \text{for } h_x \leq a \quad (6)$$

We recognize a pentaspheric model in the direction orthogonal to that of the partial derivative and a model with a hole effect in the direction of derivation.
In the other directions, the graph of the variogram is comprised between these two envelopes (Renard and Ruffo, 1993).

3.3 VARIOGRAM FITTING

For the Limousin case study, since the topography is rather smooth, the variograms have been computed in the horizontal plane only. Figure 3 shows sample variograms for the Limousin dataset.
The first remark is the difference of scale for the sill value between the vertical component and the horizontal ones. The reason is that the mean of the vertical gradient is significantly larger than zero due to the sub-horizontality of the layers, which results in a smaller variance for the vertical gradient component than for the horizontal ones. We also notice a large nugget effect for all components (nearly half of the total variability).
This difference of sill is modelled with a zonal anisotropy. The final model for the potential covariance is thus a nested cubic model:

$$K_P(\mathbf{h}) = K_3\left(\sqrt{h_x^2 + h_y^2 + h_z^2}\right) + K_2\left(\sqrt{h_x^2 + h_y^2}\right) + K_1(h_y) \quad (7)$$

The ranges are 25000m, 17000m and 55000m, respectively, for K_3, K_2, and K_1.
The corresponding sills are 781000, 1700000, 10800000, respectively.
In comparison, the default values previously proposed by the software correspond to a single isotropic component with a range of 98000m (the size of the domain) and a sill of 229×10^6.
These two covariance models lead to rather different geometric models. For example, the depth of the LGU interface is up to 450m deeper with the default covariance model than with the fitted covariance. However, we must not forget, that we are extrapolating from data sampled on the topographic surface only.

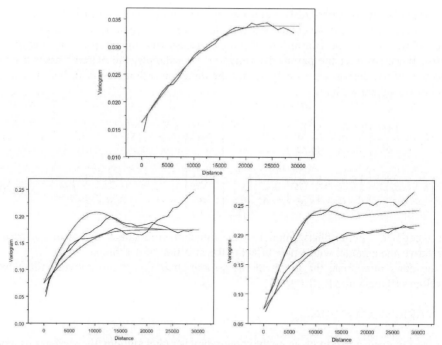

Figure 3. Experimental and fitted variograms for the components of the gradient. G^z (top), $G^{x//}$ and $G^{x\perp}$ (bottom left) and $G^{y//}$ and $G^{y\perp}$ (bottom right). The symbol // (resp. $\perp$) corresponds to the variogram in the direction of differentiation (resp. in a direction orthogonal to that of the differentiation).

In order to make the software easy to use for non-geostatisticians, an automatic procedure of variogram fitting based on the Levenberg-Marquardt method (Marquardt, 1963) has been implemented. The aim is the minimisation of the weighted metric distance between the sample variograms and the variogram model in the vectorial space of the fitted parameters (nugget effect, sill, range). It is a non linear regression method optimally using two minimisation approaches: quadratic and linear. A factor allows the use of one or another.

4 Determination of uncertainty

4.1 "REDUCED POTENTIAL" CARTOGRAPHY

When the covariance was chosen *a priori*, without consideration to a sample variogram, the method could not claim for optimality and the cokriging variance had no precise meaning. But now, since the model is well defined, determining the uncertainty on the position of the interface in depth makes sense and to get a better idea of the degree of uncertainty for the drawing we define a "reduced potential".

Let Z_0 be the value of the potential for a point on the considered interface, $Z^*(x)$ the value estimated at a point x and $\sigma_{CK}(x)$ the cokriging standard deviation at the same point.

The reduced potential $\Phi(x)$ is given by:

$$\Phi(x) = \frac{Z^*(x) - Z_0}{\sigma_{CK}(x)} \quad (8)$$

For a given point, this variable represents the reduced estimation of the potential deviation from Z_0. It can be shown that the larger this value, the less chance the point has to be on the interface. With a Gaussian assumption for the potential field, Φ is a standardized normal variable, so that for example, the area inside the curves $\Phi = \pm 2$ includes the interface in about 95% of the cases. Figure 4 shows the interpolated LGU interface (black line) and the value of Φ in blue. In short, the yellow zone, which corresponds to $|\Phi| < 3$, is like a forbidden area for the drawing of the interface.

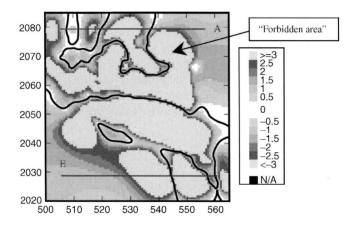

Figure 4. Limousin dataset. Map of the reduced potential.

Likewise, Figure 5 shows two cross-sections in the north (A) and the south (B) of the field. Of course, when the number of data is large, the position of the interface is well constrained, whereas in extrapolation there is a lot of uncertainty.

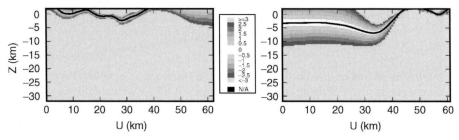

Figure 5. Limousin dataset. Cross-section A (left) and B (right) of the reduced potential.

4.2 UNCERTAINTY ON MODEL PARAMETERS

The covariance fitting has some part of uncertainty too. Thanks to a Bayesian approach it is possible to determine the uncertainty of the model parameters (Goria, 2004).
The aim is to simulate these parameters according to a posterior distribution, which is proportional to an *a priori* distribution and a likelihood function:

$$\pi(\theta \mid Z) \propto \pi(\theta)\pi(Z \mid \theta) \quad (9)$$

where $\pi(\theta)$ is the *a priori* distribution of the parameters vector θ and $\pi(Z|\theta)$ is a likelihood function. The vector θ includes the coefficients of the drift basis function, and the parameters of the covariance.
We assume a normal distribution for the coefficient of the drift and a gamma distribution for the precision (inverse of the sill). For the range and the relative nugget effect, a discrete uniform prior is used. The results show a large uncertainty on the model parameters. If we use the maximum values of the estimated parameters for the covariance model, we see some differences in the geometry of the interface. For example, the depth of the LGU interface is around 200m deeper with this "Bayesian" covariance model than with the classical fitted covariance.
The posterior distribution can also be incorporated in the cokriging or conditional simulation process.

5 Other practical implementation issues

5.1 SEVERAL INTERFACES

When there are several geological layers, some rules must be respected to avoid crossing the boundaries. If the interfaces are not sub-parallel, several potential fields are used. Two rules, "erode" and "onlap", as well as a stratigraphic column make it possible to solve all the issues facing us. The column defines the chronological order of the interfaces and the rules define the priority between the layers. The rule "erode" has always the priority and is used to mask the eroded part of the previous formations or to model an intrusive body. For example, on Figure 7 right, we can see the interface (1) which is in onlap on the interface (2).

5.2 FAULTS

Discontinuities are taken into account too. Faults are defined as external discontinuous drift functions in the cokriging system (Maréchal, 1984). The method requires the knowledge of the fault planes and the zones of effect of the faults. The discontinuity can be "infinite" and then crosses the whole field, dividing it in two subzones D and D'. The fault induces a discontinuity in the potential field, taken into account by a drift function such as:

$$f(\mathbf{x}) = 1_D(\mathbf{x})$$

This function complements the classical polynomial drift functions used for the non-stationarity in the cokriging system.

With a finite fault (Figure 6), the throw is determined with an influence area. As with an infinite fault, the discontinuity divides the delimited area in two sub-zones. In this case, the drift function has a bounded support and the function reaches its maximum at the centre of the fault. Outside the area the fault has no effect.

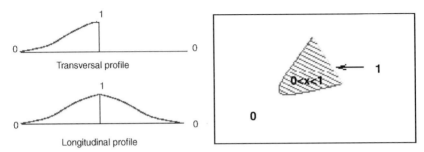

Figure 6. Finite fault. Left, transversal profile (top) and longitudinal profile (bottom) of the drift function. Right, area of influence of the discontinuity in the horizontal plane.

5.3 BOREHOLE ENDS

The last term in equation (1) is normally equal to zero, but could be strictly positive or negative if the points are not on the interface, which is the case when dealing with borehole ends. For example, the increment of potential is positive when the borehole end is above the considered interface with the following convention: the potential grows from the oldest geological formation to the most recent one.

Incomplete drillings can lead to a bad interpolation if a pre-processing of these soft data is not implemented. We use an iterative technique method based on the Gibbs sampler (Geman and Geman, 1984; Gilks *et al.*, 1996) to replace these soft data by hard data honouring both the inequalities and the spatial structure. That method, developed for stationary random functions (Freulon and de Fouquet, 1993) has been extended to the nonstationary case.

Figure 7 shows a synthetic example with two drill-holes (A and B) and two interfaces to be reconstructed (higher (1) with "onlap": 2 points and 1 gradient; lower (2) with rule "erode": 3 points and 1 gradient).

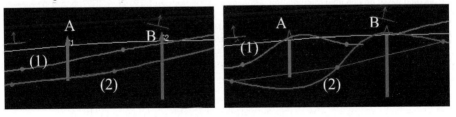

Figure 7. Interpolation of two interfaces. Left, without pre-processing, right with pre-processing.

The borehole A is only filled up with the facies whose interface (1) is the top and borehole B only with the facies whose interface (2) is the top.
On one hand, if borehole ends are not taken into account, the interpolation does not respect them as shown on Figure 7 (left).
On the other hand, Figure 7 (right) displays the result after pre-processing, interface (1) is in onlap as expected.
If P_1 and P_2 are respectively the iso-potential values for interface (1) and (2), and $P_b(A)$ and $P_e(A)$ respectively the values of potential at the beginning and at the end of the borehole, the result of the simulation gives values which respect $P_b(A)<P_1$ and $P_e(A)>P_2$.

6 Conclusion and future works

The potential-field method used in 3D geological modelling makes it possible to create models, even in complex situations, that combine different types of data, especially structural data. Thanks to the variography of these data it is possible to specify a sensible model of covariance and then to produce maps of uncertainty for the position of geological interfaces.
In this method we consider orientation data as gradient data, namely unit vectors. Only in specific cases, we know the structural intensity. The objective of ongoing work is to show, with simulations of actual situations, the impact on the covariance when actual gradient vectors are replaced by unit vectors.
Other improvements are planned like a better fault processing or the use of geophysical data.

References

Chauvet, P., Pailleux, J. and Chilès, J.-P., Analyse objective des champs météorologiques par cokrigeage, *La Météorologie, Sciences et Techniques*, 6e série, no. 4, 1976, p. 37-54.
Chilès, J.-P. and Delfiner, P., *Geostatistics: Modeling Spatial Uncertainty*, Wiley, 1999.
Freulon, X. and de Fouquet, C., Conditioning a Gaussian model with inequalities, *Geostatistics Troia '92*, vol. 1, A. Soares (Ed.), Kluwer, 1993, p. 201-212.
Geman, S. and Geman, D., Stochastic relaxation, Gibbs distributions, and the Bayesian restoration of images. *IEEE Transactions on pattern analysis and machine intelligence*, PAMI-6, no. 6, 1984, p. 721-741.
Gilks, W.R., Richardson, S. and Spiegelhalter, D.J., *Markov Chain Monte Carlo in practice*, Chapman and Hall/CRC, 1996.
Goria, S., *Evaluation d'un projet minier: approche bayésienne et options réelles*, Ph.D. Thesis, Ecole des Mines de Paris, 2004.
Lajaunie, C., Courrioux, G. and Manuel , L., Foliation fields and 3D cartography in geology : principles of a method based on potential interpolation, *Mathematical Geology*, vol. 29, no. 4, 1997, p. 571-584.
Maréchal, A., Kriging seismic data in presence of faults, *Geostatistics for Natural Resources Characterization*, Part 1, G. Verly et al. (Eds), Reidel, p. 271-294.
Marquardt, D., An algorithm for least-squares estimation of non-linear parameters. *SIAM, Journal on Applied Mathematics*, vol. 11, no. 2, 1963, p. 431-441.
Renard, D. and Ruffo, P., Depth, dip and gradient, *Geostatistics Troia '92*, vol. 1, A. Soares (Ed.), Kluwer, 1993, p. 167-178.

ACCOUNTING FOR NON-STATIONARITY AND INTERACTIONS IN OBJECT SIMULATION FOR RESERVOIR HETEROGENEITY CHARACTERIZATION

DENIS ALLARD[1], ROLAND FROIDEVAUX[2] and PIERRE BIVER[3]
1 INRA, Unité de Biométrie, Site Agroparc, 84914 Avignon, France
2 FSS Consultants SA, 9, rue Boissonnas, 1227 Geneva, Switzerland
3 Total, 64018 Pau, France

Abstract. This paper proposes an algorithm for simulating object models based on an underlying Markov point process able to reproduce attraction or repulsion between objects in a non stationary setting based on workable approximations for computing the local intensity, taking into account: i) non stationary proportions and non stationary object parameters; ii) erosion rules in the case of multi-type objects; iii) attraction or repulsion between objects.

1 Introduction

Modeling heterogeneity is the first, and possibly the most important, step of a reservoir characterization study. Depending on the geological context, several simulation techniques can be envisioned to perform this first step: sequential indicator simulation (Alabert, 1987; Goovaerts, 1997), transition probability simulation (Carle and Fogg, 1996), sequential simulation using multi-points statistics (Strebelle, 2002), truncated Gaussian or plurigaussian simulation(Le Loc'h and Galli, 1997), or Boolean simulation (Haldorsen and Macdonald, 1987; Lantuéjoul, 2002). An important feature of object simulation, which sets it apart from the other techniques, is the fact that it is not pixel-based, i.e it does not generate values at the nodes of a pre-defined grid. Rather, it generates geometric shapes in space according to some probability laws.

Although Boolean model simulation has been widely used during the last two decades to simulate sedimentary bodies (especially in fluvio-deltaic environments), several non trivial issues have remained and require scrutiny. Any general purpose object simulation program for reservoir characterization should (i) allow for the simulation of multiple object types, (ii) respect user-defined erosion rules between object of different types, (iii) reproduce specified a priori proportions, after erosion, for each object type, (iv) account for non-stationary object dimensions and orientations, (v) be conditional to existing hard-data, (vi) account for inter-actions (attraction or repulsion) between objects.

A key issue is that the single most important parameter for object models simulation, namely the (non stationary) intensity of the underlying object process, is not a parameter provided by the end user but must instead be inferred from the other input parameters listed above. A second very important issue is the fact that Boolean models traditionally used assume independence between objects. As such they are inadequate for reproducing interactions between objects. Lantuéjoul (1997, 2002) proposed a birth and death process for simulating conditional Boolean models with non stationary intensity. This algorithm considers the intensity as known and does not address the problem of making a bridge between intensity and local proportion. Recently, Benito García Morales (2003) proposed a method based on Wiener filter to estimate a non stationary intensity from non stationary proportions. This method assumes stationary distribution function of the object parameters. None of the methods described above consider multi-type objects.

This paper proposes an algorithm for simulating multi-type object models based on an underlying Markov point process able to reproduce attraction or repulsion between objects in a non stationary setting.

2 Boolean Models

2.1 GENERAL OVERVIEW

A single object type Boolean model (Stoyan, Kendall and Mecke, 1995) is made of two parts:

- A set of points (seeds), denoted $\mathbf{X} = \{\mathbf{x}_1, \ldots, \mathbf{x}_n\}$, which follow a Poisson distribution characterized by its intensity θ describing the expected number of object centroïds per unit volume. This intensity may be varying in space. As a consequence of the Poisson assumption, object centroïds are independent to each other.
- Random variables, independent of the Poisson process, that attach to each of these points random marks describing the shape, dimensions and orientation of objects A. These random variables are described by their joint probability density ψ. The random marks are independent one from the other.

The key parameters for Boolean models simulation is the point intensity parameter θ which describes how many object centroïds are expected per volume unit. This parameter, however, is not readily available: geologists have good ideas about the proportion for each geological object they want to simulate, but they have no feel for the number of such objects. Hence the need to estimate the intensity θ from the proportion p. In stationary conditions, it is well known that for Boolean models the proportion is related to the intensity according to the following relationship (Lantuéjoul, 2002; Stoyan et al., 1995):

$$p = 1 - \exp\left\{-\theta \int_{\mathbf{R}^3} E_\psi[\mathbf{1}_{0 \in A(\mathbf{v})}] d\mathbf{v}\right\} = 1 - \exp\{-\theta V\},$$

where $\mathbf{1}$ is the indicator function, 0 is the origin, $A(\mathbf{v})$ is a random object centered in $\mathbf{v}$ and V is the expectation of the volume of a random object A whose mark density is ψ. Inverting this relationship yields to

$$\theta = -\frac{1}{V}\ln(1-p). \tag{1}$$

In practice, this congenial situation is the exception rather than the rule: a priori proportions and mark densities are not stationary, objects of different types do not overlap each other randomly but according to erosion rules, objects of a given type may show a tendency to attract each other or, conversely, to repulse each other. In all these situation equation (1) cannot be used directly but requires adjustments.

2.2 ACCOUNTING FOR EROSION RULES

For each object type $k = 1, \ldots, K$, there is a corresponding proportion p_k, intensity θ_k and mark density ψ_k. Although equation (1) already accounts for the fact that several objects of the same type may overlap, it requires adjustment to ensure that, in case of multiple object type simulation, the target proportion of each type is correctly reproduced. In practice, this is done by substituting in equation (1) the proportion p_k by a corrected proportion p'_k. This correction depends on the "erosion rule" determining which type of object erodes the other. Among all the possible rules, three are commonly used: *random overlapping*, the *vertical erosion* rule (the object with the highest centroïd erodes the others) and the *hierarchical erosion* rule whereby the object type 1 always erodes the object type 2 which, in turn, always erodes object type 3, etc.

- In the case of an hierarchical erosion rule, the proportion of the type 1 object does not need to be corrected. Type 2 objects will be partly eroded by objects of type 1. Hence, for a visible proportion p_2, a corrected proportion $p'_2 = p_2/(1-p_1)$ needs to be simulated. Recursively, for the type k, the corrected proportion is given by:

$$p'_k = \frac{p_k}{1 - \sum_{i=1}^{k-1} p_i}. \tag{2}$$

- Derivation of a corrected proportion for a vertical erosion rule is somewhat more complicated. A second order approximation of this corrected proportion is given by:

$$p'_k = p_k \left(1 + \frac{(1+p_{tot})(p_{tot}-p_k)}{2}\right), \tag{3}$$

where $p_{tot} = \sum_k p_k$ is the total proportion of objects. This approximation relies on the idea that in the case of a vertical erosion, it is equally likely that an object of type k erodes an object of type l than the opposite. It leads to a corrected proportion $p'_k < 1$.
- In the case of random overlapping between object types, it is also equally likely that an object of type k overlaps an object of type l than the opposite. Hence, the same corrections as those used for vertical erosion are used.

2.3 ACCOUNTING FOR NON STATIONARITY

If object proportions or object parameters are non stationary the relationship between the non stationary proportions and non stationary parameters is more complex. Dropping, for ease of notation, the subscript referring to the object type, this relationship is expressed locally at a point $\mathbf{u} \notin \mathbf{X}$ as:

$$p(\mathbf{u}) = 1 - \exp\left\{-\int_{\mathbf{R}^3} \theta(\mathbf{v}) E_{\psi(\mathbf{v})}[\mathbf{1}_{\mathbf{u} \in A(\mathbf{v})}] d\mathbf{v}\right\}, \qquad (4)$$

where the expectation is computed with respect to the mark density with local parameters $\psi(\mathbf{v})$. This expression is extremely difficult (if not impossible) to invert. If $\psi(\mathbf{u})$ and $\theta(\mathbf{u})$ are smooth and slowly varying functions, then first order expansion in equation (4) can be used locally to approximate the local intensity $\theta_k(\mathbf{u})$ from the local corrected proportion $p'_k(\mathbf{u})$:

$$\theta_k(\mathbf{u}) = -\frac{1}{V_k(\mathbf{u})} \ln(1 - p'_k(\mathbf{u})), \qquad (5)$$

where $V_k(\mathbf{u})$ is the local expectation computed using the local mark density $\psi_k(\mathbf{u})$ and $p'_k(\mathbf{u})$ is the proportion corrected to account for erosion as described above.

3 Markov object models

It is sometimes necessary to impose that objects of a given family are attracted to each other or on the contrary that there is some sort of repulsion between objects. The general idea is to consider that repulsion or attraction is a feature of the underlying point processes, but that marks are still independent from each other. The appropriate framework for such point processes is the Markov point processes (MPP). Poisson point processes on which are built Boolean models is a particular case of MPP, for which there is no repulsion and no attraction. A comprehensive presentation of MPP can be found in Stoyan *et al.* (1995) or van Lieshout (2000).

3.1 GENERAL PRESENTATION OF MARKOV POINT PROCESSES

Markov point processes are point processes for which points are no longer independent from each other but are dependent on the configuration of the other points. According to the Hammersley-Clifford theorem, the probability density function (pdf) of a MPP depends only on functions of *cliques*. Cliques are set of points such that each point of this set is a neighbor of all other points of the set. The neighborhood relationship, used to define cliques must be symmetrical. Usually, the points $\mathbf{x}$ and $\mathbf{y}$ are neighbors (denoted $x \sim y$) if their distance $d(\mathbf{x}, \mathbf{y})$ is less than R for some $R > 0$.

The simplest possible clique to consider is a clique consisting of a single point. In this case there is no interaction and we are back to the classical Poisson process framework. In order to account for interaction, cliques of more than one point need to be considered. In practice, two point cliques will be considered and pairwise interaction functions, denoted $\beta(\mathbf{x},\mathbf{y})$, will be used to define the density of a configuration $\mathbf{X}$:

$$f(\mathbf{X}) \propto \prod_{\mathbf{x} \in \mathbf{X}} \theta(\mathbf{x}) \prod_{\mathbf{x},\mathbf{y} \in \mathbf{X}\,:\,\mathbf{x} \sim \mathbf{y}} \beta(\mathbf{x},\mathbf{y}),$$

where $\theta(\mathbf{x})$ is the intensity function. Among all pairwise interaction point processes, the simplest one is the Strauss process (Strauss, 1975; Kelly and Ripley, 1976) for which the interaction function is constant:

$$\beta(\mathbf{x},\mathbf{y}) = \beta \text{ if } \mathbf{x} \sim \mathbf{y} \text{ and } \beta(\mathbf{x},\mathbf{y}) = 1 \text{ otherwise}, \qquad (6)$$

with $0 \leq \beta \leq 1$. The pdf of a Strauss process is thus

$$f(\mathbf{X}) = \alpha \beta^{n(\mathbf{X})} \prod_{\mathbf{x} \in \mathbf{X}} \theta(\mathbf{x}),$$

where α is the normalizing constant and $n(\mathbf{X})$ is the number of neighbor pairs $\mathbf{x},\mathbf{y} \in \mathbf{X}$ with respect to the neighborhood relationship.

- If $\beta = 1$, there is no interaction whatsoever, and we are back to the non stationary Poisson point process with intensity $\theta(\mathbf{u})$.
- If $0 \leq \beta < 1$, there is some repulsion. Configurations with a high number of neighbors have a smaller density than configurations with a low number of neighbors. As a result the point process is more regular than a Poisson point process. In particular, if $\beta = 0$, configurations with neighbors have a null density and are thus impossible.
- The case of an attraction would correspond to $\beta > 1$, but without additional constraints it is mathematically not admissible because the associated density does not integrate to a finite quantity (Kelly and Ripley, 1976). However, the interaction function

$$\beta(\mathbf{x},\mathbf{y}) = \beta \text{ if } r \leq d \leq R,\ \beta(\mathbf{x},\mathbf{y}) = 0 \text{ if } d < r \text{ and } \beta(\mathbf{x},\mathbf{y}) = 1 \text{ if } d > R, \quad (7)$$

where $0 < r < R$ and d stands for $d(\mathbf{x},\mathbf{y})$, is an admissible model. In practice, the restriction introduced by r is not important because r can be chosen arbitrarily small. A typical choice is the mesh of the grid on which the simulation is represented.

In the following we will consider Strauss models for both repulsion and attraction, with the additional condition on r for attraction. The conditional density of adding to the configuration $\mathbf{X}$ a point in $\mathbf{u}$ is

$$f(\mathbf{X} \cup \{\mathbf{u}\} \mid \mathbf{X}) = f(\mathbf{X} \cup \{\mathbf{u}\})/f(\mathbf{X}) = \theta(\mathbf{u})\beta^{n(\partial \mathbf{u})}, \qquad (8)$$

where $n(\partial \mathbf{u})$ is the number of neighbors of $\mathbf{u}$. Hence, the parameter β can be interpreted as a factor multiplying locally the intensity for each point in the neighborhood of $\mathbf{u}$.

3.2 DERIVING THE INTENSITY FOR OBJECT MODELS BASED ON STRAUSS PROCESSES

Objects are now attached to the Markov point process, and for sake of simplicity, we first consider the stationary case. The proportion of objects attached to Markov point processes is not easily related to the intensity: there is no relationship comparable to (4). In the case of a Strauss model, a workable approximation for the intensity was found to be:

$$\theta(\mathbf{u}) = -\frac{p'(\mathbf{u})}{V(\mathbf{u})}\left(1 + c\frac{p'(\mathbf{u})}{2}\right), \qquad (9)$$

where $p'(\mathbf{u})$ is the proportion corrected for the erosion (as described in Section 2.2) and c is approximately the conditional probability that $\mathbf{u}$ is in an object A' given that it is already in an object A. In reservoir simulations, the objects have generally random size to account for the natural variability of geological objects. For the direction i, let us denote X_i the dimensions of an object, R_i the dimension of the interaction box and r_i the minimal distance in case of attraction. Conditional on R_i, $i = 1, 2, 3$, it can be shown that

$$c = \beta(1 - \frac{r_1 r_2 r_3}{X_1 X_2 X_3}) + (1-\beta)(1 - \frac{R_1 R_2 R_3}{X_1 X_2 X_3})[1 - \mathbf{1}_B(X_1, X_2, X_3)], \qquad (10)$$

where $\mathbf{1}_B(X_1, X_2, X_3)$ is the indicator function of the vector (X_1, X_2, X_3) being in the box B defined by the dimensions (R_1, R_2, R_3).

Markov point processes are usually defined for fixed interaction distances, as in Section 3.1 and objects are usually random with probability functions F_i. Taking the expectations of Equation (10) leads to

$$c = \beta g(r_1, r_2, r_3) + (1-\beta)g(R_1, R_2, R_3), \qquad (11)$$

with

$$\begin{aligned}g(R_1, R_2, R_3) &= 1 - F_1(R_1)F_2(R_2)F_3(R_3) - R_1 R_2 R_3 h_1(a_1)h_2(a_2)h_3(a_3) \\ &\quad - [h_1(a_1) - h_1(R_1)][h_2(a_2) - h_2(R_2)][h_3(a_3) - h_3(R_3)]),\end{aligned} \qquad (12)$$

where a_i is the smallest dimension of the object in the direction i and $h_i(r_i) = \int_{\max\{a_i, r_i\}}^{\infty} f_i(x)/x\, dx$.

4 Simulation using birth and death processes

Non conditional Boolean models (i.e. corresponding to $\beta = 1$) can be simulated directly: for each type of object k, first draw the number of objects from a Poisson random variable with parameter $\Theta_k = \int_D \theta_k(\mathbf{u})\, d\mathbf{u}$, then locate randomly the objects according to the intensity $\theta_k(\mathbf{u})$.

In all other cases (presence of conditioning data and/or Markov object models) simulation must be performed using a birth and death process. Birth and death processes are continuous time Markov Chains belonging to the family of Markov Chain Monte Carlo (MCMC) methods (van Lieshout, 2002; Lantuéjoul, 2002).

Starting from an initial configuration, an object is either removed or added according to some transition probability that depends on the current configuration of the simulation at each time step. This transition probability is chosen in such a way that the stationary distribution of the Markov chain is precisely the density we wish to simulate from. According to standard results of Markov chain theory, if the birth and death process is ergodic, there exists a stationary spatial distribution and the convergence to the stationary distribution will always occur independently on the initial configuration. Ergodicity holds if the detailed balance equation is verified at each iteration (see e.g. van Lieshout, 2002 p. 79).

It can be shown that the detailed balance is verified for the following choices: the probability of choosing a birth is $q(\mathbf{X}) = \Theta/(\Theta + n(\mathbf{X}))$ where Θ is the sum of $\theta(\mathbf{u})$ on $\mathcal{D}$. Then, $1 - q(\mathbf{X})$ is the probability of choosing a death. For a Strauss point process with $\beta \neq 1$, it is convenient to introduce an auxiliary field $\sigma(\mathbf{u})$ defined in the following way: for a repulsion (i.e., $\beta < 1$), $\sigma(\mathbf{u}) = \beta^{n(\partial \mathbf{u})}$; for an attraction (i.e., $\beta > 1$), $\sigma(\mathbf{u}) = \beta^{\min\{(n(\partial \mathbf{u}) - n_{max}), 0\}}$, where n_{max} is the maximum number of neighbors of each object. Its main effect is to stabilize the algorithm by avoiding a large quantity of objects piling on each other without increasing the proportion of this object type. In case of birth, a new object is proposed in $\mathbf{u}$ proportionally to an intensity $b(\mathbf{X}, \mathbf{u}) = \theta(\mathbf{u})\sigma(\mathbf{u})$. In case of death, the object to be removed is chosen with a uniform probability among the list of objects.

For performing conditional multi type conditional simulations, the conditioning taking into account the erosion rules must be checked each time a new object is added or removed.

5 Implementation

The implementation of the algorithm described in the previous section raises some critical issues.

- *Border effects :* It is important to ensure that objects intersecting the domain D but whose centroïds are outside this domain can be simulated. A practical way consists in considering a bigger domain D^s whose dimension is the dimension of the domain under study, D, increased by the dimension of the largest conceivable object. There is one such domain D_k^s for each type of object and the expected number of objects of type k to be simulated must be computed on D_k^s. Intensities yust be extrapolated on the domain D_k^s not in D.

  For models with interaction, care must be taken to simulate correctly the Markov point process near the borders. By construction there cannot be any neighbors outside D^s. For a point $\mathbf{u}$ located near the border the number of neighbors $n(\partial \mathbf{u})$ will therefore be underestimated as compared to points located in the center of D^s. As a consequence, the field $\sigma(\mathbf{u})$ accounting for the interaction will be biased towards less interaction near the borders. In the case of repulsion for example, this bias results in an accumulation of objects near the border of D^s. Because the border of the augmented domain D^s is usually outside the true domain D, the ultimate bias is less object than expected in D and a proportion under the target. To account for this bias, the number of neighbors is corrected by $n(\partial \mathbf{u})^* = n(\partial \mathbf{u})/v(\mathbf{u})$ where $v(\mathbf{u})$ is the proportion of the volume of the interaction box contained in D. Note that in case of mutiple object types, there is one such correction per object type.

- *Initial configuration :* In case of conditional simulation, an initial simulation is performed which will honor all the hard data. This is achieved by defining a new domain D^i which guarantees that any new simulated object, whatever its location, dimensions or orientation, will intersect at least one conditioning data. There is no birth and death process in this initial phase: new objects are added until all hard data are intersected. To avoid possible endless iterations, a maximum number of iterations is specified for this phase.
- *Convergence :* The question of finding a criterion for deciding if the algorithm has reached convergence is a very difficult one. There is no general rule for evaluating the number of iterations necessary to reach a pre-specified distance between the theoretical stationary distribution, and the actual distribution after n iterations. A considerable amount of literature has been devoted to this subject, see e.g. Meyn and Tweedie (1993) for a survey on this subject. Most of the proposed methods are either limited to some very simple cases or difficult and time consuming to implement. As a result, for practical purpose, the stopping rule will be a combination of a maximum number of iterations and a monitoring of some important output parameters (number of simulated objects, number of conditioning objects that have been replaced, average number of neighbors).

6 Illustrative example

To illustrate the proposed algorithm, let us consider tow examples. In both cases two types of objects are considered: dunes (fan shaped sedimentary bodies) and sinusoidal channels. In the first example, vertical proportion curves are imposed. For the dunes the proportion decreases steadily from a maximum of 30% at the top of the reservoir to a minimum value of 1% at the bottom. For the sinusoidal channels the trend is reversed: 30% at the bottom and 1% at the top. There are no interaction between the objects, neither for the channels nor for the dunes, and a vertical erosion rule is enforced. Figure 1 shows a typical cross-section of one realization. As can be seen, the trends in proportions are correctly reproduced. On average, the simulated proportion is 16% for the dunes and 14.9% for the channels, almost identical to the target proportion which were 15.5% in both cases. In the second example the target proportions are stationary, 10% for both the channels and the dunes, but object interactions are imposed: the dunes will repulse each other (the interaction box is 10% percent larger than the size of the objects) whereas the channels will attract each other. The interaction box is twice as large as the object width and height but has the same length, which means that the attraction operate only laterally and vertically. In this case a hierarchical erosion rule is applied. Figure 2 shows a horizontal and vertical cross-section. The simuated proportions are 10.5% for the fans and 10.0% for the channels. The results conform to the constraints which have been imposed: dunes are all distinct one from the other with no overlap and they systematically erode the channels, which tend to cluster together. Again, the average simulated proportions match almost exactly the target proportions.

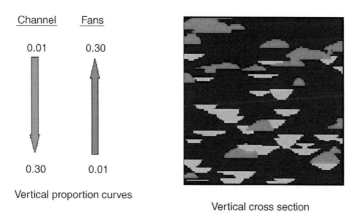

Figure 1. Vertical cross section of a simulation with vertical erosion rule

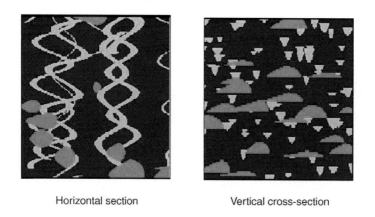

Figure 2. Horizontal and Vertical cross section of a simulation with hierarchical erosion rule

7 Discussion

The algorithm which has been presented offers a lot of flexibility and has proven effective for producing realistic simulations of reservoir heterogeneity in non-stationary situation. However, as is the case with all simulation algorithms, it is not universal and its limits of application should be respected. A critical issue, when performing conditional simulation, is the consistency between hard data, object parameters (shape and dimensions) and target proportions. This consistency becomes even more important if some or all parameters are non-stationary. The size of the object to be simulated is a critical issue. The larger the object the more difficult it will be to reproduce a target proportion and to honour conditioning data. Consistency between the object size (in particular its thickness) and the resolution at which facies are coded in well data must imperatively be verified. The approximations presented above are valid for not too high proportions. It is recommended that each proportion does note exceed 50% and that there is at least 20% of matrix, even locally. Although this algorithm can accommodate non-stationarity care should be taken that this non stationarity describes a smooth variation. Discontinuities must be avoided. The concept of neighborhood is not an intuitive one. Selecting too large a neighborhood may prove self-defeating: every point is the neighbour of every other point (they all belong to the same clique!). In case of attraction the points will not show any tendency to group in cluster, and in case of repulsion the process may never converge since it will be impossible to reach the target proportion and remain consistant with the repulsion constraint.

References

Alabert, F., 1987, Stochastic Imaging of Spatial Distribution Using Hard and Soft Information. M. Sc. Thesis, Stanford University.

Benito García-Morales, M., 2003, Non stationnarité dans les modèles de type Boléen: application à la simulation d'unités sédimentaires. PhD Thesis from the Ecole Nationale Supérieure des Mines de Paris, Centre de Géostatistique.

Carle, S.F. and Fogg G.E ., 1996, Transition probability-based indicator geostatistics: Math. Geology, v. 28, no. 4, p. 453–476.

Goovaerts, P., 1997, Geostatistics for natural resources evaluation: Oxford University Press, Oxford, 483 p.

Haldorsen, H.H. and Macdonald, C.J., 1987, Stochastic modeling of underground reservoir facies (SMURF). SPE 16751, p. 575–589.

Kelly, F. and Ripley, B.D., 1976, A note on Strauss's model for clustering: Biometrika, v. 63, no. 2, p. 357–360.

Lantuéjoul, C., 1997, Iterative algorithms for conditional simulations, in Baafi E.Y. and Schofield, N.A., eds, Geostatistics Wollongong '96: Kluwer Academic Publishers, Dordrecht, p. 27–40.

Lantuéjoul, C., 2002, Geostatistical simulation; models and algorithms: Springer-Verlag, Berlin, 256 p.

Le Loc'h, G. and Galli, A., 1997, Truncated plurigaussian method: theoretical and practical points of view, in Baafi E.Y. and Schofield, N.A., eds, Geostatistics Wollongong '96: Kluwer Academic Publishers, Dordrecht, p. 211–222.

Meyn, S.P. and Tweedie, R.L., 1993, Markov Chains and Stochastic Stability: Springer-Verlag, London, 550 p.

Stoyan, D., Kendall, W. and Mecke J., 1995, Stochastic Geometry and its Applications, 2nd Edition: John Wiley, Chichester, 436 p.

Strauss, D.J., 1975, A model for clustering: Biometrika, v. 62, no. 2, p. 467–475.

Strebelle, S., 2002, Conditional simulation of complex geological structures using multi-points statistics: Math. Geology, v. 34, no. 1, p. 1–22.

van Lieshout, M.N.N., 2002, Markov Point Processes: Imperial College Press, London, 175 p.

ESTIMATING THE TRACE LENGTH DISTRIBUTION OF FRACTURES FROM LINE SAMPLING DATA

CHRISTIAN LANTUÉJOUL, HÉLÈNE BEUCHER, JEAN-PAUL CHILÈS, CHRISTIAN LAJAUNIE and HANS WACKERNAGEL
Ecole des Mines, 35 rue Saint-Honoré, 77305 Fontainebleau, France

PASCAL ELION
ANDRA, 1-7 rue Jean Monnet, 92298 Châtenay-Malabry, France

Abstract. This paper deals with the estimation of the length distribution of the set of traces induced by a fracture network along an outcrop. Because of field constraints (accessibility, visibility, censorship, etc...), all traces cannot be measured the same way. A measurement protocol is therefore introduced to systematize the sampling campaign. Of course, the estimation procedure must be based on this protocol so as to prevent any bias. Four parametric procedures are considered. Three of them (maximum likelihood, stochastic estimation-maximization and Bayesian estimation) are discussed and their performances are compared on 160 simulated data sets. They are finally applied to an actual data set of subvertical joints in limestone formations.

1 Introduction

Fractures such as faults and joints play a key role in the containment of nuclear waste in geological formations, in the oil recovery of a number of petroleum reservoirs, in the heat recovery of hot dry rock geothermal reservoirs, in the stability of rock excavations, etc. The fracture network is usually observable through its intersection with boreholes or through its traces on outcrops (see Fig. 1). An important parameter of a fracture network is the fracturation intensity, i.e. the mean area occupied by the fractures per unit volume, which is experimentally accessible even from unidimensional observations such as boreholes. The same fracturation intensity can however correspond to very different situations, whose extremes are a network of few large well-connected fractures and a network with a large number of small disconnected fractures. Getting geometrical and topological information about the fractures - size, orientation, aperture, connectivity - is therefore very important. Despite substantial work, this remains an arduous task, mainly because of the geometrical or stereological biases resulting from this limited observability (Chilès and de Marsily, 1993).

Figure 1.   Example of an outcrop and its traces (vertical outcrop in a quarry, Oxfordian limestones, East of France)

This paper deals with the estimation of the length distribution of a set of traces along an outcrop. (The link between trace length and fracture size is briefly discussed at the end of the paper.) The difficulty lies in that the traces are usually not entirely visible. Their lower part is often buried. Their upper part may not be available either if the region around the outcrop has been eroded or mined out. In the practical case considered, the outcrop is a vertical face in a limestone quarry and all traces are sub-vertical.

In order to reduce risks associated with the sampling of such traces, it is convenient to resort to a sampling protocol that says what traces should be effectively measured and how. In the practical case considered, all traces are sub-vertical, which simplifies the protocol as well as its presentation (see Fig. 2):

1. All traces hitting a horizontal reference line - and only those traces - are selected for measurement;
2. Only the part of a selected trace above the reference line is actually measured;
3. A measurement is achieved even if the upper part of the trace is incomplete.

In other words, not all traces are sampled. Moreover, sampling a trace consists of measuring its residual - and sometimes censored - length above the reference line. The approach presented here is applicable to more general situations (the assumption that the traces are vertical or parallel is not really required), and can be easily generalized to other sampling schemes, including areal sampling.

In this paper, four parametric procedures[1] are proposed to estimate the trace length distribution starting from residual length data. These are the maximum likelihood estimation (MLE), its estimation-maximization variation (EM), the

---

[1] A non-parametric procedure based on the Kaplan-Meier estimation can also be designed. It is not described here to simplify the presentation.

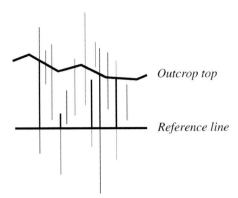

Figure 2. Protocol for sampling the outcrop

stochastic estimation-maximization algorithm (SEM) and the Bayesian estimation (BE). The performances of three of them (MLE, SEM and BE) are compared on data sets simulated from a Weibull distribution. Finally, the same three procedures are applied to a fracturation data set coming from an underground research laboratory of ANDRA (French national radioactive waste management agency).

2 Consequences derived from the protocol

At first, it should be pointed out that the selection of the traces is biased. The longer a trace, the more chance it has to hit the reference line. Quantitatively speaking, if $g(\ell)$ denotes the probability density function (p.d.f.) of the traces with length ℓ, then that of the selected traces is $\ell g(\ell)/m$. In this formula, the mean m of g acts as a normation factor.

Now a selected trace hits the reference line at an arbitrary location. In probabilistic terms, this amounts to saying that a residual length can be written as a product LU, where L is the length of a selected trace and U is a random variable uniformly distributed on $]0,1[$ and independent of L. Accordingly, the cumulative distribution function (c.d.f.) R of the residual lengths satisfies

$$1 - R(\ell) = P\{LU \geq \ell\} = \int_{\ell}^{+\infty} \frac{xg(x)}{m} P\left\{U \geq \frac{\ell}{x}\right\} dx = \frac{1}{m}\int_{\ell}^{+\infty}(x-\ell)g(x)\,dx.$$

By a first differentiation, the p.d.f. r of the residual lengths is obtained as a function of the c.d.f. G of the actual traces

$$r(\ell) = \frac{1 - G(\ell)}{m}, \tag{1}$$

and by a second differentiation both p.d.f.'s turn out to be related by the formula

$$g(\ell) = -\frac{r'(\ell)}{r(0)}. \tag{2}$$

3 Four estimation procedures

We turn now to the problem of estimating the p.d.f. g of the actual traces starting from the available residual traces, namely the complete ones $\ell_I = (\ell_i, i \in I)$ and the censored ones $t_J = (t_j, j \in J)$. Equation (2) suggests to concentrate at first on the estimation of r, and then on that of g. The four procedures presented hereunder have been designed along that line.

3.1 MAXIMUM LIKELIHOOD ESTIMATION (MLE)

In this procedure, the trace length p.d.f. g is supposed to belong to a parametrized family $(g(\cdot|\theta), \theta \in T)$. For each p.d.f. $g(\cdot|\theta)\}$, a residual p.d.f. $r(\cdot|\theta)$ can be associated. The MLE procedure consists of finding a parameter θ that maximizes the likelihood of the data

$$L(\ell_I, t_J, \theta) = \prod_{i \in I} r(\ell_i|\theta) \prod_{j \in J} [1 - R(t_j|\theta)]$$

or

$$L(\ell_I, t_J, \theta) = r(\ell_I|\theta)[1 - R(t_J|\theta)] \qquad (3)$$

for short (Laslett, 1982). It should be pointed out that this procedure is not universal. For instance, the likelihood may have no maximum[2]. Moreover, even if a maximum does exist, its determination by differentiation of the likekihood may turn out to be ineffective.

3.2 EXPECTATION-MAXIMIZATION PROCEDURE (EM)

A possible approach for estimating the maximum likelihood is to resort to the EM algorithm (Dempster et al., 1977). This is an iterative algorithm that produces a sequence of parameter values in such a way that the likelihood of the data increases at each iteration. To present this algorithm, it is convenient to introduce the residual random lengths $L_J = (L_j, j \in J)$ that have been censored to $t_J = (t_j, j \in J)$. Of course $L_J \geq t_J$.
(i) let θ be the current parameter value;
(ii) calculate the conditional distribution r_θ of L_J given $L_J \geq t_J$;
(iii) find θ_m that maximizes $\theta' \longrightarrow F_{r_\theta} \ln[r(\ell_I|\theta')r(L_J|\theta')]$;
(iv) put $\theta = \theta_m$, and goto (ii).

It should be pointed out that step (iii) of this algorithm also includes a maximization procedure. However the functions to be maximized do not depend on R and are therefore simpler to maximize than the likelihood of the censored data.
Nonetheless, this algorithm has some drawbacks. Calculating the expectation of step (iii) may be problematic. On the other hand, convergence may take place only to a local maximum that depends on the initial parameter value. Finally, the rate of convergence may be quite slow.

---

[2] However, the family of p.d.f.'s is usually designed so as to warrant a maximum whatever the data set considered.

3.3 STOCHASTIC EXPECTATION-MAXIMIZATION PROCEDURE (EM)

All these difficulties prompted Celeux and Diebolt (1985) to introduce the SEM algorithm that consists of replacing the calculation of the expectation by a simulation:
(i) let θ be the current parameter;
(ii) generate $\ell_J \sim r_\theta$;
(iii) find θ_m that maximizes $\theta' \longrightarrow r(\ell_I|\theta')r(\ell_J|\theta')$;
(iv) put $\theta = \theta_m$, and goto (ii).
Once again, this algorithm requires a maximization procedure. But what has to be maximized is the likelihood of pseudo-complete data instead of that of the censored data. As mentioned by Diebolt and Ip (1996), the outcome of such an algorithm, after a burn-in period, is a sequence of parameter values sampled from the stationary distribution of the algorithm. Its mean is close to the MLE result. Its dispersion reflects the information loss due to censoring.

3.4 BAYESIAN ESTIMATION (BE)

Now that the expectation step has been avoided, the tedious part of the SEM algorithm is the maximization step. It can also be avoided by putting the estimation problem into a Bayesian perpective. More precisely, assume that θ is a realization of a random parameter Θ with *prior* distribution p. Then the *posterior* distribution of Θ is

$$q(\theta|\ell_I, t_J) \propto p(\theta) r(\ell_I|\theta)[1 - R(t_J|\theta)]$$

The following algorithm has been designed so as to admit q for stationary distribution:
(i) generate $\theta \sim p$;
(ii) generate $\ell_J \sim r_\theta$;
(iii) generate $\theta \sim p(\cdot)r(\ell_I|\cdot)r(\ell_J|\cdot)$, and goto (ii).
This algorithm is nothing but a Gibbs sampler on (L_J, Θ). Step (ii) updates L_J while step (iii) updates Θ.

4 Weibull distribution

In order to implement MLE, EM, SEM and BE, an assumption must be made on an appropriate family of p.d.f. for g. Many choices are possible (Gamma, Pareto, Weibull etc...These distributions are described in full detail in Johnson and Kotz (1970)). In this paper, the actual trace lengths are supposed to follow a Weibull distribution with (unknown) parameter α and index b ($\alpha, b > 0$)

$$w_{\alpha,b}(\ell) = \alpha b \exp\left\{-(b\ell)^\alpha\right\} (b\ell)^{\alpha-1} \qquad \ell > 0 \qquad (4)$$

If $L \sim w_{\alpha,b}$, then $bL \sim w_{\alpha,1}$. In other words, b is nothing but a scale factor. In contrast to this, the parameter α determines the shape of the distribution. If $\alpha < 1$, then $w_{\alpha,b}$ is monotonic decreasing and unbounded at the origin. If $\alpha > 1$, then

$w_{\alpha,b}$ is a unimodal distribution that is similar in shape to a normal distribution at large α values. In the intermediary case $\alpha = 1$, $w_{1,b}$ is an exponential distribution. The Weibull distribution has finite moments at all positive orders

$$E(L^n) = \frac{\Gamma(n\alpha^{-1} + 1)}{b^n} \qquad (5)$$

A Weibull distribution can be simulated either by inverting its distribution function or by considering a standard exponential variable U and then delivering $U^{1/\alpha}/b$.

The residual p.d.f. associated with the Weibull distribution is

$$r_{\alpha,b}(\ell) = \frac{b}{\Gamma(\alpha^{-1} + 1)} \exp\{-(b\ell)^\alpha\} \qquad \ell > 0$$

This p.d.f. is monotonic decreasing whatever the values of α and b. The moments are equal to

$$E(R^n) = \frac{1}{b^n(n+1)} \frac{\Gamma((n+1)\alpha^{-1} + 1)}{\Gamma(\alpha^{-1} + 1)} \qquad (6)$$

It can be noted that $E(R) < E(L)$ when $\alpha > 1$ as well as $E(R) > E(L)$ when $\alpha < 1$. The equality $E(R) = E(L)$ that takes place in the case $\alpha = 1$ stems from the lack of memory of the exponential distribution.

A simple way to generate a residual trace is to put $R = UV^{\frac{1}{\alpha}}$ where U an V are two independent variables respectively uniformly distributed on $]0,1[$ and gamma distributed with parameter $\alpha^{-1} + 1$ and index b.

5 A simulation test

In order to test the efficiency of three of the procedures presented (MLE, SEM and BE), they have been applied to populations of residual traces sampled from $r_{0.5,1}$ with mean[3] $6m$ and variance $84m^2$. Each population can have 4 possible sizes $(100, 200, 500$ or 1000 traces), as well as 4 possible censoring levels $(0.924m, 2.817m, 7.250m$ and ∞, in accordance with the percentiles $75\%, 50\%, 25\%$ and $0\%)$. For each of the $4 \times 4 = 16$ types considered, 10 populations have been simulated.

Figure 3 shows the influence of the size of the population, of the censorship proportion and of the type of estimator on the estimation of both parameters α and b. Several observations can be made:

1. The estimated points (α, b) are organized as elongated clouds;
2. those clouds tend to shorten as the size of the population increases;
3. the target point $(0.5, 1)$ is not offset;
4. the censorship proportion is mainly influential for large populations sizes;
5. the Bayesian clouds are shortest.

---

[3] To give a comparison, the mean and the variance of the Weibull distribution $w_{0.5,1}$ are respectively $2m$ and $20m^2$.

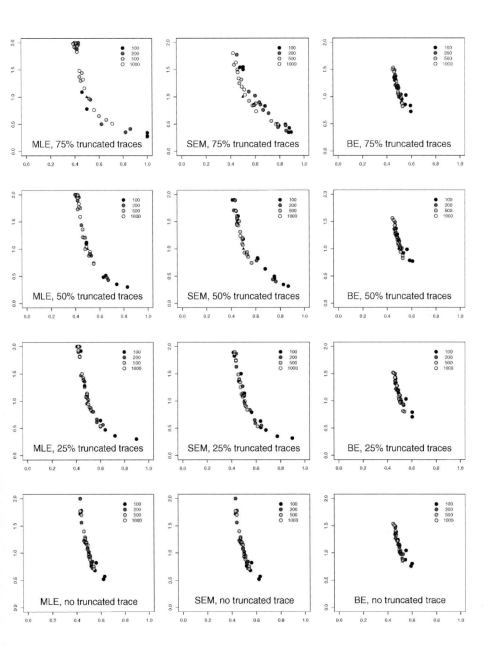

Figure 3. This figure plots the estimation of b versus that of α as a function of the estimator chosen, the number of traces in the population and the proportion of traces censored

The first observation suggests that both estimates $\hat{a}$ and $\hat{b}$ are functionally dependent. To fix ideas, consider for instance the case of the MLE. By taking the partial derivative of the log-likelihood of the simulated data w.r.t. b, one can end up with an equation of the form

$$\hat{b} = f\left(\#I, \#J, \frac{\sum_{i \in I} \ell_i^{\hat{\alpha}}}{\#I}, t, \hat{\alpha}\right)$$

where f is a deterministic function, $\#I$ (resp. $\#J$) denotes the number of elements of I (resp. J) and t is the censoring level. In other words, the complete residual traces act in the estimation process only via their number and their empirical moments. In particular, if $\#I$ and $\#J$ have been fixed, all variability that can be expected in the parameter estimation derives from the statistical fluctuations of those empirical moments. When $\#I$ is large, they are not significantly different from the moment of order α of $r_{0.5,1}$ (see (6)) and the relationship between $\hat{\alpha}$ and $\hat{b}$ becomes deterministic.

The second observation is standard. The estimators have less and less variability as the population increases. In the case of large populations, the third observation indicates that the estimators tend to concentrate around the target point. In other words, the estimators are asymptotically unbiased.

The fourth observation is not surprising either. For large population sizes, the only factor that can affect the variability of the estimators is the censorship threshold. The fifth observation suggests that the Bayesian procedure gives better results than MLE or SEM. This observation should be mitigated by the fact that the results obtained are highly dependent on the prior distribution chosen for (α, b). Here it has been supposed to be uniform over $]0, 1[\times]0, 2[$. If the range of uniformity of only one of the parameters had been extended, then the variability of both estimators would have been substantially increased.

6 Case study

The same three estimation procedures have been applied to a population of 419 traces taken from different outcrops embedded in the same geological formation, the Oxfordian limestones which overlie the Callovo-Oxfordian argilite formation of the underground research laboratory of ANDRA in the East of France. Fractures are subvertical and comprise faults and joints. A detailed structural and statistical study of the various fracture sets has been carried out (Bergerat et al, 2004). Here a directional set of subvertical joints is considered. The trace lengths range between $0.2m$ and $15m$ with a mean of $2.4m$ and a standard deviation of $2.5m$. Only 62 of the traces are censored (15%). Preliminary experiments suggested that one should certainly have $\alpha < 2$ as well as $b < 1$. This motivated us to apply the BE procedure with (α, b) *a priori* uniformly distributed on $]0, 2[\times]0, 1[$. On the other hand, the SEM and BE procedures have been resumed during 5000 iterations including a burn-in period of 1000 iterations. The 4000 pairs of values (α_n, b_n) generated by each procedure have been averaged to obtain the estimates of α and b of Table 1. This table gives also estimates of the mean and of the standard deviation of the

Weibull distribution. They have been obtained by calculating at first the mean m_n and the standard deviation σ_n associated to each (α_n, b_n), and then averaging them.

	$\hat{\alpha}$	$\hat{b}$	$\hat{m}$	$\hat{\sigma}$
MLE	0.860	0.370	2.919	3.407
SEM	0.837	0.464	2.366	2.841
BE	1.094	0.374	2.582	2.362

Table 1. Estimates obtained from the three procedures

The three estimated values for α and b cannot be considered as similar. Nonetheless, they give reasonably comparable estimations for the mean trace length. In contrast to this, the differences between the estimated standard deviations are more pronounced[4]

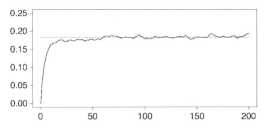

Figure 4. Variogram of the estimates of the mean along iterations

A potent advantage of Bayesian estimation is that it delivers a posterior distribution for the parameters under study, from which variances, quantiles as well as confidence limits can be deduced. For instance, it is possible to assign a variance to the estimate of the mean. As the values generated are dependent a simple approach is to consider the sill of the experimental variogram of the m_n's (see Fig. 4). We arrive at a variance of $0.183m^2$ (or a standard deviation of $0.43m$). Using similar approaches, it is also possible to attribute a variance to the standard deviation estimate $(0.050m^2)$ or even a covariance between the mean and the standard deviation estimates $(0.048m^2)$.

7 Discussion

In this paper, the traces have been considered as independent. This is of course a simplifying but not always appropriate assumption. In the case where joints tend to cluster or when they abut to the border of sedimentological layers, dependence relationships must be introduced between traces.

---

[4] It can also be mentioned that the trace length distribution was estimated in a previous exercise (Bergerat *et al*, 2004) using a MLE based on the gamma family. The estimated mean (2.67m) and the estimated standard deviation (2.84m) obtained are perfectly compatible with the results of this paper.

Another simplification has been made by assuming that the joints have their trace length independent of their orientation. If this assumption is not valid, Terzhagi's correction must be applied to compensate for the fact that a joint has more chance to be observable when its orientation is orthogonal to the outcrop (see Chilès and de Marsily (1993) and references therein).

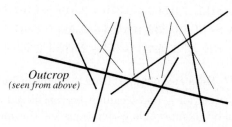

Figure 5. The more elongated the joint in the direction orthogonal to the outcrop, the more chance it has to be observed as a trace

One also may wonder what is the relationship between the trace length and the joint height distributions? To fix ideas, suppose that the joints are rectangles. Then a random joint has its statistical properties specified by the trivariate distribution of its width W, its height H and its dihedral angle Θ with the outcrop. The p.d.f. g of the trace lengths is related to that of the joint heights f by the formula

$$g(h) \propto f(h)E\{W\sin\Theta|H=h\}$$

Simplifications occur in the following cases:

1. If W and H are proportional, then $g(h) \propto hf(h)E\{\sin\Theta|H=h\}$;
2. If W and H are independent, then $g(h) \propto f(h)E\{\sin\Theta|H=h\}$;
3. If Θ is uniform on $]0,\pi/2[$, then $g(h) \propto f(h)E\{W|H=h\}$.

8 References

Bergerat F., Chilès J.P., Frizon de Lamotte D. and Elion P. (2004) - Analyse des microstructures tectoniques du Dogger et de l'Oxfordien calcaires. Bilan des Etudes et Travaux de l'Andra - 2003, chapitre II, fiche technique 2.4.2.
Chilès J.P. and de Marsily G. (1993) - Stochastic models of fracture systems and their use in flow and transport modeling. In Bear J., Tsang C.F. and de Marsily G. (eds) *Flow and contaminant transport in fractured rocks*, Academic Press, San Diego, pp. 169-236.
Dempster A.P., Laird N.M. and Rubin D.B. (1977) - Maximum likelihood from incomplete data via the EM algorithm. *J. Roy. Statist. Soc. Ser. B*, **39**, pp. 1-38.
Diebolt J. and Ip E.H.S. (1996) - Stochastic EM: method and application. In Gilks W.R., Richardson S. and Spiegelhalter D.J. (eds) *Markov chain Monte Carlo in practice*, CRC Press, Boca Raton, pp. 259-273.
Johnson N.L. and Kotz S. (1970) - Distributions in statistics (Volume 1). Wiley, New York.
Laslett G.M. (1982) - Censoring and edge effects in areal and line transect sampling of rocks joint traces. *Math. Geol.*, **14-2**, pp. 125-140.

ON SOME CONTROVERSIAL ISSUES OF GEOSTATISTICAL SIMULATION

L.Y. HU and M. LE RAVALEC-DUPIN
Department of Reservoir Engineering
Institut Français du Pétrole
1 et 4 avenue du Bois-Préau
92852 Rueil-Malmaison – France

Abstract. In this paper, we intend to clarify some conceptual issues of geostatistical simulation such as reproduction of the model covariance, equi-probability, independence, etc. We also introduce discussions on the confusion between simulating ergodic random functions and sampling random vectors. Our focus is on the interpretation of these probabilistic concepts in terms of realizations rather than the precision of simulation algorithms.

1 Introduction

Should conditional simulations reproduce the model covariance? This is one of the many controversial issues in geostatistics. Some argue that conditional and unconditional realizations are realizations of the same random function model and that they must reproduce the model covariance due to the ergodicity of the random function model. Of course, this reproduction is up to statistical fluctuations, i.e. fluctuations from the model parameters because of the limited size of a realization. Others argue that conditional realizations must respect the conditional (or posteriori) covariance but not the model (prior) covariance, and that this conditional covariance is different from the model prior covariance and even non-stationary whatever the model prior covariance.

Another famous controversial issue relates to the equi-probability of independently generated realizations. For ones, as random seed numbers are equi-probable, the resulting realizations are equi-probable too. For others, when in the Gaussian framework for instance, realizations (discretized as vectors) in the neighborhood of the mean vector are more likely to happen than the others. Therefore, realizations of a multi-Gaussian vector are not equi-probable even when independently generated.

The analysis of these contradictory points of view leads to other issues like
- Is there any (numerical) criterion to say that two (or several) realizations of a random function (in a large-enough domain) are independent?
- Can we generate an infinity of "independent" realizations of a random vector of finite dimension?

- Is geostatistical simulation of random functions consistent with the sampling of multivariate distributions?

Understanding what is behind these issues is not only of philosophical interest, but also of great importance in the application of geostatistical methods to model calibration, model sampling and uncertainty estimation etc.

In this paper, we intend to clarify some of the above issues and to introduce discussions about some others. Our discussions are limited to the stationary (multi-)Gaussian random function. We focus on conceptual issues of numerical simulations rather than numerical precision of simulation algorithms. We also explore the significance of some well established concepts of probability (Feller, 1971) in terms of an individual realization or a set of realizations. We always assume that the simulation domain is large enough with respect to the covariance range.

2 Regional covariance and covariance matrix

2.1 REGIONAL COVARIANCE

We study physical properties that are unique and defined in a field. With the geostatistical approach, a physical property is considered as a realization of an ergodic random function. The ergodic property is necessary for the inference of the structural parameters (regional mean, variance and covariance etc.) of the random function model from a single realization. When a random function model is adopted to represent the physical property, we use the measurements (data) at some locations of the field to infer the structural parameters that specify the random function. Then we build realizations of the random function and each of these realizations should honor, up to statistical fluctuations, the inferred structural parameters due to ergodicity.

Assume that we have enough data to infer correctly the regional covariance. The uncertainty in the inference of the regional covariance is an important, but different issue. The reproduction of the regional covariance in geostatistical simulations is essential because this covariance is inferred from physical data and not just only prior idea. We believe that it is methodologically inconsistent to infer the covariance from a data set and then to build a realization, conditioned to the same data set, that has a covariance, i.e., the posterior covariance in the Bayesian terminology (Tarantola, 1987; de Marsily et al., 2001), conceptually different from the inferred one.

Let us examine how conditioning an unconditional realization by kriging preserves the regional covariance. Consider a stationary standard Gaussian random function $Y(x)$. Let $(Y(x_1), Y(x_2),...,Y(x_n))$ be a standard Gaussian vector and $Y^*(x)$ the simple kriging of $Y(x)$ using the covariance function $C(h)$ and the data set $(Y(x_1), Y(x_2),...,Y(x_n))$. Let $S(x)$ be a standard Gaussian random function with $C(h)$ as covariance function but independent of $Y(x)$, and $S^*(x)$ the simple kriging of $S(x)$ using the data set $(S(x_1), S(x_2),...,S(x_n))$. Then, $Y_c(x)$ defined by

$$Y_c(x) = Y^*(x) + S(x) - S^*(x)$$

is a standard Gaussian random function with $C(h)$ as covariance function, and $Y_c(x)$ is conditioned to the random vector $(Y(x_1), Y(x_2),..., Y(x_n))$.

We note that the proof of the above result in Journel and Huijbregts (1978) or in Chilès and Delfiner (1999) assumes that the conditioning data set is a random vector. When the conditioning data set is fixed, $Y_c(x)$ becomes actually non-stationary. In particular, the mean values of $Y_c(x)$ at the data locations $x_1, x_2,..., x_n$ equal respectively the data values, and the variances of $Y_c(x)$ at these locations are zero. In general, the covariance of the random function $Y_c(x)$ with a fixed conditioning data set is non-stationary (dependent on the location x) and therefore different from $C(h)$.

However, the fact that the covariance of the random function $Y_c(x)$ with a fixed conditioning data set is non-stationary does not necessarily mean that the regional covariance of an individual realization of $Y_c(x)$ would not reproduce the model covariance $C(h)$. Indeed, because $Y_c(x)$, conditioned to the random vector $(Y(x_1), Y(x_2),..., Y(x_n))$, is a stationary ergodic random function, all realizations of $Y_c(x)$ should reproduce the covariance function $C(h)$ up to statistical fluctuations. Consequently, for any fixed data set $(y(x_1), y(x_2),..., y(x_n))$, i.e. a realization of $(Y(x_1), Y(x_2),..., Y(x_n))$, and any realization $s(x)$ of $S(x)$, $y_c(x)$ defined by

$$y_c(x) = y^*(x) + s(x) - s^*(x)$$

is a Gaussian realization conditioned to $(y(x_1), y(x_2),..., y(x_n))$ and provides $C(h)$ as its regional covariance up to statistical fluctuations.

The difference and the relation between the regional covariance and the covariance of a random function should become clearer by examining the concept of covariance matrix.

2.2 COVARIANCE MATRIX

In practice, we often need to discretize a random function over a finite grid. So we deal with random vectors and we can define their covariance matrixes. In the literature, the covariance matrix in the Bayesian framework is related to a set of realizations, instead of a single realization. For instance, the posterior covariance matrix of a random vector (after being conditioned to a data set) is related to the set of conditional realizations (not to a single conditional realization). Similarly, the prior covariance matrix of a random vector (before being conditioned to a data set) is related to the set of unconditional realizations (not to a single unconditional realization).

It is important not to confound the regional covariance of an individual realization with the covariance matrix of an ensemble of realizations. Any conditional simulation of an ergodic random function (discretized over a grid) must preserve the regional covariance but not the prior covariance matrix that will certainly change after conditioning.

The above discussion can be further clarified through the following example. Let $(y(x_1), y(x_2),..., y(x_N))$ be an unconditional realization of the Gaussian random vector $(Y(x_1), Y(x_2),..., Y(x_N))$. We use, for instance, the sequential Gaussian simulation method for generating realizations and we assume that all conditional distributions are computed without any approximation. When N is large enough and when the grid nodes $(x_1, x_2,..., x_N)$ covers a domain much larger than the area delimited by the covariance range, the experimental (regional) covariance should reproduce the theoretical covariance. Now, consider $y(x_1)$ as a conditioning datum, and we generate realizations of $(Y(x_2),..., Y(x_N))$ conditioned to $y(x_1)$. By using the same random numbers for sampling the conditional distributions at the nodes $(x_2,..., x_N)$ as in the case of the above unconditional simulation, we obtain the realization $(y(x_1), y(x_2),..., y(x_N))$ conditioned to $y(x_1)$. This conditional realization is identical to the unconditional realization and therefore has the same experimental (regional) covariance. However, if we generate a set of realizations conditioned to $y(x_1)$, their covariance matrix will be different from the model prior covariance.

In general, an unconditional realization $(y(x_1), y(x_2),..., y(x_N))$ of a random vector $(Y(x_1), Y(x_2),..., Y(x_N))$ can always be seen as a realization conditioned to $(y(x_1), y(x_2),..., y(x_I))$ for $I < N$. Evidently, this suggests that the conditioning does not necessarily change the regional covariance.

2.3 SUMMARY

The regional covariance and the covariance matrix (in the Bayesian terminology) are two different concepts in geostatistical simulation. A conditional simulation method should guaranty that the regional covariance of each conditional realization reproduces, up to statistical fluctuation, the model covariance function. However, the covariance matrix of a set of conditional realizations is conceptually (not because of statistical fluctuations) different from that of a set of unconditional realizations.

The covariance reproduction in geostatistical simulation means the reproduction of the regional covariance, not the (prior) covariance matrix. The reproduction of a covariance matrix is a much stronger requirement than that of a regional covariance. Reproduction of a covariance matrix requires generating a large-enough number of realizations that represent correctly the multivariate probability distribution, while the regional covariance is related to a single realization.

Because of the uniqueness of the physical property under study, the regional covariance has a physical sense, while the covariance matrix is a model concept.

3 Equi-probability and likelihood

"Are realizations of a stochastic model equi-probable?" is another controversial issue that still troubles practitioners of geostatistics. Considering a set of realizations as equi-probable or not can change completely the way we evaluate uncertainties from these realizations.

3.1 EQUI-PROBABILITY

Consider, for instance, the numerical simulation of a stationary Gaussian random function of order 2 over a finite grid of the simulation field. Namely, we simulate a Gaussian vector Y of N components $(Y(x_1), Y(x_2),..., Y(x_N))$. Now if we generate K realizations of Y: $y_1, y_2,..., y_K$, starting from K independent uniform numbers (random seeds issued from a random number generator), these realizations are equi-probable. This is because the uniform seeds can be considered as equi-probable, and for a given seed, a simulation algorithm produces a unique realization of Y after a series of deterministic operations.

3.2 LIKELIHOOD

However, the probability density values of the random vector Y at $y_1, y_2,..., y_K$ are different. Consider two realizations y_1 and y_2 and assume that $g(y_1) > g(y_2)$, where g stands for the probability density function of Y. Thus, we are more likely to generate realizations in the neighborhood of y_1 than in that of y_2. In other words, for a given small-enough domain $\delta(y)$ located at y and a large-enough number of realizations of the vector Y, there are more realizations in $\delta(y_1)$ than in $\delta(y_2)$. But this does not mean y_1 is more probable to happen than y_2. Consequently, when evaluating uncertainty using a set of independently generated realizations, they must be equally considered with the same weight.

3.3 SUMMARY

Before generating realizations, there is a larger probability to generate a realization in the neighborhood of the realization of higher probability density. But once a set of realizations is generated independently between each other, they are all equi-probable.

4 Independence, correlation and orthogonality

4.1 INDEPENDENCE

When performing numerical simulation of the random function model, one often needs to generate more than one realization. These realizations are said independent because they are built by using the same simulation procedure but with different random seeds. These random seeds are said statistically independent. This is meaningful when a large number of random seeds are generated. Now, if we generate only a few realizations, say only two realizations, we need then only two random seeds. Because it does not make sense to talk about statistical independence with only two fixed numbers, does it make sense to talk about independence of two realizations of a random function model?

But geostatisticians are used to build models with only two "independent" realizations. This is the case when building realizations of the intrinsic model of coregionalization (Matheron, 1965; Chilès and Delfiner, 1999), when perturbing a realization by substituting some of its values with some other "independent" values (Oliver et al., 1997), when performing a combination of two independent realizations within the gradual deformation method (Hu, 2000), when modifying a realization using the probability perturbation method (Caers, 2002), etc.

If it does not make sense to check the independence between two realizations, it is nevertheless meaningful, at least for large realizations, to evaluate the degree of their correlation.

4.2 CORRELATION

Consider again the N-dimensional standard Gaussian vector $Y = (Y(x_1), Y(x_2), ..., Y(x_N))$. Let $y_i = (y_i(x_1), y_i(x_2), ..., y_i(x_N))$ ($i = 1, 2, ..., I$) be I independent realizations of Y. For each realization y_i, we compute its mean and its variance:

$$m_i = \frac{1}{N} \sum_{n=1}^{N} y_i(x_n)$$

$$\sigma_i^2 = \frac{1}{N} \sum_{n=1}^{N} [y_i(x_n) - m_i]^2$$

When N is large enough and when the grid $(x_1, x_2, ..., x_N)$ covers a domain whose dimension in any direction is much larger than the covariance range, we have $m_i \approx 0$, $\sigma_i^2 \approx 1$. For any two realizations y_i and y_j, we usually compute their correlation coefficient as follows:

$$r_{ij} = \frac{1}{N} \sum_{n=1}^{N} \left[\frac{y_i(x_n) - m_i}{\sigma_i} \right] \left[\frac{y_j(x_n) - m_j}{\sigma_j} \right]$$

We have $r_{ij} = 1$ for $i = j$, and because of the "independence" between y_i and y_j for $i \neq j$, we have $r_{ij} \approx 0$. In practice, if the correlation coefficient r_{ij} ($i \neq j$) is significantly different from zero, then the two realizations y_i and y_j are considered as correlated, and therefore dependent.

4.3 ORTHOGONALITY

Now for any two realizations y_i and y_j, we define the following inner product:

$$\langle y_i, y_j \rangle = \frac{1}{N} \sum_{n=1}^{N} y_i(x_n) y_j(x_n)$$

We have $\langle y_i, y_j \rangle \approx r_{ij} \approx 0$ for $i \neq j$ and $\langle y_i, y_j \rangle \approx r_{ij} = 1$ for $i = j$. Therefore, when $I = N$, the N vectors $y_i = (y_i(x_1), y_i(x_2), ..., y_i(x_N))$ ($i = 1, 2, ..., N$) constitute an orthonormal basis (up to statistical fluctuations) of an N-dimensional vector space V_N furnished with the above inner product. All other realizations of the random vector Y can be written as linear combinations of these N independent realizations $y_1, y_2, ..., y_N$. In other words, we cannot generate more than N realizations of an N-dimensional random vector so that the above usual correlation coefficient between any two of these realizations equals zero.

4.4 CONSEQUENCE

The above remark has an unfortunate consequence for many iterative methods that involve the successive use of independent realizations. For instance, the gradual deformation method requires generating, at each iteration, a realization independent from all realizations generated at previous iterations. When the number of iterations of the gradual deformation method becomes equal to or larger than the number of grid nodes, the optimized realization at iteration l ($l \geq N$) and a new realization at iteration $l + 1$ are linearly dependent (not because of statistical fluctuations). Consequently, the condition for applying the gradual deformation method with combination of independent realizations is no longer satisfied when the number of iterations is larger than the number of grid nodes. This explains why when applying the gradual deformation method with a large number of iterations (much larger than the number of grid nodes), it is possible to progressively force the optimized realization to have a regional covariance different from the initial one (Le Ravalec-Dupin and Noetinger, 2002).

Nevertheless, in practice, the number of iterations is hopefully much smaller than the number of grid nodes. Otherwise, the method is not applicable when the calculation of the objective function requires heavy computing resources.

5 Random function or random vector?

Up to now, we have assumed that an ergodic random function can be represented by a random vector. This seems questionable. Consider the following experimentation of thought. Let $y(x)$ be an ergodic realization of the standard Gaussian random function $Y(x)$. Assume now $k(x) = \exp[y(x)]$ represents a physical property distributed in a field, say rock permeability in an oil field. The "real" permeability field is then completely known. Starting from a large-enough data set of $k(x)$, we can infer the covariance of the random function model $Y(x)$.

Now because we generate realizations over a finite grid $(x_1, x_2,..., x_N)$, we deal with a Gaussian vector $Y = (Y(x_1), Y(x_2),..., Y(x_N))$. As discussed before, there is a non-negative probability to generate realizations in a given domain $\delta(y)$ located at y. For a domain of fixed size, this probability is maximal when y equals the mean vector. If we sample correctly the random vector Y, there is a non-negative probability to generate realizations in the neighborhood of the mean vector. These realizations have small regional variances (smaller than the model variance 1) and their regional covariance will not respect the model covariance inferred from the "reality": $y(x) = \ln[k(x)]$!

The above reasoning (if it makes sense) leads to the following consequences:

- Geostatistical simulations (over a finite grid) cannot honor both the regional covariance and the multivariate probability density function.

- The sequential simulation algorithm (Johnson, 1987; Deutsch and Journel, 1992) is related to random vectors, and therefore it cannot generate realizations of ergodic random functions (Lantuéjoul, 2002), even in the Gaussian case where we can compute the conditional distribution without approximation by using the global neighborhood.

- The use of an exact sampling method based on the Markov iteration or the acceptation/rejection (Omre, 2000) will make it possible to generate a set of realizations representative of the multivariate probability density function. Due to the maximum likelihood of the mean vector, we must expect some realizations close to the smooth mean vector. This is not compatible with the foundation of geostatistics whose aim is to model spatial variability inferred from real data.

6 Conclusions and further discussions

If the theory of geostatistics based on simulating ergodic random functions is compatible with that based on sampling random vectors, then we have the following conclusions:

- The regional covariance of each conditional or unconditional simulation should reproduce, up to statistical fluctuations, the model covariance function inferred from real data.
- A large-enough set of conditional (or unconditional) simulations should respect, up to statistical fluctuations, the conditional (or unconditional) covariance matrix.
- A set of independently generated realizations are equi-probable but can have different probability density values.
- We cannot generate more then N orthogonal realizations of an N-dimensional random vector, if we use the usual correlation coefficient as the measure of correlation between realizations.

However, it seems that the theory based on the exact sampling of random vectors is contradictory with that based on the simulation of ergodic random functions. If this is true, there are then two possible theories of geostatistics: one based on random functions and the other based on random vectors. In the framework of the random function based geostatistics, we can talk about ergodicity, regional covariance and its inference from a single (fragmentary) realization (i.e., the real data set). In the framework of the random vector based geostatistics, we can talk about covariance matrix, but not ergodicity (that is not defined). The inference of the model parameters from a single realization is then questionable. These two theories are self-consistent and but they seem not compatible between each other. To avoid, at least, terminological confusion, it is necessary to choose one of the two frameworks: random vectors or random functions. In practice, we should expect that these two theories converge to each other with huge random vectors and random functions in large field.

Note finally that, in most real situations, the primary concern in geostatistical modeling remains the choice of a physically realistic random function (or set) model. For instance, a multi-Gaussian model is in contradiction with many geological settings such as fluvial channel or fractured system (Gomez-Hernandez, 1997). Then comes the difficulty of building realizations that preserve the spatial statistics inferred from data and that are calibrated to all quantitative (static and dynamic) data. The further evaluation of uncertainty is meaningful only under the following conditions: first the probability density function (pdf), conditioned to all quantitative data, covers correctly the range of uncertainty and second enough samples (realizations) of this conditional pdf can be obtained within an affordable time. The second condition depends on the efficiency of the sampling algorithms and the computer resources. But the first condition depends on the degree of objectivity of the pdf model. Because of the uniqueness of the reservoir property of interest, a pdf model should largely be subjective. We can evaluate uncertainty only within an subjective model (Matheron, 1978; Journel, 1994). The preservation of the model spatial statistics and the model calibration to data are objective problems, while the uncertainty evaluation is a subjective one (although mathematically meaningful and challenging within a pdf model).

References

Caers, J., 2002, Geostatistical history matching under training-image based geological model constraints, Paper SPE 77429.

Chilès, J.P. and Delfiner, P.,1999, Geostatistics - Modeling spatial uncertainty, Wiley, New York, 695p.

Deutsch, C.V. and Journel, A.G., 1992, GSLIB - Geostatistical software library and user's guide, Oxford University Press, New York, 340p.

Feller, W., 1971, An introduction to probability theory and its applications, Vol.I and Vol.II, John Wiley & Sons, New York, 509p. and 669p.

Gomez-Hernandez, J.J., 1997, Issues on environmental risk assessment, In E.Y. Baafi and N.A. Schofield (eds.), Geostatistics Wollongong '96, Kluwer Academic Pub., Dordrecht, Vol.1, p.15-26.

Hu, L.Y., 2000, Gradual deformation and iterative calibration of Gaussian-related stochastic models, Math. Geology, Vol.32, No.1, p.87-108.

Johnson, M., 1987, Multivariate statistical simulation, John Wiley & Sons, New York, 230p.

Journel, A.G. and Huijbregts, C. J., 1978, Mining geostatistics, Academic Press, London, 600p.

Journel, A.G., 1994, Modeling uncertainty: some conceptual thoughts, In Dimitrakopoulos (ed.), Geostatistics for the next century, Kluwer Academic Pub., Dordrecht, p.30-43.

Lantuéjoul, C., 2002, Geostatistical simulation - Models and algorithms, Springer-Verlag, Berlin, 256p.

Le Ravalec-Dupin, M., and Noetinger, B., 2002, Optimization with the gradual deformation method, Math. Geology, Vol.34, No.2, p.125-142.

Marsily, G. de, Delhomme, J.P., Coudrain-Ribstein, A., Lavenue, A.M., 2001, Four decades of inverse problems in hydrogeology, In Zhang and Winter (eds.), Theory, Modeling and Field Investigation in Hydrogeology, Geophysical Society of America, Special Paper 348, p.1-17.

Matheron, G., 1965, Les variables régionalisées et leur estimation - Une application de la théorie des fonctions aléatoires aux sciences de la nature, Masson, Paris, 305p.

Matheron, G., 1978, Estimer et choisir, Fascicule 7, Les Cahiers du Centre de Morphologie Mathématique de Fontainebleau, Ecole des Mines de Paris, 175p.

Oliver, D.S., Cunha, L.B. and Reynolds, A.C., 1997, Markov chain Monte Carlo methods for conditioning a permeability field to pressure data, Math. Geology, vol. 29, no. 1, p. 61-91.

Omre, H., 2000, Stochastic reservoir models conditioned to non-linear production history observations, In W. Kleingeld and D. Krige (eds.), Geostatistics 2000 Cape Town, Vol.1, p.166-175.

Tarantola, A., 1987, Inverse problem theory - Methods for data fitting and model parameter estimation, Elsevier, Amsterdam, 613p.

ON THE AUTOMATIC INFERENCE AND MODELLING OF A SET OF INDICATOR COVARIANCES AND CROSS-COVARIANCES

EULOGIO PARDO-IGÚZQUIZA[1] and PETER A. DOWD[2]
[1]*Department of Mining and Mineral Engineering,*
University of Leeds, Leeds LS2 9JT, UK
[2]*Faculty of Engineering, Computer and Mathematical Sciences,*
University of Adelaide, Adelaide, SA 5005, Australia

Abstract. The indicator approach to estimating spatial, local cumulative distributions is a well-known, non-parametric alternative to classical linear (ordinary kriging) and non-linear (disjunctive kriging) geostatistics approaches. The advantages of the method are that it is distribution-free and non-parametric, is capable of dealing with data with very skewed distributions, provides a complete solution to the estimation problem and accounts for high connectivity of extreme values. The main drawback associated with the procedure is the amount of inference required. For example, if the distribution function is defined by 15 discrete thresholds, then 15 indicator covariances and 105 indicator cross-covariances must be estimated and models fitted. Simplifications, such as median indicator kriging, have been introduced to address this problem rather than using the theoretically preferable indicator cokriging. In this paper we propose a method in which the inference and modelling of a complete set of indicator covariances and cross-covariances is done automatically in an efficient and flexible manner. The inference is simplified by using relationships derived for indicators in which the indicator cross-covariances are written in terms of the direct indicator covariances. The procedure has been implemented in a public domain computer program the use of which is illustrated by a case study. This technique facilitates the use of the full indicator approach instead of the various simplified alternatives.

1 Introduction

The general estimation problem can be stated as the estimation at unsampled locations of the most probable value of a variable together with a measure of the uncertainty of the estimation (e.g. estimation variance). A more complete and interesting solution to the problem, however, is to estimate at each unsampled location the local cumulative distribution function (cdf) conditioned to the neighbouring data. Point estimates, interval estimates, measures of uncertainty and probabilities (e.g. probability of the unknown value being greater than a specified threshold) can be easily obtained from the estimated distribution function,. The indicator approach (Journel, 1983; Goovaerts, 1997) offers a non-parametric solution to the problem of estimating such local cdf. A discrete representation of the cdf is defined by K thresholds from which K indicator random functions can be derived by applying the thresholds to the continuous variable $Z(u)$:

$$I(u; z_k) = \begin{cases} 1 & \text{if } Z(u) \leq z_k \\ 0 & \text{otherwise} \end{cases} \quad (1)$$

Each indicator random function is assumed to be second-order stationary, i.e. unbounded semi-variograms are not allowed and covariances and semi-variograms are equivalent statistical tools. The expected values of the indicator variables are interpreted as the distribution function of the continuous variable for the respective thresholds:

$$E\{I(u; z_k)\} = F(u; z_k) = \Pr\{Z(u) \leq z_k\} \quad (2)$$

Estimating the different indicators provides estimates of the cdf, $F(u; z_k)$, for the different thresholds $\{z_k; k = 1,...,K\}$. The complete cdf is estimated by assuming a form of the cdf between the thresholds and for the tails (Deustch and Journel, 1992; Goovaerts, 1997).

Journel and Alabert (1989) argue that indicator cokriging is theoretically the best estimator (in a least squares sense) of the cdf using indicators from the experimental data for all the thresholds simultaneously. This is, however, an onerous procedure requiring the inference of K^2 indicator covariances and cross-covariances. In practice, the cross-covariances are assumed to be symmetric and the number of models is reduced to K indicator covariances and $K(K-1)/2$ indicator cross-covariances. An asymmetrical cross-covariance implies (from the non-centred cross-covariance) that:

$$P\{Z(u) \leq z_k, Z(u+h) \leq z_{k'}\} \neq P\{Z(u) \leq z_{k'}, Z(u+h) \leq z_k\} \quad (3)$$

and:

$$P\{Z(u) \leq z_k\} \neq P\{Z(u+h) \leq z_k\} \quad (4)$$

If the indicator random functions are second-order stationary, the random function $Z(u)$ is distribution-ergodic (Papoulis, 1984), the restriction in (4) no longer holds and thus symmetrical cross-covariances are justified.

Even with the assumption of symmetric cross-covariances with, say, $K = 10$ there are 10 indicator covariances and 45 indicator cross-covariances to infer and model. The inference of direct indicator covariances is not particularly difficult and can be done more or less automatically by using maximum likelihood (Pardo-Igúzquiza, 1998) to infer the parameters (e.g., range, sill, nugget, anisotropy angle) without the need to estimate the covariance for a number of lags and fit a model. Even using maximum likelihood the inference and modelling of indicator cross-covariances is more difficult because, *inter alia*, of the order relations (Journel and Posa, 1990) which impose restrictions on the types of models that can be fitted to the indicator cross-covariances. A much more efficient and routine procedure would be to express the cross-covariances in terms of the direct indicator covariances.

2 Methodology

Given a continuous variable, $Z(u)$, and its cdf, the range of values of the variable is represented by K discrete thresholds. Given any pair of these thresholds, z_k and $z_{k'}$, (with the convention $z_{k'} > z_k$) a class, or categorical, variable, c_k, can be defined with an associated indicator random function $I(u;c_k)$:

$$I(u;c_k) = \begin{cases} 1 & \text{if } Z(u) \in c_k \\ 0 & \text{otherwise} \end{cases} \quad (5)$$

or, equivalently:

$$I(u;c_k) = \begin{cases} 1 & \text{if } z_k < Z(u) \leq z_{k'} \\ 0 & \text{otherwise} \end{cases} \quad (6)$$

with

$$I(u;c_k) = I(u;z_{k'}) - I(u,z_k) \quad (7)$$

The covariance of the indicator random function for the class on the left hand side must be equal to the covariance of the difference between the indicator random functions for the thresholds on the right hand side:

$$\text{Cov}\{I(u;c_k)\} = \text{Cov}\{I(u;z_{k'}) - I(u,z_k)\} \quad (8)$$

or $\text{Cov}\{I(u;c_k)\} = \text{Cov}\{I(u;z_{k'})\} + \text{Cov}\{I(u,z_k)\} - 2\text{Cov}\{I(u;z_k), I(u;z_{k'})\} \quad (9)$

then $C_I(h;z_k,z_{k'}) = \frac{1}{2}\{C_I(h;z_{k'}) + C_I(h;z_k) - C_I(h;c_k)\} \quad (10)$

which expresses the indicator cross-covariance of each pair of thresholds as a function of the direct indicator covariances at the thresholds and for the class that they define. There are $K(K+1)/2$ models, all defined by indicator covariances, which can be efficiently modelled by maximum likelihood and where:

$$C_I(h;z_k,z_{k'}) = \text{Cov}\{I(u;z_k), I(u+h;z_{k'})\} = E\{I(u;z_k)I(u+h;z_{k'})\} - F(z_k)F(z_{k'}) \quad (11)$$

$$C_I(h;z_k) = \text{Cov}\{I(u;z_k)I(u+h;z_k)\} = E\{I(u;z_k)I(u+h;z_k)\} - F^2(z_k) \quad (12)$$

$$C_I(h;c_k) = \text{Cov}\{I(u;c_k)I(u+h;c_k)\} = E\{I(u;c_k)I(u+h;c_k)\} - (F(z_{k'}) - F(z_k))^2 \quad (13)$$

Note that, in general, Equation (10) defines a composite model for the indicator cross-covariance even if the direct indicator covariances are simple models. There is no need to fit a specific model to the indicator cross-covariance as it is defined by the indicator covariances and Equation (10).

Journel and Posa (1990) give the order relations for indicator covariances and indicator cross-covariances. The order relations follow from the fact that the non-centred indicator covariances $K_I(h;z_k,z_{k'})$ are bivariate cumulative distribution functions, with

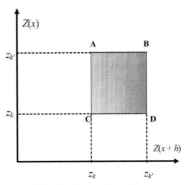

Fig. 1. General order relation

$$K_I(h;z_k,z_{k'}) = C_I(h;z_k,z_{k'}) + F(z_k)F(z_{k'}) \quad (14)$$

The general order relation can be written as (Journel and Posa, 1990):

$$C_I(h; z_k) + C_I(h; z_{k'}) + E \geq 2C_I(h; z_k, z_{k'}) \quad (15)$$

with $E = F^2(z_k) + F^2(z_{k'}) - 2F(z_k)F(z_{k'})$. This order relation is easily derived from Figure 1 in which the shaded area represents a bivariate cumulative distribution function, i.e. a probability which must be non-negative. Then $B - A - D + C \geq 0$, where each of the corners is a non-centred indicator covariance and

$$K_I(h; z_{k'}) - K_I(h; z_k, z_{k'}) - K_I(h; z_{k'}, z_k) + K_I(h; z_k) \geq 0$$

from which (15) can be easily derived taking account of (14) and the assumption of a distribution-ergodic random function $Z(u)$. Thus the non-centred indicator cross-covariance is symmetric with respect to the thresholds.

Substituting (10) into (15) gives:

$$C_I(h; c_k) \geq 2F(z_k)F(z_{k'}) - F^2(z_k) - F^2(z_{k'}) \quad (16)$$

which can be written as:

$$C_I(h; c_k) \geq -(F(z_k) - F(z_{k'}))^2 \quad (17)$$

As the term on the right hand side is always negative this inequality will be satisfied if the covariance of the class indicator is positive. This inequality shows that (10) conforms to the general order relation given by (15).

3 Case study

A realization of a non-Gaussian random function was generated on an 80 × 80 grid by sequential indicator simulation using the program sisimm (Deutsch and Journel, 1992). The thresholds, quantiles and covariance models used for generating the realization are given in Table 1 and a plot of the realization is shown in Figure 2. The range increases as the quantile increases, implying that

Threshold number	z_k	$F(z_k)$	Range of isotropic spherical covariance model (length units)
1	0.1	0.1	6
2	0.5	0.3	8
3	2.5	0.5	10
4	5.0	0.7	12
5	10.0	0.9	24

Table 1. Thresholds and indicator models used in generating the realization shown in Figure 2 using sisim (Deutsch and Journel, 1992).

there is greater connectivity of high values than low values - the semi-variogram range for the 0.9 quantile is four times larger than that for the symmetrical quantile with respect to the median (0.1).

A sample of 60 randomly located data was drawn from the realisation; the values of the samples are represented in Figure 3. Maximum likelihood inference does not require $Z(u)$ to be Gaussian; in fact in this application it is applied to binary indicator data. Nevertheless, using the likelihood of the multivariate normal distribution provides an efficient estimator of the indicator covariance parameters (Pardo-Igúzquiza, 1998). The procedure is also useful for model selection as shown hereafter.

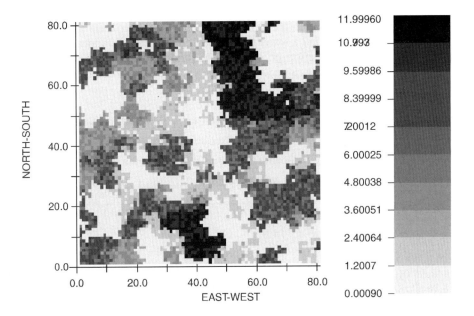

Fig. 2. Simulated random field using indicator sequential simulation with the parameters given in Table 1.

For a specified type of model (spherical, exponential or Gaussian) the program infers the model parameters of the indicator covariance for each threshold and all classes. Eight models were assessed: (1) one isotropic model with no nugget; (2) one isotropic model with nugget; (3) one anisotropic model with no nugget; (4) one anisotropic model with nugget; (5) two nested isotropic models with no nugget; (6) two nested isotropic models with nugget; (7) two nested models: one isotropic, one anisotropic and no nugget; (8) two nested models: one isotropic, one anisotropic and a nugget.

The number of parameters ranges from two for model (0) to seven for model (7) - nugget, sill of isotropic model, range of is otropic model, sill of anisotropic model, long range, short range and anisotropy angle. The most appropriate model could simply be chosen by inspection of the method of moments estimate of the semi-variogram (Figure 4) for the direct indicator semi-variograms at the thresholds. Any of the models can be tried and the quality of the fit can be assessed by the value of the

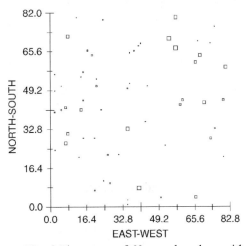

Fig. 3 Pictogram of 60 sample values with locations selected at random from the 80 × 80 grid of the simulated field.

negative log-likelihood function (NLLF). However, the model becomes more flexible as more parameters are used, and a lower NLLF may be achieved by a meaningless over-specification. A model selection criterion, such as the Akaike information criterion (AIC) (Akaike, 1974), provides a trade-off between simple models and more exact fits:

$$AIC(\ell) = 2L(\ell) + 2k(\ell)$$

where: $AIC(\ell)$ is the Akaike information criterion value for the ℓ-th model,

$L(\ell)$ is the value of the negative log-likelihood function for the ℓ-th model, and

$k(\ell)$ is the number of independent parameters fitted in the ℓ-th model.

The model with the lowest $AIC(\ell)$ value is chosen.

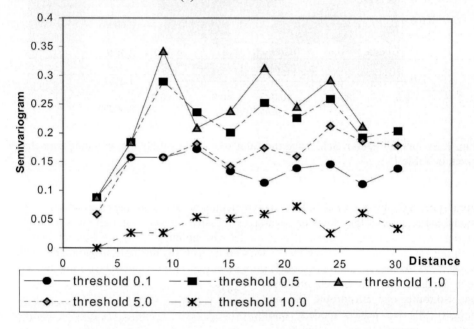

Fig. 4 Experimental indicator semi-variograms for the five thresholds

For few observations simple models should be chosen as, in general, there is insufficient evidence in the data for a model with a large number of parameters. In such cases using a large number of parameters amounts to modelling the fluctuations generated by sampling variability.

Table 2 shows the spherical model fitted by the program using the 60 observations shown in Figure 3 for model 1. In terms of the AIC values the best model is one isotropic structure with no nugget, which is the model used to generate the simulated realization. When comparing estimated parameters with those used in the simulation it should be remembered that only 60 randomly located data were used for the estimates and, as a consequence, they are subject to a high degree of sampling variability. Nevertheless, the ranges of the spherical models are quite well estimated.

Indicator covariances

Threshold I	AIC	C_0	C	Range 1	Range 2	Anisotropy angle	Model type
1	57.440	0.0	0.149	5.883	5.883	0.0	1
2	78.399	0.0	0.226	8.818	8.818	0.0	1
3	77.549	0.0	0.226	9.336	9.336	0.0	1
4	63.744	0.0	0.204	13.996	13.996	0.0	1
5	-27.800	0.0	0.058	22.627	22.627	0.0	1

Indicator covariances for the indicator classes

I_1	I_2	AIC	C_0	C	Range 1	Range 2	Anisotropy angle	Model type
1	2	44.997	0.0	0.136	6.228	6.228	0.0	1
1	3	79.625	0.0	0.249	8.473	8.473	0.0	1
1	4	83.529	0.0	0.249	12.098	12.098	0.0	1
1	5	68.098	0.0	0.185	11.752	11.752	0.0	1
2	3	64.314	0.0	0.214	7.782	7.782	0.0	1
2	4	80.712	0.0	0.233	6.401	6.401	0.0	1
2	5	82.137	0.0	0.241	9.336	9.336	0.0	1
3	4	26.942	0.0	0.056	10.371	10.371	0.0	1
3	5	74.151	0.0	0.202	9.336	9.336	0.0	1
4	5	55.506	0.0	0.173	13.996	13.996	0.0	1

Table 2. Results for one isotropic structure and no nugget, i.e. two parameters: variance (sill) and range. I is indicator number, model type 1 is spherical.

From Table 2 and using (10) the models shown in Figure 5 are fitted to the indicator cross-covariances. The example has been restricted to five thresholds to limit the size of tables and number of figures, but even for large numbers of thresholds the procedure is computationally efficient, e.g. the program generates the 120 models for 15 thresholds in a few minutes.

4 Conclusions

The indicator cokriging of local cumulative distributions is often avoided because of the burden of modelling a large number of indicator covariances and cross-covariances. The authors have described a procedure that bases the modelling of the indicator cross-covariances on direct models of indicator covariances for the thresholds and for the classes defined by pairs of thresholds. The direct indicator covariances can be efficiently inferred and modelled by maximum likelihood which can be applied without assuming that the continuous random variable is multivariate Gaussian.

A public domain program, available on request from the authors, allows the modelling of a wide range of structures, with or without nugget, isotropic or anisotropic, and with up to two nested models. Each structure may be spherical, exponential or Gaussian. The AIC, provided by the program for each indicator covariance, can be used for model

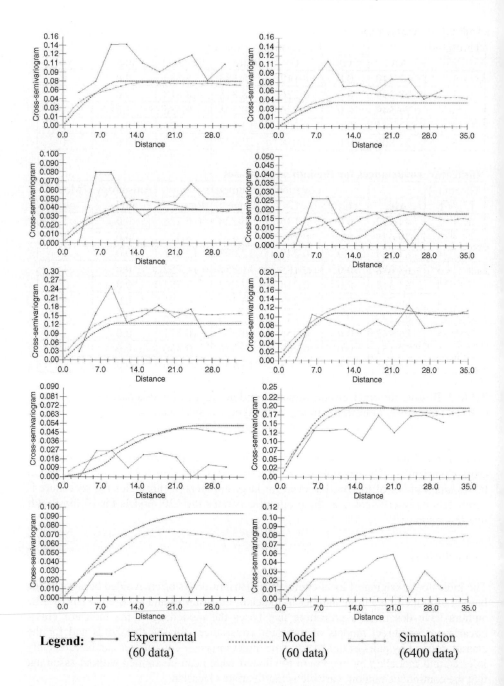

Fig. 5 Indicator cross-semivariograms and models fitted by (10) and the maximum likelihood estimates given in Table 2. Left to right and top to bottom: (a) thresholds 1 and 2; (b) thresholds 1 and 3; (c) thresholds 1 and 4; (d) thresholds 1 and 5; (e) thresholds 2 and 3; (f) thresholds 2 and 4; (g) thresholds 2 and 5; (h) thresholds 3 and 4; (i) thresholds 3 and 5; (j) thresholds 4 and 5.

selection. A case study illustrated the methodology on a simulated realization of a random field.

Acknowledgements

The work reported in this paper was funded by UK Engineering and Physical Sciences Research Council Grant number GR/M72944.

References

Akaike, H., A new look at the statistical model identification. *IEEE Transactions on Automatic Control*, vol. AC-19, no. 6, 1974, p. 716-723.

Deutsch, C.V. and Journel, A.G., GSLIB. Geostatistical Software Library and User's Guide. Oxford University Press, New York, 1992, 340 p.

Goovaerts, P., Geostatistics for Natural Resources Evaluation. Oxford University Press, New York, 1997, 483p.

Journel, A. G., Nonparametric estimation of spatial distributions. *Mathematical Geology*, vol. 15 no. 3, 1983, p. 445-468.

Journel, A. G. and Alabert, F., Non-Gaussian data expansion in the Earth Sciences. *Terra Nova*, vol. 1, 1989, p. 123-134.

Journel, A. G. and Posa, D., 1990. Characteristic behavior and order relations for indicator variograms. *Mathematical Geology*, vol. 22, no. 8, 1990, p. 1011-1025.

Papoulis, A., Probability, Random Variables and Stochastic Processes. McGraw-Hill International Editions, Singapore, 1984, 576 p.

Pardo-Igúzquiza, E., Inference of spatial indicator covariance parameters by maximum likelihood using MLREML. *Computers and Geosciences*, vol. 24, no. 5, 1998, p. 453-464.

ON SIMPLIFICATIONS OF COKRIGING

JACQUES RIVOIRARD

Centre de Géostatistique, Ecole des Mines de Paris, Fontainebleau - France

Abstract. Due to the number of variables or of data, cokriging can be a heavy operation, requiring simplifications. Two basic types of simplications, with no loss of information, are considered in this comprehensive paper. The first type of simplifications consists, in the isotopic case, in reducing cokriging to kriging, either of one or several target variables, or of spatially uncorrelated factors. The example of variables linked by a closure relation (e.g. constant sum, such as the indicators of disjoint sets) is in particular considered. The other type of simplifications is related to some particular models that, in given configurations, screen out a possibly large part of data. This results in simplified and various types of heterotopic neighborhoods, such as collocated, dislocated, or transferred.

1 Introduction

Given a set of multivariate data, cokriging theoretically improves the linear estimation of one variable by taking into account the other variables. In particular, it ensures consistency between the estimates of different variables: the cokriging of a linear combination of variables equals the linear combination of their cokriging, which is in general not the case for kriging. For instance, the cokriged estimate of the difference between the top and the bottom elevations of a geological layer is the same as the difference of their cokriged estimates. Similarly, if we consider sets that correspond for instance to classes of values of a random function, cokriging ensures the relation:

$$[1_{Z(x)<z_2}]^{CK} = [1_{Z(x)<z_1}]^{CK} + [1_{z_1 \leq Z(x)<z_2}]^{CK}$$

which is clearly desirable. In addition cokriging is central in multivariable simulation in a Gaussian framework, where regressions are linear and where the absence of correlation is equivalent to independence.

However, due to the ever increasing number of data (in term of data points or of measured variables), cokriging can become rapidly a heavy operation, hence the interest in looking for simplifications, either in the isotopic cases (when all variables are known at each data point), or in the heterotopic cases. In this comprehensive paper, two types of simplifications without loss of information (i.e. the estimation coinciding with full cokriging) are considered: isotopic cases where cokriging of some variable reduces to its kriging (other variables being screened out), and cokriging neighbourhood being simplified by screening out some types of data.

In the following, we will consider p variables $Z_1(x)$, ..., $Z_i(x)$, ..., $Z_p(x)$. For simplicity the variables are supposed second order stationary with simple and cross-covariances:

$$C_i(h) = C_{ii}(h) = Cov\left[Z_i(x), Z_i(x+h)\right]$$
$$C_{ij}(h) = Cov\left[Z_i(x), Z_j(x+h)\right]$$

or (in particular when cross-covariances are even functions) intrinsic with simple and cross-variograms:

$$\gamma_i(h) = \gamma_{ii}(h) = \frac{1}{2}E\left(\left[Z_i(x+h) - Z_i(x)\right]^2\right)$$
$$\gamma_{ij}(h) = \frac{1}{2}E\left[Z_i(x+h) - Z_i(x)\right]\left[Z_j(x+h) - Z_j(x)\right]$$

The difficult problem of estimating consistently multivariate structures when all variables are not known at all data points is not addressed in this paper. Note that different choices may be possible for the variables (e.g. the indicators of disjoint classes of values of a random function, or the indicators of accumulated classes), that are theoretically equivalent. However one choice may be found to be easier to detect possible simplifications of cokriging.

2 Reduction of cokriging to kriging in isotopic cases

2.1 SELF-KRIGEABILITY

Given a set of p variables ($p \geq 2$), one of the variables, for instance Z_1, is said to be self-krigeable, if its cokriging coincides with its own kriging in any isotopic configuration (Matheron 1979, Wackernagel 2003). The necessary and sufficient condition for this is its cross-structure with the other variables being identical (more exactly, proportional) to its own structure, which is denoted (a little abusively when the proportionality factor is zero) as:

$$C_{1j}(h) \equiv C_1(h)$$

or

$$\gamma_{1j}(h) \equiv \gamma_1(h)$$

for all j, and can possibly be checked using sample simple and cross-structures. (Note that the concept of self-krigeability is relative to a given set of variables: for instance a self-krigeable variable may not remain self-krigeable if new variables are added.)

Suppose now that, among a set of p variables ($p \geq 2$), two of them are self-krigeable, say Z_1 and Z_2: $\gamma_{12}(h) \equiv \gamma_1(h)$ and $\gamma_{12}(h) \equiv \gamma_2(h)$. Then two cases can be distinguished. Either they have the same structure (the cross-structure being proportional to these, not excluding it being zero):

$$\gamma_{12}(h) \equiv \gamma_1(h) \equiv \gamma_2(h)$$

in which case they are intrinsically correlated, in the sense that the correlation coefficient between $Z_1(v)$ and $Z_2(v)$ within a domain V is "intrinsic", not depending on the support v nor on the domain V (Matheron, 1965; this should not to be confused with the intrinsic model based on increments). Or they have different structures, in which case the variables (or their increments in the intrinsic model) are necessarily spatially uncorrelated:

$$\gamma_{12}(h) = 0 \quad \forall h$$

Suppose now that all p variables are self-krigeable. We can group the variables that have the same simple structure (up to a proportionality factor). Then all variables within a given group are intrinsically correlated. Moreover two variables from different groups are spatially uncorrelated. So a set of self-krigeable variables can be partitioned into groups of intrinsically correlated variables, each group having a different structure, and with no correlation between groups (Rivoirard, 2003; another proof is given in Subramanyam and Pandalai, 2004). In particular we have the two typical cases of reduction of cokriging to kriging:
- when all variables are intrinsically correlated, all simple and cross-structures being proportional to a common structure, and kriging weights being the same for all variables;
- when they have no cross correlation.

In the particular case of p self-krigeable variables being linked by a closure relation (such as concentrations, or such as the indicators of disjoint sets):

$$Z_1(x) + Z_2(x) + ... + Z_p(x) = 1$$

or more generally being linearly dependent, separating the different groups yields:

$$\text{var}[\sum Z_i(x)] = \text{var}[\sum_{group} \sum_{i \in group} Z_i(x)] = \sum_{group} \text{var}[\sum_{i \in group} Z_i(x)] = 0 \Rightarrow \text{var}[\sum_{i \in group} Z_i(x)] = 0 \text{ for}$$

each group.

Hence each group of intrinsically correlated variables is closed. Note that, in the case of the indicators of disjoint sets partitioning the space, there can be only one group, since these indicators are necessarily correlated. So if the indicators of disjoint sets are self-krigeable, their cross and simple structures are proportional to a unique structure (which corresponds to the mosaic model with independent valuations, Matheron (1982)).

Linear dependency between variables is not a problem for defining cokriging (e.g. as a projection). The cokriged value is perfectly determined, but in the isotopic case, the cokriging system is singular and weights are not all uniquely defined. So it is interesting to remove (at least) one of the variables. Providing that linear dependency vanishes, cokriging does not depend on the choice of which variables are removed. However some choices may be better than others, in term of simplification of cokriging. Suppose that $p-1$ out of a set of p variables are self-krigeable, say Z_1, ..., Z_i, ..., Z_{p-1}. If these $p-1$ variables are intrinsically correlated, then all p variables are intrinsically correlated and for each variable, cokriging reduces to kriging of that variable. But suppose now that there is more than one group of intrinsically correlated variables in the $p-1$ variables (which cannot be the case for the indicators of disjoint sets). Then the last variable:

$$Z_p = 1 - Z_1 - Z_2 - ... - Z_{p-1}$$

which is necessarily correlated to at least some of the $p-1$ variables since:

$$\text{var} Z_p = -\sum_{i<p} \text{cov}(Z_p, Z_i) > 0$$

cannot be self-krigeable: as groups are uncorrelated, the cross-structure with each group is proportional to the structure of the group, and so these cross-structures cannot all be identical (they are either different, or possibly equal to zero but only for some of them). Then cokriging is simplified by kriging the $p-1$ self-krigeable variables, and deducing the estimation of the last one.

2.2 FACTORIZATION

Up to now, we have considered simplification of isotopic cokriging resulting from initial variables being self-krigeable, which can be detected directly from the observation of simple and cross-structures. We will consider now an extension through the use of factors.

2.2.1 Model with residual

If we consider a set of $p = 2$ variables Z_1 and Z_2, with Z_1 being self-krigeable:

$$\gamma_{12}(h) = a\, \gamma_1(h)$$

the residual of the linear regression of $Z_2(x)$ on $Z_1(x)$ at same point x:

$$R(x) = Z_2(x) - a\, Z_1(x) - b$$

(by construction with mean zero and uncorrelated to $Z_1(x)$ at same point x) has no spatial correlation with Z_1. In this model of residual, or "Markov" type model (Journel, 1999; Chilès and Delfiner 1999; Rivoirard, 2001), the variable Z_2 is subordinated to the master variable Z_1:

$$Z_2(x) = a\, Z_1(x) + b + R(x)$$

The model is factorized into Z_1 and R, which (being spatially uncorrelated) are self-krigeable. So we have:

$$Z_1^{CK} = Z_1^{K}$$
$$R^{CK} = R^{K}$$
$$Z_2^{CK} = a\, Z_1^{K} + b + R^{K}$$

for any isotopic configuration. This model was illustrated in a mining case study by Bordessoule et al. (1989): this included the deduction of the self-krigeability of a variable from the observed simple and cross-variograms, the analysis of the residual, and the cokriging, showing in particular that cokriging can be significantly different from kriging in practice, if there was any doubt.

Note that if the structure of the residual is identical to this of Z_1, all simple and cross-structures are identical, so that Z_1 and Z_2 are intrinsically correlated. Then any variable can be taken as master.

2.2.2 Intrinsically correlated variables

If variables have simple and cross-structures proportional to a common structure (say a correlogram $\rho(h)$), so do their linear combinations. So intrinsic correlation between a set of variables extends to their linear combinations. It follows that if two variables, or two linear combinations of variables, are uncorrelated at same point ($C_{ij}(0) = 0$), their cross-covariance is identically zero:

$$C_{ij}(h) = C_{ij}(0)\rho(h) = 0$$

So, for intrinsically correlated variables, the absence of statistical correlation implies the absence of geostatistical, or spatial, correlation. As a consequence, any statistical factorization (eigen vectors, successive residuals, etc) of intrinsically correlated variables gives spatially uncorrelated factors (Rivoirard, 2003). Note that, while arbitrary or conventional in the sense they depend on the choice of the factorization method, these factors are objective, in the sense their values at a point where the variables are known, are determined. While factorisation is always possible when the model is admissible (exhibiting factors ensures the variances-covariances matrix to be valid), it does not provide simplification in isotopic cokriging, as all variables and linear combinations are self-krigeable anyway.

2.2.3 Linear model of coregionalization

The linear model of coregionalization corresponds to a decomposition of structures into a set of basic structural components (e.g. describing different scales). The corresponding components of the variables for a given scale are intrinsically correlated (Journel and Huijbregts, 1978), and a factorisation of these ensures the validity of the model. However there is usually more factors than variables, and these factors have not an objective meaning. Factorial (co-)kriging, or kriging analysis, allows estimating consistently these, but does not simplify cokriging.

2.2.4 Objective factors

Cokriging can be greatly simplified when the variables are factorized into objective factors, for the knowledge of the variables at a data point is equivalent to the knowledge of factors. And since the factors are spatially uncorrelated, they are self-krigeable, yielding cokriging of all linear combinations, and in particular of the initial variables. Note that some factors can share the same structure (in which case these factors are intrinsically correlated and their choice is conventional).

Of course a question is how to determine, if possible, such factors in practice. In general a statistical factorization (non-correlation at same point) does not yield zero cross-correlation. The technique of min-max autocorrelation factors (Desbarats and Dimitrakopoulos, 2000; Switzer and Green, 1984) allows building factors that are not only uncorrelated at distance 0, but also at a distance chosen from sampling: then the lack of correlation must be assumed or checked for other distances. This reduces cokriging to kriging of factors, and in the gaussian case, multivariate simulation to separate simulations of factors. Another approach, seeking the absence of correlation simultaneously for all lags of variograms, is proposed by Xie and Myers (1995) and Xie et al. (1995).

The absence of cross-correlation between factors for all distances also corresponds to the isofactorial models of non-linear geostatistics (disjunctive kriging, i.e. cokriging of indicators, being obtained by kriging the factors).

3 Simplifications of cokriging neighbourhood in heterotopic cases

3.1 DATA EXPRESSED AS INDIVIDUAL VALUES OF VARIABLES OR FACTORS

Simplifications coming from non spatially correlated variables are still valid in heterotopic cases, when data consist of individual values of these variables. For instance, if Z_1 and Z_2 are spatially uncorrelated, Z_2 is screened out when cokriging Z_1, whatever the configuration. An exception must be noted, when uncorrelated variables have means, or drifts, that are unknown but related (Helterbrand and Cressie, 1994). The screen is deleted by the estimation of means which is implicitly performed within cokriging. We do not consider this case in this paper. Screen can also be deleted by data that do not consist of individual values of variables. If Z_1 and Z_2 are spatially uncorrelated for instance, the knowledge of e.g. the sole sum $Z_1 + Z_2$ at a data point can delete the screen.

Similarly, the simplifications due to factorized models hold in heterotopic cases where data consist in values of individual factors. In the model of residual, with Z_2 subordinated to master variable Z_1, factors are Z_1 and the residual, and simplifications hold providing that data can be equivalently expressed in term of Z_1 and Z_2, or of Z_1 and R. This occurs when Z_1 is known at each data point, for the possibly additional knowledge of Z_2 at this point is equivalent to that of the residual.

More generally the cokriging of a self-krigeable variable reduces to kriging of that variable, in heterotopic cases where it is known at every data point (Helterbrand and Cressie, 1994), for the possibly available other variables at these points are screened out. Of course it is not necessarily so in other heterotopic cases (this is why the definition of a self-krigeable variable assumes isotopy).

If the target variable is Z_2, subordinated to one (or possibly several) master variable Z_1 informed at all data points, its cokriging at any target point is simplified:

$$Z_2^{CK} = a\, Z_1^{K} + b + R^{K}$$

where Z_1 is kriged from all data points and R from only the data points where Z_2, or equivalently R, is known. If Z_1 is known at any desired point, so in particular at target point, we have:

$$Z_2^{CK} = a\, Z_1 + b + R^{K}$$

and cokriging reduces to kriging of the residual (Rivoirard, 2001). Then cokriging is collocated, making use of the auxiliary variable only at target point and at points where the target variable is known, not at other data points. In other models, the residual R is spatially correlated to the auxiliary variable Z_1. Then, the auxiliary variable Z_1 still being supposed known at all desired points, the advantage of collocated cokriging is to be more precise (in term of estimation variance) than kriging the residual, for it corresponds to cokrige this residual from the same data. However, by using a collocated neighbourhood, both collocated cokriging and kriging of the residual result in a loss of information compared to full cokriging, when the residual is spatially correlated to the auxiliary variable, i.e. when the cross-structure is not proportional to this of the auxiliary variable.

3.2 OTHER SIMPLIFICATIONS OF NEIGHBOURHOOD

Consider for instance the case where the cross-structure is proportional to that of the target variable, not of the auxiliary variable, i.e. the target variable is the master variable Z_1. In the case the residual is pure nugget, knowing additionally Z_2 at a Z_1 data point corresponds to knowing the residual. Being uncorrelated to all data and to the target, this is screened out in any Simple Cokriging configuration, and the neighbourhood is dislocated, making use of the auxiliary variable only at target point (if available) and where Z_1 is unknown. In Ordinary Cokriging, the Z_2 values where Z_1 is known are not screened out for they participate to the estimation of the Z_2 mean, but all receive the same weight, which also simplifies cokriging.

Such simplifications of the cokriging neighbourhood are studied in detail by Rivoirard (2004), where a number of other simplifications are listed (with possible extensions to more than two variables). In particular (assuming target is unknown):
- If the target variable is master and has a pure nugget structure, it is screened out from Simple Cokriging where known alone, in the case the auxiliary variable is available at target point; else all data are screened out.
- If the target variable is subordinated to an auxiliary nugget and master variable, this last is screened out where known alone, except at target point if available, in any Simple Cokriging configuration.
- If the target variable is subordinated to an auxiliary master variable with a nugget residual, it is screened out from Simple Cokriging where the auxiliary variable is known (neighbourhood being transferred to the auxiliary variable for common data points); if additionally the auxiliary variable is available at target point, all data except this are screened out from Simple Cokriging.

4 Conclusions

In this paper, different simplifications of cokriging have been finally considered:
- cokriging reduced to kriging for a self-krigeable variable, in isotopic cases or when it is known at all data points;
- cokriging obtained from the kriging of spatially uncorrelated variables or factors (e.g. residuals), in isotopic cases or when data consist of individual values for these factors;
- screening out of some type of data in the neighbourhood, in some particular models with residual where master variable or residual are pure nugget.

Some of these simplifications, in given configurations, can be directly deduced from the observation of the simple and cross-structures of the variables (self-krigeability, intrinsic correlation, model with residual). Other simplifications depend on the possibility of building objective factors that are not cross-correlated. An open question is how to measure the efficiency of the simplifications for a possible departure from the assumptions.

References

Bordessoule, J. L., Demange, C., and Rivoirard, J., 1989, Using an orthogonal residual between ore and metal to estimate in-situ resources, *in* M. Armstrong, ed., Geostatistics: Kluwer, Dordrecht, v. 2, p. 923-934.
Chilès, J.-P., and Delfiner, P., 1999, Geostatistics: Modeling spatial uncertainty, Wiley, New York, 695 p.
Desbarats, A. J. and Dimitrakopoulos, R., 2000, Geostatistical simulation of regionalized pore-size distributions using min/max autocorrelation factors. *Math. Geol.* Vol. 32 No 8, p. 919-942.
Helterbrand, J. D., and Cressie, N., 1994, Universal cokriging under intrinsic coregionalization, Math. Geol.ogy, v. 26, no. 2, p. 205-226.
Journel, A.G. and Huijbregts, C.J., *Mining Geostatistics*, Academic Press, 1978.

Journel, A., 1999, Markov models for cross-covariances: Math. Geology, v. 31, no. 8, p. 955-964.

Matheron, G., 1965, Les variables régionalisées et leur estimation, Masson, Paris, 306 p.

Matheron, G., 1979, Recherche de simplification dans un problème de cokrigeage, Technical Report N-628, Centre de Géostatistique, Fontainebleau, France, 19 p.

Matheron, G., 1982, La destruction des hautes teneurs et le krigeage des indicatrices, Technical Report N-761, Centre de Géostatistique, Fontainebleau, France, 33 p.

Rivoirard, J., 2001, Which Models for Collocated Cokriging?: Math. Geology, v.33, no. 2, p. 117-131.

Rivoirard, J., 2003, Course on multivariate geostatistics, C-173, Centre de Géostatistique, Fontainebleau, France, 76 p.

Rivoirard, J., 2004, On some simplifications of cokriging neighborhood: Math. Geology, v.36, no. 8, p. 899-915.

Subramanyam A. and Pandalai, H. S., 2004, On the equivalence of the cokriging and kriging systems, Math. Geology, v. 36, no. 4.

Switzer, P. and Green, A. A., 1984, Min/max autocorrelation factors for multivariate spatial imaging. Technical report no 6, Department of Statistics, Stanford University, 14 p.

Wackernagel, H., 2003, Multivariate geostatistics: an introduction with applications, 3rd ed., Springer, Berlin, 387 p.

Xie, T. and Myers, D.E., 1995. Fitting matrix-valued variogram models by simultaneous diagonalization (Part I: Theory), Math. Geology, v. 27, p. 867-876.

Xie, T., Myers, D.E. and Long, A.E. 1995, Fitting matrix-valued variogram models by simultaneous diagonalization (Part I: Application), Math. Geology, v. 27, p. 877-888.

EFFICIENT SIMULATION TECHNIQUES FOR UNCERTAINTY QUANTIFICATION ON CONTINUOUS VARIABLES:
A process preserving the bounds, including uncertainty on data, uncertainty on average, and local uncertainty

BIVER P.
TOTAL SA
Centre Scientifique et Technique Jean Feger
Avenue Larribau
64000 Pau, France

Abstract. In the petroleum industry, the standard Monte Carlo technique applied on global parameters (rock volume, average petrophysics) is often used to evaluate hydrocarbon in place uncertainty. With the increasing power of computers, these methodologies have become old fashioned compared to geostatistics. However care must be taken in using the latter blindly; multiple geostatistical realisations do not cover a reasonable uncertainty domain because of undesirable effects. For instance: a net to gross map with a small variogram range tends to reproduce the prior mean over the domain of interest even if this prior mean is established on a small data set; uncertainty on the data themselves may produce values outside the prior distribution range; more over the alternative data set may be biased comparing to initial prior distribution.

In this paper, we present a fully automatic process based on the classical normal score transformation which is able to handle, in a non-stationary model, the following characteristics:
- uncertainty on the data set (with systematic biases)
- uncertainty on the prior mean
- local uncertainty
- preservation of bounds defined on the prior model

An example is given on a real field case in the framework of net to gross modelling. The beta law is used in order to provide high frequencies observed at the bounds (0,1); the robustness of an automatic fit for this distribution type is highlighted in order to adjust a non-stationary model on the data set. All aspects described above have been handled successfully in this non-stationary context; and the ensemble of realisations reproduces rigorously the prior distribution. The balance between local uncertainty and global uncertainty is provided by the user; consequently the volumetrics distribution are easily controlled. A final comparison with the classical geostatistical workflow is provided.

1 Introduction

A continuous variable has to be depicted with a complete probability density function; it could be a non parametric density function issued from smoothed histogram (when a large number of data are available) or a parametric model fitted to the noisy experimental histogram; for instance: a porosity histogram is often calibrated with a Gaussian model, but for a net to gross histogram the beta distribution can be more appropriate.

Except for the Gaussian model, the assumption of kriging requires a normal score transform to process the data and build a random field; the back-transform is subsequently applied to come back to the real variable.

The motivation in the framework of multiple realisations is to incorporate the uncertainty on data and the uncertainty on the mean; and at the same time to retrieve the probability density function over all the possible realisations especially in the case of bounded models (uniform, triangular and beta distribution). This goal is achieved with the use of the normal score transformation, in a generalized context, coupled with a kriging of the mean.

2 Uncertainty on the mean

The uncertainty evaluation exercise may be performed in the appraisal phase of a reservoir; at this stage sparse data are available and it is not obvious that the prior mean (derived from data analysis) is perfectly well known. More frequently, this mean is uncertain and this uncertainty has a direct impact on hydrocarbon in place distribution. The quantification of the relative uncertainty on the mean can be assessed with a declustering formula.

Let us define, for the variable to simulate $Z(x)$,

a relative dispersion of the mean in the real space $r = \sigma_m / \sigma_t = 1/\sqrt{n}$

with   - σ_m the uncertainty on the mean in the real space

    - σ_t the global uncertainty on $Z(x)$

    - n the number of independent data

Let us assume a covariance model $C(h)$ for $Z(x)$; n can be derived from the ordinary kriging system if the covariance vector is set to zero (ordinary kriging of the mean m);

$$m = \sum_{i=1}^{n} \lambda_i . Z(x_i) \tag{1}$$

we have : $\sum_{j=1}^{n} C_{ij} . \lambda_j = \mu$

$\sum_{i=1}^{n} \lambda_i = 1$

with λ_i the kriging weights and μ the Lagrange parameter

It can be shown that: $\sigma_{OK}^2 = \sigma_m^2 = C(0).\mu$ hence $\mu = 1/n$ as $C(0) = \sigma_t^2$ (2)

If we draw a new mean value in the real space and if we want to preserve the distribution shape of the residuals, it is necessary to shift the distribution; as a consequence, the bounding values are not preserved from one realisation to another. An alternative procedure is to work in the normal score space and to consider that the standard normal distribution is in fact a sum of two random variables:

- the first random variable Δ characterises the uncertainty of the mean; it has an infinite spatial correlation, it is centred on zero and has a standard deviation of $r = \sqrt{\mu} = 1/\sqrt{n} = \sigma_m/\sigma_t$ (relative dispersion of the mean in the real space);

- the second random variable characterises the uncertainty of the residuals; it has a limited spatial correlation, it is centred on Δ and has a standard deviation of

$$\sigma_r = \sqrt{1 - \frac{1}{n}}$$ (3)

Each realisation (characterised by its Δ) is then back transformed with the global cumulative distribution of the variable. As a consequence, each realisation has a specific distribution according to the value of Δ; it is different from the initial global distribution. However, if we consider the ensemble of all these different realisations, the global distribution is retrieved. Despite of this property, we have to check that the uncertainty on the mean in the real space (induced by the variation of Δ) is consistent with the relative dispersion we want to impose.

The procedure is illustrated on Figure 1. Two initial global distributions are considered; a beta distribution between 0 and 1 with shape parameters p = 0.7 and q = 0.4 (this model could be appropriate to represent the distribution of a net to gross) and a Gaussian distribution with mean m = 0.2 and standard deviation σ = 0.06 (this model could represent a porosity uncertainty). The number n of equivalent independent data is set to 4 ($r = 0.5$). The updating of the distribution for each value of Δ is represented on Figure 1; for the beta distribution, the asymmetry is gradually modified with Δ; for the Gaussian case, the shape is preserved and the influence of Δ is only a resizing of the Gaussian curve.

By construction, the global distribution is perfectly reproduced. The histograms of the mean values in real space are depicted on Figure 2. Concerning the mean, the distribution in the Gaussian case is of course an exact reproduction of the reference; the distribution in the beta case is slightly skewed comparing to the reference. Even with this high value of $r = 0.5$, the approximation of the mean dispersion is excellent.

If more asymmetric distributions are envisioned, the uncertainty on the mean could be asymmetric for high values of r but when r is decreased, it converges rapidly to the reference Gaussian distribution; this property is a consequence of the central limit theorem. Concerning the variance, the drawback of considering the variation of the mean in the normal score domain is that the variance is not constant for all realisations but perfectly correlated to the mean value.

The methodology can be extended in non-stationary cases. In these situations, the global cumulative distribution is substituted by a local cumulative distribution and, as a consequence, the normal score transform is locally adapted; however, in the Gaussian space, the simulation strategy with Δ is unchanged and still stationary. Its effect in real space is different, the uncertainty on the mean will be higher in the location where local cumulative distribution corresponds to a larger dispersion. In this case, the influence of the variable mean and variable trend dip (that can be major sources of uncertainty, see Massonnat, Biver, Poujol) are considered in one single step.

3 Uncertainty on the data

The data used in practical applications are not always considered as "hard" data. Today, log analysts are able to quantify uncertainties on their data; two kinds of uncertainties can be considered:
- measurements uncertainties linked to the resolution of logging tools,
- interpretation uncertainties linked to the parameters of the interpretation law chosen to derive interpreted logs from raw logs (for instance the exponents of the Archie's laws or the resistivity of connate water used to derive water saturation).

The first is mainly a noise on the data set, the second introduces a systematic bias from one data set to another; they are treated differently in the data set simulation procedure handled by the log analysts.

However, if systematic biases are suspected, an up date of the distribution is needed for each data set. Let's assume that we have multiple realizations of the data set of interest, and that a histogram and a prior distribution model are derived from the ensemble of data set realizations. The suggested procedure of updating can be described as following:

- compute normal score transform $G(x_i)$ of the current data set realization x_i $i=1,n$
- compute the potential bias distribution in the normal score space with ordinary kriging of the mean for $G(x)$, using previously mentioned declustering formula with the normal score transform variogram model ;
- draw a value of the bias Δ' in its distribution with mean $m_{\Delta'}$ and standard deviation $\sigma_{\Delta'}$;
- add this bias Δ' to Δ corresponding to the uncertainty on the mean with a fixed data set ;
- compute a random field for residuals using the mean $(\Delta + \Delta')$ and the variance $(1-\sigma^2_{\Delta}-\sigma^2_{\Delta'})$ (4)

With this procedure, the global unit variance of the normal score is split in the different categories of uncertainties that could affect the final map (residual uncertainty, uncertainty on the mean, uncertainty on data with bias).

The graphical illustration of the corresponding multiple realizations loop is depicted on Figure 3.

4 Specific aspects of beta distributions

The beta model distribution (second type) is often used to describe volume ratios as net to gross, but also potentially effective porosities and effective irreducible water saturations. This model is interesting for the following reasons:

- It has bounding values *(min, max)* ;
- the Gaussian model can be seen as a particular case of the beta model
- depending of the shape parameters *(p, q)*, a large variety of behaviour can be described (see Figure 4)

The corresponding distribution law is given by:

$$f(y) = \frac{1}{B(p,q)} \cdot \frac{y^{p-1}}{(1+y)^{p+q}} \quad \text{with} \quad B(p,q) = \frac{\Gamma(p).\Gamma(q)}{\Gamma(p+q)} \qquad (5)$$

$$y = \frac{x - \min}{\max - \min}$$

In this model, Γ is the gamma function; p and q the shape parameters; they can be derived from the mean *m* and variance σ^2, assuming that bounding values *(min, max)* are fixed and known:

$$p = (m - \min)\left[\frac{(\max - m).(m - \min)}{\sigma^2} - 1\right] \qquad (6)$$

$$q = (\max - m)\left[\frac{(\max - m).(m - \min)}{\sigma^2} - 1\right]$$

This relationship allows us to estimate a first guess for fitting the beta law with experimental value of mean and variance. Moreover, in a non-stationary case, a local updating of the shape parameters can be performed from the local mean and variance.

5 Practical case study

All the previous concepts have been used to simulate petrophysical attributes (net to gross, porosity, irreducible water saturation, log Kh, and Kv/Kh ratio) in a carbonate platform reservoir sampled in the hydrocarbon pool with 19 wells.

In this reservoir, five environments of deposition have been distinguished; moreover, the statistics of the net to gross values are different in each of the 15 layers of the model. The corresponding number of fit to achieve is large (450) and cannot be done manually.

The fit of a non-stationary beta law model is performed in two steps:
- computing a vertical trend of mean and standard deviation for each facies,
- derive the shape parameters p and q from local mean and standard deviation using formula (6).

An example of a geostatistical simulation based on this non-stationary model is depicted on Figure 5 (net to gross visualization regarding facies map) and Figure 6 (comparison of statistics between simulation and data)

Hence, the relative uncertainty on the mean values are estimated from the declustering formula (2) using a variogram of 500 meters. This relative uncertainty depends on the facies (some facies are more frequently observed than other) and varies from one variable to another (some variables are less sampled that other, for instance effective water saturation is less sampled than net to gross); for net to gross $r = \sigma_m/\sigma_t$ is between 0.07 and 0.17 (from offshore to upper shoreface facies), for water saturation $r = \sigma_m/\sigma_t$ is between 0.08 and 0.22 (from offshore to upper shoreface facies).

Alternative data set have been produced from a log data uncertainty study. Unfortunately, the systematic bias which exists from possible alternative interpretations have not been correctly represented; to illustrate however the procedure, another example with alternative data set have been generated.

The complete simulation loop process (remember Figure 3) has been used to define uncertainty on volumetrics; it has been compared to the standard case (multiple realizations without uncertainty on data and without uncertainty on means) and to an intermediate case (multiple realizations without uncertainty on data and with uncertainty on the mean).

The volumetrics results are provided on Figure 7. It tends to illustrate that the uncertainty on data are the key uncertainty for this practicle case. This is a consequence of the systematic bias affecting the data; on this mature field, the hydrocarbon pool is controlled by wells and, as a consequence, the uncertainty on mean and the uncertainty on residuals have a small impact on volumetrics. This conclusion is not obvious for reserves and production profiles which are more dependent of local heterogeneities.

6 Conclusions

Uncertainty quantification for hydrocarbon in place evaluation is a frequent exercise in oil industry, this paper has illustrated the possibility of using a geostatistical workflow to achieve this goal. This is not the standard multiple realisation loop frequently observed in commercial software; the suggested procedure involves uncertainty on the distribution model itself, coupled with an uncertainty on conditioning data. Through the case study, it has been shown that this data and mean uncertainties aspect can be a key issue.

It may be argued that the well known technique of experimental design can be used as an alternative approach to treat uncertainty on data and mean. However, multiple realizations are needed to assess local uncertainties on hydrocarbon location; more over, hydrocarbon in place evaluation is not a CPU intensive computation; for all these reasons, it seems more appropriate to use Monte Carlo simulation to explore exhaustively the uncertainty domain instead of focusing on a limited number of cases with experimental design.

Acknowledgements

The author would like to acknowledge J. Gomez-Hernandez (University of Valencia), R. Froidevaux (FSS International), and A. Shtuka (Shtuka Consulting) for stimulating discussions and implementation aspects.

References

Goovaerts, P., *Geostatistics for Natural Resources Evaluation*, Oxford University Press, 1997.
Journel A.G., *Fundamentals of geostatistics in five lessons*, p.39, Short course in geology: vol.8, American Geophysical Union, Washington DC, 1989
Massonnat G., Biver P., Poujol L., 1998, Key sources of uncertainty from reservoir internal architecture when evaluating probabilistic distribution of OOIP, SPE paper no 49279, *Proceedings of the SPE fall meeting 1998*.
Norris R., Massonnat G., Alabert F., 1993, Early quantification of uncertainty in the estimation of oil in place in a turbidite reservoir, SPE paper no 26490, *Proceedings of the SPE fall meeting 1993*, pp. 763-768.

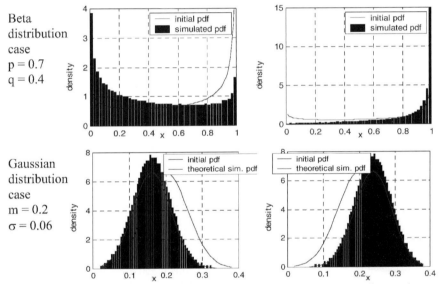

Figure 1: histogram of local values for extreme realisations according to the mean.

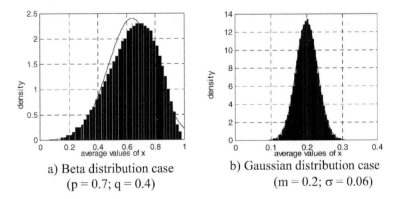

a) Beta distribution case
(p = 0.7; q = 0.4)

b) Gaussian distribution case
(m = 0.2; σ = 0.06)

Figure 2: histograms of the means for the ensemble of realizations.

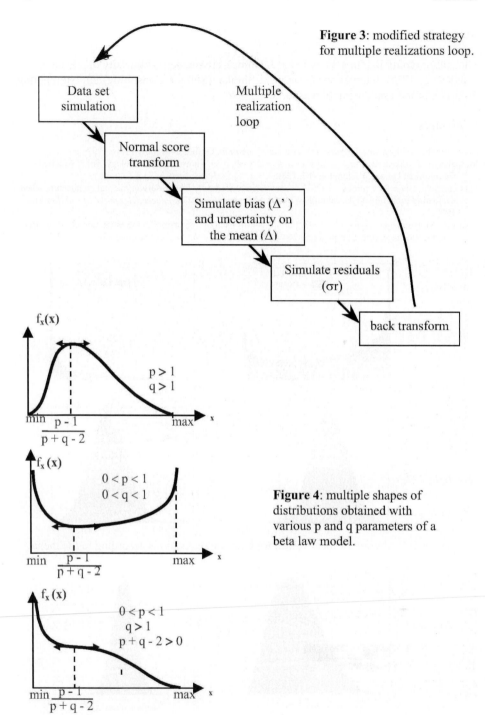

Figure 3: modified strategy for multiple realizations loop.

Figure 4: multiple shapes of distributions obtained with various p and q parameters of a beta law model.

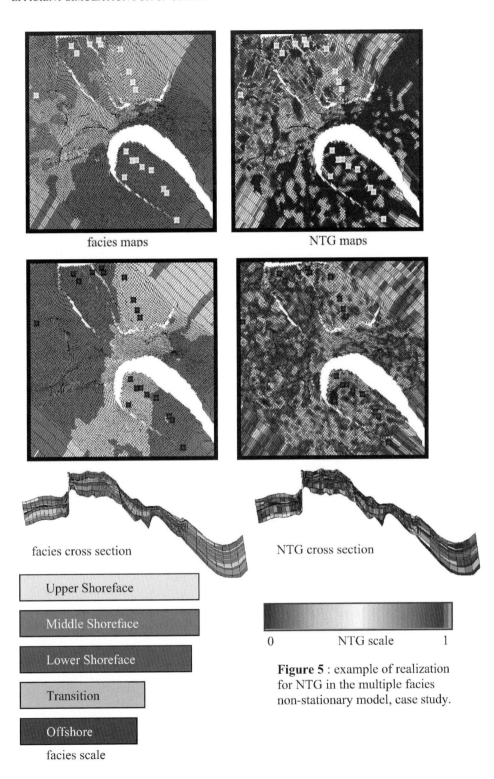

Figure 5: example of realization for NTG in the multiple facies non-stationary model, case study.

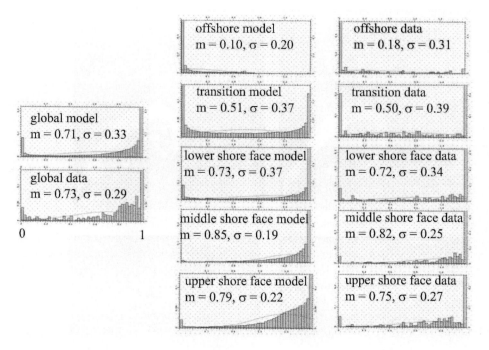

Figure 6: realization statistics versus data statistics, case study, quality control of the model.

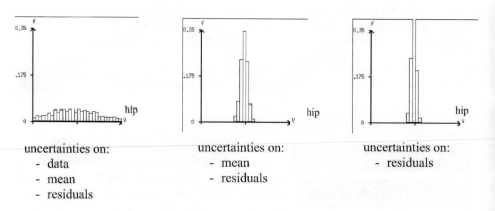

Figure 7: hydrocarbon in place distributions, comparison of runs with different uncertainty sources.

HIGHER ORDER MODELS USING ENTROPY, MARKOV RANDOM FIELDS AND SEQUENTIAL SIMULATION

COLIN DALY
Roxar Limited. Pinnacle House, 17-25 Hartfield Rd., Wimbledon London SW19 3SE, U.K.

Abstract. Approaches to modelling using higher order statistics from statistical mechanics and image analysis are considered. The mathematics behind a very general model is briefly reviewed including a short look at entropy. Simulated annealing is viewed as an approximation to this model. Sequential simulation is briefly introduced as a second class of methods. The unilateral model is considered as being a member of both classes. It is applied to simulations using learnt conditional distributions.

1 Introduction

For certain problems the entity to be predicted is a non linear function of some poorly known control variable. An example is prediction of flow behaviour in an oil reservoir which involves solving differential equations which are highly non-linear with respect to the spatial distribution of permeability. When the control variable may be modelled stochastically then the accepted practice is to generate several realizations and make predictions by using the distribution of results found by applying the function to each. Assuming that the nonlinear function adequately captures the physics, then our attention turns to seeing if the stochastic modelling adequately captures the geology and to inquire which parts of the geology are relevant for the entity to be predicted.

Two principle methods have been used for simulation, methods based on variograms and object models (although see Tjelmeland and Besag, 1998, for a Markov Random Field approach). The former has tended to focus on simplicity and ease of well conditioning, the latter on geological realism and connectivity. Recently there has been an increase in interest in methods using higher order statistics. There has been some controversy about whether a random function necessarily exists which satisfies these higher order statistics. This paper briefly reviews relevant results from statistical mechanics and image analysis, stating existence results and considering a rigorous general model capable of handling complex higher order statistics. A popular method for using higher order statistics is based on a sequential simulation algorithm (e.g. Strebelle, 2002). This is briefly considered in section 3, which, it should be noted is largely independent of section 2 apart from questions of existence. A unilateral model is a simple special case which is both a Markov random field and yet may be simulated sequentially. For a seemingly non symmetric model it gives surprisingly good results.

2 Review of results from Statistical Mechanics

2.1 THREE VIEWPOINTS

Suppose we are interested in building stochastic models of a discrete variable (e.g. facies for a petroleum reservoir model) satisfying some vector of statistics η (with given explicit values of η_I - possibly calculated from a training image). Therefore the result of a calculation on a realisation of the model should match these statistics in some sense. The types of calculations (called the *interactions*) that may be used are now defined. They are very wide in scope. Let Λ the grid on which we want to simulate. First we allow *potentials* φ acting on finite subsets of Λ by

(1) $\varphi(X) = \varphi(X + a)$ so that the potential is invariant under translation
(2) $\varphi(\phi) = 0$ where ϕ is the empty set
(3) $\varphi(X) = 0$ for all X such that $diam(X) > R$

The third point says that there is a range R such that the potential acting on a subset bigger than R takes the value zero. The sets for which φ takes non-zero values are called cliques. This restriction can be weakened and is weakened in many of the references given in this section. Then *an interaction* for φ is defined as

$$\Psi = \sum_{X \subset \Lambda} \varphi(X) \qquad (1)$$

The *statistics* that we calculate and compare to the given statistics are just a normalised version of this calculated on a realisation $I(x)$

$$\eta = \frac{1}{|\Lambda|} \sum_{X \subset \Lambda} \varphi(X) \qquad (2)$$

A couple of examples of interactions are

a) Consider Θ to be the set of pairs of points x_1 and x_2 separated by a vector h. In a two facies case, let $I(x)$ be the indicator function for one of the facies. Define a potential by $\varphi_0(X) = \tfrac{1}{2}$ if $X \in \Theta$ and $I(x_i) \neq I(x_i + h)$, $\varphi_0(X) = 0$ for any other type of set X. The associated statistic is then just the variogram at lag h.

$$\eta_0 = \frac{1}{|\Lambda|} \sum_{\substack{x_i, x_i+h \\ I(x_i) \neq I(x_i+h)}} \tfrac{1}{2} = \frac{1}{2|\Lambda|} \sum_{x_i} (I(x_i) - I(x+h))^2$$

b) Suppose $I(x)$ is a facies model and a permeability value is assigned for each facies. Let X be a set of radius R and let k be the effective permeability calculated in X. Let the histogram of effective permeability for sets of size X be split into n bins. Define potential $\varphi_i(X)=1$ if k falls in bin i, and $\varphi_i(X)=0$ otherwise. The set of interactions $\{\Psi_i\}$ simply records the observed histogram of effective permeability for the realisation I.

Our objective will be to be able to do simulations using interactions like $\Psi_0,..., \Psi_n$ by trying to condition to prescribed values of the statistics (where possible). As we can see, the types of interaction and hence statistics that we can work with are very general. From now on we will assume that we are working with vectors of interactions and when we refer to an interaction we will generally mean a vector $\Psi = (\Psi_0, ..., \Psi_n)$. We now look at three formulations of this problem which will turn out to be equivalent.

Microcanonical ensemble: Assume that an interaction Ψ is given. This gives rise to statistics η when calculated on a realisation. Let us think of Ω as being the set of all possible realisations on the grid Λ. So if there are n possible facies, then $|\Omega| = n^{|\Lambda|}$ (Actually, we think of the grid as infinite in size in the microcanonical perspective). Let $\Omega(\eta)$ be the subset of Ω taking the statistics η. So this subset contains all the possible realisations with calculated statistics equal to η. Some values of η give large subsets, while others might give small or even empty subsets. The latter correspond to cases where the interaction is not capable of producing the desired statistics. As a simple example, suppose we tried to match $\gamma(h) = -1$ in the first example above. Since the variogram can only be positive, the set $\Omega(-1)$ must be empty. More generally, use of many components in an interaction could lead to contradictions between the components and so to empty subsets. The ideal simulation method for a given set of statistics η would be to sample uniformly from the set $\Omega(\eta)$. Each realisation from $\Omega(\eta)$ takes the same probability but the probability may change for different outcomes n. As such Ω splits into equivalence classes based on the statistics η.

The Gibbs Distribution: In most cases we need to work on a finite grid. The statistics that we calculate on a finite realisation $\eta_\Lambda(I)$ will therefore have some statistical fluctuations. We should not expect to match our input statistics η exactly. The next best thing would be to match them in expectation. The probability distribution that we will choose to work with, call it p, should satisfy

$$E_p[\eta_\Lambda(I)] = \eta \quad (3)$$

Defining $s(q) = -\int q(I) \log q(I) dI$ to be the entropy of a distribution q, we will *choose* the distribution p with maximum entropy, that is, $p = \underset{q}{\operatorname{argmax}}(s(q))$ subject to the constraint (3). Then p follows a Gibbs distribution (see Jaynes 1957, Zhu et al. 1997)

$$p(I; \lambda, \Psi, \Lambda) = \frac{1}{Z_\Lambda(\lambda)} \exp(-\langle \lambda, \eta_\Lambda(I) \rangle) \quad (4)$$

where λ are the Lagrange multipliers and $Z_\Lambda(\lambda)$ is a normalisation to ensure the probabilities add to 1. The angle brackets signify scalar product. The λ are found by satisfying equation (3). This is usually a complicated and iterative process.

Markov Random Field: Suppose that we have a neighbourhood system $\{N_i\}_{i \in \Lambda}$, then $I(x)$ is a Markov Random Field (MRF) for the neighbourhood system if

$$P[I_i | \hat{I}_i] = P[I_i | N_i] \quad (5)$$

where $\hat{I}_i$ means all points on the grid except point i. That is to say, the conditional distribution at point i given all other points only actually depends on the values in the neighbourhood of i.

2.2 EQUIVALENCE OF THE VIEWPOINTS

These three viewpoints give an equivalent mathematical formulation for the problem of trying to produce realisations matching statistics η. We don't attempt to demonstrate the results here as the proofs are long and fairly technical. However we will try to convey the flavour of some of the concepts involved. The question of existence of a random function for an interaction Ψ is treated rigorously in several texts by showing the existence of a *Gibbs measure* (e.g. Ruelle 1968, Georgii 1988). This Gibbs measure has the property that on any finite subgrid, given the values on the exterior of the subgrid, it reduces to the Gibbs distribution on the subgrid. The equivalence of the first and second perspective is called the equivalence of ensembles. The equivalence of the second and third is the Hammersley-Clifford theorem (see e.g. Moussouris 1974). One statement of existence of the random function is given in terms of a new definition of entropy. This definition appears different to the one previously given which was defined as an integral over all realisations using the probability measure of the random function. It is defined below in terms of the size of the set $\Omega(\eta)$ honouring the statistics. Again we take a lax approach to rigor in the following discussion. If $\Omega_\Lambda(\eta)$ is the set of realisations on the finite grid Λ taking the statistics η, we define

$$s(\eta;\Psi) = \lim_{|\Lambda|\to\infty} \frac{\log|\Omega_\Lambda(\eta)|}{|\Lambda|} \qquad (6)$$

to be the entropy function for the interaction Ψ. The dependence on both Ψ and η might appear confusing, but by the latter we mean the type of calculation that is performed to the realisation while the former are the resulting statistics. We will usually drop the explicit reference to the interaction and just refer to $s(\eta)$. The existence theorem states that for the interactions that we have defined, this limit exists and takes values greater than $-\infty$ for at least some values of η (Ellis 1985). Moreover s is a concave function of η. The definition of s shows us that the number of possible realisations taking statistic η acts like $|\Omega(\eta)| \sim e^{|\Lambda|s(\eta)}$ for large grids. The two definitions of entropy turn out to be equivalent (e.g. Ellis 1985). In fact the entropy $s(\eta)$ is the maximum value that $s(p)$ attains when comparing over all distributions p that satisfy equation 3 (as $\Lambda \to \infty$)

It is important to note how the numbers of realisations depend on the entropy. Suppose we have two possible statistics η_1 and η_2 for the same interaction (concretely, if we think of our calculation of the variogram at lag h giving us two different numbers on different realisations). Suppose that $s(\eta_1) > s(\eta_2)$. Then the proportion of realisations that take value η_1 is $\dfrac{e^{|\Lambda|s(\eta_1)}}{e^{|\Lambda|s(\eta_1)} + e^{|\Lambda|s(\eta_2)}} = \dfrac{1}{1+e^{|\Lambda|(s(\eta_2)-s(\eta_1))}} \to 1$. So the number of possible realisations taking higher entropy values grows exponentially higher than those taking lower entropy values.

However this does not yet account for the role of the interaction which supplies a probability distribution on Ω. To see this we use another result (Ruelle, 1968). Starting from the Gibbs distribution perspective (equation 4), it can be shown that a convex function $\rho(\lambda)$ defined below exists. It is called the density function in statistical mechanics

$$\rho(\lambda) = \lim_{|\Lambda| \to \infty} \frac{\log Z_\Lambda(\lambda)}{|\Lambda|} \qquad (7)$$

We can now calculate the probability of a statistic set $\Omega(\eta)$. Notice that the Gibbs distribution assigns the same probability to all realisations taking the same statistics η. Thus we have $P[\Omega_\Lambda(\eta)] = \frac{|\Omega_\Lambda|}{Z_\Lambda(\lambda)} \exp(-\langle \lambda, \eta \rangle)$ By taking logs and passing to the limit and using the density definition we get *the probability rate function, r_λ* (Wu et al. 1999)

$$r_\lambda(\eta) \stackrel{def}{=} \lim_{|\Lambda| \to \infty} \frac{\log P[\Omega_\Lambda(\eta)]}{|\Lambda|} = s(\eta) - \langle \lambda, \eta \rangle - \rho(\lambda) \qquad (8)$$

We can see that the probability of taking statistic η is asymptotically $P[\Omega(\eta)] \sim e^{|\Lambda| r_\lambda(\eta)}$ so that for large grids the Gibbs distribution samples uniformly from the set $\Omega(\eta_{max})$, where $\eta_{max} = \max_\eta (s(\eta) - \langle \lambda, \eta \rangle - \rho(\lambda))$. We know that the entropy function is concave. If, furthermore, it is strictly concave then this implies that only one set of statistics attains the maximum of the probability rate function (Lanford, 1973 gives a sufficient criteria due to Dobrushin based on having sufficiently low values of λ). In other words, the probability concentrates on one particular set of statistics (there is no *phase transition* in the language of statistical mechanics). It turns out that s and ρ are convex conjugate pairs (Lanford, 1973) and using this it is easy to show that for low values of λ (high 'temperature') the probability concentrates on the class which has the highest entropy consistent with the interaction as measured by the entropy function s.

2.3 TWO CONSEQUENCES AND A SIMPLE EXAMPLE

Example: This is a simple example due to Lanford, 1973 in the context of a digression on sums of independent variables. It was instrumental in the reinvigoration of work on the theory of large deviations. Consider a random function that assigns 0 or 1 with equal probability to each point on the grid independently of the value at other points and a potential function that assigns a value 1 to single points, 0 to anything else. Then the statistic of interest is $\eta = 1/|\Lambda| \sum I(x)$, the mean value of the realisation. By applying the binomial theorem and Stirling's formula and taking limits, it can be shown that the entropy is

$$s(\eta) = \begin{cases} -\eta \log \eta - (1-\eta)\log(1-\eta) - \log 2 & 0 \leq \eta \leq 1 \\ -\infty & \text{otherwise} \end{cases}$$

This takes its maximum value, 0, at $\eta = 0.5$ as expected. So 'virtually all' realisations (elements of Ω) have a mean value of 0.5. Other mean values are possible (between 0 and 1), but in the case of iid random variables they will almost never occur. Since s takes the value $-\infty$ outside $[0,1]$ we get the (obvious) result that the mean cannot take values outside this interval.

Next we look at two consequences:

1) Calculations made on realisations from the same equivalence class $\Omega(\eta)$ give the same results
2) A quick look at 'Simulated Anncaling'.

In the petroleum industry it is quite common to build a reservoir model based on some input parameters and test the results or make predictions based on calculations that were not explicitly controlled by the input set. For example we might build a model based on variogram information and validate the model by comparing upscaled permeability values with the observed histogram of upscaled permeability (found by interpreting well test for example). For our purposes we will call such a calculation an *observable o* and demand that the observable satisfies the same constraints as an interaction (e.g. effective permeability calculation over finite regions is an observable as in the earlier example). We know that the realisations of our model will generally sample from $\Omega(\eta_{max})$ where the statistics maximise the probability rate function. We ask how this set decomposes according to the possible values that the observable o might take. Well, for each value of o there is a subset of $\Omega(\eta_{max})$, call it $\Omega(\eta_{max}, o)$. Consider the entropy function for the extended interaction (η_{max}, o). It must attain it's maximum for some value of o_{max}. With no phase transition this value is unique. By the exponential growth argument used earlier, the volume $\Omega(\eta_{max}, o_{max})$ is far larger than that for other values of o, so $\Omega(\eta_{max})$ is dominated by $\Omega(\eta_{max}, o_{max})$ and $s(\eta_{max}) = s(\eta_{max}, o_{max})$. In other words for any interaction and an arbitrary observable the realisations will produce the same statistics for the observable with very high probability. These *typical* realisations have maximum entropy with respect to the extended interaction. Other *atypical* elements of $\Omega(\eta_{max})$ have lower values of the entropy of the extended interaction ('by chance' they have some extra information about o) and are comparatively rare so are unlikely to be sampled. So, if two realisations do not have the same statistics on some o then it is unlikely that they are from the same equivalence class.

As mentioned before, the Gibbs distribution, given by (4), offers a rigorous method of sampling from a distribution taking a set of statistics. It does so by first finding a set of weights λ for the interaction satisfying the constraints (3). Hence the statistics are satisfied on average. This can be an involved and computer intensive calculation. We now compare this to an application of simulated annealing that has been made regularly in the geostatistics literature, e.g. Deutsch and Journel, 1998. In this method a new Gibbs distribution is proposed (for each value of T),

$$p(v) = \frac{1}{Z} e^{-\frac{1}{T}\|v-\eta\|} \qquad (9)$$

Here $\|.\|$ is some distance measurement from v to η. For a fixed value of T, an MCMC technique like the Metropolis algorithm can sample from this distribution. As T→0 the distribution becomes concentrated on those realisations taking the value η, in other words on the set $\Omega_\Lambda(\eta)$. Simulated annealing reduces T slowly to ensure that the sampling is uniform on $\Omega_\Lambda(\eta)$. For large grids, where the statistical fluctuations on the statistics η are small, we can see that this can be viewed as a direct attempt to sample from the microcanonical ensemble. (Technically, we would have to prove that a limit exists as $\Lambda \to \infty$. This is not done here, but it appears to resemble typical statistical mechanical proofs.) A few comments can be made.

1) This will give essentially the same result as sampling from the Gibbs distribution for large grids
2) It can be viewed as an approximate sampling for large grids from the Gibbs measure whose local conditional distributions are the Gibbs distributions (4). In other words, it makes an approximate sample from a regular, well defined random function.
3) For a particular set of data η, which we consider to be associated to an interaction Ψ, there is no guarantee that $\Omega_A(\eta)$ is nonempty (we have a theorem which says that for any linearly independent set Ψ, there are *some* statistics α for which $\Omega(\alpha)$ is nonempty). If empty, $s(\eta)$ will take the value $-\infty$ in equation (6) and we say that the statistics are incompatible with Ψ. In this case $\Omega_A(\alpha) \to 0$ faster than exponentially. This will manifest itself as an inability to match the input statistics to a reasonable degree of accuracy, even for relatively small images. So if the method is producing realisations matching the statistics, the grid is large enough and the annealing schedule is slow, then $\Omega(\eta)$ is non empty and this method should give reasonable results.

3 Sequential Simulation

3.1 GENERAL METHOD

A more detailed analysis of the algorithms in this section is currently in preprint form. The sequential simulation algorithm as proposed in the geostatistics literature, e.g. Deutsch and Journel, 1998, works by assuming known the conditional distribution of a point, x, given any number of neighbours that have been observed within a fixed *search neighbourhood* n_x of the point. The available neighbours have either been previously simulated or are initial conditioning data. We will call these available neighbours the *parents* of x and label the parents as ∂_x and refer to the conditional distribution as $f_d(x|\partial_x)$. The subscript d alludes to the fact that we are *driving* the simulation by claiming knowledge of some conditional distributions. These distributions may come from an analytic model such as the Gaussian, or they may be empirical distributions coming from some training data. We do not assume, and it is not generally the case, that the simulated model will reproduce all of these statistics. The sequential method then does a simulation on a finite grid by following a path p through the points on the grid. The result follows the distribution (the resultant model does not have the subscript d however it is labelled by p to indicate dependence on the path)

$$f^p(x_1,\ldots,x_N) = f_d(x_1|\partial_{x_1})f_d(x_2|\partial_{x_2})\ldots f_d(x_N|\partial_{x_N}) = \prod_{i=1}^{N} f_d(x_i|\partial_{x_i}) \quad (10)$$

Of course, a restriction on this model is that each parent set ∂_x must be known before simulation of x. It is straightforward to show that the conditional distribution is of the form in equation (11) and so depends on the parents of x_i, $pa(x_i)$, the children of x_i, $ch(x_i)$, and the other parents of the children of x_i, $pa(ch(x_i))$

$$f^p(x_i|\hat{x}_i) \propto \prod_{\{j:i\in\partial_j \text{ or } j=i\}} f_d(x_j|\partial_{x_j}) \quad (11)$$

This set of dependency points is always contained within the *dependency neighbourhood* of x_i, defined as $d(x_i) = pa(x_i) \cup ch(x_i) \cup pa(ch(x_i))$. Thus it is always possible to write

$$f^p(x_i | \hat{x}_i) = f^p(x_i | d(x_i)) \qquad (12)$$

So that we have a Markov Random Field. The dependency neighbourhood changes for each x_i, so the result is not a stationary MRF. There are two straightforward ways to reintroduce stationarity; using a raster scan path which is the topic of the next section or using a random path as follows. Let P be a random path, that is, a random variable which 'picks' from the set $\wp$ of permutations of $\{1,...,N\}$ uniformly. We now consider the simulation strategy of first picking a path at random and then doing sequential simulation. The distribution f^p is a randomisation of that given in (10)

$$f^P(x_1,...,x_n) = \frac{1}{n!} \sum_{p \in \wp} f^p(x_1,...,x_n) \qquad (13)$$

The distribution is now the same for all internal points on the grid. The model still does not reproduce all the input statistics of the driving distribution.

3.2 THE UNILATERAL MODEL

This is a simple model (Picard, 1980) which has the advantage of being both a sequential algorithm and a readily parametrizable MRF at the same time. It reproduces its input statistics and can be simulated in one pass. This gives the advantage of good results (less 'speckle') and a fast algorithm. However, it does depend on initial conditions as we shall see. Let us consider a 2d example to simplify notation. For this model the parent set of any point $x=(x_1,x_2)$ is chosen to be some subset of $\{(a,b); b < x_2 \text{ or } (b = x_2 \text{ and } a < x_1)\}$. We assume that the same subset is always chosen so that the parent set $\partial_x = \partial_{x+h}$ for any h, a translation on the grid. A sequential simulation is made with these parent sets by starting at the top left and finishing in the bottom right. Equation (10) still holds to define the decomposition of the probability distribution. As before, the conditional distribution depends on the parents of x_i, the children of x_i and the other parents of the children of x_i. This time however the dependency is the same for all points and the model may be represented as a stationary MRF (see the example below). Simulation of a unilateral model may be made by choosing some initial values along 'the top' of the grid and then simulating in a raster scan order. The fact that we have had to choose some initial values means that the method should be run for a while before it starts sampling correctly from the conditional distribution. An exact sampling, such as that proposed by Propp and Wilson, 1996, appears to be possible but in practice the 'run in' seems to be very short.

An example shows how we get from a representation of the type given by (10) to the Gibbs equivalent.

Example: Consider a 2 facies model where each point x has three parents whose co-ordinates relative to the point are (-1,-1), (-1,0) and (0,-1). Figure 2 labels these points as well as the children and parents of children of x. Applying (11) we get (drop the d)

$$f(x|\hat{x}) \propto f(x|1,2,4)f(6|2,3,x)f(9|4,x,7)f(8|x,6,9) \qquad (14)$$

Consider one of the 4 terms on the right, call it $f(y|a,b,c)$, the distribution of y given 3 conditioning points. Apart from being a conditional distribution this is arbitrary. We now write a formula for its most general form. Since there are 2 facies, the number of configurations of the conditioning points is $2^3=8$. For each configuration we have to specify the probability that y takes the value 1 (p(0) = 1-p(1) gives the other value). Let us assume that $f(y|a,b,c)>0$ for all combinations of variables (this is not strictly necessary for unilateral process but is for general MRF). Then we can write f in the form $f(y|a,b,c) = \exp(y\Psi(a,b,c))$ where Ψ is an arbitrary function (because y can only take the values 0 or 1 and we only have to concern ourselves with 1 – note this number is not between 0 and 1, but that will be fixed by the normalisation). The most general function of 3 binary variables can be rewritten as

Figure2. Labels for neighbours of x; red = parents; blue = children; yellow = other parents of children

$$\Psi(a,b,c) = \alpha + \beta a + \gamma b + \delta c + \varepsilon ab + \eta ac + \kappa bc + \upsilon abc \qquad (15)$$

by identification of terms, for example $\beta a = \Psi(a,0,0) - \Psi(0,0,0)$. Substituting these into (14) and throwing out terms that do not contain an x gives the final MRF result.

$$f(x|\hat{x}) \propto e^{x\{\alpha+\beta(1+9)+\gamma(2+8)+\delta(4+6)+\varepsilon(12+48+69)+\eta(14+26+89)+\kappa(24+36+78)+\upsilon(124+236+478+689)\}} \qquad (16)$$

This example shows that the unilateral model retains some anisotropy, for example there is no term with (3+7). However, by using larger neighbourhoods any desired terms can be included into the model and we can adequately model complex behaviour. Figure 3 shows a training image and two unilateral simulations of a channel system. The image on the right used a neighbourhood consisting of 60 points. This would appear to necessitate learning 2^{60} configurations for the conditional distribution. While this is true in principle, entropic reasoning tends to suggest that the number of configurations that actually occur is a very small fraction of this. This is not to say that the inference issue is easy. In fact it is the major problem facing techniques trying to use higher order statistics, but in this case we do get a reasonable result (the speckle on the right image is the onset of problems owing to neighbourhood size). Wei and Levoy, 1999 use a unilateral approach with image pyramids to reduce dimensionality.

Conditioning to data introduces some nonstationarity for all sequential simulation methods (conditioning data have no parents – and so have a different conditional distribution to other points). The unilateral method can be made to condition rigorously by using a MCMC on its equivalent MRF formulation. This reduces the efficiency of the unilateral method to that of a typical MRF. The unilateral method offers the possibility of starting with an approximate technique (for example, the algorithm looks to see if any conditioning data are children of the current point being simulated. If so, they are switched and used as parents of the current point). This approximate simulation may be improved by using several iterations of an MCMC algorithm if needs be. Figure 4 shows a conditional simulation using only the approximation technique and no MCMC.

  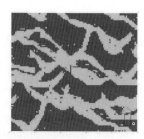

Figure 3. On the left a slice through a boolean channel model. The center and right images are unilateral models with statistics learnt from the left image. The centre model uses a small neighborhood (24 points) while the right one uses 60 points.

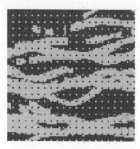 

Figure 4. On the left, a slice through a Boolean model and 400 'observed wells' (black and white dots) sampled from the model. The right hand model is a conditional unilateral simulation using the approximation method.

References

Deutsch, C.V. and Journel, A.G., *GSLIB: Geostatistical Software Library and User's Guide*, 2nd Edition, Oxford University Press, 1998.
Ellis, R.S., *Entropy, large deviations and statistical mechanics*. Springer Verlag. New York 1985
Georgii, H.-O., *Gibbs measures and phase transitions*. Walter de Gruyter & Co,. Berlin-New York, 1988
Jaynes, E.T., Information Theory and Statistics, *The Physical Review*, vol.106, no. 4, 1957, pp620-630
Lanford, O.E., Entropy and equilibrium states in classical statistical mechanics *in Statistical Mechanics and Mathematical Problems*, Springer-Verlag, Berlin. Lecture Notes in Physics vol. 20., 1973 pp1-113
Moussouris, J., Gibbs and Markov random systems with Constraints, *Journal of Statistical Physics*, vol. 10, no. 1, 1974, pp11-33.
Pickard, D.K., Unilateral Markov fields, *Adv. Appl. Probab.*, vol. 12, 1980, pp. 655–671.
Propp, J.G and Wilson D.B., Exact sampling with coupled Markov chains and applications to Statistical Mechanics, *Random Structures and Algorithms*, vol. 9, 1996 pp223-252
Ruelle, D. *Statistical Mechanics: Rigorous results*. W.A. Benjamin Inc. New York-Amsterdam, 1969
Strebelle, S., Conditional Simulation of Complex Geological structures using Multiple-Point Geostatistics, *Math. Geol.*, vol. 34, no. 1, 2002
Tjelmeland H. and Besag J., Markov random fields with higher order interactions, *Scandinavian Journal of Statistics*, vol. 25, 1998, pp 415-433
Wei, L.Y. and Levoy, M., Fast Texture Synthesis using Tree-structured Vector Quantization in *Proceedings of SIG-GRAPH 2000* (July 2000), pp 479–488.
Wu, Y.N, Zhu, S.C and Liu, X., Equivalence of Julesz ensembles and FRAME models, *Int' Journal of Computer Vision*, vol. 38, no. 3, 2000, pp. 245-261,
Zhu, S.C., Wu, Y., Mumford, D., Filters, Random Fields and Maximum Entropy (FRAME), *Int'l journal of Computer Vision*, vol. 27, no. 2, 1998, pp1-20

BEYOND COVARIANCE:
THE ADVENT OF MULTIPLE-POINT GEOSTATISTICS

ANDRE G. JOURNEL
Department of Geological and Environmental Sciences, Stanford CA 94305.

Abstract. In any estimation or simulation endeavor there are two types of information, the conditioning data typically numerical, most often location-specific, and the structural model which relates deterministically or stochastically the conditioning data to the unknown(s). Adding conditioning data is valuable only inasmuch as the structural model that links them to the unknowns is accurate and reflects data redundancy.

Traditional (cross) covariances/variograms, being only 2-point statistics, are limited in the amount of prior structural information they can carry, in addition in 3D they are notoriously difficult to infer and model. In many applications, particularly those related to mapping of categorical variables, facies or rock types distributions, critical structural information can be obtained from training images drawn from prior expertise, outcrops or similar deposits. From such training images complex statistics involving jointly values at multiple locations can be extracted. Using these statistics may be preferable to letting the estimation algorithm impose its arbitrary and likely inappropriate version of the same statistics.

The recently introduced concept of multiple-point (mp) geostatistics allows a fresh methodological look at the general problem of numerical modeling under data conditioning, where the concept of "data" is now open to include structural information much beyond variogram models. Such structural data are often soft and represent a major source of uncertainty, which must and can be appraised through the consideration of alternative training images all consistent with the numerical data available. Training images return more of the modeling responsibility to the geologist, more generally to the physicist who can add interpretation and valuable expertise to the numerical data.

1 From kriging to simulation

Conditional simulation was introduced in the 1970's, interestingly enough not to address uncertainty but as a remedy for the smoothing effect of kriging. It was understood that for many applications reproduction of the patterns of spatial variability, as reflected by the data then modeled through a variogram, was more important than local accuracy. Most of the ensuing developments in the next three decades related to faster and more flexible simulation algorithms, most notably sequential simulation algorithms and object-based (Boolean) algorithms and their conditioning to diverse data, both hard and soft.

The practice of simulations quickly revealed the practical limitations of the otherwise extraordinarily congenial Gaussian random function (RF) underlying most simulation algorithms. A single covariance model or matrix was not enough to characterize even the simplest curvilinear structure, any structure that did not display the maximum entropy characteristic of the Gaussian model, a property extraneous to the very concept of a geological structure. Any geostatistician could spot visually and immediately a map generated from Gaussian-related model, any geologist would judge it as relevant only for homogeneously heterogeneous spatial distributions within well-defined homogeneous zones. The concept of indicator RFs extended a little the practicality range of simulation: different categories or classes of a continuous variable could have different variograms. However, in addition to the burden of multiple variograms inference, indicator geostatistics suffered from embarrassing order relation problems, yet did not deliver the flexibility required.

Then came object-based simulations algorithms whereby parametric objects mimicking geological structures were dropped onto the simulation field then moved and morphed iteratively to honor the data. But parametric objects do not offer full flexibility: not all natural shapes can be approximated by simple parametric shapes, and strict conditioning to dense and diverse data was difficult. A new simulation paradigm was needed.

2 From variogram to multiple-point statistics

One reason for the staying power of variogram/covariance-based random function models could be the sense of objectivity one felt at estimating the structural model from actual data. Unfortunately, that sense is more illusion than reality. First, except for data-rich fields, variograms are notoriously difficult to infer then model, to a point that, in petroleum applications for example where hard data are scarce, that task is often not even attempted: variograms are synthesized from expertise, distant outcrops or loosely related ancillary data such as provided by seismic surveys. Second, what really matters for simulation is the random function model adopted, more precisely its multivariate or multiple-point (mp) distribution, not its variogram which is but a two-point statistics; one notable exception is (precisely!) the Gaussian model. Different RF models sharing the same variogram model and honoring the same data values at the same locations could yield drastically different simulated realizations, see Figure 1. The variogram has little structural resolution, cannot distinguish vastly different patterns of heterogeneity. Consequently the variogram is also an incomplete measure of uncertainty; the major source of uncertainty does not lie in fluctuations of various conditional realizations sharing the same variogram model but in the choice of the generating RF model much beyond its variogram, see Caumon and Journel (2004).

But the consideration of mp statistics raises the issue of inference. If variograms are already difficult to infer from actual data, there is no hope to infer even an elementary 3-point statistics, let alone multiple-point statistics. Under stationarity, that inference would require availability of multiple replicates of triplets of data sharing the same geometric configuration, that is the same pair of separation vectors ($\mathbf{h}_1$, $\mathbf{h}_2$), as opposed to a variogram requiring replicates of only doublets of data sharing the same single separation vector $\mathbf{h}$. One had to overcome the illusion of objectivity provided by direct inference of statistics and accept that multiple statistics could be inferred from training images (TIs), which in many practices was already done for the variogram. Inferring a variogram from a noisy experimental variogram cloud then accepting blindly the higher order statistics implicit to, say, a Gaussian RF model is no more objective than inferring the same high order statistics from a visually explicit training image. A training image can be refuted by an expert geologist; the same cannot be said about a geologically non-significant variogram model.

3 From mp statistics to direct probability inference

Introduction of the concept of mp statistics and accepting that they could be inferred from training images led to the very demise of these statistics. In the first instance, why do we need a variogram? The variogram model is used to build the various conditional probability distributions from which simulated values are drawn. If 3-point, 4-point, mp statistics were available, one could derive from them better conditional probabilities, better in the sense that the resulting simulated values would reflect these higher order statistics in addition to the variogram. But as to infer these high order statistics from a training image why not infer from the same training image and directly the required conditional probabilities? Identifying conditional probabilities to conditional proportions read directly from the training image would shortcut completely the step of moments inference and modelling and that awkward step of kriging which relates the variogram model to the conditional probability. In practice, one would scan the training image for replicates of the multiple-point conditioning data event; these replicates would provide a distribution of the corresponding central training values; that distribution is then taken as the conditional probability. It can be shown that if a training image is considered a representation of an ergodic random function, its training proportions identify exactly the conditional probabilities that would be calculated from the experimental moments lifted from the same training image. If the conditional probabilities are available directly, why take the indirect and painful route of inferring all the relevant high order moments to reconstitute exactly the same probabilities through some kriging?

This remarkable leap of thought was due to Srivastava (Guardiano and Srivastava, 1992). Srivastava's original implementation required, however, to scan the training image repetitively for each new unsampled location, an overwhelming cpu task if large fields with 10^6 to 10^8 nodes are to be simulated. Then came faster desktop computers with larger RAM and the contribution of Strebelle (2000, 2002). Strebelle suggested to scan the training image only once with a specific data template (size and geometry), record in RAM all training data events together with their central training values. The search tree data structure used for that record allows a fast retrieval of all required conditional probabilities. Strebelle and large RAM availability made the new paradigm of Srivastava practical.

Figure 1 gives three very different spatial distributions, yet sharing the same histogram (here a proportion) and, most significantly, approximately the same variogram. Thus any solely variogram-based simulation algorithm would fail to resolve the structural difference between these training images. Figure 2 gives three conditional simulations using Strebelle's snesim code, each drawn from a different training image (as given in Figure 1), but conditioned to the same 30 data values and the same global proportion. Each mp-based conditional simulation reproduces fairly well its training image and, most significantly, the common variogram model even though no such model was ever input into the mp simulation code. The mp simulation algorithm also never used any kriging, although one could argue that the identification of the conditional probability to the training proportion amounts to solve a kriging system with a single (normal) equation, hence the name snesim (single normal equation-based simulation) given by Strebelle to his algorithm.

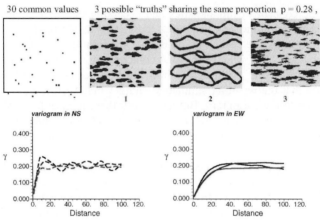

Figure 1. The need to go beyond the variogram. The variogram model cannot resolve the 3 possible "truths"

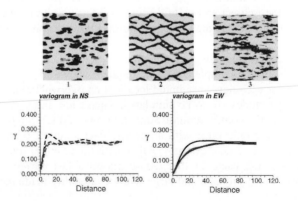

Figure 2. mp simulations using the training images of figure 1. They are conditioned to the same 30 samples and global proportion p=.28

Figure 3 gives a mp simulation generated from a training image which is a realization of a variogram-based sequential Gaussian algorithm. Again no variogram model was ever considered, yet the mp realization generated succeeded to reproduce the training image variogram, thus proving that mp simulation could replace traditional variogram-based algorithms as long as the relevant training image is available. Because no kriging system had to be built and solved, the mp generation of that Gaussian realization was faster cpu-wise than the generation of the training image using the traditional Gaussian sequential simulation algorithm. This remark points to the idea of a "universal" catalog of training images that would include all typical geological structures, a class of which being that of maximum entropy Gaussian-type structures. Once an appropriate training image is retrieved from that catalog, mp simulation with data conditioning could be lightning fast without the encumbrance of variogram modelling and kriging.

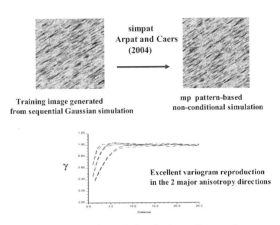

Figure 3. Pattern-based simulation of a continuous variable

4 From point simulation to pattern simulation

Change of paradigm breeds accelerated advances. Since the introduction of the snesim algorithm by Strebelle, within a period of a few years, many other mp simulation algorithms have been developed, some already in a state of beta testing, Arpat and Caers (2004), Zhang et al. (2004). As for lifting from a training image probability distributions of central point values conditioned by a mp data event, why not lift probabilities of whole multiple-point patterns conditioned to the same mp data event? A pattern, that is a mp event, would be drawn from a certain class of training patterns and patched onto the simulated field with due consideration to conditioning data.

Pattern simulation is based on the same concept: infer distributions directly from training images shortcutting all steps of elementary statistics inference and reconstruction of conditional probabilities. Stochastic simulation becomes an exercise of image construction drawing from a set of training puzzle pieces:

1) training patterns are first classified in bins according to some similarity/distance criterion. Each bin is characterized by an average pattern called prototype.
2) define a path (typically random) visiting all unsampled nodes of the field to be simulated.
3) at any location along that path, collect its mp conditioning data event, find the training prototype most similar to that mp data event, and draw a training pattern from that prototype bin.
4) patch that pattern onto the simulation field overriding any non hard data, a hard data being either original data or previously simulated values marked as hard. All values of that pattern becomes conditioning data for calculating distances but only the central part of that pattern is marked as hard data never to be changed.
5) move to the next non hard data location along the simulation path and repeat the pattern simulation procedure until the path is completed. A simulated conditional realization has been generated.

Figure 4 gives an example of such mp pattern simulation. At the top of the Figure is the training image, a binary image of dry soil cracks; 50 hard data are taken from that image. At the bottom of the figure are given 3 conditional simulations using the recently developed mp code filtersim (Zhang et al., 2004). These simulations honor exactly the 50 data at their locations, and reproduce reasonably the training image patterns although with shorter less connected cracks, a generic problem associated with the Nyquist frequency limitation.

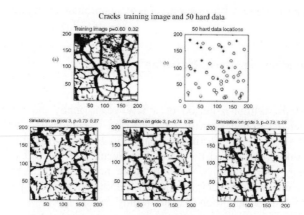

Figure 4. Three pattern-based conditional simulations

5 Issues and challenges

mp geostatistics started by recognizing the limitation of the variogram/covariance tool which are mere two-point statistics. Next the value of prior structural information lifted from training images was recognized. mp higher order statistics much beyond the variogram could now be inferred, but then also and directly the conditional probabilities needed for simulation. Direct lifting of these conditional probabilities voids the need for inference of any other statistics, whether a variogram or else, and also voids the need of reconstructing the same probabilities by indicator kriging or any other algorithm.

One could see a training image as the representation of an ergodic RF, or be bold enough to shed all probabilistic reference and take the training image for what it is truly, a repository of training patterns from which similar looking images anchored to prior data can be built. The puzzle reconstruction draws from a box with many bins, each bin filled with "similar-looking" patterns, the rule being that any pattern used is immediately replaced by a twin in its bin. From a given training image, for a given set of conditioning data, there are many alternative reconstructions possible: this is the within-model uncertainty. Many alternative training images could be considered, which amounts to retaining different puzzle boxes and results in different sets of simulated images: this is the model uncertainty, typically much larger than any within-model uncertainty, Caumon and Journel (2004).

There are definite issues in how to select a training image type then building that training image if it is not already available in a catalog, but this is no different from choosing a variogram model and the RF spatial law implicit to any specific simulation algorithm. There is no escaping from adopting a RF model the very moment one draws a map or contour a line, even if this is done by hand. A training image allows utilizing much richer prior structural information, as could be obtained from experience, outcrops or analog fields. It would be foolish to ignore such valuable information, information that a mere variogram cannot carry.

Then there is the issue of model and data consistency. If the training image is inconsistent with the local conditioning data, and if the mp simulation algorithm freezes the latter, discontinuities will be simulated next to these data. Such discontinuities may not be spotted if there are few conditioning data. Indeed just about any structural model could be anchored to sparse data, the model uncertainty is then overwhelming, as should be expected. Figure 5 gives a large (5×10^5 nodes) 3D pattern-based simulation of a channel reservoir conditioned to a large number of well data: the result is impressive because the training image reflects accurately the heterogeneity patterns of the actual reservoir, in particular the vertical stacking of the channels. Figure 6 repeats that exercise but now using a poorer training image which fails to display the channel vertical stacking: that inconsistency with the well data leads to a much poorer simulated realization: the realization honors the well data but with discontinuities.

A maximum entropy structural model a la Gaussian or a la high nugget effect would be much more permissive as for data inconsistency thanks, precisely, to its high entropy/disorder which crowds data discontinuities. Such tolerance is, however, dangerous because it masks problems.

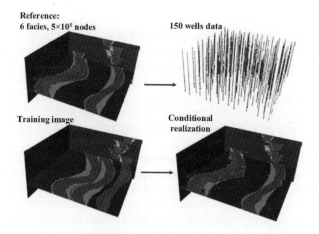

Figure 5. 3D pattern-based simulation with consistent data

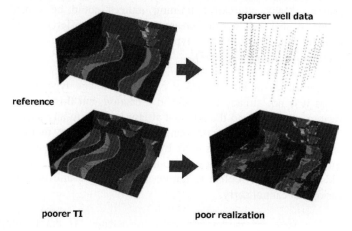

Figure 6. Data inconsistent with the training image

6 Conclusions

The inception of the multiple point concept has brought major changes into the way one sees spatial modelling. With the remarkable processing power now available on desktop computers, there is no more reason to ignore critical prior structural information brought by human expertise under the argument that it is subjective or "messy", not delivered concisely through a few statistics. Just like the original development of geostatistics in the 1960's was made possible by the availability of digital computing (variogram calculation and solving kriging systems), now is the time to graduate into massive processing of expert structural information which will tie the discipline of geostatistics increasingly more to image and computer sciences.

A model should never take precedence on data, or analytical convenience on completeness of the model. Real phenomena can rarely be summarised into a few simple statistics (e.g., a histogram and a variogram), any complexity that matters must be faced outright calling on all sources of information available to model it. It is counterproductive to ignore expert interpretative data because they are soft. A variogram model delivers little, yet its inference from typically too few data makes it as soft as a well documented training image that carries much more valuable information. As any other data, training images are uncertain and can be considered as random variables; a distribution of alternative training images might be considered, thus adding an essential component of uncertainty to the final prediction. The implicit high entropy training image built in the traditional mapping algorithms is no less uncertain than the visually explicit training image provided by an expert geologist.

References

Arpat, B. and Caers, J., *A multiple-scale pattern-based approach to sequential simulation*, In Proc. of the 2004 Banff Geostatistics Congress, Kluwer publ., (this volume), 2004

Caumon. G. and Journel, A.G., *Early uncertainty assessment: Application to a hydrocarbon reservoir development*, in ibid, 2004

Zhang, T., Switzer, P. and A. G. Journel, *Sequential conditional simulation by identification of training patterns*, in ibid, 2004

Guardiano, F. and Srivastava R. M., *Multivariate geostatistics: Beyond bivariate moments*. In Geostat Troia 1992, ed. Soares, Kluwer publisher, 1992

Strebelle, S., *Sequential simulation drawing structures from training images*, unpublished PhD Thesis, Stanford University, 2000

Strebelle, S., Conditional *simulation of complex geological structures using multiple-point statistics*, Mathematical Geology, Vol. 34, no. 1, 2002, p. 1-22.,

NON-STATIONARY MULTIPLE-POINT GEOSTATISTICAL MODELS

SEBASTIEN STREBELLE[1] and TUANFENG ZHANG[2]
[1] ChevronTexaco Energy Technology Company,
6001 Bollinger Canyon Road, San Ramon, CA 94583, USA
[2] Department of Geological and Environmental Sciences
Stanford University, Stanford, CA 94305, USA

Abstract. During the last few years, the use of multiple-point statistics simulation to model depositional facies has become increasingly popular in the oil industry. In contrast to conventional variogram-based techniques such as sequential indicator simulation, multiple-point geostatistics enables the generation of facies models that capture key depositional elements (e.g. curvilinear channels) characterized by unique and predictable shapes. In addition, multiple-point geostatistics is more intuitive because the complex mathematical expression of the variogram is replaced with an explicit three-dimensional training image that depicts the geometrical characteristics of the expected facies.

In multiple-point geostatistics, the stationarity assumption that underlies the inference of a variogram model from sparse sample data is extended to infer facies joint-correlation statistics from the training image. A consequence of this assumption is that patterns extracted from the training image can be reproduced in any region of the reservoir model where the training image is thought to be representative of the geological heterogeneity. Yet actual reservoirs are generally non-stationary: topographic constraints, sea-level cycles, or changes of sedimentation sources lead to spatial variations of facies deposition directions and facies geobody dimensions. Three-dimensional fields of location-dependent facies azimuth/dimensions representing those spatial variations are commonly estimated from well log and seismic data, or from geological interpretations based on analogs. This paper proposes a modification of the multiple-point statistics simulation program **snesim** to account for such non-stationary information.

In the original **snesim**, prior to the simulation, the multiple-point-point statistics inferred from the training image are stored in a dynamic data structure called a search tree. In the presence of a locally-varying azimuth field, the range of possible azimuths over the study field is first discretized into a small number of classes. Then, the training image is successively rotated by the average value of each azimuth class and a search tree is built for each resulting rotated training image. During the simulation, at each unsampled node, multiple-point statistics are retrieved from the search tree built for the class in which the local azimuth falls, enabling the local reproduction of patterns similar to those of the corresponding rotated training image. A similar process is proposed to account for a field of location-dependent facies geobody dimensions. The new modified **snesim** program is applied to the simulation of a fluvial reservoir with locally-variable channel orientations and widths.

1 Introduction

Multiple-point geostatistics has emerged recently as a practical approach to characterize and model facies at reservoir scale (Strebelle *et al*, 2002). The first step of this approach is the construction of a three-dimensional training image describing the facies thought to be present in the study area. The training image captures the geometrical characteristics of each facies, as well as the complex spatial relationships among multiple facies. The training image is a purely-conceptual geological model; it contains no absolute location information and in particular, is not conditioned to any actual field data. In reservoir modeling applications, non-conditional object-based modeling techniques appear to be well-suited to create such three-dimensional conceptual models. The second step of this approach consists of inferring from the training image statistics on the joint-correlation of facies at multiple locations, and using these statistics to reproduce patterns similar to those of the training image while honoring hard and soft conditioning data.

The theoretical framework of multiple-point geostatistics was developed as early as 1989 by Journel and Alabert and was revisited by Guardiano and Srivastava in 1993. The first practical implementation was proposed by Strebelle (2000), who introduced a dynamic data structure called a search tree, to efficiently store and retrieve all multiple-point statistics inferred from the training image. During the last few years, multiple-point geostatistics has been shown to overcome the major limitations of traditional facies modeling technologies:

- Multiple-point statistics (MPS) simulation enables improved modeling of curvilinear and large-scale continuous facies patterns, such as sinuous channels, relative to variogram-based techniques (Strebelle *et al*, 2002). In addition, the training image is much easier to analyze/discuss than a variogram model.
- In contrast to object-based modeling techniques (Holden *et al*, 1996; Viseur, 1997; Lia *et al*, 1998), MPS simulation is a very flexible data integration tool. In particular, MPS models honor all conditioning well data, i.e. reproduce at all well data locations the facies connectivity/geometry observed in the training image, with no limitation on the number of wells (Strebelle and Journel, 2001).

One important assumption underlying the inference of multiple-point statistics from the training image and their reproduction in the MPS model is the stationarity of the field under study: facies relative proportions, geometries, and associations are expected to be reasonably homogeneous over the field. Yet, most actual reservoirs are not stationary. Local topographic constraints such as the presence of a salt dome, sea level cycles, or changes of sedimentation sources, lead to significant spatial variations of facies deposition directions and facies geobody dimensions.

In this paper, we first review the implications of the stationarity assumption in multiple-point geostatistics. Then we propose modifying the MPS simulation program **snesim** (Strebelle, 2000) to reproduce pre-defined non-stationary information such as locally-varying facies azimuth and/or facies geobody dimension data.

2 Stationarity

Geostatistics relies on the concept of Random Function. The Random Function represents the statistical model of spatial variability of some property over some study field. In traditional geostatistics, the Random Function model is generally limited to some one-point and two-point statistics moments, namely a cumulative distribution function and a variogram model. In multiple-point geostatistics, the Random Function model consists of the multiple-point facies joint-correlation moments that can be inferred from the training image. The inference of statistics representing the Random Function model requires some repetitive sampling. For example, a porosity cumulative probability distribution is typically inferred from the histogram of porosity data collected from all well logs available over the study field. However, when pooling sample data together into a single histogram, the modeler makes an assumption of stationarity: all porosity sample values are assumed to originate from the same unique population, regardless of their location in the reservoir. Another stationarity decision is commonly taken whenever a variogram is computed by pooling information at similar lag distances together into a single scatter plot.

In multiple-point geostatistics, the stationarity assumption carries over to higher order statistics: multiple-point statistics moments are inferred from training patterns present in the training patterns regardless of the location of these patterns in the training image. As a consequence, non-stationary features of the training image cannot be preserved in MPS models. Figure 1 shows a clearly non-stationary training image wherein ellipses are South West-North East-oriented in the left half of the image, and North West-South East-oriented in the right half. The resulting model generated by the MPS simulation program **snesim** displays a mix of ellipses oriented in both directions over the whole field.

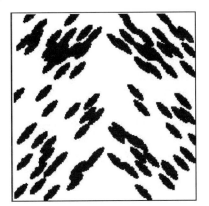 

Figure 1. Non-stationary training image (left), and resulting MPS model (right). The specific locations of the South West-North East and North West-South East-oriented ellipses in the training image are not preserved in the MPS model.

The non-stationary features of the training image are not captured in MPS models. Therefore, we propose using a stationary training image and applying rotation and affinity transforms to the training image to reproduce non-stationary features to MPS models. Prior to that, the implementation of the original MPS simulation program **snesim** is briefly recalled.

3 Multiple-point statistics simulation implementation

The MPS simulation program **snesim** proposed by Strebelle (2000) is a pixel-based direct sequential simulation algorithm: all simulation grid nodes are visited only once along a random path and simulated node values become conditioning data for cells visited later in the sequence. Let S be the categorical variable (depositional facies) to be simulated, and s_k, $k=1\ldots K$, the K different states (facies types) that the variable S can take. At each unsampled node $\mathbf{u}$, d_n denotes the data event consisting of the n conditioning data $S(\mathbf{u}_1)=s(\mathbf{u}_1)\ldots S(\mathbf{u}_n)=s(\mathbf{u}_n)$, closest to $\mathbf{u}$. The conditional probability distribution function (cpdf) at $\mathbf{u}$ is inferred by scanning the training image to find all training replicates of d_n (same geometric configuration and same data values as d_n), and identifying the conditional facies probabilities as the facies proportions obtained from the central values of the training d_n–replicates.

Instead of repeatedly scanning the whole training image at each unsampled node to search for training replicates of the local conditioning data event, Strebelle (2000) proposed storing ahead of time all conditional facies probabilities that can be inferred from the training image in a dynamic data structure called a search tree. More precisely, given a conditioning data search window W, which may be a search ellipsoid defined using GSLIB conventions (Deutsch and Journel, 1998), τ_N denotes the data template (geometric configuration) constituted by the N vectors $\{\mathbf{h}_\alpha, \alpha=1\ldots N\}$ corresponding to the N relative grid node locations included within W. Prior to the simulation, the training image is scanned with τ_N, and the numbers of occurrences of all training data events associated with τ_N are stored in the search tree. During the simulation, at each unsampled node $\mathbf{u}$, τ_N is used to identify the conditioning data located in the search neighborhood W centered on $\mathbf{u}$. d_n denoting the data event consisting of the n conditioning data found in W (original sample data or previously simulated values, $n \leq N$), the local probability distribution conditioned to d_n is retrieved directly from the above search tree; the training image need not be scanned anew.

Theoretically, a large data template τ_N should be used to capture the large-scale features of the training image. However, such large template would increase dramatically the memory used to build the search tree and the cpu-time needed to retrieve conditional probabilities from it. One practical solution to capture large-scale structures while keeping the size of the data template τ_N reasonably small ($N \leq 100$) is to use a multiple grid simulation approach (Strebelle, 2000). In **snesim**, this approach consists of simulating a series of G increasingly-finer grids, the g-th ($1 \leq g \leq G$) grid comprising each 2^{G-g}-th node of the final (finest) simulation grid. After the data template $\tau_N=\{\mathbf{h}_\alpha, \alpha=1\ldots N\}$ has been defined on the finest grid, its components $\mathbf{h}_\alpha$ are rescaled proportionally to the node spacing within the grid being simulated. Thus the rescaled data template $\tau_N^g=\{\mathbf{h}_\alpha^g=2^{G-g}.\mathbf{h}_\alpha, \alpha=1\ldots N\}$ is used to build the search tree and search for conditioning data when simulating the g-th grid.

In the next two sections, we show how rotation and affinity transformations can be applied to the data template τ_N prior to building the search tree, to integrate location-dependent azimuth and geobody size information into MPS models.

4 Integration of azimuth data

In this section, we first study the simple case in which the main direction of continuity of the facies geobodies is assumed to be constant over the field under study, but possibly different from the main direction of continuity of the training facies. Then we extend this technique to handle 2D or 3D fields of location-dependent azimuths. Only azimuths defined in the xy-plane are considered in this section because, in practice, dip is typically taken into account by the layering of the stratigraphic grid in which the facies model is built.

4.1 CONSTANT AZIMUTH

Consider the case in which the facies geobodies in the MPS model should have a constant principal direction of continuity, yet possibly different from that of the training image. Let θ be the difference in degrees counter-clockwise between those two directions.

Given a training image and a data template $\tau_N = \{h_\alpha, \alpha=1...N\}$, Zhang (2002) proposed modifying the **snesim** algorithm as follows. First the search tree is built from the training image using τ_N. Then, a new data template $\tau_N(\theta)$ is created from τ_N by the following method:

1. Rotate by θ each single component h_α of τ_N.
   In 2D, the coordinates $(x_\alpha(\theta), y_\alpha(\theta))$ of the rotated component $h_\alpha(\theta)$ are computed from the coordinates (x_α, y_α) of the original component h_α as:
   $x_\alpha(\theta) = x_\alpha \cos\theta + y_\alpha \sin\theta$
   $y_\alpha(\theta) = - x_\alpha \sin\theta + y_\alpha \cos\theta$
2. Relocate the rotated components $h_\alpha(\theta)$ to the nearest nodes of the simulation grid currently simulated.

During the simulation, at each unsampled node, the rotated data template $\tau_N(\theta)$ is used to search for nearby conditioning data, and the corresponding conditional probability distribution function (cpdf) is retrieved from the search tree, which was built using the original data template τ_N.

However, as described in the previous section, **snesim** uses a multiple-grid simulation approach that consists of simulating a series of increasingly-finer grids. Thus, at the early stage of the simulation, the rotated components $h_\alpha(\theta)$ are relocated to the closest nodes of some coarse grids, entailing drastic approximations regarding the actual locations of the conditioning data. Such approximations lead to the inaccurate estimation of facies probability distributions and the poor reproduction of training patterns. However, because the simulation grids used in **snesim** are regular Cartesian grids and the distance between nodes is the same along both x and y-directions, the components $h_\alpha(\theta)$ of the rotated data template match exactly existing grid nodes for $\theta=0$, 90, 180, or 270 degrees. This is a property that we will use in the next sub-section.

An alternative approach consists of keeping the original data template τ_N to search for conditioning data, but rotating that data template to build the search tree prior to the simulation. In this case, the rotated data template is built in a slightly different way than in Zhang's method:
1. Rotate by $-\theta$ each single component $\mathbf{h}_\alpha$ of the original data template τ_N.
   In 2D, the coordinates $(x_\alpha(-\theta), y_\alpha(-\theta))$ of the rotated component $\mathbf{h}_\alpha(-\theta)$ are computed from the coordinates (x_α, y_α) of the original component $\mathbf{h}_\alpha$ as:
   $x_\alpha(-\theta) = x_\alpha \cos\theta - y_\alpha \sin\theta$ and : $y_\alpha(-\theta) = x_\alpha \sin\theta + y_\alpha \cos\theta$
2. Relocate the rotated components $\mathbf{h}_\alpha(-\theta)$ to the nearest nodes of the training image grid. When using **snesim**, the training image is assumed to have the same node spacing as the (finest) simulation grid.

The resulting rotated data template $\tau_N(-\theta)$ is used to build the search tree from the training image. The exact same result can be obtained by rotating the training image by θ, then building the search tree from that rotated training image using the original data template τ_N. During the simulation, at each unsampled node, τ_N is used to search for nearby conditioning data, and the corresponding cpdf is retrieved from the above search tree.

The most critical advantage of that new technique over Zhang's original method is that the relocation of the rotated components $\mathbf{h}_\alpha(-\theta)$ to the training image grid entails only minor approximations of the actual locations of the conditioning data, thus resulting in a reasonably good reproduction of the training patterns. As an application, this modified **snesim** program was used to model a horizontal 2D section of a fluvial reservoir. The training image depicts the prior conceptual geometry of the sinuous sand channels expected to be present in the subsurface (Figure 2). The size of that image is 250*250=62,500 pixels, and the channel proportion is 27.7%. A non-conditional simulated realization was generated using the same direction of continuity as that of the training image (Figure 2), then two additional models were created using different arbitrary main directions of continuity: 20 and 50 degrees (Figure 3). All models reproduce equally well the patterns displayed in the training image.

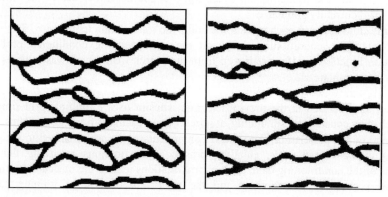

Figure 2. Training image used for the simulation of a 2D horizontal section of a fluvial reservoir (left), and reference MPS model (right).

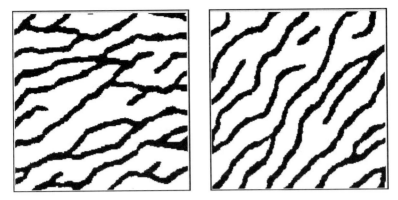

Figure 3. Fluvial reservoir MPS models obtained with a 20 degree counter-clockwise azimuth difference with the training image (left), and a 50 degree difference (right).

4.2 LOCATION-DEPENDENT AZIMUTHS

In reservoir facies modeling applications, it commonly is observed that the principal direction of continuity of the facies varies from one region of the reservoir to another. For example, topographic constraints, such as changes in the slope gradient, may lead to the formation of several sand fairways with different depositional directions. Data regarding such variations can be derived from different sources. In particular, local depositional directions can be obtained from geological interpretation (Harding *et al*, this meeting), or can be computed from seismic data (Strebelle *et al*, 2002).

Suppose that local azimuths can be estimated at each location **u** of the reservoir, and that $\theta(\mathbf{u})$ denotes the difference in degrees counter-clockwise between the azimuth value estimated at node **u** and the azimuth of the (stationary) training image. Given a data template τ_N, the method previously presented for a constant azimuth field can be extended to the location-dependent azimuth field $\theta(\mathbf{u})$ as follows:

1. Consider the range $[\theta_{min}, \theta_{max}]$ of all azimuth values estimated over the entire study field. Discretize that range into a small number L of classes, using regularly-spaced threshold values: $\theta_i = \theta_{min} + i*(\theta_{max} - \theta_{min})/L$, $i=0...L$.
2. Using the method described in the previous sub-section, compute for each class $[\theta_i, \theta_{i+1}]$ the search tree corresponding to the rotated data template $\tau_N(-\theta)$ where θ is the central value of the class: $\theta = (\theta_i + \theta_{i+1})/2$
3. During the simulation, at each node **u** to be simulated, use the original data template τ_N to search for nearby conditioning data, and retrieve the local cpdf from the search tree corresponding to the class of azimuth angles to which $\theta(\mathbf{u})$ belongs.

If the range $[\theta_{min}, \theta_{max}]$ of azimuth angles is greater than 90 degrees, Zhang's original technique can be used to decrease the range of the individual discretized classes $[\theta_i, \theta_{i+1}]$. For example, consider the simulation of node **u** where $\theta(\mathbf{u}) = \theta_{min} + 100°$. The rotated data template $\tau_N(90°)$ can be used to search for conditioning data (recall that 0, 90, 180, and 270 degrees are the only rotation angles for which Zhang's method requires no data relocation). Then the resulting cpdf can be retrieved from the search tree corresponding to the class of azimuth angles to which $(\theta_{min}+100°)-90°= \theta_{min}+10°$ belongs. Therefore, in any case, the maximum range of azimuth values to discretize is 90 degrees. The number L of azimuth classes should depend then on the uncertainty about the local azimuth values. With $L=5$ classes, the range of each class is 18 degrees. This is equivalent to estimating local azimuth values with an error of ± 9 degrees.

One limitation of the above technique may be the memory demand because one search tree per azimuth class needs to be built. However, one can consider one azimuth class after the other, i.e. build the search tree corresponding to a given azimuth class, simulate all grid nodes corresponding to that class, then delete that search tree prior to considering the next azimuth class. Building, then deleting search trees is a relatively fast process compared to the actual grid simulation process.

Figure 4 shows a 2D azimuth field and a resulting simulated realization using the fluvial reservoir training image of Figure 2. The reproduction of the training patterns is similar to that in the reference simulated realization of Figure 2. Note also that, although only five azimuth classes were used, the discretization of the range of possible azimuths did not create any artifact in the simulated realization.

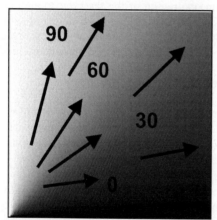 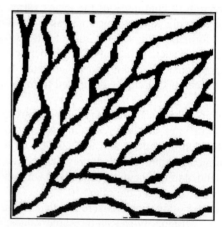

Figure 4. 2D location-dependent azimuth field (left), and resulting MPS model obtained using the fluvial reservoir training image of Figure 2 (right).

5 Integration of geobody dimensions data

Facies geobody dimensions that may depend, for example, on the distance to the sedimentation source, represent another traditional non-stationary feature of hydrocarbon reservoirs. A technique similar to that presented in the previous section to impose locally-varying azimuths is proposed to integrate geobody dimensions data, using some affinity transform of the data template used to build the search tree. For the sake of simplicity, we assume in this section that an isotropic rescaling factor (same affinity ratio in x, y, and z-directions) is sufficient to describe the variations of geobody dimensions in the volume under study.

Consider first the case in which the facies geobodies should have constant dimensions over the study field, yet possibly different from the dimensions of the training geobodies. Let λ be the ratio between target and training facies dimensions. Given a training image, and a data template $\tau_N = \{\mathbf{h}_\alpha, \alpha=1...N\}$, a new data template $\tau_N(1/\lambda)$ is obtained from τ_N by the following method:

1. Rescale by $1/\lambda$ each component $\mathbf{h}_\alpha$ of τ_N. In 2D, the coordinates $(x_\alpha(1/\lambda), y_\alpha(1/\lambda))$ of the rescaled component $\mathbf{h}_\alpha(1/\lambda)$ are computed from the coordinates (x_α, y_α) of the original component $\mathbf{h}_\alpha$ as: $x_\alpha(1/\lambda) = x_\alpha/\lambda$ and: $y_\alpha(1/\lambda) = y_\alpha/\lambda$
2. Relocate these rescaled components to the nearest nodes of the training image grid.

The resulting rescaled data template $\tau_N(1/\lambda)$ is used to build the search tree from the training image. The exact same result can be obtained by rescaling the training image by λ, then building the search tree from that rescaled training image using the original data template τ_N. During the simulation, at each unsampled node, the original data template τ_N is used to search for nearby conditioning data, and the corresponding cpdf is retrieved from the above search tree.

The extension of that technique to integrate location-dependent geobody dimensions data is straightforward and similar to the integration of location-dependent azimuth data. If $\lambda(\mathbf{u})$ denotes the ratio between target and training facies dimensions at the grid node location $\mathbf{u}$, then MPS simulation using locally-varying geobody rescaling factors consists of dicretizing the range of $\lambda(\mathbf{u})$ values into a smaller number of classes, and building a search tree for the average rescaling factor value of each class.

Figure 5 shows a 2D rescaling factor field and a resulting simulated realization using the fluvial reservoir training image of Figure 2. The reproduction of the training patterns is similar to that in the reference simulated realization of Figure 2.

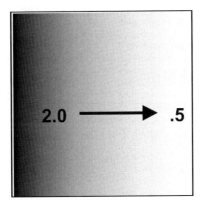

Figure 5. 2D field of location-dependent geobody dimension rescaling factors (left), and resulting MPS model obtained using the training image of Figure 2 (right).

6 Conclusion

In multiple-point geostatistics, statistics on facies joint-correlation at multiple locations are inferred from patterns displayed by a training image regardless of the location of these patterns in the training image. As a consequence, non-stationary features, such as spatial variations of facies azimuths or geobody dimensions that the training image may contain are not preserved in the multiple-point statistics simulated realizations.

To integrate variable azimuth/dimensions data, we propose applying a series of rotation/affinity transforms to a stationary training image, and building a search tree to store the multiple-point statistics inferred from each rotated/rescaled training image. During the simulation, multiple-point statistics are retrieved from the search tree corresponding to the class where the local azimuth/rescaling factor occurs. The application of that process to a 2D horizontal section of a fluvial reservoir indicates that the reproduction of the training patterns in non-stationary MPS models is similar to that observed in stationary models.

This technique can be easily generalized to create non-stationary models using several different training images thought to be representative of the geological heterogeneity in different areas of the reservoir provided that there is a smooth transition between the different training images.

References

Deutsch, C.V. and Journel, A.G., *GSLIB: Geostatistical Software Library and User's Guide*, 2nd edition, Oxford University Press, 1998.

Guardiano, F. and Srivastava, R.M., Multivariate Geostatistics: Beyond Bivariate Moments, in Soares, A., editor, *Geostatistics-Troia*, vol. 1, p. 133-144. Kluwer Academic Publications, 1993.

Holden, L., Hauge, R., Skare, Ø. and Skorstad, A., Modeling of Fluvial Reservoirs with Object Models, *Mathematical Geology*, vol. 30, no. 5, 1998, p. 473-496.

Journel, A. and Alabert, F., Non-Gaussian Data Expansion in the Earth Sciences, *Terra Nova*, no. 1, 1989, p. 123-134.

Lia, O., Tjelmeland, H. and Kjellesvik, L.E., Modeling of Facies Architecture by Marked Point Models, in *Geostatistics-Wollongong*, p. 386-398, Kluwer Academic Publications, 1997.

Strebelle, S., *Sequential Simulation Drawing Structures from Training Images*, Ph.D. Thesis, Department of Geological and Environmental Sciences, Stanford University, 2000.

Strebelle, S., and Journel, A., Reservoir Modeling Using Multiple-point Statistics, paper SPE 71324 presented at the 2001 SPE Annual Technical Conference and Exhibition, New Orleans, Sept. 30-Oct. 3, 2001.

Strebelle, S., K. Payrazyan, and J. Caers, Modeling of a Deepwater Turbidite Reservoir Conditional to Seismic Data Using Multiple-Point Geostatistics: paper SPE 77425 presented at the 2002 SPE Annual Technical Conference and Exhibition, San Antonio, Sept.29-Oct. 2, 2002.

Viseur, S., Stochastic Boolean Simulation of Fluvial Deposits: a New Approach Combining Accuracy and Efficiency, paper SPE 56688 presented at the 1999 SPE Annual Technical Conference and Exhibition, Houston, Oct. 3-6, 1999.

Zhang, T. Program 2001 snesim version 5.0. In *Report 15, Stanford Center for Reservoir Forecasting*, Stanford, CA, 2002.

A WORKFLOW FOR MULTIPLE-POINT GEOSTATISTICAL SIMULATION

YUHONG LIU
ExxonMobil Upstream Research Company, P. 0. BOX 2189, Houston, TX 77252, USA. E-mail: yuhong@pangea.stanford.edu

ANDREW HARDING, RUSTY GILBERT
ChevronTexaco Energy Technology Company, San Ramon, CA 94583, USA

ANDRE JOURNEL
Geological and Environmental Science Department, Stanford University, Stanford, CA 94305, USA

Abstract. There are presently two main avenues in the stochastic modeling of depositional facies: pixel-based and object-based geostatistics. They both have strengths and weaknesses: traditional pixel-based geostatistics is good at data conditioning, but it depends on variograms to capture spatial structures and hence fails to reproduce definite patterns common to most geological facies; while object-based geostatistics is good at reproducing crisp facies shapes but is difficult to condition to dense well data or exhaustive 3D seismic data. Multiple-point simulation, a newly developed pixel-based technique, integrates the strengths of both: it keeps the flexibility of pixel-based techniques for data conditioning, while allowing pattern reproduction through consideration of multiple-point statistics. In this paper, a workflow for multiple-point stochastic simulation is discussed in details. This workflow is applied to an industry project. The results show reproduction of the prior geological knowledge and honoring of both well and seismic data.

1 Introduction

Integrating all available information when building a geological model is a recurrent and difficult problem. One challenge lies in how to condition the model to different types of measured reservoir data, such as wells and 3D seismic data. Another challenge lies in how to account for prior geological knowledge, a fuzzy yet important conceptual information. Geostatistics provides an ensemble of tools for data integration and uncertainty evaluation (Deutsch and Journel, 1992; Goovaerts, 1997). Through computer-based data integration algorithms, multiple equi-probable numerical models of the reservoir properties are built, whose difference reflects uncertainty.

There are two main avenues in geostatistical modeling: pixel-based and object-based. Object-based geostatistics performs simulation by "dropping" different geological bodies one after another onto the simulation field (Deutsch and Wang, 1996). Any added geological body is accepted, rejected, or modified through evaluating some objective function measuring the match to local data. Hence by its nature, it is good at reproducing the crisp shapes of geological bodies. But often it is CPU demanding, particularly when extensive hard data must be honored; also their capability in integrating 3D seismic data is limited: typically only 2D aerial proportion maps derived from seismic can be accounted for. By contrast, pixel-based algorithms simulate each grid node of the reservoir model one pixel at a time (Deutsch and Journel, 1992; Goovaerts, 1997). All the unsampled nodes are sequentially visited along a random path. The probability distribution function (cpdf) at any given node is estimated conditional to all data (both hard and soft) found in its neighborhood. A value is drawn from that cpdf using Monte Carlo simulation. This simulated value is frozen as hard data for simulation of the subsequent unknown nodes. Because pixel-based geostatistics performs simulation one pixel at a time, it is very flexible and easy to condition to most conditional data. However, the traditional pixel-based geostatistics uses the variogram, a two-point statistic, to capture the spatial structures of facies. It has been found that it is difficult for a simple variogram model to capture the curvilinear structures of geological bodies. Higher order statistics are required. For example, Figure 1 shows three distinct images with similar variograms (Strebelle, 2000).

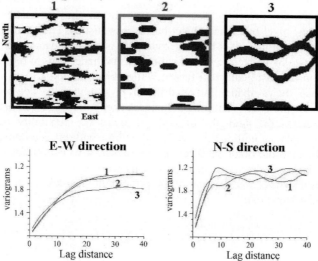

Figure 1. Three distinct images with similar variograms. (Source: Strebelle, 2000, slightly modified.)

Hence none of the traditional geostatistical simulation techniques are ideal for depositional facies modeling integrating both geology and 3D seismic data. Multiple-point geostatistics (Journel, 1992; Guardiano and Srivastava, 1993; Strebelle, 2000; Strebelle and Journel, 2000; Strebelle, et al, 2002; Liu, 2003; Liu, et al, 2004) is called upon to address this problem. In the following sections two case studies are presented to illustrate the application of multiple-point geostatistics. Next, a workflow of multiple-point simulation is proposed.

2 Why multiple-point geostatistics?

Multiple-point geostatistics aims at reproducing complex statistics involving two and more points at a time. This allows capturing information much beyond the reach of a mere variogram model. Being an advanced pixel-based technique, multiple-point geostatistics inherits the advantage of building the numerical model one pixel at a time, allowing easy data conditioning. Yet compared with the traditional variogram-based algorithms, it has the enhanced capability of reproducing curvilinear shapes of geological bodies, a feature traditionally reserved to object-based algorithms. Two examples are presented in this section to respectively illustrate these two strengths.

2.1 BETTER INTEGRATION OF GEOLOGY

This first example is a synthetic case study of building a numerical reservoir property model from limited well data. Figure 2a shows a photo of the Wagon Caves Rock outcrop (Anderson et al, 1998). From this outcrop, rock properties such as sand/shale indicators, grain size, porosity and permeability, are measured along the two marked vertical columns. They are taken as known well data, while rock properties at all other grid nodes are assumed unknown and are simulated using different geostatistical algorithms.

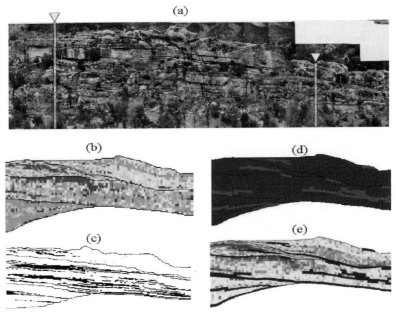

Figure 2. Integration of geology using a multiple-point simulation algorithm. (a) the Wagon Rock Caves outcrop, from which, two vertical columns are taken as well data; (b) one realization of permeability by the two-point model; (c) training image of mud layers, used for multiple-point simulation; (d) one multiple-point realization of mud layers; (e) one permeability realization including simulated mud layers.

First a variogram-based two-point geostatistical simulation is performed. Figure 2b shows one realization of permeability using sequential Gaussian simulation (Deutsch and Journel, 1998): it displays high-entropy spatial patterns typical of Gaussian simulations. One major problem with any two-point model is that it can not reproduce the elongated thin mud layers, which can act as flow barriers due to their very low permeabilities.

From the outcrop, two types of mud layers are observed: continuous mud layers spanning across the two wells and discontinuous mud layers that pinch out before reaching either one of the two wells. Multiple-point simulation is performed to reproduce these different types of mud layers. Figure 2c depicts the shapes of the mud layers over a vertical section, which can serve as a training image for multiple-point simulation of these mud layers. For sensitivity analysis purpose, we build three models: one without mud layers, one with discontinuous (pinched-out) mud layers, and one with continuous mud layers crossing all the way between the two wells. Figure 2d shows one realization of the simulated mud layers using the multiple-point simulation *snesim* algorithm (Strebelle, 2000). Figure 2e shows one realization of permeability including the simulated mud layers.

To analyze the flow response of these different models, we perform a single-phase steady-state upscaling over the whole model to get a single upscaled permeability tensor for the whole simulation field. A high effective permeability along a certain direction means easy flow along that direction. Figure 3a shows the effective permeability along the vertical direction (denoted as Kz) for three different variogram-based models, each model with a different range and represented by 10 equi-probable realizations. When the range is decreased from 1200 to only 150, Kz changes within a small range of 140-220 md. In contrast, when a multiple-point algorithm is used to incorporate different types of mud layers, Kz changes dramatically within a range of 220-30-1 md (Figure 3b).

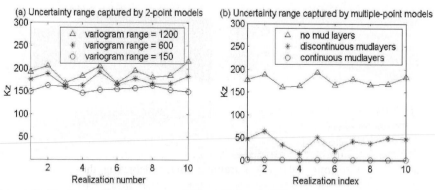

Figure 3. Vertical effective permeability (Kz) of different models. (a) Kz of three different variogram-based models; (b) Kz of three different multiple-point models.

This example illustrates the importance of simulating correctly the crisp shapes and continuity of geological bodies: they can have significant impact on flow response, a major concern of petroleum engineers. Traditional variogram-based geostatistics fails in this aspect, while multiple-point geostatistics achieves it at little additional CPU cost, provided a conceptual training image of the type of Figure 2c is available.

2.2 EASY DATA CONDITIONING

Another advantage of multiple-point simulation is easier data conditioning. Object-based algorithms can reproduce crisp shapes and continuity of geological bodies. A major problem with object-based algorithms, however, is difficult data conditioning, especially in presence of dense well data and diverse types of soft data, such as 3D seismic data or production data. Figure 4 illustrates such a real reservoir study (Liu, 2003; Liu, et al, 2004).

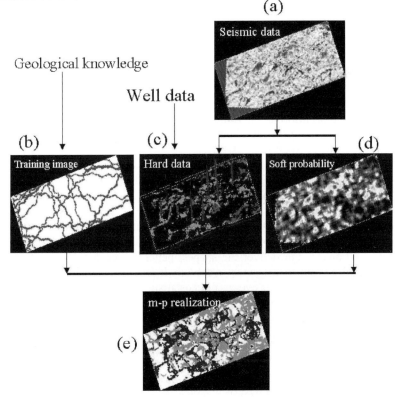

Figure 4. Multiple-point simulation integrating diverse types of information (Note all the following data/information/realizations are in 3D, although only horizontal slices are shown here). (a) seismic data; (b) a training image depicting the prior geological concepts; (c) hard data (well data + seismic imaged channel pieces); (d) seismic-derived soft probability for sand; (e) one multiple-point realization honoring the information shown in (b), (c), (d).

In this reservoir characterization study, there are four different sources of information:
- Geological knowledge: the depositional environment of this reservoir is interpreted as a fluvial channel system, which can be summarized by a training image (see Figure 4b).
- Well data: they are considered as hard data and must be reproduced exactly by all simulated realizations (see the vertical columns in Figure 4c).
- Geobodies extracted from seismic data: good quality 3D seismic data can clearly image some characteristic geobodies (channel segments in this case study). These geobodies (see the two types of clusters in Figure 4c) are deemed certain and need to be reproduced exactly by all realizations. A PCA clustering technique (Scheevel and Payrazyan, 1999), capable of recognizing such characteristic facies pieces, is used to extract these geobodies from the 3D seismic data.
- Soft information from seismic data: in areas not clearly imaged by seismic data, a soft probability data cube for presence of sand can be derived from the seismic data (see Figure 4d). All alternative simulated models for sand/shale should be all consistent with the seismic data in that probabilistic sense.

While it would be extremely difficult to constrain object-based simulation to all these different sources of information, multiple-point simulation can easily achieve this, because it operates one pixel at a time. Figure 4e presents one such a multiple-point realization.

3 A workflow for multiple-point simulation

In multiple-point geostatistics, a training image is used to deliver prior geological concepts about the geometry of reservoir heterogeneities. This training image should be reasonably stationary, and deliver the shapes, patterns and distributions of geological objects deemed present in the actual reservoir. It essentially plays the same role as a variogram model in traditional two-point geostatistics: it provides statistics relating the unsampled value to conditioning data involving jointly multiple locations. The simulation process amounts to take the training image patterns, "morphing" and anchoring them to location-specific reservoir data.

Strebelle (2000) developed a *snesim* algorithm, which significantly speeds up the original multiple-point simulation algorithm proposed by Guardiano and Srivastava (1993). In this program, the training image is scanned only once to retrieve the frequency of occurrences of observed outcomes for the central nodal value given a template of neighboring conditioning data. These probabilities are then stored into a search tree data structure, which allows fast storage and retrieval of probabilities corresponding to the actual hard conditioning data events encountered during sequential simulation. Conditioning to hard sample data (e.g. well data) in multiple-point simulation is done the same way as in two-point algorithms. The hard data values are frozen at their nodal locations and never changed; each unknown node is sequentially visited and simulated conditional to the original hard data and previously simulated values. As for soft data conditioning (e.g. seismic data or production data), a Bayesian-type paradigm is applied in the multiple-point simulation workflow. Instead of using some cokriging algorithms as in two-point geostatistics (Goovaerts, 1997), soft data conditioning is performed in two steps. The first step is to extract the useful information from the soft data. For example, a prior facies conditional probability is derived from the seismic data. The second step is to perform simulation integrating hard and soft data, which can be further decomposed into three sub-steps. At each unknown node, first the probability conditioned to the current multiple-point hard data event is read from the search tree. Then it is combined with the previously derived soft conditional probability to get the final or posterior probability conditioned to both hard and soft data. Finally, a facies indicator value is drawn from that posterior probability.

This multiple-point simulation workflow can be subdivided into three parts, each of which has a different conditional probability involved (Figure 5). All three parts are later explained in more detail.

Part 1, P(A|B): modeling with hard data and conceptual geology
Part 2, P(A|C): seismic data analysis.
Part 3, P(A|B,C): data integration.

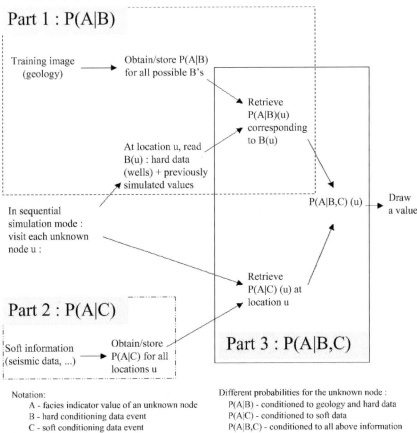

Figure 5. A multiple-point simulation workflow, decomposed into three parts.

3.1 P(A|B): MODELING WITH HARD DATA AND CONCEPTUAL GEOLOGY

P(A|B) denotes the conditional probability of the value to be simulated given a multiple-point hard conditioning data event B, with A representing, e.g., a facies indicator value. This part aims at capturing and reproducing the geological information provided by the training image, conditional to the hard data event B.

First a data template, composed of multiple nodes with any user-specified configuration, is used to scan the training image, and the number of replicates of each different multiple-point data event is retrieved. These numbers are stored in a search tree data structure (Strebelle, 2000), which allows an easy retrieval of information.

Next, in a pixel-based simulation mode, each uninformed node **u** is sequentially visited. Its neighboring conditioning data event B (including both the original hard sample data and previously simulated values) is collected. Two numbers are then retrieved from the search tree: number of replicates of the joint data event (A,B) and the number of replicates of the conditioning data event B. The multiple-point conditional probability P(A|B) is then easily calculated as:

$$P(A|B) = \frac{P(A,B)}{P(B)} = \frac{\text{number of replicates for } (A,B)}{\text{number of replicates for } (B)} \qquad (1)$$

This conditional probability P(A|B) is used to either directly draw a value for the node **u**, if no soft information is available, or is combined with any co-located soft data conditional probability P(A|C) to get a joint conditional probability P(A|B,C). The value at location **u** is then drawn using the updated probability P(A|B,C), see hereafter.

3.2 P(A|C): SEISMIC DATA ANALYSIS

As discussed above, when there is soft information, such as seismic data, it is necessary to determine the conditional probability P(A|C), denoting the facies probability given the soft data C (say, seismic) alone. This part tries to establish the relationship between seismic patterns, and geological patterns, enabling prediction of rock properties from the measured seismic data. Many different techniques can be used to retrieve this probabilistic information from the seismic data. They can be divided into two categories: supervised vs. unsupervised techniques. Supervised techniques are used when there exist calibration geological data associated with the seismic data, for example, well data versus corresponding seismic data, or interpreted geological facies versus corresponding seismic data. The pattern recognition from seismic data is then "supervised" by the known geological data. In contrast, unsupervised techniques try to directly identify patterns from seismic data without any prior geological constraints. Techniques in both categories have been used by geostatisticians: supervised or unsupervised neural network (Caers, 1999), principal components clustering (Scheevel and Payrazyan, 1999; Strebelle et al., 2002; Liu, 2003), maximum message length technique (Arroyo, 2000), the latter uses entropy to measure the dispersion of different seismic patterns.

3.3 P(A|B,C): DATA INTEGRATION

After retrieving useful information separately from the geology + hard data, and from seismic, that is, obtaining the two individual conditional probabilities P(A|B) and P(A|C), the next step is to combine them into one single posterior probability, P(A|B,C), conditioned to all available information. This is the data integration part.

Journel (2002) proposed a "Permanence of Updating Ratios" paradigm to integrate P(A|B) and P(A|C) into P(A|B,C). The basic assumption of this algorithm is that the relative contribution of data event C is the same before and after knowing B:

$$\frac{x}{b} = \frac{c}{a} \qquad (2)$$

where, a, b, c and x represent distances to the event A occurring defined as:

$$a = \frac{1-P(A)}{P(A)}, \quad b = \frac{1-P(A|B)}{P(A|B)}, \quad c = \frac{1-P(A|C)}{P(A|C)}, \quad x = \frac{1-P(A|B,C)}{P(A|B,C)}$$

All these distances are bounded within $[0,\infty)$. They reach 0 if the probability of A occurring is 1, infinity if that probability is 0.

From the permanence relation (Eq.2), P(A|B,C) is calculated as:

$$P(A|B,C) = \frac{1}{1+x}$$

Zhang and Journel (2003) later showed that the previous permanence assumption is equivalent to a Bayesian updating under conditional independence of B and C given A. To account for dependence between B and C data, Journel proposed the following generalization using a power parameter $\tau > 0$:

$$\frac{x}{b} = \left(\frac{c}{a}\right)^\tau \qquad (3)$$

Setting $\tau > 1$ increases the impact of seismic data, conversely, setting $\tau < 1$ decreases the impact of seismic data.

These three parts establish a general workflow for multiple-point geostatistical simulation.

4 Conclusions

In this paper, a multiple-point simulation workflow is proposed and discussed. Anchored in the pixel-based category, which allows an easier conditioning to a variety of data, the multiple-point approach aims at identifying and reproducing the spatial patterns typically displayed by geological bodies. Hence it incorporates the advantages of both pixel-based techniques and object-based techniques. The proposed multiple-point simulation workflow is composed of three parts:

- Modeling with hard data and conceptual geology, i.e., obtaining P(A|B): The prior geological knowledge is represented by a training image, which is scanned to obtain the probability P(A|B), namely, the probability of

presence/absence of a facies A given its multiple-point hard conditioning data event B.
- Seismic data analysis, i.e., obtaining P(A|C): This part establishes the relationship between seismic patterns (C) and facies patterns (A). The result is a probability P(A|C) field, denoting the facies probability given the neighboring multiple-point seismic data C.
- Integration of different sources of information, i.e., obtaining P(A|B,C): This part integrates the two previous individual conditional probabilities, P(A|B) and P(A|C), into one single posterior probability P(A|B,C), conditioned to both geology and seismic data. The facies indicator (A) at each unsampled node is drawn from this updated posterior probability.

Acknowledgements

We would like to thank ChevronTexaco Energy Technology Company and its affiliates for the permission to publish one of the case studies. The helps provided by Sebastien Strebelle, William Abriel and staffs of ChevronTexaco are especially acknowledged.

References

Anderson, K.; Hickson, Thomas A.; Crider, G and Graham, S.: *Integrating teaching with field research in the Wagon Rock Project. Journal of Geoscience Education*, vol.47, no.3, May 1999, pp.227-235.
Caers, J.: *Modeling Facies Distributions from Seismic Using Neural Nets.* SCRF Annual Report No.13, vol.1. 2000.
Deutsch, C.V. and Journel, A.G., *GSLIB: Geostatistical Software Library and User's Guide*, Oxford University Press, 1992.
Deutsch, C. and Wang, L.: *Hierarchical Object-Based Stochastic Modeling of Fluvial Reservoirs.* Mathematical Geology, vol. 28, no. 7, 1996, p 857-880.
Gilbert, R., Liu, Y., Abriel, W. and Preece, R.: *Reservoir Modeling Integrating Various Data and at Appropriate Scales*, The Leading Edge, Vol.23, no. 8, Aug. 2004, p784-788.
Goovaerts, P., *Geostatistics for Natural Resources Evaluation*, Oxford University Press, 1997.
Guardiano, F. and Srivastava, R.M.: "Multivariate Geostatistics: Beyond Bivariate Moments", Geostatistics-Troia, A. Soares (ed.), Kluwer Academic Publications, Dordrecht, 1993, vol 1, p 113-114.
Journel, A: *Geostatistics: Roadblocks and Challenges*. In A. Soares (Ed.), Geostatistics-Troia, Kluwer Academic Publ., Dordrech, 1992, p 213-224.
Journel, A: *Combining knowledge from diverse sources: an alternative to traditional data independence hypotheses*, Mathematical Geology, vol. 34, no. 5, 2002.
Liu, Y.: *Downscaling Seismic Data into A Geological Sound Numerical Model*, Ph.D. dissertation, Department of Geological and Environmental Science, Stanford University, Stanford, 2003, pp. 202.
Liu, Y., Harding, A., Abriel, W. and Strebelle, S.: *Multiple-Point Simulation Integrating Wells, 3D Seismic Data and Geology*, AAPG Bulletin, Vol. 88, No.7, 2004.
Scheevel, J. R., and Payrazyan, K.: *Principal Component Analysis Applied to 3D Seismic Data for Reservoir Property Estimation*, SPE 56734, SPE Annual Technical Conference and Exhibition, Houston, 1999
Srivastava, M., 1995: *An Overview of Stochastic Methods for Reservoir Characterization*, in Yarus, J., and Chambers, R., eds., Stochastic modeling and geostatistics: principles, methods, and case studies, v.3: AAPG Computer Applications in Geology, p. 3-16.
Strebelle, S.: *Sequential Simulation Drawing Structures from Training Images*, Ph.D. dissertation, Department of Geological and Environmental Sciences, Stanford University, Stanford, 2000.
Strebelle, S., and Journel, A.: *Reservoir Modeling Using Multiple-point Statistics*, SPE 71324, SPE Annual Technical Conference and Exhibition, 2001.
Strebelle, S., Payrazyan, K. and Caers, J.: *Modeling of a Deepwater Turbidite Reservoir Conditional to Seismic Data Using Multiple-Point Geostatistics*, SPE 77425, SPE Annual Technical Conference and Exhibition, San Antonio, Texas, 2002.
Zhang, T. and Journel, A. G.: *Merging Prior Geological Structure and Local Data: the mp Geostatistics Answer*, SCRF Annual Report No.16, vol. 2, 2003.

A MULTIPLE-SCALE, PATTERN-BASED APPROACH TO SEQUENTIAL SIMULATION

G. BURC ARPAT and JEF CAERS
Department of Petroleum Engineering, Stanford University
367 Panama St., Stanford, CA 94305-2220, USA

Abstract. In the context of multiple-point geostatistics, a new algorithm (SIMPAT) is presented. This algorithm relies on several image processing concepts, such as image similarity, to borrow and reproduce patterns from training images constrained to hard and soft data. The method makes use of a new multiple-grid approach by which the scale relations between the training image patterns are better captured and reproduced.

1 Introduction

Sequential simulation is one of the most widely used stochastic imaging techniques within the Earth Sciences. The theory is well understood (Daly, 2004; Goovaerts, 1997) and many practical, fast and robust algorithms have been developed (Deutsch and Journel, 1998) such as sequential Gaussian simulation (SGSIM) and sequential indicator simulation (SISIM).

However, realizations generated by SGSIM (and also SISIM) are often deemed too 'synthetic' looking, not reflecting the actual variability of Earth Science phenomena such as facies distributions in oil reservoirs or sedimentary deposits in aquifer systems. The limitations of SGSIM lie in the assumption of a multi-Gaussian distribution that requires knowledge of a histogram and a variogram. The variogram, as a two-point statistics, is not capable of modeling complex, connected and curvilinear spatial variation. To overcome these limitations, multiple-point geostatistics (MPS) was introduced together with the concept of "training image" (Guardiano and Srivastava, 1993). A training image is an exhaustive 3D picture containing patterns believed to be similar to the actual field under investigation. The training image serves as a concept, a vision of what spatial variability of the study area should look like. As a mere concept, the training image need not be constrained to any hard or soft data.

Based on the original MPS idea, Strebelle (2000) proposed a practical algorithm (SNESIM) that generates realizations mimicking the 3D patterns of the training image while constraining to hard and soft data. The algorithm works in the same way as many other sequential algorithms do: (1) visit each node of the simulation grid randomly; (2) at each node, estimate the conditional probability given the neighboring data and previously simulated nodes (called a "data event"), and (3) draw from that probability distribution and assign the value to the node. In SNESIM, the conditional probability is sampled from the training image by looking for replicates of the data event.

In this paper, an alternative pattern-based algorithm is proposed by redefining the problem of pattern reproduction as an image processing problem. In image processing, one generally tackles complex images by finding common patterns in the image and working on these patterns (Palmer, 1999). A similar approach can be devised for geostatistical modeling where one finds all the patterns of a training image. These patterns correspond to multiple-pixel configurations within a user-defined template and capture meaningful pieces of geological shapes known to exist in field of study. Such patterns exist at different geological scales and patterns at various scales interact with each other. The idea is to generate realizations that reproduce these multiple-scale patterns on the simulation grid. A new practical algorithm SIMPAT (SIMulation with PATterns) is implemented to achieve this goal. The paper shortly describes the inner workings of this algorithm, presents some 3D examples and discusses how SIMPAT complements the already existing sampling-based algorithms such as SNESIM.

2 A New, Pattern-based Sequential Simulation Method

2.1 NOTATION

Define $z(\mathbf{u})$ as the realization of a random variable $Z(\mathbf{u})$ modeling the variable of study where $\mathbf{u} = (x,y,z) \in \mathbf{G}$ and $\mathbf{G}$ is the regular Cartesian grid discretizing the field of study. $Z(\mathbf{u})$ can be a model of either a continuous or a categorical variable. The random function itself is denoted as $\mathbf{Z} = \{Z(\mathbf{u}), \forall \mathbf{u} \in \text{study area}\}$ and a realization as $\mathbf{z}$.

$\mathbf{z}_T(\mathbf{u})$ indicates a location-specific vector of $z(\mathbf{u})$ within a template $\mathbf{T}$ centered at $\mathbf{u}$, i.e.:

$$\mathbf{z}_T(\mathbf{u}) = \{z(\mathbf{u}+\mathbf{h}_0), z(\mathbf{u}+\mathbf{h}_1), \ldots, z(\mathbf{u}+\mathbf{h}_\alpha), \ldots, z(\mathbf{u}+\mathbf{h}_{n_T-1})\} \quad (1)$$

where $\mathbf{h}_\alpha$ vectors are the vectors defining the geometry of the n_T nodes of the template $\mathbf{T}$ and $\alpha = 0, \ldots, n_T - 1$ with the special vector $\mathbf{h}_0 = 0$ identifying the node $\mathbf{u}$. A flag notation $z(\mathbf{u}) = \chi$ is used for 'unknown' nodes, i.e. nodes still to be informed by the sequential simulation and hence that do not have an assigned value yet.

To distinguish the training image, the hard and the soft data from the simulated realization $\mathbf{z}$, the notations $\mathbf{ti}$, $\mathbf{hd}$ and $\mathbf{sd}$ are used. For example, a multiple-point event scanned from the training image $\mathbf{ti}$ at location $\mathbf{u}'$ is denoted by $\mathbf{ti}_T(\mathbf{u}')$, i.e.:

$$\mathbf{ti}_T(\mathbf{u}') = \{ti(\mathbf{u}'+\mathbf{h}_0), ti(\mathbf{u}'+\mathbf{h}_1), \ldots, ti(\mathbf{u}'+\mathbf{h}_\alpha), \ldots, ti(\mathbf{u}'+\mathbf{h}_{n_T-1})\} \quad (2)$$

where location $\mathbf{u}' \in \mathbf{G}'$ and $\mathbf{G}'$ is the regular Cartesian grid discretizing the training image. The training image grid $\mathbf{G}'$ need not be the same as the realization grid $\mathbf{G}$.

A pattern $\mathbf{pat}_T^k$ is the particular k-th configuration of the above vector of values $\mathbf{ti}_T(\mathbf{u}')$ defined by the template $\mathbf{T}$ where $k = 0, \ldots, n_{pat} - 1$ and n_{pat} is the number of total available patterns. Each k-th configuration is assumed to be location-independent and, thus, the vector $\mathbf{pat}_T^k$ is written as:

$$\mathbf{pat}_T^k = \{pat_T^k(\mathbf{h}_0), pat_T^k(\mathbf{h}_1), ..., pat_T^k(\mathbf{h}_\alpha), ..., pat_T^k(\mathbf{h}_{n_T-1})\} \qquad (3)$$

where all patterns are defined on the same template **T**.
In sequential simulation, a data event $\mathbf{dev}_T(\mathbf{u})$ is defined as the set of hard data and previously simulated values neighboring the visited location **u** within the template **T**, i.e. $\mathbf{dev}_T(\mathbf{u}) = \mathbf{z}_T(\mathbf{u})$.

The dissimilarity (distance) between a data event and a pattern is calculated using a node-based distance function:

$$d\langle \mathbf{dev}_T(\mathbf{u}), \mathbf{pat}_T^k \rangle = \sum_{\alpha=0}^{n_T-1} |dev_T(\mathbf{u}+\mathbf{h}_\alpha) - pat_T^k(\mathbf{h}_\alpha)| \qquad (4)$$

where $d\langle\rangle$ denotes the distance function (Manhattan distance; Duda et al., 2001). When for a certain node $\mathbf{u} + \mathbf{h}_\alpha$, $dev_T(\mathbf{u} + \mathbf{h}_\alpha) = \chi$ (unknown), the value is ignored in the distance calculation. For other distance functions that can be used with the SIMPAT algorithm, the reader is referred to Arpat (2004).

2.2 THE SINGLE-GRID, UNCONDITIONAL SIMPAT ALGORITHM

The algorithm starts by scanning the training image using a template **T** to acquire all patterns of **ti**. A filter can be applied to discard undesirable patterns. Remaining patterns are stored in a pattern database and such patterns are denoted by $\mathbf{pat}_T^k$ where the size of the pattern database is n_{pat} as defined in the previous section.

The simulation part of the algorithm follows the sequential simulation framework. During simulation, nodes are randomly visited and the data event $\mathbf{dev}_T(\mathbf{u})$ is extracted. Then, $\mathbf{dev}_T(\mathbf{u})$ is compared to all available patterns in the pattern database using a predefined similarity criterion. The aim is to find the 'most similar' pattern to the data event, denoted by $\mathbf{pat}_T^*$. In other words, the algorithm minimizes $d\langle\rangle$ of Equation 4 for all patterns $\mathbf{pat}_T^k$ and labels the minimum as $\mathbf{pat}_T^*$. Once this most similar pattern is found, the data event $\mathbf{dev}_T(\mathbf{u})$ is replaced by $\mathbf{pat}_T^*$, i.e. the values of $\mathbf{pat}_T^*$ are pasted on to the simulation grid at the current node **u**.

The above outlined algorithm can be divided into two main parts:

(1) Pre-processing of the training image:

P-1. Scan the training image using the template **T** to obtain all existing patterns $\mathbf{pat}_T^k$ that occur over the training image.

P-2. Reduce the number of patterns to n_{pat} by applying filters to construct the pattern database. Typically, only unique patterns are taken, i.e. repetitions (frequency) of patterns are ignored.

(2) Simulation on the simulation grid:

S-1. Define a random path on the simulation grid to visit each node **u** only once.

S-2. At each node **u**, retain the data event $\mathbf{dev}_T(\mathbf{u})$ and find the $\mathbf{pat}_T^*$ that minimizes $d\langle \mathbf{dev}_T(\mathbf{u}), \mathbf{pat}_T^k \rangle$ for $k = 0, ..., n_{pat} -1$, i.e. $\mathbf{pat}_T^*$ is the 'most similar' pattern.

S-3. Once the most similar pattern $\mathbf{pat}_T^*$ is found, assign $\mathbf{pat}_T^*$ to $\mathbf{dev}_T(\mathbf{u})$, i.e. for all the n_T nodes $\mathbf{u} + \mathbf{h}_\alpha$ within the template $\mathbf{T}$, $dev_T(\mathbf{u} + \mathbf{h}_\alpha) = pat_T^*(\mathbf{h}_\alpha)$.

S-4. Move to the next node of the random path and repeat the above steps until all the grid nodes along the random path are exhausted.

On large simulation grids, a practical problem occurs due to the finite size of the template $\mathbf{T}$. To capture the large scale correlations of the training image, a large template would need to be used. Figure 5b of Section 4 (Examples) demonstrates this problem. Yet, using a large template would make the minimization step (Step S-2) of the SIMPAT algorithm too CPU demanding. To overcome this problem, SIMPAT employs a modified version of the multiple-grid approach as proposed by Tran (1994). The idea is to use a set of cascading multiple-grids and sparse templates instead of a single grid and one large dense template. The simulation is first performed on the coarse grid and then these coarse values are passed to the subsequent finer grids as conditioning information. This idea is elaborated below.

2.3 THE MULTIPLE-GRID, UNCONDITIONAL SIMPAT ALGORITHM

On a Cartesian grid, the multiple-grid view of a grid $\mathbf{G}$ is defined by a set of cascading coarse grids $\mathbf{G}^g$ and templates $\mathbf{T}^g$ instead of a single fine grid and one large dense template where $g = 0, ..., n_g - 1$ and n_g is the total number of multiple-grids. The g-th coarse grid ($0 \leq g \leq n_g - 1$) is constituted by each 2^g-th node of the final grid ($g = 0$) in each direction. If $\mathbf{T}$ is a template defined by vectors $\mathbf{h}_\alpha$, then the template used for a coarser grid $\mathbf{T}^g$ is defined by $\mathbf{h}_\alpha^g = 2^g \times \mathbf{h}_\alpha$ and has the same configuration of n_T nodes as $\mathbf{T}$ but with spacing 2^g times larger. Figure 1 illustrates this concept.

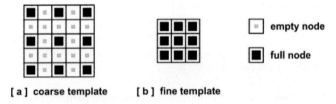

[a] coarse template   [b] fine template

Figure 1. A 3x3 fine template (b) and its corresponding coarse template (a) obtained by expanding the fine template with 2^g spacing where $g = 1$ (the first coarse grid).

The multiple-grid simulation of a realization is achieved by successively applying the single-grid algorithm explained above to the multiple-grids starting from the coarsest grid. After each multiple-grid simulation, the values calculated on the current grid are transferred to the one finer grid and g is set to $g - 1$. This succession of multiple-grid simulations continues until $g = 0$. On a multiple-grid, the previously calculated coarser grid values contribute to the distance calculation of Step S-2, i.e. if on node $\mathbf{u}$ a previous coarse grid value exists, this value is taken into account when minimizing the distance.

Different from the classical multiple-grid approach (Tran, 1994), SIMPAT does not 'freeze' the coarse grid values when they are transferred to a finer grid, i.e. such values are still allowed to be updated and visited by the algorithm in the subsequent multiple-grid simulations. In other words, the coarse grid nodes are always included in the finer grid simulations.

The above multiple-grid approach allows the values determined on the coarser grids to be modified by the finer grids, i.e. coarse to fine scale interaction. For complex training images, fine to coarse interaction might also be desired to fully capture the scale relations of training image patterns. SIMPAT utilizes a feedback mechanism, termed "dual template simulation", that allows such fine to coarse scale interaction.

Consider the pattern $\mathbf{pat}_T^k$ scanned from the training image at location $\mathbf{u}$ using the template $\mathbf{T}$ for a coarse grid simulation. Another, fine template $\mathbf{T'}$ can be used at the same location to obtain the corresponding fine pattern such that template $\mathbf{T'}$ covers the same area (volume in 3D) as template $\mathbf{T}$ but with all the nodes of the finest grid. $\mathbf{T'}$ is called the "dual template" of $\mathbf{T}$. The relation between $\mathbf{T}$ and $\mathbf{T'}$ is shown in Figure 2.

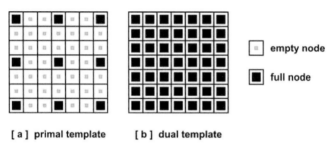

Figure 2. Illustration of the primal (a) and the dual (b) template concepts.

The two patterns scanned using $\mathbf{T}$ and $\mathbf{T'}$ (called the "primal pattern" and the "dual pattern") are linked to each other in the pattern database. Then, during the simulation, whenever a coarse $\mathbf{pat}_T^*$ is found using the distance calculations and pasted on to the coarse simulation grid, the corresponding fine pattern (scanned from the same location $\mathbf{u}$ but with template $\mathbf{T'}$) retrieved from the pattern database is simultaneously pasted on to the fine grid. In essence, the multiple-grid simulation of the realization is performed in parallel on all grids but using only the similarity criterion of the current coarse grid. Figure 3 illustrates the steps of this approach for a single node $\mathbf{u}$ on the coarse grid of a 2 multiple-grid simulation.

The values simulated using the dual templates will affect the results of the subsequent distance calculations on the finer grids, thus allowing the desired feedback from the finer grids to the coarse grids. Consider the case of 3 multiple grids. When the coarsest grid ($g = 2$) simulation is completed, due to the use of the dual templates, the simulation grid will be completely full. Then, during the middle grid simulation, the values previously pasted by the dual template on to the finest grid will affect the distance calculations of the middle grid, hence providing the fine to coarse feedback. In essence, the algorithm places the large scale patterns on the coarsest grid and 'roughly' decides on small scale patterns and then 'corrects' for finer details on the subsequent grids.

Figure 5d of Section 4 (Examples) demonstrates the effect of the modified multiple-grid approach as used in SIMPAT.

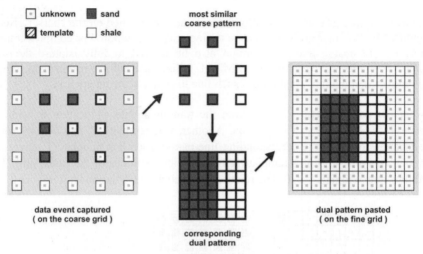

Figure 3. Illustration of using dual templates with a binary (sand/shale) variable. First, the data event **dev**$_T$(**u**) on the coarse grid is captured. The most similar pattern **pat**$_T$* to this data event is found by minimizing the distance between **dev**$_T$(**u**) and **pat**$_T^k$. Then, the corresponding dual pattern of **pat**$_T$* is retrieved from the pattern database and pasted on to the finest grid.

3 Data Conditioning

3.1 HARD DATA CONDITIONING

In SIMPAT, conditioning to hard data is performed in Step S-2 of the algorithm, during the search for the most similar pattern. If conditioning data exists on any $dev_T(\mathbf{u} + \mathbf{h}_\alpha)$, the algorithm first checks whether $pat_T^k(\mathbf{h}_\alpha)$ is equal to this data. If the pattern **pat**$_T^k$ does not fulfill this condition (i.e. there is a mismatch), it is skipped and the algorithm searches for the next most similar pattern until a match is found. If none of the available patterns fulfill the condition, the algorithm selects a pattern such that only the nodes of the data event that has conditioning information are considered during the distance calculations and other nodes are ignored. If several patterns fulfill this condition, then a second minimization is performed on the non-conditioning nodes of the data event using only these patterns. In essence, a two-stage similarity check is performed: first, only for the conditioning data; then, for the previously simulated nodes.

3.2 SOFT DATA CONDITIONING

In Earth Sciences, soft data typically refers to data obtained from indirect measurements acquired using some form of remote sensing (e.g. geophysical methods). Thus, soft data is nothing but a 'filtered' view of the original field of study, where the filtering is performed by some forward model **F**. In general, the forward model **F** is not known exactly and is approximated by a known model **F***.

In SIMPAT, conditioning to soft data calls for a soft training image. This soft training image can be obtained by applying the approximate forward model **F*** to the (hard) training image (See Figure 7c and 7d of Section 4 for an example). The patterns of the soft data (for example, a response from a seismic survey) are related to the geological (hard) patterns of the realization through the above mentioned filter model. The pair of hard and soft training images provides a basis for modeling the multiple-point relationship between hard and soft patterns. In fact, any joint statistics or pattern pairs extracted from the two training images can be considered as the multiple-point alternative of a cross-variogram in variogram-based geostatistics.

Once the soft training image is obtained, SIMPAT explicitly relates the patterns in the hard training image and the soft training image by creating a joint pattern database. In other words, Step P-1 of the pre-processing part of the algorithm is modified such that, for every $\mathbf{pat}_T^k$ of the hard training image, a corresponding soft pattern is extracted from the soft training image from the same location **u**. Another modification is done to the Step S-2 of the simulation part of the algorithm, i.e. the search for the most similar pattern. Instead of minimizing the distance between $\mathbf{dev}_T(\mathbf{u})$ and $\mathbf{pat}_T^k$, the algorithm now minimizes,

$$d^{1,2}\langle \cdot, \cdot \rangle = \omega \times d^1 \langle \mathbf{dev}_T(\mathbf{u}), \mathbf{pat}_T^k \rangle + (1-\omega) \times d^2 \langle \mathbf{dev}_T^2(\mathbf{u}), \mathbf{pat}_T^{2,k} \rangle \quad (5)$$

i.e., the summation of two distances where $\mathbf{dev}_T^2(\mathbf{u})$ denotes the soft data event obtained from the soft data grid **sd**, $\mathbf{pat}_T^{2,k}$ is a soft pattern and ω is a weight that is attached to the combined summation to let the user of the algorithm give more weight to either the hard or the soft values, reflecting the 'trust' of the user to the soft data. The flowchart of these modifications when conditioning only to soft data is given in Figure 4.

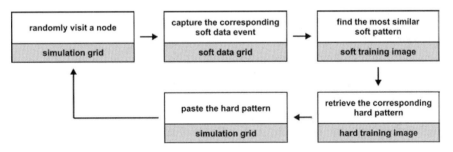

Figure 4. The flowchart for the conditional search for the most similar pattern $\mathbf{pat}_T^*$ when the hard data event is not informed. When there is conditioning data or previously simulated nodes within the hard data event, a joint search is performed instead.

The net result of the above modifications is that, for every node **u**, the algorithm now finds the most similar pattern not only based on the previously calculated nodes but also based on the soft data. When there is also hard data available, this minimization is performed only after the patterns that condition to the hard data are found as explained in the previous section, i.e. hard data has priority over soft data.

4 Examples

Figure 5a is a 7 facies training image depicting a tidal channel system in an oil reservoir. A notable property of Figure 5a is that, the image is highly non-stationary, especially the large scale variation: note how one facies appears only on the front part of the cube. Figure 5b is an unconditional SIMPAT realization obtained using a single-grid simulation. Figure 5c shows the application of the traditional multiple-grid approach (Tran, 1994). Section 2.3 explains two modifications done to this traditional approach where (1) the coarse grid values are not 'frozen' on the finer grids and (2) the dual template simulation technique is employed. Figure 5d is an unconditional SIMPAT realization obtained using these modifications. As these final figure illustrate, the algorithm successfully captures the non-stationary behavior of the training image, while adequately reproducing the facies relations of the training image.

Figure 6a shows a synthetic reference case with 6 facies in an oil reservoir. A dense data set is sampled from this reference to test conditioning to hard data (Figure 6b). The training image used is shown in Figure 6c. The final conditional SIMPAT realization is in Figure 6d. The training image used in this example is highly representative of the reference case; both the reference and the training image contain stacked channels. This agreement keeps the number of conflicting patterns to a minimum during the simulation. For a more realistic case, the reader is referred to Arpat (2004).

Figure 7 demonstrates the application of soft data conditioning using SIMPAT. In this case, soft data is obtained by applying a seismic forward model **F*** to the binary reference case (Wu, Mukerji and Journel, 2004). The same model is applied to the training image to obtain the soft training image. The final SIMPAT realization (Figure 7f) conditions to soft data relatively well but pattern reproduction is somewhat degraded as made evident by the disconnected channel pieces. This issue, along with possible solutions, is further discussed in Arpat (2004).

5 Conclusion

The sequential simulation method of SIMPAT replaces the traditional probability framework of drawing from conditional probability distributions (for example, as used in the SNESIM algorithm of Strebelle, 2000) with calculations of similarity between patterns. This entirely new approach to stochastic simulation has the advantage that it focuses directly on one of the core purposes of stochastic simulation: reproduction of patterns (Be it two-point or multiple-point, stationary or non-stationary). The similarity approach does not share many of the restrictions of a probabilistic approach, which often calls for a rather strong assumption of stationarity in the inference or modeling of patterns via probabilities. While the initial results are promising, the downside of this approach is that the purely algorithmic formulation of the method to stochastic pattern reproduction and data conditioning is not yet well understood. Future research will therefore focus on understanding better the advantages and limitations of the various new concepts (similarity, dual templates, similarity-based data conditioning, etc.) presented in this paper.

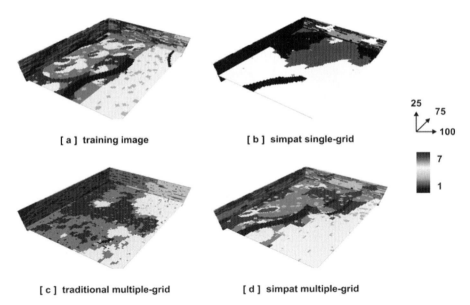

Figure 5. Unconditional SIMPAT. (b) - (d) all use 11×11×5 templates. In (b), only a single grid is used. (c) utilizes the traditional multiple-grid method with 3 multiple-grids (where simulated nodes are frozen and stay constant for the rest of the simulation) and (d) is obtained using the new multiple-grid approach that employs the dual template simulation technique.

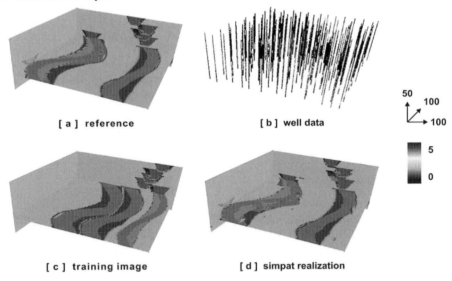

Figure 6. Hard data conditioning using SIMPAT. (b) is sampled from the reference (a) and constitutes 2% of all nodes. (c) is the training image and (d) is the final conditional SIMPAT realization obtained using a 11×11×5 template and 3 multiple-grids.

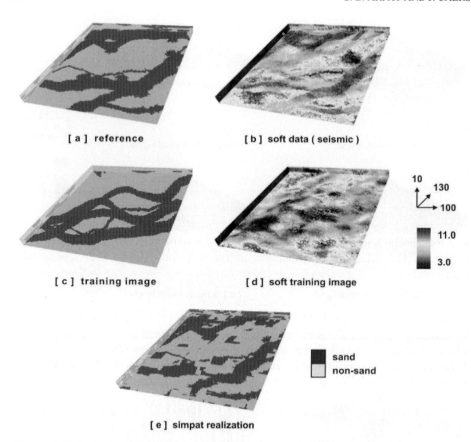

Figure 7. Soft data conditioning using SIMPAT. The soft training image (d) is obtained by applying an approximate model **F*** to (c). The final conditional SIMPAT realization (e) is obtained using a 11×11×3 template and 3 multiple-grids.

References

Arpat, G. B., SIMPAT: stochastic simulation with patterns, 17th SCRF Meeting, Stanford Center for Reservoir Forecasting, Stanford University, 2004.
Daly, C., Higher order models using entropy, Markov random fields and sequential simulation, *Geostatistics-Banff*, Kluwer Academic Publications, 2005.
Deutsch, C. and Journel, A., *GSLIB: Geostatistical Software Library*, Oxford University Press, 2nd ed., 1998.
Duda, O., Hart, P. and Stork, D., *Pattern Classification*, John Wiley & Sons, Inc., 2nd ed., 2001.
Goovaerts, P., *Geostatistics for Natural Resources Evaluation*, Oxford University Press, 1997.
Guardiano, F. and Srivastava, R., Multivariate geostatistics: Beyond bivariate moments, *Geostatistics-Troia*, Kluwer Academic Publications, 1993, pp. 133 - 144.
Palmer, S.E., *Vision Science: Photons to Phenomenology*, MIT Press, 1999.
Strebelle, S., *Sequential Simulation Drawing Structures from Training Images*, Ph.D. thesis, Stanford University, 2000.
Tran, T., Improving variogram reproduction on dense simulation grids, *Computers and Geosciences*, vol. 20 no. 7, 1994, pp. 1161–1168.
Wu, J., Mukerji, T. and Journel, A., Prediction of spatial patterns of saturation time-lapse from time-lapse seismic, *Geostatistics-Banff*, Kluwer Academic Publications, 2005.

SEQUENTIAL CONDITIONAL SIMULATION USING CLASSIFICATION OF LOCAL TRAINING PATTERNS

T. Zhang, P. Switzer, A. Journel
Department of Geological and Environmental Sciences
Stanford University, CA, 94305, U.S.A

Abstract Local spatial structures, as depicted by a training image, can be summarized by a few general linear filter scores. Local training patterns are then classified according to these scores. Sequential simulation proceeds by associating each conditioning multiple-point data event with a score class and then patching a pattern from this class onto the simulation grid. This procedure can handle both binary and continuous variable training images as illustrated by several diverse training images.

1 Introduction

Multiple point (mp) simulation aims to capture local patterns from a training image (TI) and anchor them to actual data. A training image reflects only general aspects of spatial structure or texture. It should display stationary patterns, which can be transported to the actual simulation space, see Figure 1a for an horizontal 2D fault training image example.

The original mp simulation concept was introduced by Srivastava, in Guardiano and Srivastava (1993). The original algorithm was very CPU-demanding in that the training image had to be rescanned for each node being simulated. Strebelle (2002) traded the CPU problem for a greater RAM demand by scanning the TI only once and storing all required information in a search tree data structure. Strebelle's program *snesim* made mp simulation feasible for 3D applications but limited to the joint simulation of no more than 4 or 5 categories.

In this paper, we propose a new mp simulation approach that can deal with both categorical and continuous variable training images with reasonable CPU and memory demand. Dimension reduction, hence RAM saving, is obtained by classifying the training patterns according to a few linear filters. A neighborhood template is passed over the training image. At each pixel location of the training image the template records the local pattern as an array of values. The array is reduced to a low-dimensional set of scores by applying a few general linear filters, see Figure 2. Patterns with similar scores are then grouped together into pattern classes.

To capture patterns at different scales, the template can be rescaled to scan the same training image. For example, if the finest template comprises N pixels at horizontal spacing 1x1, a coarser template with the same topology would comprise again N pixels but with spacing 2x2, or 4x4. This corresponds to the concept of multiple grids commonly used in sequential simulation (Tran, 1994).

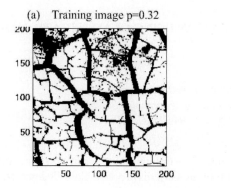
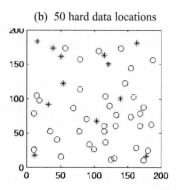

Figure 1. (a) training image with p=32% of faults, (b) 50 data locations sampled from (a) (star-faults; circle-background)

2 Pattern scoring

Let $X(i,j)$ denote the value at location (i,j) in the training image. A score $S_f(i,j)$ for the pattern in the neighborhood of (i,j) is defined for a filter $f(u,v)$ as follows:

$$S_f(i,j) = \sum_{v=-n}^{n} \sum_{u=-n}^{n} X(i+u, j+v) f(u,v)$$

where the dimension of the local neighborhood or template is $(2n+1) \times (2n+1)$. We define six different filters $f_1, ..., f_6$ as follows.

(1) f_1 : N-S average

$$f_1(u,v) = 1 - \frac{|v|}{n}, v = -n, ..., n$$

see Figure 2a.

(2) f_2 : E-W average

$$f_2(u,v) = 1 - \frac{|u|}{n}$$

see Figure 2b.

(3) f_3 : N-S gradient

$$f_3(u,v) = \frac{v}{n}$$

see Figure 2c.

(4) f_4 : E-W gradient

$$f_4(u,v) = \frac{u}{n}$$

see Figure 2d.

(5) f_5 : N-S curvature

$$f_5(u,v) = \frac{2|v|}{n} - 1$$

see Figure 2e.

(6) f_6 : E-W curvature

$$f_6(u,v) = \frac{2|u|}{n} - 1$$

see Figure 2f.

Each of these six filters is used to scan the TI. At each pixel location, the template of neighborhood data is weighted by the filters to produce a series of 6 scores. If the six scores are assigned to the pixel at the center of the template, we thus obtain score maps of the training image itself. In Figure 3, we see the score maps for the training image in Figure 1a. The size of the template used is 27x27 pixels (n=13), while the training image is 200x200.

The first two score maps S_1 and S_2 are weighted moving averages of the 27x27=729 template values. They highlight the object center locations. The next two scores S_3 and S_4 come from gradient filters; they provide edge detection, and highlight the object boundary contrast. The last two scores S_5 and S_6 are derived by curvature filters, they provide gradient changes. Note that these 6 filters privilege the NS and EW directions; appropriate rotations should be applied to either the training image or the filter weight maps if one wishes to emphasize different directions yet with the same total number of filters.

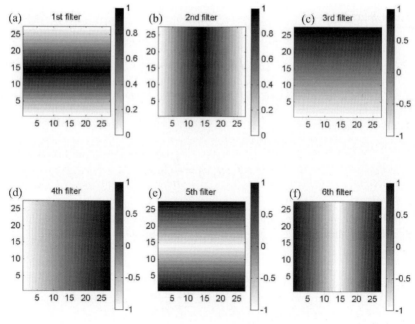

Figure 2. Six general filters (27x27)

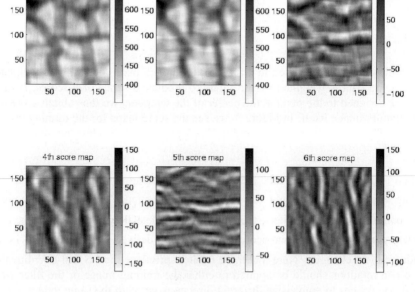

Figure 3. Six score maps at the finest grid

3 Pattern classification

Scanning the TI with each of the six filters produces a frequency distribution for each of the scores $f_1, ..., f_6$. Each of these marginal frequency distributions is discretized into 5 equal frequency bins according to their respective quintiles. This results in a partition of the 6-dimensional score space into $5^6 = 15625$ cells. [For binary data on a fine grid it may happen that many templates consist of all zeros or all ones. Therefore it is possible for some quintiles to be the same, resulting in fewer effective bins].

Even though each of the six scores has been divided into equal frequency bins, the 6-component joint cell frequencies are not equal. Many cells are empty because there are no local training patterns having such filter score combinations. Training patterns whose filter scores fall into the same cell are thus grouped into pattern classes. For each non-empty score cell, a "prototype" is obtained by averaging all patterns falling into that class, which can be seen as the aggregate of similar training patterns. Figure 4 shows the first 8 prototypes with the most training pattern replicates taken from Figure 1a on the finest scale with a template size 27x27.

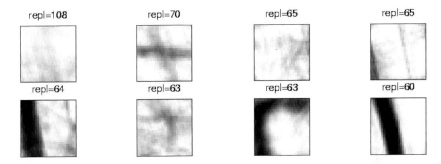

Figure 4. 8 prototypes with the most replicates from the training image of Fig. 1a at the finest grid (27x27)

4 Pattern simulation

Based on the previous classification of local training patterns, sequential simulation with multiple-grids can be utilized to generate pattern simulations that together mimic structural features of the training image.

At each node to be simulated, conditioning data are searched within a data template centered at this node. This data template has the same dimensions as that used to scan the training image at the current grid level.

If there are no conditioning data within the data template, we choose the template prototype closest to the target global mean value and pick a training pattern from this prototype class, the pattern whose mean is closest to the target mean. A target mean value is specified before the simulation and it is expected that the averaged value of each realization should be close to this value. If there are conditioning data in the data template, calculate the distance between this data event (DEV) and each training prototype (PROT) template recorded at the current grid level.

There are three types of conditioning data, $k = 1,2,3$:
(1) hard original data
(2) previously frozen simulated nodes
(3) non-frozen previously simulated nodes from pattern patches, see below.
The distance expression is written as

$$d(DEV, PROT) = \sum_{k=1}^{3} \omega(k) \frac{\sum_{i_k=1}^{n_k} |x^k(i_k) - y^{(k)}(i_k)|}{n_k}$$

where i_k are the pixel locations of information of type k and $\omega(k)$ are weights for the three respective information types with $\omega(1) \geq \omega(2) \geq \omega(3)$.

Once we identify the local template prototype closest to the conditioning template information, we sample a specific pattern from the prototype pattern class. We patch the sampled pattern at the current simulation node but retaining hard data and previously frozen simulated locations. The "inner" part of the patch is frozen and is not revisited in the sequential simulation. A larger inner patch area makes simulation faster, but may cause discontinuity. The outer part of the patch will be revisited, hence re-simulated. The concept of using a patch instead of a single node can be found in texture synthesis (Liang et al., 2001).

We use multi-grids to capture pattern structure and texture at different scales. The training image is scanned using local templates at several grid scales. Separately, at each grid scale, local patterns are converted to 6-dimensional scores using the filters described earlier. Thus, at each grid scale we get a classification of local patterns at that scale. In our examples, we used two coarser scales that are 2 times and 4 times the dimension of the finest grid scale.

Simulation proceeds from the coarsest grid to the finest grid. Simulated values from the preceding coarser grid are used as conditioning information at the finer grid simulation. However, all coarser grid simulated values are revisited and re-simulated at the finer grid.

5 Illustrations

The illustrations shown here exhibit the ability of local filter scores to capture spatial patterns when used to simulate categorical and continuous patterns from training images. First, we investigate fault structural simulation using the 200x200 binary categorical training image of Figure 1a. The area proportion covered by faults in the training image is p=0.32.

Three grid scales are used; the local template size is 27x27 pixels; the patch size is 19x19 pixels. 50 hard data, as shown in Figure 1b, are sampled from the training image. After training image classification, sequential simulation proceeds by randomly visiting grid nodes (pixels) and identifying local pattern prototypes that match the currently available information in the neighborhood of the simulation node. The target proportion of fault area was set to 0.30. Figures 5a-5c display the same conditional simulated realization at the different grid scales, the final simulated image being at the right of the figure. Figure 6 displays three additional conditional simulations at the final grid. It can be seen that the fault structures, including small fractures, are reasonably reproduced, although with less large scale continuity than displayed on the training image of Figure 1a. The 50 hard conditioning data are honored exactly by all simulated realizations.

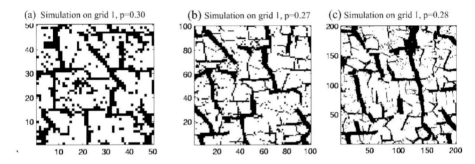

Figure 5. (a)-(c) The same conditional simulation at different scales

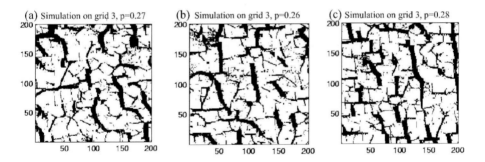

Figure 6. (a)-(c) Three additional conditional simulations at the final grid

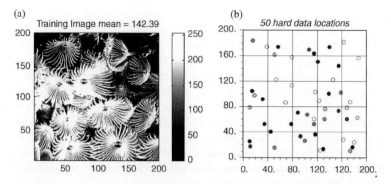

Figure 7. (a) Texture training image, (b) 50 data location map
The same grey scale is used for all maps of Fig.7-8

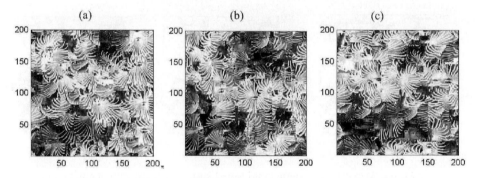

Figure 8. (a)-(c) 3 conditional simulations at the final grid

A continuous variable training image is displayed in Figure 7a, it is a picture of sea anemones. It contains visible gray scale textures with curvatures. Figure 7b shows 50 hard data locations. These samples were generated by sampling a non-conditional simulation from this training image; we used 27x27 pixel templates to classify local training patterns over 3 grid scales. The patch size is 19x19; Figures 8a-8c show three conditional simulations based on the training image of Figure 7a. The simulated anemones can be recognized as such, however with square discontinuities corresponding to the template size. This is the price to pay for working with patches instead of points. This problem calls for future tuning of the algorithm.

It is important to specify correctly both the template size and the patch size. The template size depends on the complexity and the scale of training patterns. The guiding rule is that the template should be large enough that on the coarsest grid it can capture the pattern objects and their interaction. For example, for the sea anemones training image, the size of the largest template should be at least equal to the size of the average anemone object. The patch size can be up to 2/3 of the template size in each direction. A larger patch speeds up the simulation and improves the pattern reproduction, but at the cost of generating discontinuities.

It is suggested to test different template and patch sizes using the full training image but simulating only part of the required field.

6 Conclusions

In this paper, we apply a set of filters to scan training images. Local patterns and textures in training images are classified by a set of filter scores. This leads to a significant dimension reduction of the space of training patterns. Drawing from classes of training patterns allows us to simulate whole patterns as opposed to point values. The simulation proceeds by sequentially visiting each simulation node and identifying the closest training pattern class to the local template data centered at the simulation node. We sample a specific pattern from the identified pattern class, and patch the sampled pattern at the simulation node. Freezing the inner part of patched pattern not only makes the simulation faster but also ensures better pattern reproduction. Multi-grid simulation is implemented, allowing for pattern reproduction at different scales. Although all illustrations are given in 2D with six filters, the algorithm can be extended to 3D using correspondingly 9 filters.

References

Guardiano, F., and Srivastava, R. M., 1993, Multivariate geostatistics: Beyond bivariate moments, *in* Soarses A., ed. Geostatistics-Troia Vol. 1: Kluwer Academic, Dordrecht, pp.133-144

Liang, L., Liu, C., Guo, B., and Shum, H.-Y., 2001, Real time texture synthesis by patch-based sampling, ACM Transactions on Graphics, vol.20, Issue 3, p.127-150

Strebelle, S., 2002, Conditional simulation of complex geological structures using multiple-point statistics: *Math. Geol.*, v.34, no1, p.1-21.

Tran, T. 1994, Improving variogram reproduction on dense simulation grids, *Computers & Geosciences*, vol.25, no.7, pp.1161-68

ra
A PARALLEL SCHEME FOR MULTI-SCALE DATA INTEGRATION

OMER INANC TUREYEN and JEF CAERS
Stanford University, Department of Petroleum Engineering, Stanford, CA
94305-2220, USA

Abstract. In this paper, we propose a parallel modeling approach for solving large and complex inverse problems involving multiple data sources each with different scale of observation. This parallel approach relies on building property models on multiple grids with different resolution at the same time, rather than selecting a single modeling grid. By keeping a high resolution model and its upscaled, coarsened model in constant consistency with each other during the inversion process, a fully consistent integration of all data sources is achieved.

1 Introduction

With the advance of CPU power, numerical models have become an essential part of most engineering applications, be it a finite difference code of flow in the subsurface or a boundary element code modeling the geo-mechanical behavior of faulting and folding structures. Any model, analytical or simulated using a finite element/difference code, is as good as the input material properties on which the physical model is applied. In an Earth Science context, the modeling of such properties is subject to a large degree of uncertainty due to lack of exhaustive access and due to often strong heterogeneity of the medium under study. Instead, a wide variety of indirect data is available to construct various realizations of the media in question. Moreover the various data sources have a different "area of coverage" and "scale of observation". Fine-scale data, for example obtained by drilling a well or by taking a sample at the surface provide direct measurements but are typically sparse in coverage. Remote sensing methods cover a large area but provide only indirect evidence of the properties to be modeled. For the purpose of this paper, the data is subdivided into two parts:

- Static data: refers to direct or indirect observations of the material or rock properties being modeled as input to the physical model. For example a rock/soil type observed in a well, a 3D seismic survey.
- Dynamic data: refers to all data that are direct observations of the physical phenomenon being studied. For example a measurement of pressure/head in a well, a stress or strain measurement.

The goal pursued in this paper is a method for building property models that honor these two types of data. To address this problem, some difficult challenges will need to be addressed

- The relationship between the dynamic data and the modeled property is in general non-linear, often provided through partial differential equations simulated using a finite element/difference code.
- Static data is often of smaller scale than dynamic data, which provides integrated or convoluted information about the modeled property.

2 Solution using a parallel modeling approach

The purpose of numerical modeling is to predict a response based on a numerical model, e.g. the degree of fracturing in a structure, the production of water or oil in a well. Physical laws on which such prediction rely are generally of the form:

$$f(\mathbf{q}, \mathbf{z}) = 0 \qquad (1)$$

where $\mathbf{q}$ are the physical quantities (e.g. pressure, stress), as function of space and time and $\mathbf{z}$ are the material properties, which, in this paper is only a function of space (e.g. porosity, Poisson's ratio). In most cases a 2D or 3D regular or irregular grid of properties $\mathbf{z}$ is generated. The physical law Eq.(1) defines a forward model between the material properties and physical quantities

$$\mathbf{q}^{res} = g(\mathbf{z}, \mathbf{q}^{in})$$

for some initial state and boundary conditions, $\mathbf{q}^{in}$. Static data consist of direct or indirect information related to $\mathbf{z}$, while dynamic data consists of information $\mathbf{q}^{res}$ (e.g. pressure in a well). The non-linear nature of g forces the modeler to use iterative methods for solving the inverse problem (finding $\mathbf{z}$, for given $\mathbf{q}^{res}$), hence requires multiple evaluations of g. When g is a numerical simulation model this may be CPU demanding (Caers, 2004). In that regard the problem of modeling $\mathbf{z}$ calls for a decision on the resolution (dimension) of the modeling grid that is a trade-off between two constraints:

- The grid size should be small enough to include the static fine-scale information, particularly any direct property data.
- The grid size should be coarse enough for finite element/difference simulation of the forward model g to be feasible within reasonable CPU-time. This is important for including the dynamic data on $\mathbf{q}$

In this paper, we propose a parallel modeling approach for $\mathbf{z}$ that avoids this trade-off by working on two grids at the same time: a high resolution grid that allows including any fine-scale static information on $\mathbf{z}$ and a coarsened grid that allows running multiple evaluations of g and thereby including the dynamic data on $\mathbf{q}$. The key idea presented in this paper is to keep the two grids in constant consistency with each other both in terms of the property $\mathbf{z}$ and the responses on $\mathbf{q}$.

The proposed parallel modeling approach follows the following basic steps.

Step 1: High resolution model generation

A high resolution geostatistical realization (**z**) that honors the static data is generated. This initial high resolution model does not yet match the dynamic data, hence will need to be perturbed. Any perturbation method can be used (gradient-based, Metropolis samplers, rejection methods etc,...) In this paper, such perturbations are represented by a set of parameters **r** that change a realization **z** into a perturbed realization **z(r)**, the magnitude of perturbation given by **r**. When **r=0**, no perturbation is performed hence **z=z(0)**.

Step 2: Optimized gridding and upscaling

The next step consists of upgridding and upscaling the high resolution realization **z(r)** to a coarsened realization $\mathbf{z}^{up}(\mathbf{r})$: relationship:

$$\mathbf{z}^{up}(\mathbf{r}) = S_\theta(\mathbf{z}(\mathbf{r})) \tag{2}$$

Here S_θ represents the upscaling/upgridding technique applied on **z(r)** and θ the set of upgridding parameters (grid dimensions, averaging type, etc..) Upgridding refers to the construction of the coarse grid, which could be Cartesian or irregularly gridded. The dimension of the coarse grid is defined by the number of grid vertices in each x, y and z-direction. Upscaling refers to methods for assigning coarse grid properties given the high resolution property realization. Due to the presence of upscaling/upgridding errors, the high resolution and coarsened realization may conflict in terms of the responses when the forward model is applied on each of them. This possible inconsistency between the two grids needs to be reduced by minimizing the upscaling/upgridding errors introduced. Define as "true" but unknown upscaling error, the difference between responses of the high resolution and coarsened model:

$$\epsilon = \|g(\mathbf{z}^{up}(\mathbf{r})) - g(\mathbf{z}(\mathbf{r}))\| \tag{3}$$

This error cannot be calculated since the forward model g is too CPU-demanding to be evaluated on the high resolution realization. Hence the challenge is to reduce the upscaling error ϵ without knowing $g(\mathbf{z}(\mathbf{r}))$. To achieve this, we introduce a function g^* as an approximation to g that is less CPU demanding to evaluate. The shape of g^* is problem specific and could be a model with simplified physics or could be an analytical model (see example section for specifics). The approximate forward model allows to approximate the true upscaling error

$$\epsilon^*(S_\theta) = \|g^*(S_\theta(\mathbf{z}(\mathbf{r}))) - g^*(\mathbf{z}(\mathbf{r}))\| \tag{4}$$

The key idea is to reduce ϵ by reducing ϵ^*. ϵ^* can be reduced by adjusting the parameters θ of the upscaling/upgridding method until Eq.(4) is minimized. At the same time Eq.(3) will be minimized if the ranking of models **z** provided by the forward model g^* is the same as the ranking provided by g. In other words, ϵ and

ϵ^* need not be the same in absolute magnitude, ϵ^* must decrease when ϵ decreases and vice versa.

Step 3: Mismatch calculation and perturbation

Once an optimized coarsening of the high resolution model is determined, a model response is obtained by evaluating the forward model (g) on the coarsened realization

$$\mathbf{RP}_{\mathbf{z}^{up}}(\mathbf{r}) = g(\mathbf{z}^{up}(\mathbf{r}))$$

the $\mathbf{r}$ parameters can be optimized by minimizing the difference between the coarsened model response $\mathbf{RP}_{\mathbf{z}^{up}}(\mathbf{r})$ and the dynamic data $\mathbf{D}$:

$$\min_{\mathbf{r}} O(\mathbf{r}) = \min \| \mathbf{RP}_{\mathbf{z}^{up}}(\mathbf{r}) - \mathbf{D} \|$$

Some important properties of this approach are

- All data, fine-scale static and coarse-scale dynamic are integrated simultaneously.
- The upscaling/upgridding optimization forces the high resolution and coarsened model to be consistent with each other during the complete inversion process.
- The properties are modified on the high resolution grid, hence any model perturbation $\mathbf{z}(\mathbf{r})$ can be kept consistent with the static data and prior geological information.

3 Parallel modeling in reservoir characterization

3.1 RESERVOIR MODELING

In this section we present in greater detail how the parallel modelling methodology is applied to the problem of reservoir characterization for hydrocarbon reservoirs along with example applications.

Reservoir modelling calls for the integration of various data into a single reservoir model. Such data sources can be classified as follows:

1. Geological interpretation of reservoir architecture at all scales ranging from major faults to facies and bedding configurations. In geostatistics, such information can be quantified through the variogram or through 3D training images.
2. Well-log and core measurement information is often the most direct type of information, however is only telling of the near well-bore reservoir heterogeneity and provides information at a small (cm) scale.
3. 3D seismic information is probably most exhaustive, yet often at a scale larger than the reservoir modelling scale. In geostatistics this type of data is treated as soft, static data.

4. Reservoir dynamic data, most particularly from pressure and flow measurements, or increasingly common, 4D seismic. The scale of information provided by this kind of data set is largely unknown. It is spatially varying and dependent on boundary conditions and configurations of wells.

The current practice of reservoir modelling consists of modelling the reservoir first using the static data (sources one to three) on the high resolution grid, then upscaling the high resolution model into a coarsened model on which flow simulation (function g) is feasible. Next, the coarsened model is further adjusted to match the dynamic information (source four). In reservoir engineering this is commonly known as "history matching".

Such an approach has the following drawbacks:

- Any high resolution reservoir information (core or well-log) may be lost when the coarsened model is changed.
- Important fine and coarse scale geological information may be destroyed while history matching. Particularly when the history matching method does not take into account statistics such as variogram or multiple-point statistics that are characteristics particular to the high resolution geological model.

3.2 METHODOLOGY OVERVIEW

We provide first a broad overview of a parallel modelling scheme specific to reservoir characterization see (Figure 1).

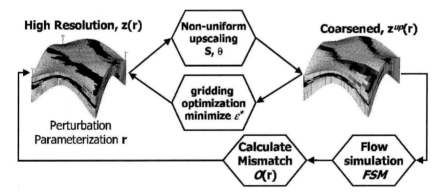

Figure 1. Parallel approach for reservoir characterization

The work-flow starts by constructing a high resolution realization ($\mathbf{z}(\mathbf{r})$) which can be perturbed using a set of perturbation parameters $\mathbf{r}$. A gridding optimization is performed in order to obtain the "non-uniformly gridded" coarse model which minimizes the mismatch between the flow responses of the high resolution model and the coarsened model. The optimization is performed on the θ vector (shown in Figure 1), which represent the various gridding parameters specific to the gridding algorithm (see next section). Once the optimally gridded coarse model is obtained, flow simulation (denoted by *FSM* in Figure 1) is performed on the coarsened model. Then the mismatch between the observed field data and the simulation results are compared. The **r** parameters are adjusted to reduce the mismatch. The entire loop is repeated until this mismatch is minimized.

3.3 PERTURBATION PARAMETERIZATION

Various perturbation methods can be used to perturb the high resolution geological model. However, perturbations should be parameterized such that any perturbation **z**(**r**) honors the same geological continuity model (variogram, object model, training image model) and the same static data as the initial high resolution model **z**. The gradual deformation (Hu and Roggero, 1997) and the probability perturbation method (Caers, 2003) methods are two examples of perturbation that honor this information.

3.4 GRIDDING - 3DDEGA

Although the above outlined parallel modeling approach is general in the type of gridding method, in this section we review an existing gridding algorithm, 3D-DEGA (3D Discrete Elastic Grid Adjustment, (Garcia, Journel and Aziz, 1992)). The algorithm is devoted to the generation of quadrilateral or hexahedric grids suitable for grid adaptation based on reservoir properties (ϕ, k), pressure fields, saturation fields or any other variable. The resulting grids are in a corner point geometry fashion and can be used with most commercial flow simulators. The main idea behind the 3D-DEGA algorithm is to generate coarse grid blocks that are as homogeneous in terms of a given input variable or variables (permeability, porosity, facies map, etc.).

3.5 GRIDDING OPTIMIZATION, APPROXIMATE FORWARD MODEL

The forward model g is a flow simulation that includes all physics necessary (gravity, capilary pressure, compressibility, multiple phases etc..) for simulating the actual reservoir flow. In real cases it may take several hours to run a full flow simulation on a grid of the order of 10^5 cells depending on the complexity of the physical model. To incorporate static information from well-logs and seismic, a typical high resolution geostatistical realization generated using stochastic simulation can be of the order of $10^6 - 10^7$ cells, a resolution on which flow simulation is not feasible. Approximate flow simulations are therefore required to minimize any upscaling errors as outlined above.

As an approximate physical model we use an incompressible single phase flow simulation (denoted by g^*). A single phase flow model can be calculated on the high resolution grid in a matter of minutes. To further aid the upgridding method we trace streamlines using the pressure (and velocity) solution of the single phase flow model. Using the streamlines simulation, approximate flow responses can be obtained, denoted by $\boldsymbol{RP^*}$.

The high resolution geostatistical model is then upscaled/upgridded to a coarser model with 3D-DEGA given some gridding parameters θ. The incompressible single phase flow solution and streamline simulations are repeated on the coarsened model, from which the same type of responses as for the high resolution model are calculated. The responses from the high resolution and the coarsened model are then compared. The gridding parameters θ are adjusted in an iterative manner until the mismatch (Eq.(4)) between the two responses is minimized.

3.6 2D EXAMPLE

3.6.1 *Definition of the problem*

A 2D, high resolution reference permeability realization is generated through a training image based technique using multiple point geostatistics. This was performed using the SNESIM (Strebelle, 2002) algorithm with the training image given in Figure 2a. The realization is representative of a channel system with a quarter five spot pattern production strategy where injector producer wells are placed on opposite corners. Water flooding for 500 days is simulated on this realization using a finite difference reservoir simulator (ECLISPE). A water cut curve is obtained (see reference permeability map and its corresponding water cut curve in Figure 2b and 2c). This curve is treated as dynamic data. A constant permeability value of 10000 md is assigned to the sand (channel) facies and 100 md is assigned to the mud (non-channel) facies. The permeability values at the well locations are assumed "known" and are treated as "hard data". This high resolution realization is composed of 100 × 100 × 1 grid blocks in the x, y and the z directions, where each block is of size 20ft × 20ft × 200ft.

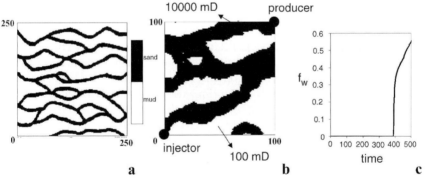

Figure 2. (a) The training image used for creating the reference permeability field, (b) The reference permeability field, (c) The flow response of the reference permeability field.

In order to demonstrate the effectiveness of the parallel modeling approach, 30 high resolution realizations were generated (using SNESIM), conditioned only to hard data (no history matching was performed). Full flow simulation was performed on these 30 realizations and Figure 3 illustrates the flow responses. As expected the flow responses of the high resolution models conditioned only to hard data provide a wide scatter of production responses.

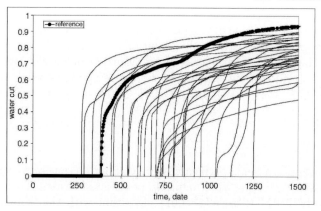

Figure 3. Flow responses of 30 realizations not matched to reference water cut data.

3.6.2 Applying the parallel modelling approach

Figure 4 illustrates the work-flow of the parallel modeling methodology specific for this example. In the first step the SNESIM algorithm is used for generating a high resolution realization (with the same dimensions of the reference realization). In generating this initial realization, the same training image given in Figure 2a is used. Streamline simulation (which acts as the approximate flow model, denoted by g^* in the previous section) is performed on this high resolution realization and a pseudo (approximate) water cut curve ($\mathbf{RP}^*$, the approximate flow response) is obtained.

Using the 3D-DEGA algorithm (with an initial guess of upgridding parameters θ), the high resolution model is upgridded and upscaling is performed by taking arithmetic averages of the permeability values from the high resolution realization. Streamline simulation is performed on the coarsened model and a pseudo water cut curve ($\mathbf{RP}^{up*}$) is calculated. An optimization step minimizing the mismatch between $\mathbf{RP}$ and $\mathbf{RP}^{up*}$ is applied, during which the upgridding parameters θ are optimized. In addition to the upgridding parameters, the number of coarse grid blocks are also optimized while keeping the total number of grid blocks constant ($n_x \times n_y = 625$).

Full flow simulation (the full flow model, denoted by g in the previous section) is performed on the optimally gridded coarsened model and the simulated water cut curve is obtained. The mismatch between the water cut curve and the reference water cut curve (given in Figure 2c) is computed. The perturbation method (probability perturbation, (Caers, 2003)) repeats the entire procedure until this mismatch is minimized by optimizing the $\mathbf{r}$ parameters.

3.6.3 Results for the 2D synthetic example

The proposed parallel modeling approach has been performed initially without the gridding optimization for 30 different random seeds in order to make a comparison with Figure 3. At the end of the inversion process, for each random seed, a coars-

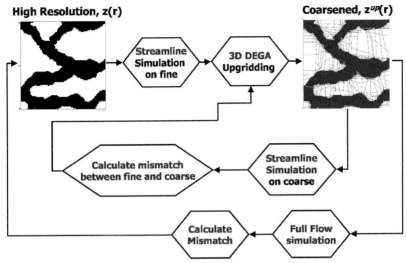

Figure 4. Schematics of the parallel approach specific to the 2D example.

ened and a high resolution flow response results as output. The coarsened models match the history well, and provide accurate future predictions for the same well configuration (see Figure 5a). Figure 5b illustrates the flow responses of the high resolution models corresponding to the history matched, coarsened models. As it is clear from the Figure 5b, the high resolution models in this case provide a match to some degree when compared with Figure 3, but not as well as the match of the coarsened models. This can be attributed to upscaling errors that introduce inconsistency between the two modeling solutions.

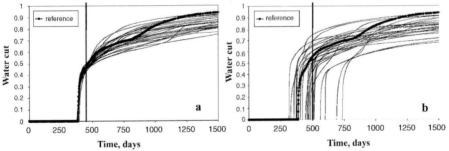

Figure 5. Flow results of the parallel modelling approach without the gridding optimization. (a) Flow responses of the non-optimally gridded coarse models, (b) Flow responses of the corresponding high resolution models.

Therefore, gridding optimization needs to be introduced each time a high resolution model coarsened. Results obtained by incorporating the gridding optimization shown in Figure 6a for the coarsened model match the history well. Figure 6b illustrates the flow responses of the corresponding high resolution models. Improvements on these responses are clear when compared with Figure 5b.

The scatter on the high resolution model response is reduced considerably through gridding optimization.

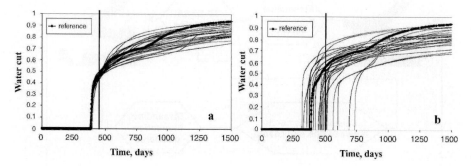

Figure 6. Flow results of the parallel modelling approach. (a) Flow responses of the optimally gridded coarse models, (b) Flow responses of the corresponding high resolution models.

4 Conclusions

In most earth sciences applications "modeling" in general is a difficult task due to lack of exhaustive data and heterogeneity of the medium under study. Fully integrating all data into a single grid is usually not plausible.

The parallel modeling approach uses multiple modeling resolutions at the same time. A (uniformly gridded) high resolution model is used for integrating static data and applying geostatistics effectively. The coarsened model may be gridded in any fashion and is used for integrating dynamic data. Consistency between these two resolutions of models is ensured through a gridding/upscaling optimization step, which forces the responses of each model resolution to be consistent.

References

Garcia, M. H. Journel, A. G. and Aziz, K. *Automatic Grid Generation for Modeling Reservoir Heterogeneities*, SPE Reservoir Engineering, May, 1992, p. 278.

Caers, J. *Geostatistical History Matching Under Training-Image Based Geological Model Constraints*, SPE Journal, September, 2003, p. 218-226.

Roggero, F. and Hu, L. Y. *Gradual deformation of continuous geostatistical models for history matching*, SPE 49004, presented at the SPE Annual Technical Conference, 27-30 September 1998, NewOrleans-Lousianna-USA.

Strebelle, B.S. *Conditional Simulation of Complex Geological Structures using Multiple-Point Geostatistics*, Math. Geol., v34, no.1, January, 2002, p1-22.

Landa, J.L., Horne, R.N. *A Procedure to Integrate Well Test Data, Reservoir Performance History and 4-D Seismic Information into a Reservoir Description*, SPE 38653, presented at the SPE Annual Technical Conference, 5-8 October 1997, San Antonio-Texas-USA.

Caers, J. *Data Conditioning With the Probability Perturbation Method*, Paper presented at the 7th International Geostatistics Congress, 26 September - 1 October 2004, Banff, CANADA.

STOCHASTIC MODELING OF NATURAL FRACTURED MEDIA: A REVIEW

JEAN-PAUL CHILÈS
Centre de géostatistique, École des mines de Paris,
35 rue Saint Honoré, 77305 Fontainebleau cedex, France

Abstract. The connectivity and the flow or mechanical properties of networks of faults and joints are key factors in a number of applications. Only a minute fraction of the fractures in the domain of interest are usually observed, so that a deterministic modeling of the fracture network is not possible. Stochastic models have been developed for a variety of fracture patterns. They can be classified in objects models, which are purely stochastic, and process-based models, which take account of the mechanical processes that rule fracturing. This presentation is focused on the geometrical and topological aspects of fracture networks.

1 Introduction

Structural discontinuities such as faults and joints occur at various scales and are now widely recognized as a key factor in a number of situations. They can act as conduits or flow barriers depending on whether they are open or sealed, and thus impact the safety of nuclear waste storage in geological formations, the oil recovery in fractured hydrocarbon reservoirs, and the heat recovery in geothermal hot dry rock reservoirs. Open fractures delimit blocks of ornamental stones or impact the stability of stopes or the safety of underground exploitations, caverns and tunnels. Veins are associated with mineralization.

Fracturing results from a number of processes and depends on the rock fabric, rock properties, geological setting, past and present mechanical constraints. Many different fracture patterns can therefore be observed and there is no general fracture network model. While large faults are usually known from geological mapping and seismic surveys, medium and small scale structures are very sparsely observed and are therefore represented by means of stochastic models. Many purely stochastic models, where the fractures are represented by objects, have been developed. The mechanical processes leading to the initiation and growth of faults and joints are better and better understood, which has lead to the development of algorithms modeling the fracturing process or at least parameters that control it. These algorithms are exploited by process-based models. We will first review the main types of object models and then examine approaches based on a modeling of geological and mechanical processes. This paper will be focused on the geometrical and topological aspects of fracture networks, but in applications this shall not be disconnected from the hydrological or mechanical aspects.

2 Object Models

2.1 BASIC RANDOM SET MODELS AND GENERALIZATIONS

The first fracture network models were deterministic, such as the model defined by three orthogonal series of parallel and regularly spaced planes. Such a network is too regular, subdividing the space like sugar cubes, so that stochastic models were soon proposed. The first stochastic models were simple prototypes corresponding to the usual models of random set theory. Priest and Hudson (1976) used random planes to represent fractures that can be considered as infinite at the scale of the domain under study. To represent finite-size fractures, Baecher et al. (1977) developed the random disk model, namely a standard Boolean model where the objects are disks with random diameters and orientations. Another way to obtain finite fractures is to start from a random plane model and define a stochastic tessellation in each plane, for example a Voronoi or Poisson polygons tessellation; each polygon is then randomly considered as a fracture or as intact rock. Depending on the method chosen for defining the tessellation, the fractures can intersect with others (Veneziano, 1978) or abut against others (Dershowitz, 1984).

These basic models represent networks with a uniform fracture intensity. In applications, however, fracture intensity is usually not uniform. This is accounted for by generalizing the Boolean model, based on Poisson points, to marked point processes based on more general point processes. The main point process models used are (e.g., Stoyan and Stoyan, 1994): (i) Poisson point process with spatial variations of intensity, these variations being either deterministic (inhomogeneous Poisson point process) or modeled as a positive stationary random function (Cox process); (ii) cluster process (also called shooting process, or parent-daughter model): primary points (targets, or parents) constitute a Poisson point process; each primary point is the center of a cluster of secondary points (shot impacts, or daughters) randomly and independently located around the primary point according to some dispersion distribution; (iii) hard-core model, to forbid the presence of too close points, for example because the relaxation of constraints in the vicinity of a fracture inhibits the creation of a new fracture. Chilès (1989) modeled fractures in a granitic site as clusters at a local scale with a regionalized intensity at a wide scale.

In practice, the orientation of the fractures is not purely random, and several fracture sets are superimposed, each one with a direction that is well defined or varies around a mean direction. Each set is linked to an event of the structural history of the site and has its own characteristics as regards to the size of the fractures, their aperture, etc.

2.2 HIERARCHICAL MODELS

Most of the models presented above were developed in view of mining engineering applications or for the study of potential nuclear-waste underground storage sites in granitic formations, namely for situations where there is no well-expressed hierarchy among the various fracture sets, despite of their chronology. The situation is very different for sedimentary rocks, so that other models were developed for them.

Hierarchical models are more complex than basic models because they must include the relationships between the various fracture sets. Conversely, fractures in sedimentary rocks are often either horizontal (bedding planes) or subvertical and confined to one or several layer (joints), so that their 3D modeling amounts to a series of 2D models. The first stochastic hierarchical model we know is a 2D model in a petroleum reservoir (Conrad and Jacquin, 1973): (i) large faults are represented by a primary network of random lines; (ii) in each Poisson polygon defined by the primary network, an independent Boolean model of segments represent finite fractures; the model is truncated by the boundaries of the polygon, so that fractures can abut against faults of the primary network but cannot intersect them.

Bourgine et al. (1995) propose another model, developed to model vertical joints at a pluridecametric scale in the Saq sandstones, which are considered as an analog to some petroleum reservoirs. That model is based on renewal processes, which offers much flexibility: (i) a primary set is composed of subparallel finite fractures; (ii) secondary sets connect fractures of the primary set, with a given proportion of terminations abutting against fractures of the primary set. A 3D model is obtained by superimposing several such models with parameters depending on the bed height. Other models can be found in the literature.

2.3 MULTISCALE BEHAVIOR AND FRACTAL MODELS

The organization of fractures, more precisely of faults, is often considered as self-similar (e.g., Turcotte, 1992) because faults can occur at all scales from cartographic faults to microfaults. This is often sustained by the fact that some variable (e.g., fracture size, fracture spacing, size of rock fragments, box counting, i.e., number of cells intersected by fractures as a function of cell size) follows a power law distribution with a noninteger parameter. In conclusion of a study of several sites in sandstones, Gillespie et al. (1993) observe that the spacings between tectonic faults follow a power law distribution because of a clustering of smaller faults in the vicinity of larger faults, whereas the spatial distribution of major joints is very regular because it is controlled by the bed thickness of the jointed unit, by the differences in mechanical properties between the jointed unit and adjacent layers and by the extensional strain. The situation is different in granitic rocks where the spacing between joints can follow a power law distribution (Barton and Zoback, 1990), whereas joint corridors and faults can be regularly spaced (Genter and Castaing, 1997). So the invariance laws (self-similarity or self-affinity ruled by a power law) which are often advocated have no universal character and shall be used only between well identified bounds, as noticed by Hatton et al. (1994) and Peacock (1996). Moreover, the observation of a power law distribution for some parameter of the fractures does not imply the fractal character of the fracture network, as shown for example by Walmann (1998) in a laboratory experiment where clayey material was submitted to mechanical tests generating cracks.

In comparison with the abundant literature about fractals, few fractal models have been proposed to represent fracture networks, probably because one or several fractal exponents are far from characterizing a fractal model. Let us mention models defined by fragmentation (Turcotte, 1986; Acuna and Yortsos, 1995): the domain of interest is subdivided in two parts by a fracture of the first generation; each part in then subdivided in two parts by a fracture of the second generation, and so on. To obtain a fractal model, the fractures are not created systematically but with a given probability; when a fracture is not created the subblock is no more subdivided. Bour and Davy (1999) developed another model, which is the transposition of the random disk model to fractals, by locating disks with a power law diameter distribution at the points of a fractal point process.

These synthetic models are often used to study the connectivity or flow behavior of the network—percolation, emergence of a continuous-medium behavior—as a function of the fractal dimension and other fractal exponents, either analytically or by simulation. The validity of the results for other models with the same fractal exponents, often assumed, is questionable.

The detailed fracture data sets studied by this author did not show evidence of a fractal behavior, either locally (Chilès, 1988, for granite) or over a wide range of scales (Castaing et al., 1996, for sandstones). In the latter case, the analysis showed a very different behavior for faults and joints, as well as characteristic scales for the joints, which could be related to the various mechanical units formed by the sedimentary beds, the sandstone formation, the sedimentary basin, and the upper crust. In such a case, stochastic models are built at a given scale, with the fractures of the finer scale incorporated with the rock matrix and the fractures at the coarser scale—usually few in number—modeled deterministically. Other models are needed for networks that are not controlled by mechanical units, for example in the sandstones studied by Odling (1997).

2.4 STATISTICAL INFERENCE

The key problem, from a geostatistical perspective, is the inference of the parameters of the stochastic models. Even for a simple model like the random disk model, this is not trivial, because fractures are 3D objects that cannot be observed directly but only through their intersections with boreholes (cores, electric logs) or outcrops (field exposures, drift walls, vertical stopes, aerial photographs). Part of the problem is also that the fractures of our models are an idealization of reality: true fractures are not planar circular disks for example. So if the choice of the model is not sound, it may be difficult for it to honor a variety of statistics about fracture density, fracture orientation, trace length, abutting fractures, connectivity, etc.

The most significant parameter is the fracture density μ, defined as the average fracture surface per volume unit. If we consider a set of fractures which are normal to a sampling line, this is simply the number of fractures per length unit; this is also the inverse of the average fracture spacing. Fractures that are oblique to a surveyed line or outcrop are underrepresented in comparison with fractures orthogonal to it (fractures parallel to the line or outcrop cannot be observed). It is therefore recommended to have several sampling lines or outcrops with different orientations, and data must be weighted according to the orientation of the fracture with respect to the line or outcrop: this is the aim of the well-known correction proposed by Terzaghi (1965) for line sampling, which can be improved in several ways (Yow, 1987) and has a variant for areal sampling (Chilès and de Marsily, 1993).

The fracture density μ is usually not the basic parameter of a fracture network model. For a random disk model for example, with disks of a fixed orientation, the parameters are the Poisson process intensity λ, namely the number of disks per volume unit, and the distribution of disk diameters F_D. It could be tempting to separately infer both parameters, as Zhang and Einstein (2000), for example, do it. These parameters, however, are not robust: it is not obvious to decide on the field that two aligned traces separated by a short intact interval are two distinct fractures or the en-échelons part of a single fracture, and structural geologists can adopt either view. The choice, however, will not affect the calculation of the total fracture length and thus the estimation of the fracture density μ, which is therefore a robust parameter. Consequently, it is preferable to infer separately the fracture density μ and the disk diameter distribution F_D, and then derive the corresponding Poisson process intensity λ by applying the relation $\mu = \pi \lambda M_2 / 4$, where M_2 is the quadratic mean of the disk diameter.

Now the diameters of the fractures cannot be measured because fractures are at best observed as fracture traces. The length of a fracture trace is shorter than the diameter of the fracture but the disk diameter distribution can be derived from the trace length distribution using the stereological formula which is appropriate for the kind of sampling used (e.g., Lyman, 2003). The determination of the trace length distribution, however, requires special care to take the various sources of bias into account: overrepresentation of long traces, truncation of the short traces, censoring when the terminations cannot be seen, etc. (see Lantuéjoul et al., 2004, and references therein).

Things are not simpler for the inference of the parameters of more complex models. For those deriving from the Boolean model, tools specific to point processes, such as Ripley's K-function and its variants for isotropic point processes and the pair correlation function can be applied to fracture centers (Stoyan and Stoyan, 1994; Wen and Sinding-Larsen, 1997). In many cases, however, the fracture trace centers are not precisely known due to censoring (at least one termination of the trace cannot be observed), which limits the use of these tools. With outcrop data for example, it could be more appropriate to use tools specifically designed for random fiber processes (Schwandtke, 1988). The spatial variability can also be studied with the variogram of the observed fracture density, defined as the cumulated fracture length in equal squares partitioning an outcrop, or as the number of fracture intersections in equal segments partitioning a borehole.

All these tools as well as simple statistical tools can be used qualitatively to assess the choice of a relevant model. For example, the distribution of the spacing between successive fractures of a given set is exponential if the fractures are randomly located, less dispersed in the case of a regular pattern, or more dispersed, with many short spacings and a long tail, if the fractures are clustered. To transform them into quantitative tools, it is necessary to know their theoretical expression as a function of the parameters of the model; these expressions must incorporate the bias sources and stereological relationships. General results are given by Pohlmann et al. (1981). A suitable approach has been developed for as complex models as the disk cluster model with regionalized intensity (Chilès, 1989) and the hierarchical model proposed by Bourgine et al. (1995).

3 Process-Based Models

3.1 IDENTIFICATION AND MODELING OF MECHANICAL PROCESSES

The models presented above place fractures in their final state. Another approach consists in modeling the fracturing process itself. That approach is theoretically more satisfactory and also allows the prediction of the future evolution of the system in response to a new tectonic event or underground works. The identification of fracturing processes motivated a detailed observation of natural fracture systems: fault zones in sandstones (Antonellini and Aydin, 1994, 1995) and granitic rock (Christiansen and Pollard, 1997), joint formation in granitic rock (Segall and Pollard, 1983), relation between faults, joints and stress (Finkbeiner et al., 1997). Laboratory experiments were also carried out on scale models to observe the nucleation and development of discontinuities: brittle varnishes for understanding the mechanical origin of joints in layered media (Rives and Petit, 1990), sand and silicones (Sornette et al., 1990) or clay and fault gouge (An, 1998; Walmann, 1998) for studying the origin of faults in the Earth's crust. Finally, numerical codes have been developed to model these mechanical processes, usually under the simplifying assumption of an elastic stress field (e.g., Thomas, 1993). Such codes have been applied to model fault growth, linkage and interaction (Aydin and Schultz, 1990; Bürgmann et al., 1994; Willemse et al., 1997; Crider and Pollard, 1998; Weinberger et al., 1999) and fault-related fracturing (Martel and Boger, 1998). Caputo and Santaroto (1998) developed a mechanical model for quantifying the ratio between extensional joints and faults.

3.2 STOCHASTIC AND PROCESS-BASED APPROACH

The numerical modeling of fracture initiation and growth is not a simple task and is carried out for rather small networks in comparison with the capabilities of flow simulators, which can handle one million fractures when the rock matrix is impervious. This has led to the development of iterative techniques that simulate the initiation of new fractures and their growth. The growth is stochastic rather than the result of a mechanical calculation but the rules that govern the direction and intensity of growth are based on mechanical principles (Takayasu, 1985; Renshaw and Pollard, 1994).

Bourne et al. (2000) are a typical example of that kind of approach. They integrate growth processes in a geomechanical model of rock deformation which is the basis for a simulation of the fracture network. Large discontinuities such as seismically visible faults are supposed to be known. The first step is to determine the stress field within the faulted reservoir; this is done by assuming that the rocks behave as a homogeneous, isotropic, and linear-elastic material, and the faults as surfaces free of shear stress. The distribution of elastic stress related to faulting is governed by the distribution of slip over the fault network. Since fault displacement observations are scarce and poorly reliable, the distribution of slip is calculated over the fault network by loading it according to the remote stress that caused the faults to slip (Jeyakumaran et al., 1992). The orientation of this remote stress is estimated from the regional tectonic history, and its magnitude according to the mean rock strength prior to faulting. The comparison of the elastic stress field with the brittle failure strength of the reservoir rock determines the areas where secondary tensile and shear fractures have occurred. The initiation, growth, and termination of these fractures are then simulated in these areas. During growth, fracture spacing and interaction are controlled: a forbidden zone is defined around each fracture, which represents an overall reduction in local stress due to the presence of the fracture; conversely, the mechanical interaction leading to the connection of fractures with neighboring tips is taken into account, leading to en-echelon structures and enhancing the connectivity of the fractures

All the elements needed by the approach above are not always available. It is however useful to account for geomechanical information. Srivastava (2002) defines a simplified approach to generate 3D simulations of a fracture network observed as lineaments at a regional scale (40 × 50 km) on well-exposed areas of the topographic surface. The fracture traces which have not been observed or whose terminations could not be seen are obtained by simulating their growth. That growth is guided by a statistical model rather than determined by a mechanical process. The statistical model rests on a detailed analysis of the distribution of the characteristics of the fractures (length distribution, dip distribution, etc.) and of their correlation (variograms). The vertical growth is simulated similarly for all fractures according to geomechanical assumptions about their shape.

3.3 INTEGRATION OF AUXILIARY INFORMATION

An intermediate solution between object models and process-based models is to model the fracture network by placement of objects but to determine the local parameters of the object model by using geomechanical rules. That approach is used for example by Cacas et al. (1997) to simulate large joints in stratified sedimentary reservoirs: the local direction of the systematic joints is deduced from a mechanical simulation of the local stress and strain tensors at the time of fracturing; the direction of fold-related joints, which is parallel to the hinge of the fold, is defined locally by a surface curvature analysis; etc.

Similarly, the fracture density is larger in the vicinity of an anticlinal axis, where the layers were submitted to an extensional regime, than on its flanks. Like in fault zones, it is then necessary to build a model of the spatial variations of fracture density. In more complex situations, the strain field can be reconstructed by finite element methods from outcrop measurements (Schultz-Ela, 1988) and its impact on the fracture network can be modeled.

The situation is more complex if the evolution of the structural setting concerns not only the fracture density but also the fracture type. For example, in the sandstones of Arches Park (Utah), joint corridors with a large permeability can be observed in the vicinity of anticlinal axes, whereas deformation bands are found in neighboring synclines, namely structures resulting from a strong compression of grains and thus with a very low permeability (Antonelli and Aydin, 1994, 1995).

4 Conclusion

With the increasing power of computers, stochastic and process-based methods shall be able to simulate realistic fracture networks. However, it will remain difficult to reproduce the exact characteristics of a given site: they depend on several factors, including the spatial variability of rock properties, which is poorly known. Intermediate approaches are therefore useful. They can integrate geomechanical rules, field data, and complementary data such as new seismic attributes brought by high resolution 3D seismic surveys (Gauthier et al., 2003).

Simulated fracture networks are usually the input of flow models. Conversely, the flow behavior of the true network can help in choosing the relevant fracture network pattern and its parameters. Inverse methods should be developed to that effect.

References

Acuna, J.A., and Y.C. Yortsos (1995). Application of fractal geometry to the study of networks of fractures and their pressure transient. *Water Resources Research*, **31(3)**, 527–540.
An, L.J. (1998). Development of fault discontinuities in shear experiments. *Tectonophysics*, **293**, 45–59.
Antonellini, M., and A. Aydin (1994). Effect of faulting on fluid flow in porous sandstones: petrophysical properties. *AAPG Bulletin*, **78(3)**, 355–377.
Antonellini, M., and A. Aydin (1995). Effect of faulting on fluid flow in porous sandstones: geometry and spatial distribution. *AAPG Bulletin*, **79(5)**, 642–671.
Aydin, A., and R.A. Schultz (1990). Effect of mechanical interaction on the development of strike-slip faults with echelon patterns. *Journal of Structural Geology*, **12**, 123–129.
Baecher, G.B., N.A. Lanney, and H.H. Einstein (1977). Statistical description of rock properties and sampling. *Proceedings of the 18th U.S. Symposium on Rock Mechanics*, A.I.M.E., 5C1:1–8.
Barton, C.A., and M.D. Zoback (1990). Self-similar distribution of macroscopic fractures at depth in crystalline rock in the Cajon Pass scientific drillhole. In: *Rock Joints*, Barton N., Stephansson O. (eds.), Balkema A.A., Rotterdam, Netherlands, 163–170.
Bour, O., and P. Davy (1999). Clustering and size distributions of fault patterns: Theory and measurements. *Geophysical Research Letters*, **26**, 2001–2004.
Bourgine, B., J.P. Chilès, and C. Castaing (1995). Simulation d'un réseau de fractures par un modèle probabiliste hiérarchique. *Cahiers de Géostatistique*, Fasc. 5, Ecole des Mines de Paris, 81–96.
Bourne, S.J., F. Brauckmann, L. Rijkels, B.J. Stephenson, A. Weber, and E.J.M. Willemse (2000). Predictive modelling of naturally fractured reservoirs using geomechanics and flow simulation. *Paper ADIPEC-0911*, Society of Petroleum Engineers, 10 p.
Bürgman, R., D.D. Pollard, and S.J. Martell (1994). Slip distribution on faults: effects of stress gradients, inelastic deformation, heterogeneous host-rock stiffness, and fault interaction. *Journal of Structural Geology*, **16**, 1675–1690.
Cacas, M.C., J. Letouzey, and W. Sassi (1997). Modélisation multi-échelle de la fracturation naturelle des roches sédimentaires stratifiées. *Comptes Rendus de l'Académie des Sciences de Paris*, t. 324, série II a, 663–668.
Caputo, R., and G. Santarato (1998). Extensional joints and faults: A 3D mechanical model for quantifying their ratio – Part 1: Theory. Part 2: Applications. In: *Mechanics of Jointed and Faulted Rock*, H.P. Rossmanith (ed.), A.A. Balkema, Rotterdam, Netherlands, 133–144.
Castaing, C., M.A. Halawani, F. Gervais, J.P. Chilès, A. Genter, B. Bourgine, G. Ouillon, J.M. Brosse, P. Martin, A. Genna, and D. Janjou (1996). Scaling relationships in intraplate fracture systems related to Red Sea rifting. *Tectonophysics*, **261**, 291–314.
Chilès, J.P. (1988). Fractal and geostatistical methods for modeling of a fracture network. *Mathematical Geology*, **20(6)**, 631–654.
Chilès, J.P. (1989). Three-dimensional geometric modelling of a fracture network. In *Geostatistical, Sensitivity, and Uncertainty Methods for Ground-Water Flow and Radionuclide Transport Modeling*, B.E. Buxton (ed.), Battelle Press, Columbus, Ohio, 361–385.

Chilès, J.P., and G. de Marsily (1993). Stochastic models of fracture systems and their use in flow and transport modeling. In *Flow and Contaminant Transport in Fractured Rock*, J. Bear, G. de Marsily, and C.F. Tsang (eds.), Academic Press, San Diego, California, Chap. 4, 169–236.

Christiansen, P.P., and D.D. Pollard (1997). Nucleation, growth and structural development of mylonitic shear zones in granitic rock. *Journal of Structural Geology*, **19(9)**, 1159–1172.

Conrad, F., and C. Jacquin (1973). Représentation d'un réseau bidimensionnel de fractures par un modèle probabiliste. Application au calcul des grandeurs géométriques des blocs matriciels. *Revue de l'I.F.P.*, **28(6)**, 843–890.

Crider, J.G., and D.D. Pollard (1998). Fault linkage: Three dimensional mechanical interaction between echelon normal faults. *Journal of Geophysical Research*, **103(24)**, 373–391.

Dershowitz, W.S. (1984). *Rock joint systems*. Ph. D. dissertation, MIT, Cambridge, Massachusetts.

Finkbeiner, T., C.A. Barton, and M.D. Zoback (1997). Relationships among in-situ stress, fractures and faults, and fluid flow: Monterey Formation, Santa Maria Basin, California. *AAPG Bulletin*, **81(12)**, 1975-1999.

Gauthier, B.D.M., M. Garcia, and J.M. Daniel (2002). Integrated fractured reservoir characterization: A case study in a North Africa field. *Paper SPE 79105*.

Genter, A., and C. Castaing (1997). An attempt to simulate fracture systems from well data in reservoirs. *International Journal of Rock Mechanics and Mining Sciences*, **34(3/4)**, p. 448, Paper No. 44. Full length paper on CD-ROM (K. Kim, ed.).

Gillespie, P.A., C.B. Howard, J.J. Walsh, and J. Watterson (1993). Measurement and characterisation of spatial distributions of fractures. *Tectonophysics*, **226**, 113–141.

Hatton, C.G., I.G. Main, and P.G. Meredith (1994). Non-universal scaling of fracture length and opening displacement. *Nature*, **367**, 160–162.

Jeyakumaran, M., J.W. Rudnicki, and L.M. Keer (1992). Modeling slip zones with triangular dislocation elements. *Seismic Society of America Bulletin*, **82**, 153–169.

Lantuéjoul, C., H. Beucher, J.P. Chilès, H. Wackernagel, and P. Elion (2004). Estimating the trace length distribution of fractures from line sampling data. This volume.

Lyman, G.J. (2003). Stereological and other methods applied to rock joint size estimation—Does Crofton's theorem apply? *Mathematical Geology*, **35(1)**, 9–23.

Martel, J.S., and W.A. Boger (1998). Geometry and mechanics of secondary fracturing around small three-dimensional faults in granitic rock. *Journal of Geophysical Research*, **103 B 9**, 21,299–21,314.

Odling, N.E. (1997). Scaling and connectivity of joint systems in sandstones from western Norway. *Journal of Structural Geology*, **19(10)**, 1257–1271.

Peacock, D.C.P. (1996). Field examples of variations in fault patterns at different scales. *Terra Nova*, **8**, 561–371.

Pohlmann, S., J. Mecke, and D. Stoyan (1981). Stereological formulas for stationary surface processes. *Mathematische Operationsforschung und Statistik*, **12(3)**, 429–440.

Priest, S.D., and J.A. Hudson (1976). Discontinuity spacings in rock. *International Journal of Rock Mechanics and Mining Sciences & Geomechanics Abstracts*, **13**, 135–148.

Renshaw, C., and D. Pollard (1999). Numerical simulation of fracture set formation: a fracture mechanics model consistent with experimental observations. *Journal of Geophysical Research*, **99**, 9359–9372.

Rives, T., and J.P. Petit (1990). Diaclases et plissements : une approche expérimentale. *Comptes Rendus de l'Académie des Sciences de Paris*, t. 310, série II, 1115–1121.

Schultz-Ela, D.D. (1988). Application of a three-dimensional finite-element method to strain field analyses. *Journal of Structural Geology*, **10(3)**, 263–272.

Schwandtke, A. (1988). Second order quantities for stationary weighted fibre processes. *Mathematische Nachrichten*, **139**, 321–334.

Segall, P., and D.D. Pollard (1983). Joint formation in granitic rock of the Sierra Nevada. *Geological Society of America Bulletin*, **94**, 563–575.

Sornette, A., P. Davy, and D. Sornette (1990). Growth of fractal fault patterns. *Physical Review Letters*, **65(18)**, 2266–2269.

Srivastava, R.M. (2002). Probabilistic discrete fracture network models for the Whiteshell research area. Ontario Power Generation, Report No: 06819-REP-01200-10071-R00, Toronto, Canada.

Stoyan, D., and H. Stoyan (1994). *Fractals, Random Shapes and Point Fields. Methods of Geometrical Statistics*. Wiley, Chichester, England.

Takayasu, H. (1985). A deterministic model of fracture. *Progress in Theoretical Physics*, **74**, 1343–1345.

Terzaghi, R.D. (1965). Sources of error in joint surveys. *Geotechnique*, **15(3)**, 287–303.

Thomas, A.L. (1993). *Poly3D: A three-dimensional, polygonal element, displacement discontinuity boundary element computer program with applications to fractures, faults, and cavities in the Earth's crust*. Masters dissertation, Stanford University, California.

Turcotte, D.L. (1986). A fractal model for crustal deformation. *Tectonophysics*, **132**, 261–269.

Turcotte, D.L. (1992). Fractal, chaos, self-organized criticality and tectonics. *Terra Nova*, **4**, 4–12.
Veneziano, D. (1978). *Probabilistic model of joints in rock.* MIT, Cambridge, Massachusetts.
Walmann, T. (1998). *Dynamics and Scaling Properties of Fractures in Clay-Like Materials.* Ph. D. thesis, University of Oslo, Norway.
Weinberger, R., V. Lyakhovsky, and A. Agnon (1999). Damage evolution and propagation paths of en-échelon cracks. In: *Rock Mechanics for Industry*, B. Amadei, Kranz, Scott, and Smeallie (eds.), Balkema A.A., Rotterdam, Netherlands, Vol. 2, 1125–1132.
Wen, R., and R. Sinding-Larsen (1997). Stochastic modeling and simulation of small faults by marked point processes and kriging. In *Geostatistics Wollongong '96*, E.Y. Baafi and N.A. Schofield (eds.), Kluwer, Dordrecht, Netherlands, Vol. 1, 398–414.
Willemse, E.J.M., D.C.P. Peacock, and A. Aydin (1997). Nucleation and growth of faults in limestones from Somerset, UK. *Journal of Structural Geology*, 1461–1477.
Yow, J.L. Jr. (1987). Blind zones in the acquisition of discontinuity orientation data. *International Journal of Rock Mechanics and Mining Sciences & Geomechanics Abstracts*, **24(5)**, 317–318.
Zhang, L., and H.H. Einstein (2000). Estimating the intensity of rock discontinuities. *International Journal of Rock Mechanics and Mining Sciences*, **37**, 819–837.

GEOSTATISTICAL SIMULATION OF FRACTURE NETWORKS

R. MOHAN SRIVASTAVA[1], PETER FRYKMAN[2] and MARK JENSEN[3]
[1] *FSS Canada Consultants, Toronto, Canada*
[2] *Danish Geological Survey, Copenhagen, Denmark*
[3] *Ontario Power Generation, Toronto, Canada*

Abstract. A geostatistical method has been developed by Ontario Power Generation to enable creation of 3D fracture network models (FNMs) that explicitly honor detailed information on surface lineaments. This approach provides a systematic and traceable method that is flexible and that accommodates data from many different sources. The detailed, complex and realistic models of 3D fracture geometry produced by this method serve as an ideal basis for developing rock property models to be used in flow and transport studies. These models are probabilistic in the sense that they consist of a family of equally likely renditions of fracture geometry. Such probabilistic models are well suited to studying issues that involve risk assessment and quantification of uncertainty.

The geostatistical procedure for simulating FNMs is described, and tested using field data collected from the Lägerdorf chalk quarry in northern Germany.

1 Introduction

Fractures play a dominant role in fluid flow and transport; 3D models of fracture networks can therefore be very useful as inputs to flow simulators. Existing procedures for simulating fracture network geometry typically simplify the undulating and curvilinear nature of fracture surfaces, approximating them as planar facets. Though this approximation is suitable for many flow and transport studies, it is unacceptable in situations where models of fracture geometry need to exactly honor known locations of a large, detailed and geometrically complex set of fractures.

Ontario Power Generation (OPG) is developing tools to assist the Deep Geologic Repository Technology Program (DGRTP) with integrating diverse types of data into geosphere models that are consistent with increasingly detailed surface and subsurface knowledge. One specific data integration task involves construction of 3D fracture network models that honor information available at the time of preliminary site characterization: surface lineaments, general structural geology principles, regional tectonic considerations and site-specific information on geomechanical characteristics. These various pieces of information provide constraints on fracture location: very strong constraints at surface in regions of good bedrock exposure; weaker constraints at depth and in regions of poor bedrock exposure.

A geostatistical simulation method has been developed for creating complex, detailed and realistic 3D fracture network models that honor the various pieces of information available for preliminary site characterization. It can also accommodate data that become available in later years, such as borehole data and information from other subsurface investigations. This method produces a family of equally probable renditions of 3D fracture geometry, each one different in detail, but all consistent with the same constraints.

2 Overview and implementation

Sequential gaussian simulation (SGS) is a procedure that is widely used for creating data-conditioned stochastic models of spatial phenomena. In a typical SGS study, the procedure is applied to volume-averaged rock properties (such as grades or permeabilities), and is performed at the nodes of a regular grid. In the procedure discussed here, SGS is applied to geometric attributes (strike of a fracture trace, or dip of a fracture surface). Furthermore, the locations at which SGS is applied are not nodes of a regular grid; instead, the procedure is applied at the tips of iteratively propagating fractures, which entails that the locations of the simulations nodes depend on the details of what has been simulated in previous iterations.

The original motivation for using an iterative procedure like SGS was to mimic the procedure proposed by Renshaw and Pollard (1994). Their approach to 2D fracture simulation was based on geomechanical principles, propagating fracture tips when stress at fracture tips exceeds a critical threshold. Their approach created very realistic synthetic images of fractures in a variety of stress environments, including many subtle features commonly observed in the field, such as zones of small *en echelon* features that bridge gaps between larger separate fractures.

Though undeniably successful, fracture simulation based on geomechanical principles proved to be prohibitively computationally intensive and has never been extended satisfactorily to 3D. By using the same broad approach — an iterative procedure for propagating fracture tips — but replacing geomechanical principles for fracture propagation with geostatistical rules, one is able to mimic much of the realism with less computational effort.

2.1 2D PROPAGATION OF SURFACE FEATURES

Figures 1 through 6 show the major steps in the first phase of the fracture simulation: 2D propagation of surface traces.

The procedure begins with the seeding of the initial fracture segments, each one of which is assigned an initial direction of propagation and an intended final length. For deterministic fractures that can be identified from aerial photography or surface reconaissance (the grey lines in Figure 1), the initial fracture segment is seeded at the midpoint, the initial direction of propagation is determined from the lineaments orientation at its midpoint and the intended final length is the length of the known fracture. In areas of poor bedrock exposure (the darker region on the left of Figure 1), additional fractures are seeded to bring the number of fractures up to the number predicted by the user-specified model of fracture density. These hidden fractures are assigned their locations according to a clustered Poisson process; their initial orientations are drawn from the model of azimuth distribution and their intended final lengths from the model of fracture length distribution. The filled dots in Figure 1 mark "constrained" endpoints, those whose propagation will be governed by the trace of known fractures. The open dots mark the "free" endpoints, those whose propagation will be determined by sequential gaussian simulation.

Figure 2 shows how free endpoints are propagated. These are visited in a random order and each one is propagated a fixed step length. The direction of propagation is determined by simple kriging using the closest segment of each nearby fracture as conditioning data. In the example shown in Figure 2, the azimuth data used in the kriging for the free endpoint marked by the question mark are shown in black. One additional piece of data (shown in grey) is included in the kriging: a point halfway to the nearest neighbor, with an orientation parallel to the line segment that connects the endpoint being propagated to its nearest neighbor. For the example in Figure 2, the distribution of possible azimuth directions has a mean of $70°$ and a standard deviation of $12°$; this is shown by the pie-shaped slice centered on the mean and with an angular width of two standard deviations. A specific direction (the dashed line) is randomly drawn from the distribution and the endpoint is propagated in that direction.

With simple kriging being performed on the strike of the fracture traces, a variogram model is required for this strike attribute. For long features that are very nearly linear, the variogram of strike will show strong spatial continuity, typically modelled with quadratic behavior at short distances and with little or no nugget effect. For features that undulate, are kinked or meander, the variogram of strike will show less spatial continuity and may be modelled with a small nugget effect. The use of variogram information on the strike ensures realistic portrayal of the fractures' undulations.

Figure 3 shows how constrained endpoints are handled differently. For simulated fractures that track deterministic features, their propagation simply follows the trace of the identified fracture. In the example in Figure 3, the constrained endpoint marked with the question mark is propagated along the trace of the deterministic trace shown in gray. As each endpoint is propagated, a new line segment is created. As in a conventional sequential simulation procedure, each newly simulated segment is available for use in all subsequent simple krigings.

When the propagation of a free endpoint collides with a previously simulated endpoint, that endpoint is terminated with a user-specified probability. Setting this truncation probability to 1 guarantees that when fractures meet, one of them will terminate, a pattern common with faults. Using lower probabilities allows fractures to cross each other, a pattern common with joints.

If both the free endpoints of a hidden fracture are terminated before the fracture reaches its intended length, then the remaining unused length is assigned to a fracture still being propagated. This preserves total fracture length in the study area, at the expense of increasing the variance of the fracture length distribution, a result of allowing more shorter fractures, as well as more longer ones.

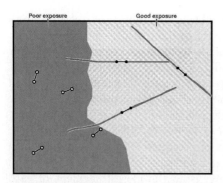

Figure 1. Initial seeding of the midpoints of the fracture traces.

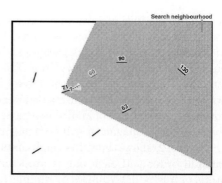

Figure 2. An example of the propagation of a free endpoint.

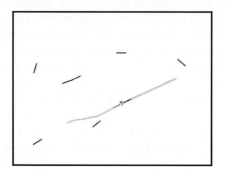

Figure 3. An example of the propagation of a constrained endpoint.

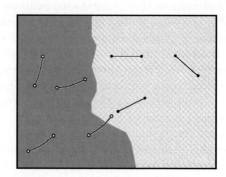

Figure 4. After one complete iteration through the endpoints.

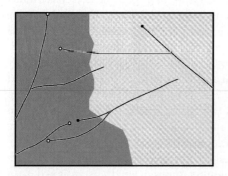

Figure 5. After five complete iterations through the endpoints.

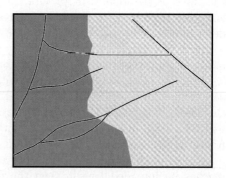

Figure 6. Final outcome of simulated fracture traces at surface.

Figure 4 shows the example after one complete iteration through all endpoints. As can be seen near the bottom of this figure, hidden fractures seeded in areas of poor exposure are allowed to propagate some user-specified distance into areas of good exposure.

Figure 5 shows the result after five complete iterations through all endpoints. Some of the fractures, such as the one whose initial propagation was shown in Figure 2, have reached their intended length and do not propagate any further. Others, such as the "hidden" one that first crossed into the area of good exposure in Figure 4, have terminated against other lineaments and do not propagate further. Some constrained endpoints, such as the eastern tip of the deterministic fracture noted in Figure 3, have been terminated because the lineament ends in an area of good bedrock exposure. Other constrained endpoints, such as the other end of the same feature, have passed into areas of poor bedrock exposure. When this happens, the endpoint becomes free and its propagation is governed by the SGS procedure since there is no longer a deterministic trace that can be followed.

Figure 6 shows the final result after several more iterations. The fractures originally identified in regions of good bedrock exposure have been honored; additional fractures have been added in regions of poor exposure; certain features have grown together into larger connected fracture systems; fractures truncate against each other in a plausible manner.

2.2 PROPAGATION TO DEPTH

Once the surface traces of the fractures have been simulated, the next step is the down-dip propagation of these traces to depth. This is accomplished using the same SGS procedure that was used to govern the 2D propagation of free endpoints. For the 2D propagation of surface traces, the geometric attribute being simulated was the strike direction of the fracture. For the down-dip propagation, the simulated attribute now becomes the dip of the fracture surface.

In the same way that each endpoint was propagated using SGS in 2D by simulating an appropriate strike direction, each line segment is now propagated down-dip, with SGS being used to simulate an appropriate dip. If the dip of a particular fracture is known (from surface reconnaisance measurements, for example), then this known dip can be used. The more common situation is that the dip is not known precisely, but only approximately. For example, regional tectonic considerations often imply that a certain set of fractures are likely the type of low-angle features commonly found in compressional environments; similarly, high-angle and sub-vertical features are usually more common in tensional environments.

When the dip of a particular fracture is not known precisely, SGS can still be guided to plausible values for the simulated dip by appropriate choices of the local mean used in the simple kriging, and by the sill of the variogram, which governs the kriging variance and the spread of the local conditional probability distributions of dip. Adjustments to the local mean can also be used to cause low-angle features to flatten with depth, a characteristic common of compressional "thrust" faults.

When there is no site-specific information on the fracture location at depth, as is typically the case for preliminary site characterization, all line segments are treated as "free" segments: they are not constrained to follow a specific down-dip trajectory. In applications where subsurface investigations do provide specific information on fracture location at depth, this type of data is easily accommodated in the same way that it was for the 2D propagation. The line segments corresponding to certain fractures can be treated as "constrained" and can be required to follow a specific trajectory.

2.3 PROPERTIES OF SIMULATED FRACTURE NETWORKS

Once the down-dip propagation of the fracture surfaces is complete, we have a complete 3D model of the geometry of the fracture network that:

- honors the surface traces of all identified fractures and, if available, subsurface information on the location of specific features;
- includes additional stochastic fractures in areas where poor exposure causes deterministic fractures to be under-represented;
- honors the fracture length distribution;
- honors the fracture orientation distribution; and,
- honors the user's assumptions about how fractures truncate against one another.

Several applications of this method to different case study examples confirm that the resulting geometry is plausible both from a geomechanical point of view, as well as from a structural geology point of view (Srivastava, 2002; Sikorsky et al., 2002); one such recent case study is discussed below.

3 Lägerdorf case study

From December 1990 to May 1992, GEO-RECON, a Norwegian consulting company, undertook a fracture mapping program at the Lägerdorf quarry in northern Germany on behalf of the Joint Chalk Research Program, a multi-company research program that focused on subjects related to North Sea chalk reservoirs. Data collection procedures are described in detail by Koestler and Reksten (1992).

The data set consists of highly detailed maps of fracture traces for 12 parallel faces of one wall of the quarry as it was advanced in small increments (roughly 1 to 1.5 metres). Each face map spans a region approximately 230 metres long by 40 metres high. All sections are inclined at approximately 50°. The faces were available for mapping during production, where the excavation process continuously scrapes a layer of 1.5 m thickness off the quarry wall by an abrading conveyor belt. During the period when the 12 faces were mapped and described structurally, the quarry wall was advanced 25 metres. The set of 12 face maps therefore represents a high resolution $2\frac{1}{2}$D fracture data set of a 230×40×25m volume.

Figure 7 shows the face maps of the fractures on walls 7 through 11. There are three major directional sets:

1. A set with steep westerly dips; these are particularly dense near two major shear zones identified as S1 and S2 on the map for Wall 10 on Figure 7.

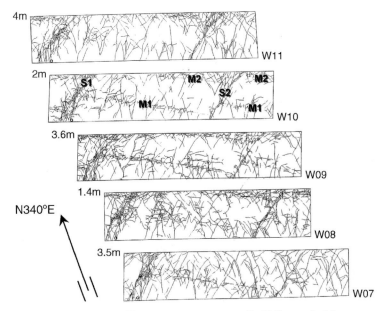

Figure 7. Fracture maps on walls 7 through 11.

2. A set with steep easterly dips; these are less common than the first set, but have similar lengths.
3. A set with shallow easterly dips; these are particularly dense along two marl layers identified as M1 and M2 on Figure 7. They tend to be shorter than the steeply dipping fractures. As shown on Figure 7, the major shear zones offset the marl layers.

To test the SGS-based geostatistical procedure for fracture simulation, the stack of parallel face maps was rotated so that the walls are essentially horizontal, with Wall 1 being at the top and Wall 12 being at the bottom. Following this rotation, we are able to treat the data from Wall 1 as a set of deterministic "surface" fractures that will be used as conditioning data for a simulation of 3D fracture geometry. In order to make the test as illuminating as possible, *all* data from Walls 2 through 12 were ignored. Once a set of simulations of 3D fracture geometry has been developed, each realization can be sliced along planes corresponding to the walls not used as conditioning data and the resulting simulated face map can be compared to the actual face map for each wall.

Figure 8 shows the conditioning data from Wall 1, the "surface" data for the test of the simulation procedure. In order to check the 2D propagation at surface,

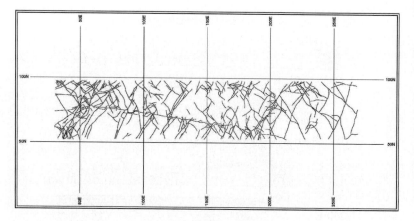

Figure 8. "Surface" fractures from Wall 1.

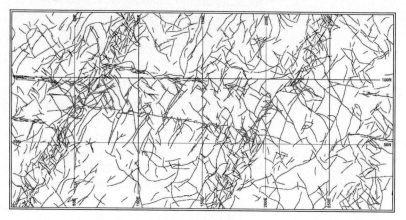

Figure 9. Realization No. 1 of simulated surface fractures.

the area covered by the mapped fractures on Wall 1 was extended to create a border, construed as a region of poor bedrock exposure where no deterministic fractures have been identified and where the simulation procedure will need to create a plausible rendition of hidden fractures.

Figure 9 shows one realization of the simulated traces at surface. The deterministic features identified by GEO-RECON are all honored, and pass quite seamlessly into the border region where the fracture traces are all simulated.

Figure 10 shows two realizations of simulated traces along the surface corresponding to Wall 11, along with the actual data from Wall 11. While the simulated fractures do have a plausible overall appearance, some of the details have been shifted slightly. The marl layers, for example, do not appear on the simulations at the same location as they do on the actual face map. The reasons for this discrepancy are well understood.

By strictly limiting the data available to the simulation to the measurements gathered from Wall 1, our model of the orientation of the marl layers (and of the

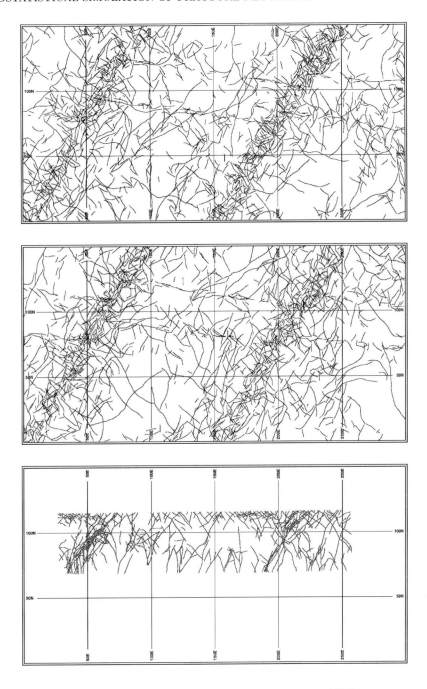

Figure 10. Simulated and actual fractures on Wall 11.

shear zones) is dependent entirely on the extrapolation of the planar orientation deduced from measurements available on Wall 1. The orientation of the marl layers deduced from Wall 1 data alone is slightly different than the orientation one can calculate from the full data set. So although the simulated fractures are definitely off in their depiction of the exact location of the densely fractured zones, this error would easily be corrected once subsurface investigations such as boreholes provided accurate 3D information on the location and orientation of the major geological features.

Future studies will compare the actual field data to the statistical and geometric characteristics of the clusters of simulated fractures. Flow simulation will also be used to compare actual and simulated flow-related characteristics, such as peak arrival times of injected tracer. For the moment, the preliminary results from the Lägerdorf case study are encouraging. The SGS-based procedure does create 3D fracture network models that are visually plausible and that honor complex and highly detailed surface information.

4 Conclusions

The proposed procedure for simulating fracture networks has now been tested on a variety of different problems at various scales. The Lägerdorf case study offers confirmation that the procedure does generate highly detailed and complex fracture patterns that mimic actual field observations.

Acknowledgements

The work presented in this paper has been funded by the Deep Geologic Repository Technology Program of Ontario Power Generation, and has benefitted greatly from the constructive criticism and encouragement of Jean-Paul Chilès, Alexandre Desbarats and Jaime Gomez-Hernàndez. The authors are also grateful to Andreas Koestler for his invaluable assistance with the data he collected and assiduously compiled from the Lägerdorf quarry.

References

Koestler, A.G., and Reksten, K., 1992, *3D geometry and flow behaviour of fractures in deformed chalk, Laegerdorf, Germany*, GEO-RECON Report no. 9012.126, GEO-RECON A.S, Oslo, Norway.

Renshaw, C., and Pollard, D., 1994, "Numerical simulation of fracture set formation: a fracture mechanics model consistent with experimental observations", *Journal of Geophysical Research*, vol. 99, pp. 9359–9372.

Sikorsky, R.I., Serzu, M., Tomsons, D., and Hawkins, J., 2002 *A GIS-based methodology for lineament interpretation and its application to a case study at AECL's Whiteshell Research Area in southeastern Manitoba*, Deep Geologic Repository Technology Program Report 06819-01200-10073, Ontario Power Generation, Toronto, Ontario.

Srivastava, R.M., 2002, *Probabilistic discrete fracture network models for the Whiteshell Research Area*, Deep Geologic Repository Technology Program Report 06819-01200-10071, Ontario Power Generation, Toronto, Ontario.

THE PROMISES AND PITFALLS OF DIRECT SIMULATION

OY LEUANGTHONG
Department of Civil & Environmental Engineering, 220 CEB,
University of Alberta, Canada, T6G 2G7

Abstract. The idea of direct simulation is to simulate in the space of the original data units, with minimal assumptions or transformations about the data distribution. A common approach to direct simulation is to proceed in a sequential fashion: direct sequential simulation (DSS). While the idea is not new, full development of the framework remains to be seen. The benefits of multiscale data integration, avoidance of the "Gaussian disease", and flexible distribution considerations are offset by problems with histogram reproduction, the pervasive influence of Gaussianity, and proportional effect reproduction.

This paper examines the promises and pitfalls of direct simulation with some illustrative examples, and also discusses the future of DSS as a practical alternative for natural resource characterization. The future of DSS calls for an engine other than kriging that accounts for possible dependency between the local variance and mean.

1 Introduction

Over the last decade, direct simulation has been proposed as a viable alternative to the venerable Gaussian simulation approaches. The idea of direct simulation is to simulate in the space of the original data units, with minimal assumptions or transformations about the data distribution. Behind this key idea is the principle of simple kriging.

Journel (1994) first showed that the covariance of simulated values reproduces the target covariance model *if* the simulated values are drawn from a distribution centred about the simple kriging (SK) mean and a variance given by the SK variance. Indeed, Bourgault (1997) showed this to be true for different distributional shapes including the uniform, dipole and of course, the Gaussian distribution. Caers (2000) also shows this for a uniform, double exponential, double exponential with a spike, and a "bootstrapped" distribution.

Covariance reproduction without relying on the Gaussian framework seeded the idea for direct simulation. The key premise for why direct simulation works lies in the orthogonality between the SK estimate, $Z^*(\mathbf{u})$, and the squared error which forms the basis for the SK error variance, $\sigma_{SK}^2(\mathbf{u})$. This can be thought of in terms of projections

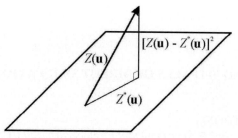

Figure 1 Kriging in terms of projection theory (redrawn from Journel and Huijbregts, 1978; Anton and Rorres, 1991).

where the squared error, $[Z(u)-Z^*(u)]^2$, is orthogonal to the space of all finite linear combinations of the random variables (RV), $Z(u_\alpha)$, $\alpha = 1,..., n$ (Journel and Huijbregts, 1978) (see Figure 1). The kriging estimate, $Z^*(u)$, lies in this space as it is a linear combination of the RVs, $Z(u_\alpha)$, $\alpha=1,..., n$:

$$Z^*(\mathbf{u}) = \sum_{\alpha=1}^{n} \lambda_\alpha Z(\mathbf{u}_\alpha) \qquad (1)$$

The squared error term, $[Z(\mathbf{u})-Z^*(\mathbf{u})]^2$, represents the distance to the unknown true value, $Z(\mathbf{u})$. Based on Projection Theory, there is a unique and exact solution that yields the linear coefficients, λ_α, $\alpha=1, ..., n$, such that this distance is minimized (Journel and Huijbregts, 1978). This solution is referred to as the projection of $Z(\mathbf{u})$ onto this space. The corollary to kriging lies in the fact that the weights, λ_α, $\alpha=1, ..., n$, are determined such that the expected squared error, $E\{[Z(\mathbf{u})-Z^*(\mathbf{u})]^2\}$, is minimum. This visual interpretation of simple kriging can also be thought of as satisfying the Generalized Theorem of Pythagoras (Anton and Rorres, 1991).

Orthogonality of the kriged estimate and the squared error leads to an error variance that is independent of the data values, commonly referred to as the homoscedascity of kriging. Under a Gaussian paradigm, this poses no problems; in fact, it would be exactly right. In reality, natural phenomena rarely possess characteristics similar to the Gaussian distribution. This is particularly evident upon examining the relationship between the local average and the local variability, which, contrary to the homoscedasticity inherent in kriging, often reveals the presence of a strong relationship between the two statistics. This heteroscedastic relationship is more specifically known as the proportional effect (Journel and Huijbregts, 1978), and poses the most significant problem for direct simulation.

This paper presents the promises and pitfalls of direct simulation with some illustrative examples. Five main areas of discussion are highlighted: (1) principle of simple kriging, (2) implementation of direct simulation, (3) multiscale data integration, (4) histogram reproduction, and (5) accounting for the proportional effect. Finally, the future of DSS is discussed.

2 The Simple Kriging Principal

Reproduction of the covariance only requires that the conditional probability distributions have a mean and variance given by simple kriging (Journel, 1994). Journel proved this by showing firstly, the detailed simulation of a variable at location **u**, then adding this simulated value to simulate the next location, **u'**, and finally checking the covariance between these two simulated variables.

Consider a stationary random variable, $Z(\mathbf{u})$, with zero mean and unit variance. Firstly, construct a simulated value such that it can be decomposed as

$$Z_S(\mathbf{u}) = m(\mathbf{u}) + R_S(\mathbf{u})$$

where $m(\mathbf{u})$ is the expected value at location $\mathbf{u} \in$ domain, A, and $R(\mathbf{u})$ is a random variable drawn from a distribution with zero mean and variance, $\sigma^2(\mathbf{u})$. The local mean is given by the kriging mean (Equation 1), and the variance is given by the SK variance:

$$\sigma_{SK}^2(\mathbf{u}) = 1 - \sum_{\alpha=1}^{N} \lambda_\alpha C(\mathbf{u} - \mathbf{u}_\alpha) \qquad (2)$$

where $C(\mathbf{u} - \mathbf{u}_\alpha)$ is the covariance between the location **u** and the data located at $\mathbf{u}_\alpha$, $\alpha=1,\ldots,n$, $\sigma_{SK}^2(\mathbf{u})$ is the simple kriging variance, and the weights, λ_α, $\alpha=1,\ldots,n$ are obtained by solving the normal equations:

$$\sum_{\beta=1}^{n} \lambda_\alpha C(\mathbf{u}_\beta - \mathbf{u}_\alpha) = C(\mathbf{u} - \mathbf{u}_\alpha), \quad \alpha = 1,\ldots,n \qquad (3)$$

This simulated value is added to the conditioning data set, and simulation is performed at the next location $\mathbf{u}'=\mathbf{u}_{n+1}$ with the following kriged mean and variance:

$$Z^*(\mathbf{u}') = \sum_{\alpha=1}^{n} \lambda_\alpha z(\mathbf{u}_\alpha) + \lambda_{n+1} Z_S(\mathbf{u}) \qquad (4)$$

$$\sigma_{SK}^2(\mathbf{u}') = 1 - \sum_{\alpha=1}^{n} \lambda_\alpha C(\mathbf{u}' - \mathbf{u}_\alpha) - \lambda_{n+1} C(\mathbf{u}' - \mathbf{u}) \qquad (5)$$

Note that the weights λ_α, $\alpha=1,\ldots,n+1$ are *not* the same as the weights λ_α, $\alpha=1,\ldots,n$ obtained from solving the system in Equation 3. The simulated value is given as

$$Z_S(\mathbf{u}') = Z^*(\mathbf{u}') + R_S(\mathbf{u}')$$

The covariance between the two simulated variables is then examined:

$$\begin{aligned}C(\mathbf{u} - \mathbf{u}') &= E\{Z_S(\mathbf{u}) \cdot Z_S(\mathbf{u}')\} \\ &= E\{Z^*(\mathbf{u}) \cdot Z^*(\mathbf{u}')\} + E\{Z^*(\mathbf{u}) \cdot R_S(\mathbf{u}')\} + \\ &\quad E\{Z^*(\mathbf{u}') \cdot R_S(\mathbf{u})\} + E\{R_S(\mathbf{u}) \cdot R_S(\mathbf{u}')\}\end{aligned} \qquad (6)$$

where $E\{Z^*(\mathbf{u}) \cdot R_S(\mathbf{u}')\}$ and $E\{R_S(\mathbf{u}) \cdot R_S(\mathbf{u}')\}$ are zero since $Z^*(\mathbf{u})$ and $R_S(\mathbf{u}')$ are independent of each other and $R_S(\mathbf{u})$ and $R_S(\mathbf{u}')$ are also independent. The remaining portions of the right hand side are non zero since the kriged mean at the second location depends on the mean and randomly drawn value at the first location.

Expanding and simplifying the remaining two terms yields

$$E\{Z^*(\mathbf{u}') \cdot Z^*(\mathbf{u})\} = \sum_{\beta=1}^{n} \lambda_\beta C(\mathbf{u}_\beta - \mathbf{u}) + \lambda_{n+1}\left[1 - \sigma_{SK}^2(\mathbf{u})\right] \quad (7)$$

$$E\{Z^*(\mathbf{u}') \cdot R_S(\mathbf{u})\} = \lambda_{n+1}\sigma_{SK}^2(\mathbf{u}) \quad (8)$$

Equations 7 and 8 are substituted into Equation 6:

$$C(\mathbf{u} - \mathbf{u}') = \sum_{\beta=1}^{n} \lambda_\beta C(\mathbf{u}_\beta - \mathbf{u}) + \lambda_{n+1}\left[1 - \sigma_{SK}^2(\mathbf{u})\right] + \lambda_{n+1}\sigma_{SK}^2(\mathbf{u})$$

$$= \sum_{\beta=1}^{n} \lambda_\beta C(\mathbf{u}_\beta - \mathbf{u}) + \lambda_{n+1}$$

$$= C(\mathbf{u}' - \mathbf{u})$$

It is by this logic that Journel (1994) proved that so long as the conditional mean and variance are provided by simple kriging, covariance reproduction could be achieved without making any assumptions about the distributional shape. This is an exciting result as it opened the way for geostatisticians to consider simulation outside of the Gaussian framework and without the inference effort required under the indicator paradigm.

3 DSS Methodology

A common approach to simulation is to proceed in a sequential fashion; thus, Direct Sequential Simulation (DSS) was coined (Xu and Journel, 1994). The sequential simulation framework is straightforward:

1. Select a random path visiting all locations.
2. At each location:
   a. Search for all nearby data of different types and/or scale and previously simulated nodes (e.g. P data types with n_p samples).
   b. Perform simple kriging to determine the parameters corresponding to the conditional distribution, $F(Z(\mathbf{u})|Z_p(\mathbf{u}_1),...,Z_p(\mathbf{u}_{n_p}))$, $p=1, ..., P$.
   c. Draw a simulated value from this conditional distribution using Monte Carlo simulation. This simulated value is added to the conditioning data set.
3. Proceed to next node and repeat Step 2, until all locations are simulated.

The virtues of simplicity cannot be understated. The sequential algorithm was proposed by Johnson (1987), and is common in most geostatistical literature (Isaaks, 1990; Goovaerts, 1997; Deutsch and Journel, 1998; Chilès and Delfiner, 1999; Sinclair and Blackwell, 2000). There are other approaches for simulation, including the matrix approach (Davis, 1987) and turning bands (Journel and Huijbregts, 1978); however, the simplicity and efficiency of sequential simulation has made it the most popular approach in practice.

Indicator and Gaussian simulation have long been the "standard" geostatistical methods of choice in modern practice. Unlike sequential Gaussian simulation and sequential indicator simulation, the promise of DSS is that neither pre- nor post-processing steps are required. There is no need for data transformation, whether it is to a Gaussian or an indicator formalism. This sequential approach is common in mainstream numerical modelling, regardless of whether that modelling is performed under a parametric or non-parametric model.

4 Multi-Scale Data Integration

The current motivation for development of the direct simulation framework is the promise of integrating multiple data types from different sources and of different scales. Integrating data of different volume supports is neither new nor difficult in theory. Cokriging using average covariance/variograms permits consideration of multiscale data that average linearly. In fact, the generalized cokriging equations are straightforward to obtain.

Consider P stationary random variables, Z_p, $p=1,...,P$ with mean μ_p defined on support V_p centred at location $\mathbf{u}_{\alpha p}$, where $\alpha = 1,..., n_p$ and n_p is the number of available data of type p. It is not necessary that the volume supports $V_p, p=1,...,P$ be constant.

$$Z(\mathbf{u}_{\alpha p}) = \frac{1}{V_p} \int_{V_p} Z_p(\mathbf{u}_{\alpha p}) du$$

Without loss of generality, consider the residual of Z_p, $Y_p = Z_p - \mu_p$. Simple cokriging of the residual yields the following simple cokriging (SCK) variance:

$$\sigma^2_{SCK} = \overline{C}(V_i(\mathbf{u}), V_i(\mathbf{u})) - \sum_{p=1}^{P}\sum_{\alpha=1}^{n_p} \lambda_{\alpha p} \overline{C}(V_i(\mathbf{u}), V_p(\mathbf{u}_{\alpha p}))$$

where

$$C(V_i, V_j) = \frac{1}{|V_i||V_j|} \int_{V_i} dy \int_{V_j} C(y - y') dy'$$

and the weights are determined by simultaneously solving the $\sum_{p=1}^{P} n_p$ equations that constitute the simple co-kriging system of equations:

$$\sum_{p'=1}^{P}\sum_{\beta=1}^{n_{p'}}\lambda_{\beta p'}\overline{C}(V_p(\mathbf{u}_{\alpha p}), V_{p'}(\mathbf{u}_{\beta p'})) = \overline{C}(V_i(\mathbf{u}), V_p(\mathbf{u}_{\alpha p})), \quad p=1,\ldots,P \qquad (9)$$

The resulting cokriging estimate and estimation variance correspond to the conditional expectation and variance of the RV $Y_p(\mathbf{u})$.

Greater efficiency can be achieved by simultaneously cokriging M multiple data types, where $M \le P$. This is simply achieved by changing the column vector of weights and right hand side covariance into an $M \times P$ matrix. An additional index is required to indicate the variable to be estimated. For this purpose, the m, $m=1, \ldots, M$, index is introduced.

$$Y_1^*(\mathbf{u}) = \sum_{p=1}^{P}\sum_{\alpha=1}^{n_p}\lambda_{\alpha p}^1 Y_p(\mathbf{u}_{\alpha p})$$

$$\vdots$$

$$Y_M^*(\mathbf{u}) = \sum_{p=1}^{P}\sum_{\alpha=1}^{n_p}\lambda_{\alpha p}^M Y_p(\mathbf{u}_{\alpha p})$$

Solving for the weights of the resulting co-kriging system requires little additional effort since the large left hand side data to data covariance matrix (in Equation 9) only has to be inverted once. Matrix multiplication of the inverted covariance matrix with the additional $M-1$ columns of the right hand side covariance will give the weights to estimate the other $M-1$ additional variables. In fact, most solvers can be modified to solve systems of simultaneous equations with multiple right hand sides without explicitly solving for an inverse. The only additional computation required in order to simultaneously estimate the collocated data types is the determination of the right hand side volume to volume covariance between the location to be estimated and the nearby data of P types.

While cokriging of one variable gives the conditional expectation and variance of the RV, simultaneous cokriging of multiple RVs gives the conditional mean vector and covariance matrix of the M RVs. Simulation using these distributional parameters must still be performed.

All this is fine in the context of estimation where cokriging can be performed in the space of the data; however, in the context of Gaussian simulation, which is the most common practical simulation method, using average statistics after a non-linear transformation and back transforming to original units, does not work. Consider three numbers: 1, 2 and 10. The average of these three numbers is 4.33. Now consider an exponential transform, e^x where x is the data. This transform gives: 2.718, 7.389 and 22026.470, respectively. The average of the transformed values is 7345.524, which after back transformation yields 8.902. This is clearly not the same as the average in original space. Thus averaging in a non-linear space, such as Gaussian space, does not provide an appropriate method of accounting for multiscale data. This provides, yet, another impetus for pursuing DSS.

5 Histogram Reproduction

The topic of histogram reproduction is quite broad. It not only encompasses the obvious global distribution reproduction, but it also addresses the challenge of inferring the local distribution based on only two parameters. While this is sufficient information for a parametric model like the Gaussian model, it is often inadequate for more realistic non-parametric distributions.

The lack of a distributional assumption requirement is an obvious benefit for DSS. Natural phenomena rarely follow a parametric form such as the Gaussian distribution, and while quantile transformation permits a change from one distribution to any another, there is nothing that says we *should* transform the data to a parametric form. That data transformation is a widely accepted part of the modelling work flow speaks volumes about our strong and continued reliance on simple, yet restrictive mathematical models.

In fact, one could argue that the effect of data transformation on the true spatial distribution of the data may be undesired. Transformation to and back-transformation from Gaussian space yields some disturbing results when applied to skewed distributions. While statistical fluctuations are an inherent property of stochastic simulation, it is expected that these deviations should be reasonable and unbiased. For any one realization, minor fluctuations from a zero mean and unit variance are expected; however, when these values are back transformed to original units a slight shift in the mean in normal space may translate to a more significant shift of the mean in original units. Similarly, the combined fluctuation of the mean and variance in normal space may translate to more noticeable shifts in original space. This is particularly true for skewed distributions, which is the case for some real phenomena. Fixes to this particular problem have been proposed (Journel and Xu, 1994); yet this can be avoided altogether if we do not perform any data transformation prior to modelling – hence direct simulation.

Although Journel (1994) showed that covariance reproduction was achievable without any distributional assumptions, histogram reproduction remained a challenge. Most of the last decade has seen the majority of research focussed on this specific issue in DSS. Soares (2001) proposed to determine the local cumulative distribution function (cdf) by sampling from part of the global cdf. Caers (2000) suggested the use of a posterior correction of the histogram originally proposed by Journel and Xu (1994), in combination with an acceptance/rejection approach to determining the local cdf. Oz et. al. (2003) proposed the prior use of a Gaussian transform to determine a table of local distributions that could be accessed during DSS.

Despite the fact that DSS permits different shapes of the local distributions, the global distribution of simulated values tend to a symmetric, bell-shaped distribution characteristic of the Gaussian distribution (see Bourgault (1997) and Caers (2000)). This is a reflection of the pervasive influence of the Central Limit Theorem, sometimes referred to as the "Gaussian disease". Of the different approaches to infer the local distribution, only the approaches proposed by Soares (2001) and Oz et.al. (2003) are successful at reproducing the global distribution without need for a post-simulation histogram correction.

While histogram reproduction is key to the success of any simulation approach, this is not a significant obstacle in the widespread consumption of DSS. Actually, the work conducted in the past decade shows that there are any number of tricks and tools that can be employed to reproduce the histogram with varying degrees of desire. Although we intuitively understand that different distributions should exist to reflect different local regions, there is nothing in the prevailing DSS algorithms that will account for the proportional effect. The practicality and hence, viability, of DSS depends heavily on the promise of honouring the proportional effect.

6 Proportional Effect

By virtue of DSS' dependence on kriging, the resulting local variance is independent of the data values and the estimate, hence it is homoscedastic. In contrast, the variance of mineral grades or petrophysical properties found in a real deposit or reservoir often changes depending on the local mean – a property called heteroscedasticity. For example, it is common to find a low variance in a low valued area, and a correspondingly high variance in a high valued area. This heteroscedastic behavior is commonly referred to as the proportional effect (Journel and Huijbregts, 1978).

Consider the well-known Walker Lake data set and the lead pollution data from Dallas. A moving average approach was used with non-overlapping windows to determine the relationship between the local mean and variance. Figure 2 shows a very strong positive correlation for both data sets, in fact its relation appears quadratic, i.e.

$$\sigma^2(\mathbf{u}) = f\left(m(\mathbf{u})^2\right)$$

Note that this relationship is characteristic of real data (alternatively, it is sometimes shown as a linear relation between the standard deviation and the mean value), and it is more pronounced for a lognormal distribution (Armstrong (1998), Chilès and Delfiner (1999)). This relationship is neither new nor surprising. Journel and Huijbregts (1978), Isaaks and Srivastava (1991), Goovaerts (1997), and Chilès and Delfiner (1999), have all discussed the importance of the proportional effect in natural resource characterization. It is precisely in this aspect that direct simulation presents its biggest promise.

Yet there is a major flaw in the foundation of direct simulation. Its basis is founded in kriging, which yields a local variance that is data-value independent. As a result, it cannot produce models that will reproduce the heteroscedastic behaviour that would otherwise be found in real mineral deposits or reservoirs. Clearly, the flaw lies in the very fact that least squares estimation is the engine behind the simulation. For it to fulfil its promise, direct simulation must be built on a method that yields dependent mean and variance.

THE PROMISES AND PITFALLS OF DIRECT SIMULATION

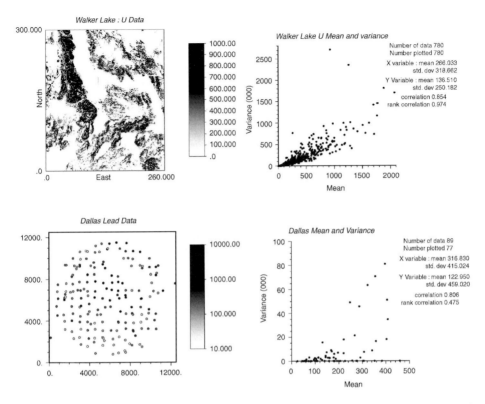

Figure 2 Illustration of proportional effect for Walker Lake data (top), and the lead pollution data from Dall (bottom). Plan view of the data is shown on the left, and crossplots of local variance vs. local mean are shown on the right.

7 Future of DSS

DSS presents one of the future avenues for geostatistics. It is among the latest in a series of simulation approaches that have been introduced in the last two decades. Whether it will rank among the "standard" approaches remains to be seen, advances in particular areas will certainly be key to its popularity. DSS promises (1) the ability to integrate multiple scale data since no transformation of the data is required, (2) reduced reliance on the multiGaussian paradigm, (3) simplicity in methodology, and (4) flexibility to consider different local distribution shapes to account for multivariate non-stationarity.

These promises, however, are balanced by the pitfalls of DSS which include (1) the unavoidable influence of multiGaussianity due to the Central Limit Theorem, (2) problems in histogram reproduction which have led to ad hoc post-processing techniques, (3) the inability to account for spatial heteroscedasticity, specifically the proportional effect, and (4) flexibility in using different distribution shapes locally has not been shown to be practically advantageous or straightforward to implement.

A number of issues must still be resolved to show a real advantage to DSS. The practical significance of accounting for the proportional effect is enormous. Resolution of this issue will lend serious credibility to DSS in construction of realistic numerical models, for application in all natural resource sectors. A second area of research lies in inference of the multivariate distribution. Many authors have expended tremendous research energies into univariate distribution inference, yet the true multiscale data integration benefits of direct simulation will never be realized if the multivariate distribution cannot be properly inferred.

Although DSS was built on the principles of simple kriging, its future cannot remain anchored to simple kriging. It does not lie in the homoscedastic kriging variance, as real data show a very strong relationship exists between the variance and the data values. For it to be of practical significance and in fact, to prevent it from simply becoming an academic exercise, the underlying principle of DSS must permit a heteroscedastic variance that is data-value *dependent*. This is contrary to its simple kriging foundations.

References

Armstrong, M., *Basic Linear Geostatistics*, Springer-Verlag, 1998.
Bourgault, G., Using Non-Gaussian Distributions in Geostatistical Simulations, *Mathematical Geology*, vol. 29, no. 3, 1997, p. 315-334.
Caers, J., Adding Local Accuracy to Direct Sequential Simulation, *Mathematical Geology*, vol. 32, no. 1, 2000, p. 815-850.
Chiles, J.P. and Delfiner, P., *Geostatistics: Modeling Spatial Uncertainty*, John Wiley & Sons, 1999.
Davis, M.W., Production of Conditional Simulations via the LU Triangular Decomposition of the Covariance Matrix, *Mathematical Geology*, vol. 19, no. 2, 1987, p. 91-98.
Deutsch, C.V., *Geostatistical Reservoir Modeling*, Oxford University Press, 2002.
Deutsch, C.V. and Journel, A.G., *GSLIB: Geostatistical Software Library and User's Guide*, 2nd Edition, Oxford University Press, 1998.
Goovaerts, P., *Geostatistics for Natural Resources Evaluation*, Oxford University Press, 1997.
Isaaks, E.H., *The Application of Monte Carlo Methods to the Analysis of Spatially Correlated Data*, PhD thesis, Stanford University, Stanford, CA, USA, 1990.
Johnson, M.E., *Multivariate Statistical Simulation*, John Wiley & Sons, 1987.
Journel, A.G., Modeling Uncertainty: Some Conceptual Thoughts, *Geostatistics for the Next Century*, R. Dimitrakopoulos, ed., Kluwer Academic Publishers, 1994a.
Journel, A.G. and Xu, W., Posterior Identification of Histograms Conditional to Local Data, *Mathematical Geology*, vol. 22, no. 3, 1994b, p. 323-359.
Journel, A.G. and Huijbregts, C.J., *Mining Geostatistics*, Academic Press, 1978.
Oz, B., Deutsch, C.V., Tran, T.T. and Xie, Y., DSSIM-HR: A Fortran 90 Program for Direct Sequential Simulation with Histogram Reproduction, *Computers & Geosciences*, vol. 29, 2003, p. 39-51.
Sinclair, A.J. and Blackwell, G.H., *Applied Mineral Inventory Estimation*, Cambridge University Press, 2002.
Soares, A., Direct Sequential Simulation and Cosimulation, *Mathematical Geology*, vol. 33, no. 8, 2001, p. 911-926.

SAMPLE OPTIMIZATION AND CONFIDENCE ASSESSMENT OF MARINE DIAMOND DEPOSITS USING COX SIMULATIONS

GAVIN BROWN
De Beers Group Services, Boundary Terraces, Mariendahl Lane, Newlands, 7725, Cape Town, South Africa

CHRISTIAN LANTUÉJOUL
Ecole des Mines, Centre de Géostatistique, 35 rue Saint-Honoré, 77305, Fontainebleau, France

CHRISTIAN PRINS
De Beers MRM R&D, Mendip Court, Bath Road, Wells, BA53DG, United Kingdom

Abstract. Conditional simulations are used to develop realistic and quantitative images of spatial variability for analysis. In particular, they are often used to evaluate the impact of uncertainty in applying economic optimization to a natural resource. Algorithms for conditional simulation of geological data and ore grades are well established. However, the majority are not applicable to placer diamond deposits which exhibit a discrete and patchy textural nature. This prompted the development of a new, enhanced suite of algorithms specifically designed for conditional simulations of diamond deposits. The Cox simulation algorithm is fast, incorporates complex mineralization structures, handles large sets of conditioning data and covers several different discrete distributions including those that are highly skewed. In addition, tests are available for model validation of both the distributional and spatial integrity. This enhanced conditional simulation tool is used, firstly, to determine the confidence limits of block estimates and secondly, to quantify and manage the risk of selective mining above an economic cut-off grade.

1 Introduction

Marine diamond evaluation is a very challenging process, balancing the sampling strategy, the estimation technique and the exploitation scenario. These challenges can be investigated by simulating a flexible model that can accommodate many specific features often observed in diamond samples, such as a long distributional tail and high occurrence of zero values.

This model is the so-called Cox process. Since its inception for modeling diamond deposits (Kleingeld, 1987), considerably more data has been collected in different marine environments, describing a large variety of distributional features for simulations to be generated. This prompted the requirement for enhancement of the Cox process. With the ability to perform conditional Cox simulations, the efficiency of sampling campaigns in different geological environments can be investigated. Furthermore, their impact on the assessment of spatial risk can be tested. This will be illustrated in two documented examples.

2 The Cox model

2.1 DEFINITION

Assume that the deposit is partitioned into a family of small congruent domains $(v_i, i \in I)$. For each index $i \in I$, let N_i denote the random number of particles falling within the domain v_i. In this paper, a model for the deposit is specified by the multivariate distribution of the random vector $(N_i, i \in I)$. The specification implies that any mineralization structure that is present at a smaller scale than the domains, is not accounted for by the model. This is compatible with the data collected, as the sample information does not comprise of the coordinates of its particles, but only the number of particles within each sample. In practical applications, the size and shape of the domains equate to those of the samples.

The number of particles in a domain v_i is affected by several factors, including the particle source, the local terrane and footwall lithology. As it is generally impossible to separately assess the contribution of each factor, it is convenient to summarize them as a single factor, say Z_i, that represents the propensity of the domain v_i to be rich. The larger the factor Z_i, the greater the chance that v_i will contain many particles. For this reason, Z_i is named the potential of the domain v_i.

For the Cox model, the number of particles within each domain is independently Poisson distributed, with the mean of each domain equal to its potential. However, as the potentials of the domains are unknown, they are considered random. Thus the numbers of particles within the domains are not independent - neighboring domains have correlated potentials - but are only conditionally independent. In this paper, we let $Z_i = \varphi(Y_i)$ for each $i \in I$, where $(Y_i, i \in I)$ is a standardized Gaussian vector.

2.2 STATISTICAL INFERENCE

The spatial distribution of the Cox model is characterized by two parameters, namely the anamorphosis φ of the potential of the domains and the covariance C of the underlying Gaussian vector Y. The inference of each parameter is considered in turn.

By definition, φ is determined by the distribution of Z_i and as there is a one-to-one correspondence between the distributions of Z_i and N_i (Feller, 1971), φ is also determined by the distribution of N_i. The statistical inference of φ can therefore be reduced to that of N_i.

In practice, the distribution of N_i is often chosen within a family of pre-specified models (negative binomial, Sichel, Cox-lognormal...), and its statistical inference consists of estimating the parameters of the selected model from experimental statistics provided by the data. Such parameters include the mean number of particles per sample, the variance and the proportion of barren samples. For example two models are presented with their corresponding potential models:

- if N_i follows a negative binomial distribution with index $\alpha > 0$ and parameter $p = 1 - q$, then Z_i is gamma distributed with the same index α and scale factor $b = q/p$:

$$p_n = \frac{\Gamma(\alpha+n)}{\Gamma(\alpha)\, n!} q^\alpha p^n \quad n \in \mathbb{N} \qquad f(z) = \frac{b^\alpha}{\Gamma(\alpha)} e^{-bz} z^{\alpha-1} \quad z \in \mathbb{R}^+$$

- if N_i follows a Sichel distribution[1] with index $\alpha > 0$ and tail parameter $0 < \theta < 1$ (Sichel, 1973), then Z_i follows an inverse Gaussian distribution with parameters $a = (1-\theta)/\theta$ and $b = \alpha^2 \theta/4$:

$$p_n \propto \frac{(\alpha\theta/2)^n}{n!} K_{n-1/2}(\alpha) \quad n \in \mathbb{N} \qquad f(z) \propto \frac{1}{z^{3/2}} \exp\left(-az - \frac{b}{z}\right) \quad z \in \mathbb{R}^+$$

Once φ has been estimated, several procedures can be performed to estimate the covariance C of Y. The simplest one rests on the fact that the covariances of N and Z differ only by their nugget effect:

$$Cov\{N_i, N_j\} = Cov\{Z_i, Z_j\} + E\{N_i\} 1_{i=j}$$

The covariance of Z can easily be derived from that of Y when φ has been expanded into Hermite polynomials:

$$Cov\{Z_i, Z_j\} = \sum_{n=1}^\infty \frac{\varphi_n^2}{n!} C_{ij}^n$$

Combining both equations and assuming that $Cov\{N_i, N_j\}$ is experimentally accessible, we can obtain C_{ij} from the equation

$$Cov\{N_i, N_j\} = \sum_{n=1}^\infty \frac{\varphi_n^2}{n!} C_{ij}^n + E\{N_i\} 1_{i=j}$$

This equation returns a single solution, due to the fact that its right-hand side member is a monotonic increasing function.
When all values for C_{ij} have been estimated, the subsequent modeling of C is necessary to ensure it is positive definite.

2.3 SIMULATION

The simulation using the Cox process with anamorphosis φ, covariance C and incorporating the available data comprising of a family of samples or blocks,

---

[1] In this formula, K_μ stands for the modified Bessel function of order μ.

is presented. Each sample and block is represented by a subset of indices of I (consisting of only one index in the case of a sample). The conditioning data can be written as $(N_A = n_A, A \in \mathcal{A})$ for any family $\mathcal{A}$ of subsets of I. There is no inconvenience in assuming that the supports of the conditioning data are pairwise disjoint[2]. In the following algorithm, I_c denotes the subset of indices affected by conditioning (e.g. $I_c = \cup_{A \in \mathcal{A}} A$):

(i) *simulate* $(Y_i, i \in I_c)$ *given* $(N_A = n_A, A \in \mathcal{A})$;
(ii) *simulate* $(Y_i, i \in I)$ *given* $(Y_i = y_i, i \in I_c)$;
(iii) *simulate* $(N_i, i \in I)$ *given* $(Y_i, i \in I)$ *and* $(N_A = n_A, A \in \mathcal{A})$.

The first step consists of simulating the distribution

$$g_c(y_i, i \in I_c) \propto g(y_i, i \in I_c) \prod_{A \in \mathcal{A}} e^{-z_A} z_A^{n_A}$$

where $z_A = \sum_{i \in A} \varphi(y_i)$ is the potential of block A. This can usually be achieved using an acceptance-rejection method unless the number of elements $\#I_c$ of I_c is large, in which case the Gibbs sampler is required (Geman and Geman, 1984; Kleingeld et al., 1997). The second step is simply a conditional simulation of a Gaussian vector. Regarding the third step, three different cases have to be considered. Each component N_i of $(N_i, i \notin I_c)$ is independently Poisson distributed with mean $z_i = \varphi(y_i)$. If $\#A = 1$, say $A = \{i\}$, then we put $N_i = n_i$. Finally, if $\#A > 1$, then the vector $(N_i, i \in A)$ is simulated according to a multinomial distribution with index n_A and parameters $(z_i/z_A, i \in A)$.

3 Application of the Cox model by examples

The south-western African coast hosts significant diamond deposits within gravels associated with paleo-shorelines, channels and transgressive lags (Kuhns, 1995). The development of the marine diamond deposits was complex and involved the interaction of fluvial, shoreline and aeolian sedimentary environments on a stable continental shelf under conditions of changing sea levels. The deep water ($> 70m$ water depth), offshore deposits comprise low-grade, aerially extensive, but thin, composite marine lag gravels (Corbett, 1996). The current mining technology requires a highly selective mine plan to facilitate profitable exploitation. Sampling costs for these offshore deposits are high and consequently, optimization of these programmes and an understanding of the confidence of the estimates are essential to ensure successful exploitation.

3.1 SAMPLE OPTIMIZATION

The objective of this study was to devise the optimum sample support size and pattern to evaluate and ultimately selectively mine a particular marine diamond deposit. At first, a stochastic model of the diamond content was created using

---

[2] Define new blocks by removing the indices of the samples within them, and update their numbers of particles.

geological information and reconnaissance sampling of the actual target area, as well as a regional understanding of the environment. Then, different sampling campaigns of varying sample sizes and patterns were simulated, from which block estimates and mine plan selections were calculated. Finally, a financial analysis of the various sampling campaigns and their expected mining revenues was carried out to determine the most optimum sampling strategy.

The topography of the deposit has favored the long-term preservation of gravel accumulations, which constitute semi-permanent trapsites that concentrate diamonds and other heavy minerals. The reconnaissance sampling data comprised only 63 widely spaced samples, and experience has shown that a limited dataset of this size creates a constrained tail to the overall stone density of the deposit. However, the stone distribution tail can be better modeled using information from analogous deposits which showed an overall Sichel distribution (Sichel, 1973) and a significantly higher variance (10.85) than the sample data (see Figure 1).

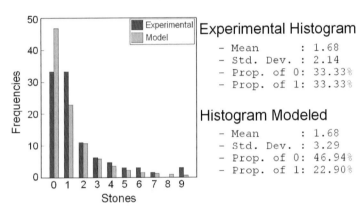

Figure 1. Modeled histogram of the stone distribution compared to the experimental stone distribution.

The variogram of the deposit was also derived from similar deposits (33% nugget effect; anisotropic spherical structure with ranges $345m$ ($305°$) and $145m$ ($215°$)). The model assumptions were validated by analyzing the statistics produced by a set of non-conditional simulations performed at the conditioning data points (see Figure 2).

Twenty conditional simulations of the diamond stone density were created over the entire target area using the Cox process. Geologists also visually reviewed the realizations and confirmed their validity against conceptual geological models. A series of sampling campaigns were then designed according to financial considerations and available sampling tools. Various sample sizes (from $12.2m^2$ to $700m^2$), sample shapes (trench, block and drill) and sample spacings ($50m \times 50m$ to $250m \times 50m$) were considered.

Ordinary kriged estimates of the stone density for $50m \times 50m$ blocks were performed on each sampling campaign and each realization. The block estimates could then be compared to their respective "actual" values in the simulated deposits

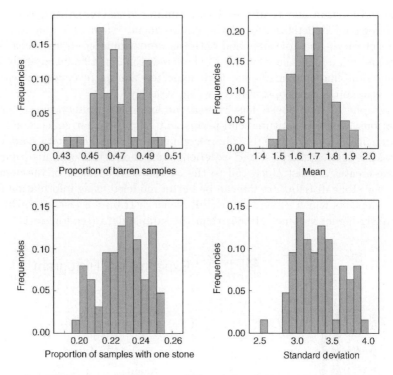

Figure 2. Histograms of the unconditional simulations at data points to validate the simulation model.

(regularized into 50m × 50m blocks). Regressions comprising of the "actual" grades to the estimates for each sampling campaign per realization are summarized in Table 1.

The results of the comparison between estimated and "actual" block grades could be related to the characteristics of the source sample data. Sampling campaigns with few large samples (trench or bulk) returned accurate but highly smoothed mean estimates (a function of poorly defined spatial structure). Sampling campaigns with numerous small samples (drills) returned less accurate mean estimates, but defined the grade variability more accurately (a function of better defined spatial structure). The accuracy of the estimates relative to the "actual" grades can be shown graphically (see Figure 3).

Selective mine plans based upon the estimated grades were then set up with a realistic cut-off grade that resulted in a high degree of selectivity (only 25% of the target area is above cut-off). The actual contribution (revenue - mining cost) for each block was then calculated using the simulation as the "actual" resource grade. Thereby, the mining profits of the selective mine plans for each sampling campaign/estimation combination were established. The cost of sampling was then deducted from the mining profit and the revenue that was generated from the sampling recovery added. Finally, an overall profit was established for each

Table 1. Statistical results of the block estimates for each sampling campaign.

Type	Size ($m \times m$)	Area (m^2)	Spacing ($m \times m$)	Mean	Standard deviation	Coefficient of variation	Regression
trench	14×50	700	250×50	0.768	0.366	0.476	0.694
trench	14×50	700	150×100	0.792	0.304	0.384	0.816
trench	14×25	350	100×100	0.775	0.464	0.599	0.894
bulk	25×25	625	150×100	0.796	0.321	0.404	0.828
bulk	14×14	196	100×100	0.776	0.426	0.548	0.872
drill	3.5×3.5	12.25	100×100	0.757	0.458	0.604	0.837
drill	3.5×3.5	12.25	100×50	0.755	0.508	0.674	0.902
drill	3.5×3.5	12.25	50×50	0.768	0.616	0.802	0.945

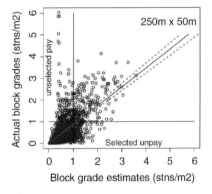
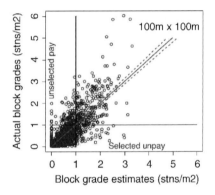

Figure 3. Scatterplots of block estimates versus "actual" block grades for two different sampling campaign realizations.

sampling campaign. Table 2 presents the financial results obtained. It shows the percentage mining and overall (assuming perfect knowledge) profits obtained by the selective mining for each of the different sampling campaigns. It appears that the trench sampling campaigns produce less optimal returns for both mining and overall profits. The high degree of selectivity penalizes these estimates despite lower net sampling costs. The bulk samples realize slightly better returns for mining profit and a reasonable overall profit because of the associated low sampling costs. The drill sampling produces good returns from mining and overall profits, a consequence of more accurate estimation, however the high cost of drill sampling penalizes the overall profit, with the $100m \times 50m$ drill sample pattern producing the optimal result.

Table 2. Summary of the selective mining of the simulated deposit based upon various sampling campaigns.

Type	Size ($m \times m$)	Area (m^2)	Spacing ($m \times m$)	Number of samples	Mean percentage of mining profit	Mean percentage of overall profit
trench	14×50	700	250×50	231	59%	52%
trench	14×50	700	150×100	180	72%	66%
trench	14×25	350	100×100	270	82%	77%
bulk	25×25	625	150×100	180	73%	68%
bulk	14×14	196	100×100	270	76%	73%
drill	3.5×3.5	12.25	100×100	270	81%	76%
drill	3.5×3.5	12.25	100×50	567	90%	82%
drill	3.5×3.5	12.25	50×50	1080	93%	81%

3.2 CONFIDENCE LEVELS OF BLOCK ESTIMATES

The objective of this second study was to determine the relative confidence of block estimates of stone density (expressed as stones per square meter) for another marine diamond deposit. The approach adopted here consisted of creating conditional simulations of the deposit, deriving the stone density distribution of each block, and deducing confidence limits for each block estimate.

The sampling data comprised 544 large diameter drill samples ($7m$ diameter), predominantly arranged on a $50m$ square grid. A section of the deposit was less densely sampled ($70m$), whereas another portion was unsampled (see Figure 4).

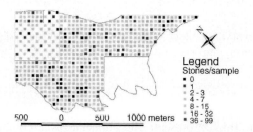

Figure 4. Locality plot showing the large diameter samples.

The stone distribution was modeled as a Sichel distribution using the experimental mean and the proportion of barren samples. The sample variogram of the deposit was determined from the sample data and inferences from regional geological structures (34% nugget effect; two isotropic nested spherical structures, 50% of the sill at $76m$ and 16% at $300m$), from which a variogram in Gaussian space was derived. The Cox model was validated by analyzing the statistics produced by a set of unconditional simulations at the locations of the conditioning data points. One hundred and twenty-five conditional Cox simulations were then created at the sample support size and averaged into $50m \times 50m$ blocks. An example of a realization at the sample support size, together with the conditioning sample data is presented in Figure 5.

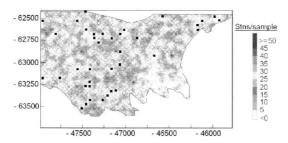

Figure 5. Conditional simulation of a marine diamond deposit.

From the simulations, statistics for each block were calculated and compared with ordinary kriged estimates, using the same dataset and the same variogram model (see Figure 6).

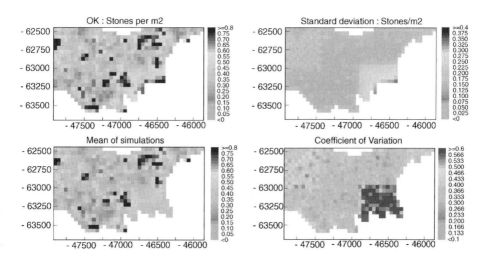

Figure 6. Comparison between kriged and simulated results.

The kriged estimate (top left) and the mean of the simulations (bottom left) of the stone density show good similarity with compatible low and high grade areas. The estimation variance (top right) provides an indication of the data density for these estimates as this parameter highlights areas of similar sample spacings. The coefficient of variation, measuring the relative variability of the distribution of possible grades, provides further information by incorporating the relevant sample grade information. The simulated distribution of each block is also used to guide mining selections when a cut-off grade is considered. Figure 7 shows a comparison between the blocks whose kriging estimate lies above a given cut-off grade, and those whose simulated grade has more than 65% chance of lying above the same cut-off grade.

In this case study, the simulation based selection, using the probability above cut-off selection criteria, selected 56% of those blocks selected above cut-off of the kriged estimates. However, the reduced "probability" selection contains 89% of the total diamonds. More generally, this approach allows the relative risk of the mining selection to be incorporated into mine-planning decisions and has proved to return better results than a single grade cut-off method.

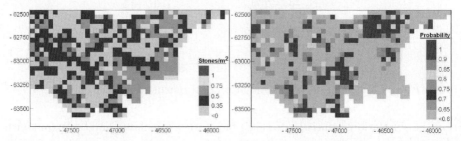

Figure 7. Mining selections above a cut-off grade based on kriged estimates (left) and 65% or higher probability of above a cut-off grade based on Cox simulations (right).

4 Conclusion

The development of the Cox process to model the spatial distribution of diamonds within marine deposits has enabled the successful optimization of sampling campaigns in terms of both sample size and pattern. The quantification of block grade uncertainty from conditional simulations has also proven to significantly assist mine planning decisions.

References

Corbett, I.B., *A Review of Diamondiferous Marine Deposits of Western Southern Africa*, Africa Geoscience Review, Vol. 3-2, 1996, pp. 157-174.

Feller, W., *An Introduction to Probability. Theory and Applications Vol. 2*, Wiley, New York, 1971.

Geman, S. and Geman, D., *Stochastic Relaxation, Gibbs Distribution and the Bayesian Restoration of Images*, IEEE Trans. Pattern Anal. and Mach. Intel., Vol. 6, 1984, pp. 721-741.

Kleingeld, W.J., *La géostatistique pour des Variables Discrètes*, Ph.D. Thesis, School of Mines of Paris, 1987.

Kleingeld, W.J., Thurston, M.L., Prins, C.F. and Lantuéjoul, C., *The Conditional Simulation of A Cox Process with Application to Deposits with Discrete Particles*, In Baafi E.Y. and Schofield N.A. (eds.), Geostatistics Wollongong '96, Vol. 2, Kluwer, Dordrecht, 1997, pp. 683-694.

Kuhns, R., *Sedimentological and Geomorphological Environment of the South African Continental Shelf and its Control on Distribution of Alluvial, Fluvial and Marine Diamonds*, Society for Mining, Metallurgy and Exploration, Proceedings of Annual Meeting, Denver, 1995.

Sichel, H., *Statistical Valuation of Diamondiferous Deposits*, In Salamon M.D.G. and Lancaster F.H.(eds.), Application of Computer Methods in the Mineral Industry, The South African Institute of Mining and Metallurgy, Johannesburg, 1973, pp. 17-25.

INVERSE CONDITIONAL SIMULATION OF RELATIVE PERMEA-BILITIES

JAIME GÓMEZ-HERNÁNDEZ
Universidad Politécnica de Valencia, Spain
CAROLINA GUARDIOLA-ALBERT
Universidad Politécnica de Valencia, Spain
currently with Instituto Geológico y Minero de España, Madrid, Spain

Abstract. The spatial variability of relative permeability curves has not attracted much attention yet. This paper addresses this issue, and extends the self-calibration technique for the generation of absolute and relative permeabilities conditioned not only to permeability data but also to saturation and pressure data.

The paper starts with a sensitivity analysis presenting a synthetic example where the spatial variability of relative permeability is relevant, then proceeds with a derivation of the algorithm used to condition a realization of relative permeabilities to pressure and saturation data (both steady state and transient) and concludes with the demonstration of the technique with one synthetic example.

The paper shows that the spatial variability of relative permeabilities is important in reservoir characterization. It also demonstrates how the self-calibrating simulation method can be used to generate realizations of spatially variable relative permeability curve parameters which are consistent with measured values of the state of the reservoir.

1 Introduction

Stochastic modeling of multi-phase flow in heterogeneous porous media is becoming common practice in petroleum engineering and subsurface hydrology. Inverse modeling theory provides a methodology to integrate both static and dynamic data in reservoir characterization. Absolute permeability is one of the parameters that are typically estimated with inverse conditional or unconditional simulations, whereas relative permeabilities are assumed to be known homogeneous functions within the reservoir. However, when studying multiphase flow, relative permeability is the parameter that controls the rate of displacement of the different phases present in the reservoir. This paper discusses the characterization of the spatial variability of relative permeabilities by inverse conditional simulation. A new inverse technique to estimate spatial distributions of both absolute and relative permeability parameters has been developed based on the self-calibrating method. Since relative permeabilities are dependent on one of the state variables (saturation), the optimization problem is highly non linear, increasing the difficulty of the inverse problem that has to be solved.

2 Sensitivity Analysis

The governing equations for immiscible two-phase flow are formulated in terms of water saturation and fluid pressure. Substituting a generalized form of Darcy's law into the mass conservation equation, and neglecting gravity effects, the diffusivity equation for the horizontal flow of each fluid is obtained:

$$\frac{\partial (\phi \rho_l S_l)}{\partial t} - \nabla \cdot \left[\rho_l \frac{\mathbf{k} k_{rl}}{\mu_l} \nabla p_l \right] = -q_l \quad \text{for } l = w, o \tag{1}$$

where subscripts w and o refer, respectively, to water (wetting phase) and oil (non-wetting phase), $\mathbf{k}$ is the absolute permeability tensor $[L^2]$, k_{rl} the dimensionless relative permeability for phase l, S_l is the saturation, ϕ is the porosity (dimensionless parameter), ρ_l is the fluid density $[M/L^3]$, μ_l is the viscosity $[ML/T]$, p_l is the fluid pressure $[M/LT^2]$, q_l is the injection or production rate per unit volume $[T^{-1}]$ and t is time $[T]$.

The above equations are solved by finite differences after neglecting gravity and capillary pressure terms and assuming the following constitutive equation relating relative permeability and saturation (Brooks and Corey, 1966):

$$k_{rl} = k_{rl}^0 \left(\frac{S_l - S_{rl}}{1 - S_{rl} - S_{rl'}} \right)^{n_l} \quad \text{for } l = w, \, l' = o \text{ or } l = o, \, l' = w \tag{2}$$

in which n_w, n_o, S_{rw}, S_{ro}, k_{rw}^0 and k_{ro}^0 are parameters defining the relationship. For the purpose of characterizing the heterogeneity of relative permeability, we will assume that exponents n_w and n_o are homogeneous, but that the remaining coefficients, i.e., the residual saturations for oil and water (S_{rw}, S_{ro}), and the end-points of the relative permeability functions may vary within the reservoir.

To demonstrate the importance of accounting for the spatial variability of relative permeabilities, two different runs are performed, both of them with the same heterogeneous absolute permeability field, but with different relative permeability fields. The first run is done assuming homogeneous relative permeabilities, the second run is done with heterogeneous values. Conditional simulations are used to construct absolute and relative permeability fields. The absolute permeability fields were generated by GCOSIM3D (Gómez Hernández and Journel, 1993), following a lognormal distribution with known mean and variance, and a spherical variogram. The relative permeability fields are constructed by generating each of the four parameters defining relation (2) as a Gaussian field with known mean and variance and a Gaussian variogram (it is assumed that these parameters vary smoothly in space).

The 2D spatial domain mimics a quarter of a five-spot, discretized in 15 × 15 grid blocks of size 10 m ×10 m. The injector well is located at the left lower corner, and the production well at the right upper corner. The initial conditions for the 2D run are $S_w = 0.1$ and $p = 6.895 \cdot 10^6$ Pa. Water is injected at a constant rate of 2.5 kg/s. Constant 4.8-hour time steps are prescribed, and the simulations are run for 120 days.

For the first run, a single relative permeability curve for each fluid is used, with parameter values equal to the mean values used for the generation of the heterogeneous field: $k_{rw}^0 = 0.7$, $S_{rw} = 0.1$, $k_{ro}^0 = 0.85$ and $S_{ro} = 0.2$. The second run uses heterogeneous values for all four parameters, maintaining the same mean values as the previous run. The saturation and pressure fields at the end of the 120 days are shown in Figures 1 and 2.

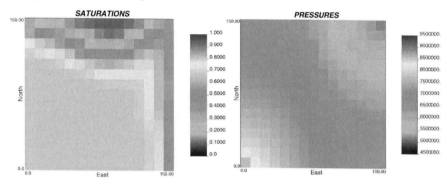

Figure 1. Saturation and pressure front at $t = 120$ days. Absolute permeability is heterogeneous, but relative permeability curves are homogeneous. The water injection well is located at the lower left grid block and the production well at the upper right grid block.

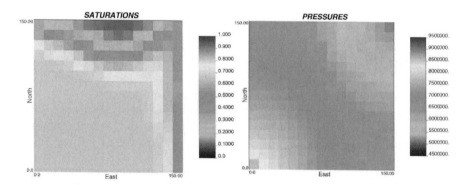

Figure 2. Saturation and pressure front at $t = 120$ days. Absolute and relative permeability are heterogeneous. Same absolute permeabilities and well configuration as in previous figure.

From these figures it can be observed that pressures are not very much affected by the heterogeneity of the relative permeability curves, but that saturations are. Similar conclusions were obtained in 1D by Guardiola-Albert and Gómez-Hernández (2002).

3 Inverse conditional simulation algorithm

The optimization algorithm is based on gradient methods, and the concept of master points is borrowed from the Sequential Self-Calibrated method (Gómez-Hernández et al., 1997; Hendricks-Franssen, 2001). Calibration of the flow model to non linearly related data is formulated as an optimization problem, which tries to minimize an objective function. A computer code was written to couple the forward two-phase flow simulator TOUGH (Pruess and Oldenburg, 1999) with an iterative inverse method. After calibration, the result is a plausible representation of the reservoir honoring historical pressure and saturation data. Calibration parameters are: absolute permeabilities, the two end-points of oil and water relative permeability functions, and the two residual saturations. For the sake of simplicity, the shape parameters (n_o and n_w) are constant and equal to 1. The main steps in the iterative process, followed by the inversion technique developed, are summarized here. The loop from step 2 to step 7 is repeated until convergence is reached.

For each iteration IT:

1. Generate a conditional or an unconditional simulation of the four parameters k_{rw}^0, k_{ro}^0, S_{rw} and S_{ro} and of the absolute permeability, k. The generated fields constitute the seed or initial input fields.
2. The two-phase flow numerical solver is run. Saturation and pressure fields are obtained for all the time steps at every grid block.
3. Evaluate the following objective function:

$$J^{IT} = \sum_{t=1}^{T_s}\sum_{i=1}^{N_s} w_{s,i}\left(S_{i,t}^{SIM,IT} - S_{i,t}^{MEAS}\right)^2 + \sum_{t=1}^{T_p}\sum_{i=1}^{N_p} w_{p,i}\left(p_{i,t}^{SIM,IT} - p_{i,t}^{MEAS}\right)^2 \quad (3)$$

where N_s and N_p are the number of saturation and pressure data points, respectively, T_s and T_p are the number of times at which saturation and pressure have been measured. Indices SIM and $MEAS$ indicate simulated and measured values, and, w_s and w_p are weighting factors.

4. If J is smaller than a pre-determined value, the simulated permeability values are said to be conditioned to the measured saturation and pressure values, and the iterative loop stops. On the contrary, the optimization continues and the k, k_{rw}, k_{ro}, S_{rw}, and S_{ro} fields are perturbed.
5. The optimization procedure determines the value of the perturbation that is applied to the initial field so that the objective function is reduced.
6. Go to step 2. The modified reservoir model is input again into the reservoir simulator.

4 Synthetic example

To check the feasibility of the inverse technique a simple synthetic example is presented. Absolute permeability is heterogeneous all over the reservoir, and relative permeability is piecewise heterogeneous as shown in Figure 3.

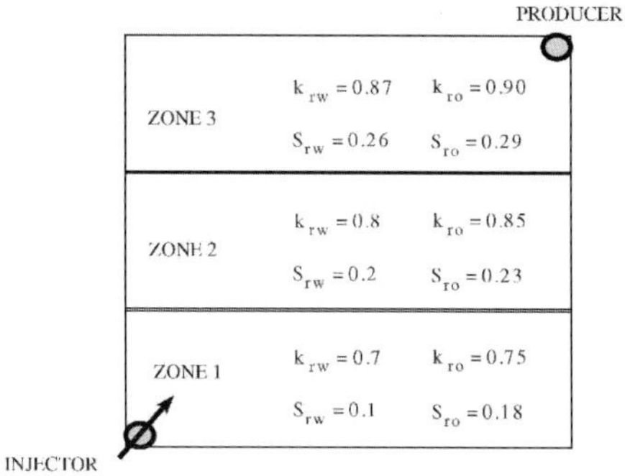

```
                                    PRODUCER
                                        ○
            ┌─────────────────────────────────┐
            │         k_rw = 0.87   k_ro = 0.90│
            │ ZONE 3                           │
            │         S_rw = 0.26   S_ro = 0.29│
            ├─────────────────────────────────┤
            │         k_rw = 0.8    k_ro = 0.85│
            │ ZONE 2                           │
            │         S_rw = 0.2    S_ro = 0.23│
            ├─────────────────────────────────┤
            │         k_rw = 0.7    k_ro = 0.75│
            │ ZONE 1                           │
            │ ↗       S_rw = 0.1    S_ro = 0.18│
            └─────────────────────────────────┘
         INJECTOR
```

Figure 3. The reservoir is divided into 3 different zones, within each of them the relative permeability parameters are assumed to be constant. The reference values for the relative permeability parameters are given in the figure.

No statistical correlation was considered between the five parameters (k, k_{rw}^0, S_{rw}, k_{ro}^0 and S_{ro}). However, there is an implicit correlation because all the parameters are calibrated to the same set of production data. A reference run as performed with the values shown in Figure 3, which was sampled at five locations and at ten time steps to be used as the conditioning information for the inverse conditional simulation.

Scatterplots for saturation and pressure at well locations for all sampling times are shown in Figure 4. They compare the degree of mismatch between the seed fields and the calibrated fields to the data (on the left, the seed fields, on the right the calibrated fields). The calibration reduces considerably the spread of the scatterplots, reflecting the effect of jointly conditioning absolute and relative permeabilities to the available saturation and pressure data. We could consider the calibrated field as a plausible representation of a reservoir for which only partial historical evolution is known.

5 Conclusions

A code has been developed and implemented for the simultaneous generation of absolute and relative permeability fields conditioned to historical production data. The heterogeneity of relative permeability is described by the heterogeneity of the parameters that describe a specific relationship between saturation and relative permeability.

Four parameters in this relationship (two end-points of the curves and the residual saturations) were chosen to characterize the relative permeability curves. Joint conditioning of absolute and relative permeabilities to pressure and saturation data considerably improved the history match in the synthetic example presented.

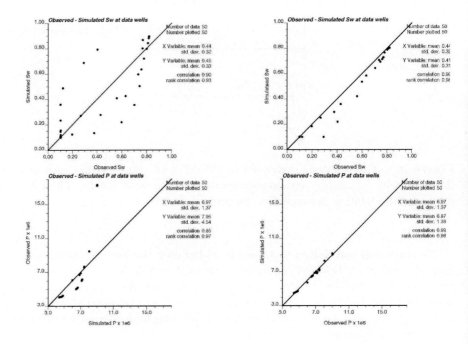

Figure 4. Simulated versus "observed" reference values are plotted before (left) and after (right) the inversion is performed. Upper scatterplots represent water saturations and lower scatter plots pressures.

Acknowledgements

The second author is grateful to the fellowship by the Repsol Foundation that financed her stay at Universidad Politécnica de Valencia during her Ph.D. studies. Partial funding from the Spanish National R&D Plan, through project REN2002-02428, and from the Spanish Nuclear Waste Management Company, ENRESA, through R&D contract 774325 are acknowledged.

References

J.J. Gómez-Hernández and A.G. Journel, *Joint simulation of multiGaussian random variables*, Oxford University Press, 1993.

R.H. Brooks and A.T. Corey, *Properties of porous media affecting fluid flow*, J. Irrigation and Drainage Division. Proceedings of the American Society of Civil Engineers, 1966.

Hendricks-Franssen, H. J., *Inverse Stochastic Modelling of Groundwater Flow and Mass Transport*, PhD dissertation. Universidad Politécnica de Valencia, Spain, 2001.

J. Gómez-Hernández and A. Sahuquillo and J.E. Capilla, *Stochastic Simulation of Transmissivity Fields Conditional to Both Transmissivity and Piezometric Data. 1. Theory*, Journal of Hydrology, vol. 203, 1997, p. 162-174.

C. Guardiola-Albert and J. Gómez-Hernández, *Inverse modelling of two-phase flow: calibration of relative permeability curves*, Calibration of Reliability in Groundwater Modelling: A Few

Steps Closer to Reality (Proceedings of ModelCARE'2002, Prague, Czech Republic. IAHS, Publ. no. 277, 2002, p. 190-195.

K. Pruess and C. Oldenburg. Tough2 user's guide. Report LBNL-43134, Ernert Orlando Lawrence Berkeley National Laboratory, Berkeley, 1999.

QUANTIFIABLE MINERAL RESOURCE CLASSIFICATION: A LOGICAL APPROACH

CHRISTINA DOHM
Mineral Resource Evaluation Department (MinRED), Exploration Division of AAplc, Johannesburg, South Africa

Abstract. In terms of the reporting codes Mineral Resource classification is a function of increasing confidence in the geoscientific information and the associated resource estimate. An overview of Mineral Resource classification approaches is given; the tendency in resource classification is to concentrate on the confidence associated with the grade estimate. Uncertainties linked to tonnage and metal estimates are rarely explicitly mentioned. As for the risk associated with the underlying geological model it is often, if at all, only considered on a global rather than a local basis. The objective is to present a quantifiable Mineral Resource classification guideline that recognises uncertainty in both geological and resource models, considers confidence in estimation of metal content for specified production periods and also takes into account both the correlation of blocks in the block model as well the change of support between an estimated block and the production period. This classification method builds on a previous publication (Dohm, 2003), where a technique for assessing the combined risk associated with both the geological and grade models was demonstrated. The final result is a succinctly classified mineral resource model, which is based on objective quantifiable classification rules that recognises the uncertainty related to subjective interpretations of the available information.

1 Introduction

The classification of Mineral Resources and Ore Reserves forms an integral part of Mineral Resource evaluation and reporting. Mineral Resource classification categories correspond to an increasing function of geoscientific knowledge and confidence. A Mineral Resource is classified as Inferred if the tonnage, grade and mineral content can be estimated with low confidence. Indicated Mineral Resources represent that part of the Mineral Resource for which tonnage, densities, shape, physical characteristics, grade and mineral content can be estimated with a reasonable level of confidence. For Measured Mineral Resources these attributes can be estimated with a high level of confidence. Only Measured and Indicated Mineral Resources can be converted to Ore Reserves. In terms of the guidelines of the reporting codes the Competent Person (JORC, SAMREC) or Qualified Person (NI 43-101) is to provide a view of the relative confidence the investment community should place on the published Mineral Resources and Ore Reserves of mining and exploration companies.

The main elements that affect the confidence in the resource estimate are the reliability of the geological model, the continuity of the mineralisation, the sampling grid configuration, the quality of the sampling data, and the reliability of the evaluation method. The most important element is the interpretation of the geology and the delineation of the resource (Stephenson, 2001). In practice the level of uncertainty in the geological model is often not easily incorporated in the Mineral Resource classification. In many cases global discount factors are applied to take account of the unpredictability of the geological features.

A number of approaches that the author encountered during project reviews presented here illustrate the evolution of Mineral Resource classification methodologies, concluding with a holistic Mineral Resource classification guideline. This guideline recognises uncertainty in both geological and resource models, considers confidence in estimation of tonnage, grade and metal content for specified production periods and also takes into account both the correlation of blocks in the block model as well the change of support between an estimated block and the production period.

2 Questionable Mineral Resource classification strategies

Since the Bre-X scandal the spotlight has been focussed on Mineral Resource classification methodologies. Two interesting but not recommended classification strategies observed during project appraisals in recent years are discussed.

2.1 NUMBER OF SAMPLES PER BLOCK OR PER UNIT AREA

The crudest set of classification rules the author has come across is: One drillhole per hectare identifies an Indicated Mineral Resource, more drillholes per hectare allow for the resource to be classified as Measured and when there are no drillholes but the area is within the mining lease it can be considered as an Inferred Mineral Resource.

This method does not take cognisance of the spatial continuity of the mineralisation, and anisotropy, if it should exist, is also ignored. Change of support is also not considered as 10000 square metres (1ha) can be achieved in a number of ways for example a rectangle (40m x 250m) or square (100m x 100m) are equivalent in this scheme.

The relative locations of the drillholes are not recognised; for example four 2x2, 1ha squares with a set of clustered drillholes close to the four touching corners will be considered Indicated as will four 1ha squares, with evenly spaced drillholes at their centroids, be classified as Indicated.

2.2 RESOURCES WITHIN A PRODUCTION PLANNING PERIOD OF RESERVES

In this two-dimensional example ordinary kriged estimates were produced for the entire mine lease area. Blocks were assigned the average grade of the deposit when their estimation became "unreliable", e.g. the criterion for the minimum number of samples in the search volume is not met. The classification rule applied here was established from time-based production planning considerations. Resources were classified as a consequence of the reserve classification and not the other way round. The argument put forward was that grade control information acquired during mining activities would be adequate to support the classification.

Planned 5-year and 20-year production period mining outlines were used to differentiate between Proved and Probable Ore Reserves. The Proved Ore Reserves incorporated the Measured Mineral Resources and the Probable Ore Reserves partially incorporated the Indicated Mineral Resources. Resources lying beyond the 20 year planning limit were also defined as Indicated if sufficient drilling information was available. The area between the 20-year plan and the lease boundary were to be considered as Probable Ore Reserves. Inferred Mineral Resources were non-existent. It is obvious that this set of classification rules was unacceptable and required revision. On recommendation the company adopted a quantifiable risk based classification strategy, still related to production periods but independent of mining lease boundaries and also including the confidence of the resource estimates.

3 Range of influence of the variogram model and Resource Classification

Methods in place for classifying resources are often based on the kriging variances of grade estimates or functions of the variogram parameters and kriging parameters. The semi-variogram of a mineral deposit reflects the spatial variability of the sample grades at fixed distances and along a given direction. Snowden (1996) suggests interpreting this spatial continuity to determine appropriate drillhole patterns to achieve various levels of confidence in resource classification.

Resources are classified as Inferred when drillholes are further apart than the range of influence of the variogram. The drill spacing at which a distinction between Measured and Indicated is made is based on a rule of thumb and is taken as the distance equivalent to two thirds of the total variability i.e. two thirds of the sill of the variogram model.

The ranges of the variogram are not sufficient for resource classification; the nugget effect will for instance have a significant influence in this classification. If the nugget effect is high and the structured component is relatively short, as is common for Witwatersrand gold deposits, this classification method will be of little use; the majority of the mineral resources will be classified as being Inferred resources.

4 Kriging variance and variations thereof in the classification scheme

One of the advantages of kriging estimation techniques is that when correctly applied these techniques produce unbiased block estimates and ensure minimum estimation variance known as the kriging variance. The kriging variance is dependent on the variogram model, and the sampling grid configuration in relation to the block that is being estimated. It is possible to calculate the kriging variance without producing the estimate. It is thus not surprising that a number of classification schemes in the past were based on the kriging variance: a few applications are given below:

4.1 INTERPOLATION, EXTRAPOLATION AND RESOURCE CONFIDENCE

This classification rule was based on the following reasoning related to the type of estimation. Measured Mineral Resources arise from interpolated blocks, which have lower kriging variances and therefore higher confidence associated with them. Indicated Mineral Resources occur when blocks are extrapolated, this means that these blocks have higher kriging variances and thus a reduced confidence is associated with them. Any blocks extrapolated beyond the range of influence of the variogram are classified as Inferred blocks.

4.2 SAMPLE VARIANCE, KRIGING VARIANCE AND NUMBER OF SAMPLES

The Mineral Resources are classified as Measured if the kriging variance of the block is less than the sample variance. If not Measured and at least 4 samples were within the maximum range of influence of the block the resource is classified as Indicated. Inferred Mineral Resources are those blocks that did not fall into the previous two categories but with a kriging variance equivalent to that of blocks in the Indicated category.

4.3 SAMPLE VARIANCE, BLOCK VARIANCE AND KRIGING VARIANCE

Blocks are classified as Measured if their kriging variance was less than the block variance. Blocks classified as Indicated have a kriging variance less than the sample variance but greater than the block variance. Blocks with an estimated mineralised proportion less than 20 % were considered to be in the Inferred category.

5 Resource classification and relative variances

Using the kriging variance, as the only measure to quantify uncertainty in block grade estimate, is questionable as the only relationship the kriging variance has to the local sample grade values is through the variogram model, which is on a global average basis rather than a local basis. This means that the kriging variance for a specific sample to block configuration is a fixed value irrespective of whether the grade values are highly variable or more uniform. It is clear that there is greater confidence in the estimate of the latter block than that of the former block. This anomaly led to the introduction of classification techniques that concentrated on relative variances that recognise the local data configuration and variability.

5.1 RELATIVE KRIGING ERRORS AND THE NUMBER OF SAMPLES

Blackwell (1998) presented an argument for introducing the Relative Kriging Variance (RKV); the ratio of the kriging variance and the kriged estimate squared. From this the Relative Kriging Standard Deviation (RKSD) is defined as the square root of the RKV. The RKSD is plotted against the number of samples used in the kriging of the block. Two threshold values for the RKSD are selected arbitrarily, but based on experience, to separate the Measured, Indicated and Inferred categories.

5.2 RELATIVE VARIABILITY INDEX

The implementation of the Relative Variability Index (RVI) as a measure of confidence in the estimate was proposed (Arik, 1999). The RVI is the ratio of the square root of the combined kriging variance and the kriged estimate of the block. The combined kriging variance is the square root of the product of the kriging variance and the local weighted average variance. The histogram of the RVI is analysed to determine thresholds for distinguishing between categories, and the proposal is to use the 50th and 90th percentile RVI values to identify the three different resource categories.

5.3 INTERPOLATION VARIANCE

Yamamoto (2000) introduced the Interpolation Variance (IV) as an alternative measure of the reliability of ordinary kriging estimates. The IV reflects the local variability as expressed by the data. It is the weighted average of squared differences between the data values and the block estimate. An advantage is that this variance recognises the proportional effect when present. This interpretation of the variance is however, only valid if and only if all the ordinary kriging weights are positive.

6 Mineral Resource classification linked to a production period

The philosophy of applying a classification rule that considers *"the % error in the estimate of the block being classified is within 15% with 90% confidence for a specific production period"* has been around for a number of years, at least since the early 1990's. This is an empirical rule that has been accepted in the mining industry. The specific production period should define the resource category: the shorter the production period, the higher the confidence category, for a longer production period the resource category is lower. Many variants of this rule are applied in practice.

The percentage error in the estimate of the mean is given by

$$\% \text{ error} = \frac{\text{Standard Error}}{\text{Estimate}} \times 100\% = \frac{s/\sqrt{n}}{\bar{z}} \times 100\% = \frac{s}{\bar{z}} \times \frac{1}{\sqrt{n}} \times 100\% = \frac{\text{CoV}}{\sqrt{n}} \times 100\%$$

CoV is the coefficient of variation; s is the standard deviation and $\bar{z}$ is the average of the samples.

Relative 90% confidence limits can be established from the product of the CoV/√n and 1.645, the standard normal deviate.

If the resource blocks can be considered independent the relative variability of a block can be converted to the equivalent of the variability in a production period by dividing the coefficient of variation of the block in the resource model by √n. Where *n* would be the number of independent blocks that would be required to represent the production period as shown below.

$$CoV_{Production\ Period} = \frac{CoV_{Block}}{\sqrt{n}}$$

The estimated resource blocks are correlated and the support of the blocks are relatively small compared to the support of for instance the annual production. In general, if the histogram of the sample support is skew and if the block support is small the histogram of estimates will be skew and as the support increases the histogram of the estimated blocks in the resource model will approach normality as per the central limit theorem. This means that the histograms of estimates of production periods, which consist of many resource blocks, are expected to approach normality and independence.

Mining, though, does not take place in independent blocks thus the effect of correlation must be brought into account when an individual estimated resource block is being classified in terms of production periods.

6.1 THE "INDEPENDENT" √n

It is necessary to modify the 90% limits of the estimated blocks in the block model to represent the equivalent variability of a production period. Therefore, a factor that takes this correlation and the production period into account has to be determined to replace the "independent" square root of "n".

$$CoV_{Production\ Period} = \frac{CoV_{Block}}{\sqrt{n}} \sim \frac{CoV_{Block}}{Factor\ Production\ Period}$$

Thus

$$Factor\ Production\ Period = \frac{CoV_{Block}}{CoV_{Production\ Period}}$$

The problem is to find estimates for a representative CoV of blocks in the block model and for the CoV of the production periods. As the resource has not yet been classified the CoV of the production periods cannot be established. It is nonetheless possible to determine estimates for these values from many realisations of a conditional simulation exercise.

6.2 CONDITIONAL SIMULATION AND UNCERTAINTY ASSESSMENT

In an endeavour to attain quantifiable Mineral Resource classifications, the trend in the mining industry has been towards the application of conditional simulation techniques. A specific set of drilling or sampling results provides one view of the resource, a different set will provide a different view, the luxury of a second campaign is however not always available. A fairly quick and inexpensive method for obtaining a spectrum of possible views of the global statistical and spatial characteristics of the orebody can be obtained through conditional simulation. The variability in realisations of the simulations can be interpreted to assess the uncertainty in the resource estimates.

It is assumed that conditional simulations are carried out to reflect both the uncertainty in the geological model as well as the uncertainty associated with the grade model. Dohm (2003) introduced a technique to combine conditional indicator simulations for geology and conditional sequential Gaussian simulations for grade to assess the combined uncertainty of the geological interpretation and the grade estimation.

It is, however, vital to realise that if this tool is applied to assess the risk associated with the Mineral Resource then it is important to establish the integrity of the simulation results. For example the number of simulations to be considered for assessment and validations of the reproducibility of both variogram model and histogram of the conditioning data are crucial. When these validations are not carried out, the results can lead to incorrect Mineral Resource classifications that could have disastrous effects on Ore Reserve classifications and investment decisions.

The task at hand is to establish from the conditional simulations, what the expected coefficient of variation for the estimated grade, tonnage and metal content of a real month's or year's mining would be.

6.3 THE 15% RULE – A LOGICAL APPROACH

The purpose is to produce a measure of confidence in the resource estimate. The specific classification rule considered here is based on two production periods, namely a monthly production period for Measured Mineral Resources and an annual production period for Indicated Mineral Resources.

Critical to this resource classification guideline are the following three CoV values:

CoV_{Local}:   a typical CoV for blocks of the same size as used in the estimation model.

$CoV_{Monthly}$:   CoV signifying the relative variability of a monthly production period.

CoV_{Annual}:   CoV signifying the relative variability of an annual production period.

To establish the monthly and annual CoV values the moving block technique is applied. Every realisation of the conditional simulation is "cookie cut" by units representing likely 'production periods' at various positions within the orebody. This process is repeated for a sufficiently large number of times, e.g. at least 120 months (10years) per realisation of the simulation exercise.

Representative CoV values for monthly and annual production periods can be calculated from the statistical analysis of these two supports.

The proposed method has been applied to a Zn deposit. The orebody comprises two superimposed mineralised horizons, which are structurally controlled and are part of an overturned fold limb. Both orebodies comprise a well mineralised massive sulphide horizon close to the footwall and an overlying iron formation containing banded and disseminated sulphides. Fan drilling is performed from hanging- or footwall drives, on 20 m spaced north–south sections. Intersection spacing on section varies between 10 m and 40 m depending on the complexity of the orebody. Deeper exploration drilling is on a 100m x 50m grid, sampled every 2m.

The three critical CoV values for classification obtained from 40 conditional simulations of the %ZN values, carried out in unfolded space were:

$CoV_{Local} = 0.754$ the relatively large CoV for a block in the resource model confirms the earlier remark that blocks are correlated.

$CoV_{Monthly} = 0.1564$ The CoV of monthly periods is less than that of the block.

$CoV_{Annual} = 0.0667$ The CoV of the annual production is as expected significantly less

The monthly adjustment factor is then calculated from

$$\text{Factor Monthly Production Period} = \frac{COV_{Block}}{COV_{Monthly}} = \frac{0.754}{0.1564} = 4.821$$

The annual adjustment factor is then calculated from

$$\text{Factor Annual Production Period} = \frac{COV_{Block}}{COV_{Annual}} = \frac{0.754}{0.0667} = 11.304$$

For a Measured Mineral Resource where the error in the monthly production estimate has to be within 15% with 90% confidence the threshold value is:

$$CoV_{Measured} = \frac{0.15 \times 4.821}{1.645} = 0.440$$

For an Indicated Mineral Resource where the error in the annual production estimate has to be within 15% with 90% confidence the threshold value is:

$$CoV_{Indicated} = \frac{0.15 \times 11.304}{1.645} = 1.030$$

Each block in the estimation model is considered in turn and its coefficient of variance, CoV $_{\text{block estimate}}$ is calculated

$$\text{CoV}_{\text{block estimate}} = \frac{\sigma_K}{Z_K}$$

Where Z_K is the kriged estimate and σ_K is the kriging standard deviation of the block.

The CoV $_{\text{block estimate}}$ value of each block in the estimation model is then compared to the above threshold values and the decision rules are:

If the CoV $_{\text{block estimate}}$ ≤ CoV $_{\text{Measured}}$ then the block is classified as Measured.

If CoV $_{\text{Measured}}$ < CoV $_{\text{block estimate}}$ ≤ CoV $_{\text{Indicated}}$ then the block is classified as Indicated.

If the CoV $_{\text{block estimate}}$ > CoV $_{\text{Indicated}}$ then the block is classified as Inferred.

Once all the blocks in the estimation model have been classified the Measured and Indicated Mineral Resources can be considered for conversion to Proved and Probable Ore Reserves. It is recommended, that as with any automated mathematical process, the classified Mineral Resource model be validated; at least visually.

The final result is a succinctly classified Mineral Resource model, which is based on objective quantifiable classification measures that recognise the uncertainty related to subjective interpretations of the available information.

7   Comments and discussion

It is essential to ensure the integrity of the simulations by validating the inherent and spatial; variability of each realisation in terms of the histogram and variogram of the conditioning data.

An advantage of applying the conditional simulation techniques is that once the resources have been converted to reserves it is possible to compare the variability of the actual mine plan with that expected from the simulations.

The author has come across two other approaches for determining "n", the number of independent blocks to use in the above classification. The first method determines "n" as the number of independent production blocks required to reach the range of the variogram. In the second method "n" is calculated as the ratio of the production period tonnage divided by the block tonnage. In both cases the square root of "n" is used as the divisor for the CoV to determine the 90% confidence limits.

The classification guideline proposed does not assume that the variance reduction factor should be in terms of a square root and uses CoV measures to address the change of support effect.

8 Conclusion

A number of Mineral Resource classification methodologies were discussed showing the development of understanding and incorporating uncertainty associated with the grade estimates. The final classification guideline presented is based on the assessment of conditional simulations that have incorporated the uncertainty in the interpretation of geological boundaries, tonnage, grade and consequently metal content estimates and likely production periods

The proposed classification technique does not replace the resource estimation; rather it serves as an additional tool to quantify confidence in the resource evaluation model.

It is further appreciated that particular Mineral Resource classification techniques are appropriate for specific situations.

A fundamental concept in Mineral Resource classification is the need for common sense and experience to prevail and it is therefore recommended that any automated mathematical technique applied be scrutinised.

9 References

Arik, A. (1999), An Alternative Approach to Ore Reserve Classification. *1999 APCOM Proceedings*, SME. Denver, p. 45-53

Blackwell, G (1998), Relative Kriging Errors – A Basis for Mineral Resource Classification, *Explor. Mining Geol.*, Vol.7 Nos 1 and 2, p. 99-105.

Dohm, C.E. (2003), Application of simulation techniques for combined risk assessment of both geological and grade models – an example. 2003 APCOM Proceedings, SAIMM, Cape Town, p. 351-354

Deutsch, C.V. and Journel, A.G., *GSLIB: Geostatistical Software Library and User's Guide*, 2nd Edition, Oxford University Press, 1998.

Goovaerts, P., *Geostatistics for Natural Resources Evaluation*, Oxford University Press, 1997.

Isaaks, E.H. and Srivastava, M.R., *An Introduction to Applied Geostatistics*, Oxford University Press, 1989.

Journel, A.G. and Huijbregts, C.J., *Mining Geostatistics*, Academic Press, 1978.

Snowden, D.V. (1996), Practical Interpretation of Resource Classification Guidelines, *AusIMM Annual Conference Procedings*, Perth March 24-28, p. 305-308.

Stephenson, P.R. and Stoker, P.T. (2001), Classification of Mineral Resources and Ore Reserves, *Mineral Resource and Ore Reserve Estimation, The AusIMM Guide to Good Practice*, Monograph 23, p. 653-659

Yamamoto, J.K. (2000), An Alternative Measure of the Reliability of Ordinary Kriging Estimates, *Mathematical Geology*, Vol32, No. 4, pp489-509.

MINING

MINING

GEOSTATISTICS IN RESOURCE/RESERVE ESTIMATION: A SURVEY OF THE CANADIAN MINING INDUSTRY PRACTICE

MICHEL DAGBERT
Consultant, Geostat Systems International Inc.
10 Blvd de la Seigneurie E., Suite 203, Blainville, Qc, Canada, J7C 3V5

Abstract. With the new NI 43-101 rules of public disclosure for exploration and mining companies listed on the Canadian exchanges, it is now possible to have access to technical reports describing in details the procedures used by those companies to estimate resources and reserves for their properties in Canada and elsewhere in the world. This paper summarises the results of a survey of such technical reports issued in the last two years. It evaluates the role of geostatistics in various aspects of the resource/reserve estimation work namely the capping of outliers sample value, the domaining according to geology, the continuity analysis, the interpolation of block grades, the evaluation of dilution factors and the categorisation of resource/reserves.

1 Introduction

Since Feb. 1st, 2001, mining and exploration companies listed on Canadian stock exchanges (Toronto, Vancouver, Calgary and Montreal) must follow the so-called National Instrument (NI) 43-101 standard whenever they disclose technical and scientific information about their properties (CSA, 2000a and 2000b). Such information includes exploration results and of course resource and reserve estimates. Like similar standards in other countries(e.g., JORC,1999), NI43-101 does not specify how the actual exploration or estimation work should be done but concentrate on the profile of the individuals who do the work (the "qualified persons") and the format of the disclosure (the "technical reports"- CSA, 2000c). On the content itself, NI-43-101 endorses the revised resource/reserve definitions of the Canadian Institute of Mining and Metallurgy (CIM, 2000).

In the last 3.5 years of application of this new regulation, several hundreds "technical reports" of NI43-101 style have been filed. They are public documents available on the web in digital form and they constitute a privileged reference to determine how mineral resources and ore reserve are currently estimated in the Canadian mining industry and, in particular, to what extent geostatistics is used in this estimation

2 Survey of technical reports

Technical reports are retrieved from the www.sedar.com site which concentrates all documents (annual reports, notices to shareholders, press releases...) in digital (PDF) form from public companies listed on Canadian stock exchanges. They are generally found in the "Other" category and are easily detected by their size of commonly several Mb since they correspond to documents of generally several hundred pages.

Exploration and mining companies are found in 3 industry groups: Gold and precious metals, Junior natural resource/Mining and Metals and Minerals. We have retrieved most of the technical reports issued since October 1, 2002 i.e. the last 2 years and all together they represent about 1200 documents.

About three quarter of those documents are uniquely concerned with so-called "exploration results" with limited drill hole information and no estimate of resource or reserve. That leaves about 300 reports dealing with properties with sufficient drilling/sampling data to warrant resource or reserve estimation (R&R reports).

The majority of the properties studied in those R&R reports is actually outside of Canada, mostly in United States, South America, Africa and Russia (with others in South Africa, China etc..). Gold is the most frequent commodity of interest (under the form of vein or disseminated type) followed by base metals (porphyry type and massive sulphide), uranium and industrial minerals. A surprisingly large fraction of properties in those R&R reports are producing mines (or have produced in the past), which can lead to some instructive reconciliation work.

Some of the issuers are well known Canadian mining houses like Barrick Gold Corp., Placer Dome Inc., Kinross Gold Corp. and Aur Resources Inc. Some of the properties described in those reports have made the headlines of mining publications in the recent years : Lac des Iles of North American Palladium Ltd., Kemess North of Northgate Minerals Corporation, Las Christinas of Crystallex International Corp.

As indicated in the introduction, the new NI 43-101 regulation puts a lot of emphasis on the "qualified person(s)" who authors the technical reports. They can be employees of the issuer but in most cases, they are outside consultants, independent of the issuer. Some of them are affiliated with large and well known consultancies from all over the world, particularly from Australia. It can be noted that there is a definite "correlation" between the approach taken in an R&R study and the background of the qualified person(s) responsible for that study and this is particularly true when it comes to the use of geostatistics.

The qualified person who authors a technical report is not necessarily the individual who has conducted the R&R work presented in the report. In many cases, the work has been done internally and the external consultant, who acts as the qualified person, after auditing and verification, simply endorses the results of that company work.

3 Use of geostatistics in resource and reserve estimation

In this age of widespread computer usage, a surprisingly high proportion (about 1/3) of the resource and reserve estimation work presented in the surveyed reports is still carried in the "manual" way with the interpretation of the limits of mineralised lenses from plans and sections and the calculation of an average grade of those lenses (or parts of them) from samples within those interpreted limits. This type of calculation seems restricted to vein type deposits to be mined by underground methods. Often there is a distinction between "geological" and "mining" resources, the later being after application of a minimum mining width to available vein intercepts and the corresponding grade dilution. Reserves are made of mining resources within limits of designed stopes and after application of dilution and recovery factors. Examples of such calculations can be found in Curtis, 2003 (for UG gold mines in Russia) and Roscoe et al, 2002 (for a UG gold mine in Canada). Needless to say, those R&R estimations do not use geostatistics.

The 2/3 balance of the R&R studies use the concept of computerized resource block model implemented in a mining package (Vulcan, Datamine, Gemcom, Mintec, Surpac...). Steps followed in such studies are invariably: geological solid modelling or "domaining", selection of block size, original sample compositing (with possibly some capping of high assays), eventually some variography of composite data in the various "domains", interpolation of block grade from surrounding composites (with search strategy and weighting scheme), categorization of estimated resources in blocks and finally conversion of resources to reserves.

Geological modelling consists of building 3D solids around material of the same "geological" nature based on lithology, alteration, degree of oxidation (the traditional leach/oxide/supergene/primary sequence of tropical or paleotropical terranes), geometry (blocks of similar general orientation in a folded structure) or simply grade. In the latter case, a low cut-off is used to delineate "potentially mineralised material" in disseminated mineralization, typically somewhere between 0.3 and 0.8 g/t in gold deposits. The resulting geological model can be fairly complex and detailed, for example 63 different solids for the Jinlonggou gold deposit in China (Fillis and Arnold, 2004). In the majority of cases surveyed, limits of those solids are "hard" limits i.e. blocks within limits are interpolated from just samples within the same limits. As a general rule, geostatistics is not used as an aid to geological modelling or to test the hard nature of the defined limits.

Resource model block size is quite variable and is linked to both the size of mineralised solids (small blocks in narrow zones) and the average spacing between samples (the old rule of the half distance between samples). It ranges from a high of 20x20x15m at Kemess North (Gray et al, 2004) to a low of 2x5x1m at the Barbrook mine (Applied Geology Service, 2004). Trend seems to have fairly small blocks especially with the sub-celling technique of mining packages (although generally all sub-cells in the same cell are given the interpolated value of that parent cell). In deposits to be mined by open-pit methods, the prevailing rule is to adjust the vertical dimension of blocks to the planned bench height of the future mine.

High grade capping of original assay data (whatever size of the corresponding interval) is the rule in gold deposits. The most popular approach to determine the cap limit is to look at changes in the slope of the cumulative frequency curve on log probability paper. In case of multiple domains, there is generally a specific cap for each domain which varies with the average grade of all samples in the domain. Unfortunately the proportion of capped samples as well as the percent gold metal lost is not always mentioned.

Composite size is generally dictated by the average size of the original assay intervals, irrespective of the dimension of blocks to be interpolated. As a result, composite size tends to be rather low, like 1m or 2m. In only a few instances, blocks for a deposit to be mined by open-pit are interpolated by bench composites (e.g. 5m composites for 10x10x5m blocks in 5m benches) thus minimizing risks of under-dilution of block grades.

Variography of composite grades in each domain is performed in only half the studies using resource block model. Description varies from an almost casual mention in the text to 120 pages of variogram plots (Belanger, 2003). Correlograms seem to prevail over regular variograms. We have not seen many indicator variograms or variograms of transformed data. The "pair wise relative variogram" is still very much popular with some consultants.

Inverse squared distance (ID2) is the most popular block grade interpolation method. We can even find cases where variograms are computed but blocks are interpolated by ID2 (Hill and Davidson, 2003). Arguments to prefer ID2 to kriging indicate that the later is not that well understood. For example : *"Inverse distance was used to interpolate gold grades into the block model for Bouroum instead of ordinary kriging due to the more significant presence of isolated high grade gold values that impacted adjacent low grade areas and the generally, more in-equidistant drill hole spacing at Bouroum"* (Vanin et al, 2003). When kriging is used, it is mostly under the form of ordinary kriging (OK) with very little indicator kriging (IK) applications.

Typical of this hesitation toward kriging is the report by Gosselin (2003) on Laronde gold deposit resources and reserves for 2003 : variograms are computed and fitted with anisotropic models in all 6 gold + base metals bearing sulphide zones but they are just used to defined search ellipsoids for the ID2 (or ID3) of blocks in the same zones.

Preference of ID2 over OK reflects some fear of diluting high-grade composites (or "smearing" high grades into adjacent low grade areas). This concern also transpires from the selection of search parameters and the fairly low maximum number of composites allowed in the interpolation (from 3 to 8 from the few studies where this information is given). Standard approach is to consider ellipsoids of increasing sizes to progressively fill the block matrix. The restricted search for high-grade composites option of some packages is sometimes used as an alternative (or in combination with) high grade capping prior to compositing.

In all the studies that we have surveyed, categorization of block model resources is strictly based on the geometry of composites with respect to blocks with no use of predicted uncertainty from variograms (the old kriging variances or the new conditional simulation). A common approach is to use the steps in the progressive search for composites around blocks to set the block category e.g. measured resources for blocks which can be interpolated with the most restrictive search (smallest ellipsoid, highest minimum number of composites..) up to inferred resources for the last blocks to be interpolated. In other cases, a specific template ellipsoid is used to test the density of composites around each block with cut-offs on number of composites within this ellipsoid corresponding to given drilling grid: for example, at Quebrada Blanca (Barr and Reyes, 2004), they use the number of 7.5m bench composites within a 75x75x18.75m ellipsoid to classify blocks with measured blocks if more than 20 composites (50x50m grid) and indicated blocks if more than 9 composites (100x50m grid).

Post-processing of block estimates is rather limited. In many cases, for deposits to be mined by open-pit methods, block values are used as-is in open-pit optimization and the calculation of reserves from resources (i.e. reserves are resources within final pit). In some instances, fixed recovery and dilution factors are applied to block estimates before pit optimization e.g. a 95%recovery in all blocks and a 10% dilution in contact blocks at Kemess North (Gray et all, 2004). In other cases, global change of support methods are used to check that block estimates have the grade variability corresponding to their size (Belanger, 2003). We have not found any study where conditional simulation is used to adjust block estimates to expected selectivity and grade control conditions of the future mining operation.

4 Conclusions

Roughly speaking, about one third of the resource and reserve studies issued by exploration and mining companies listed in Canada over the last two years use geostatistical methods in their resource estimation. Another third is also based on computerized block models but with inverse square distance. The last third is the manual approach with sectional blocks and sample averages within blocks.

Geostatistics used in those studies is fairly "classical" with composite grade variography and block estimation by ordinary kriging. Indicator kriging is not much used, even if the majority of studies deal with gold deposits. There is virtually no use of conditional simulation as an aid to resource categorization or block grade adjustment for recovery and dilution.

The general feeling that one gets when browsing those thousand pages of technical reports is that, at the moment, resource estimation is still more an art than a science with lots of subjective decisions and fudge factors (specially in high capping and categorization) which can be related to the background and past experience of the "qualified persons" who sign those reports.

Acknowledgements

Thanks are due to Gaetanne Beaulieu for her patience in retrieving technical reports on the web

References

Applied Geology Service, *Independent Qualified Person's Report. Barbrook Mines Limited, Mpumalanga Province, South Africa* [Sedar : Caledonia Mining Corp.], 67p, 2004

Barr N.C. and Reyes R., *Report on Mineral Resources and Mineral Reserve Estimates at Dec. 31, 2003. Quebrada Blanca copper mine, Region I, Chile* [Sedar : Aur Resources Inc.], 57p, 2004

Belanger M., *Technical Report, La Coipa Mine, Chile* [Sedar : Kinross Gold Corp.], 231p, 2003

Cormier M., *Technical Mana Minéral S.A..Révision des Ressources Minérales, Mana , Burkina Faso, en date du 31 Decembre 2003* [Sedar : Semafo Inc.], 46p, 2004

CSA, *National Instrument 43-101. Standard of disclosure for mineral projects*, [available as standards_disclosure_43-101-1.pdf from www.osc.gov.on.ca], 21p, 2000a

CSA, *Companion policy 43-101CP to National Instrument 43-101. Standard of disclosure for mineral projects*, [available as companion_policy_43-101.pdf from www.osc.gov.on.ca], 14p, 2000b

CSA, *Form 43-101F1. Technical report*, [available as technical_report_form_f1_43-101.pdf from www.osc.gov.on.ca], 12p, 2000c

CIM, *CIM standards on mineral resources and reserves. Definitions and guidelines*, [available as CIMdef1.pdf from www.cim.org], 26p, 2000

Curtis L., *Technical Report. Zun-Holba and Irokinda Gold Mines, Buryatzoloto for High River Gold Mines Ltd.* [Sedar : High River Gold Mines Ltd], 17p, 2003

Fillis P. and Arnold C., *Tanjanshian Gold Project, Qinghai Province, 2003 work programme and resource estimation.* [Sedar : Afcan Mining Corporation], 54p, 2004

Gosselin G., *2003 Laronde Mineral Resource and Mineral Reserve Estimate Agnico-Eagle Mines Ltd., Laronde.* [Sedar : Agnico-Eagle Mines Ltd], 81p, 2003

Gray J.H., Morris R.J., Major K.W. and Arik, A., *Revised Kemess North pre-feasibility project* [Sedar : Northgate Exploration Ltd], 138p, 2004

JORC, *Australasian Code for Reporting of Mineral Resources and Ore Reserves*, [available as JORC-code.pdf from www.jorc.org],16p, 1999

Hill A. and Davidson A., *Technical Report. The Alto Chicama property. Department of La Libertad, Peruo.* [Sedar : Barrick Gold Corp.], 50p, 2003

Roscoe Postle Associates Inc., *Review of Mineral Resources and Mineral Reserves of the Macassa Mine Property, Kirkland Lake, Ontario, prepared for Kirkland Lake Gold Inc.* [Sedar : Kirkland Lake Gold Inc.], 73p, 2002

Salmon B., *Mineral Resource and Mineral Reserve Estimates, Louvicourt Mine, January 1st, 2004*, [Sedar : Aur Resources Inc.], 230p, 2004

Vanin D., Michaud M., Thalenhorst H., *Taparko-Bouroum project, Burkina-Faso. High River Gold Mines Ltd. 43 101 F1 technical report* [Sedar : High River Gold Mines Ltd], 153p, 2003

INTEGRATION OF CONVENTIONAL AND DOWNHOLE GEOPHYSICAL DATA IN METALLIFEROUS MINES

M. KAY, R. DIMITRAKOPOULOS and P. FULLAGAR
WH Bryan Mining Geology Research Centre
The University of Queensland, Brisbane Qld 4072, Australia

Geophysical logs provide valuable data that can be linked to orebody modelling and mine planning in metalliferous mines; however, geophysical measurements provide indirect data for ore grades and require further integration with conventional assay data. Integration can be based on the generation of suitable geophysical data compositing and the use of the sequential co-indicator simulation with the Markov-Bayes approximation. A detailed study at the Kidd Creek base metal mine, Canada, shows the practical aspects of the suggested approach and the value of integrating geophysical data.

1. Introduction

Geophysical logs provide valuable, relatively inexpensive information that can be further linked to various aspects of a mining operation, including orebody modelling, mine planning, grade control and production. Although financially attractive, downhole geophysical measurements usually only provide indirect indicators of ore grades (Fullagar and Fallon, 2001) and require further analysis in order to achieve integration with conventional assay data.

Until recently few studies have examined the issue of technical integration of geophysical data in the metalliferous environment. Early attempts include grade estimation based on natural gamma logs in uranium mines, after calibration of geophysically derived grades with geochemical assays (e.g. Bryan and Roghani, 1985; David, 1988). More recently, Miller and Luark (1993) use simulation techniques to construct models of rock strength in an underground coal mines; Dimitrakopoulos and Kaklis (2001) demonstrate the use of sequential co-indicator simulation to integrate data from a radio frequency electromagnetic tomography survey and scattered geochemical assays; and Basford et al. (2001) describe refinement of blasting based on automated interpretation of natural gamma and magnetic susceptibility logs. Kay (2001) provides a detailed study on integrating and valuing downhole geophysical measurements in base metal mines.

Given the potential economic benefits (e.g. Fallon et al., 1997) that could be obtained from using downhole geophysical logs, there is a clear incentive to determine how geophysical logging data can be integrated with orebody modelling and subsequently used throughout the mining process. To do this, it is necessary to use a simulation technique that is capable of integrating geophysical logs. This is a challenge since in the mine environment there is typically insufficient geophysical logging data to allow for calibration and variogram analysis. In addition, the handling of the variable support effects associated with geophysical signals in geologically complex environments has received little attention.

This paper demonstrates the effectiveness of technically integrating downhole geophysical logging in the metalliferous environment for orebody modelling. First, the sequential co-indicator simulation method with the Markov-Bayes approximation used herein is outlined. This is followed by a brief description of the approach employed to composite downhole geophysical logs. Subsequently, a case study with copper assays and conductivity log data from Kidd Creek Copper Mine, Canada, is used to illustrate the practical aspects of the technique.

2. Sequential co-indicator simulation in brief

Sequential co-indicator simulation (ScoIS) is a suitable method for integrating 'soft' data (such as geophysical logging measurements) with 'hard' data (such as conventional assay data) in simulating orebody models. ScoIS employs indicator co-kriging (Goovaerts, 1997) to derive the cumulative distribution function of the attribute being modelled and accounts for the geospatial correlation of 'hard' and 'soft' data, as well as their spatial cross-correlation. To alleviate the tedious inference and joint modelling of indicator covariance and cross-covariance functions, a variant of ScoIS employing the so-called Markov-Bayes approximation of Zhu and Journel (1992) is convenient, particularly when 'hard' and 'soft' data measure the same attribute. The Markov-Bayes hypothesis requires (i) that for a cut-off z_k a hard indicator datum (binary transform of the original measurement) $i(u, z_k)$ at location $\mathbf{u}$ in a deposit prevails over the influence of any co-located soft indicator data $y(u, z_k)$; and (ii) that the indicator covariance function $C_{YY}(h;z)$ of the 'soft' data and cross covariance function $C_{YI}(h;z)$ between hard and soft data can be expressed as a function of the covariance of the hard data $C_{II}(h;z)$. Specifically,

$$C_{YY}(h;z) = |B(z)| \cdot C_{II}(h;z) \quad \text{for } h = 0 \tag{1}$$
$$\phantom{C_{YY}(h;z)} = B^2(z) \cdot C_{II}(h;z) \quad \forall h > 0$$

and

$$C_{IY}(h;z) = B(z) \cdot C_{II}(h;z) \quad \forall h \tag{2}$$

$B(z)$ is an accuracy index given by:

$$B(z) = m^1(z) - m^0(z) \quad B(z) \in [-1,1] \tag{3}$$

and

$$\begin{aligned} m^1(z) &= E\{Y(x;z)\,|\,I(x;z)=1\} & m^1(z) \in [0,1] \\ m^0(z) &= E\{Y(x;z)\,|\,I(x;z)=0\} & m^1(z) \in [0,1] \end{aligned} \tag{4}$$

where $m^1(z)$ and $m^0(z)$ are two conditional expectations that are obtained from calibration scatterplots of the hard versus the soft data, as shown in a subsequent section. It can be seen that $B(z)$ is equal to one when the soft data are fully equivalent to the hard data.

3. Compositing downhole geophysical data

Geophysical logging data are usually collected every few centimetres downhole with the geophysical probe's sampling volume being quite variable. This variability of support is especially pronounced for conductivity measurements in base metal sulphide deposits due to the high conductivity of the ore. As a result, it is unclear how exactly geophysical logs should be composited. Kay (2001) uses an experimental approach which consists of compositing the conductivity logs with power averaging or l_r norm (e.g. Dimitrakopoulos and Desbarats, 1993; Claerbout and Muir, 1972). The power averaging exponent ω is derived experimentally. By varying the value of ω, one can generate a continuum of 'average' values that include common averages such as the arithmetic, geometric and harmonic means, when ω is equal to 1, 0 and –1 respectively. In determining a suitable ω value, it is rational to maximise the information content of the composited downhole geophysical measurements. The procedure for maximization has the following steps:

1. Composite, for a selected length, the geophysical log data at drillhole X with a particular power averaging value ω;
2. Estimate the metal grades in drillhole X using the composited conductivity log from Step 1 and the geochemical assays available from adjacent drillholes using standardised ordinary co-kriging (e.g. Deutsch and Journel, 1992);
3. Compare the resulting grade estimates to the actual geochemical composites in drillhole X using the mean squared error μ_{e^2}, the Spearman rank correlation coefficient r' (e.g. Swan and Sandilands, 1995) and the relative differences in these measures;
4. Repeat Steps 1 to 3 for all drillholes and different values of ω; and
5. Summarize all the results from Step 3 and choose a value for ω to use.

An example of this procedure is shown in a subsequent section.

4. Case Study: Integrating downhole conductivity logs and copper assays

Data collected from the Kidd Creek base metal deposit, Canada, illustrate the use of the above methods in integrating copper assay data and downhole inductive conductivity logs. Copper assays relate to 1.5m composites. Conductivity measurements were collected every five centimetres downhole and are composited to the same length of 1.5m using power averaging.

4.1. POWER AVERAGING OF GEOPHYSICAL LOGS

The procedure for deriving the power averaging constant ω described above is applied here to a set of drillholes. Kay (2001) showed that the accuracy of the copper estimates depends on the value of ω because μ_{e^2} has a minimum and r' a maximum in [-2.0, -0.5]. This suggests that ω should be drawn from [-2.0, -0.5]. But it is less clear which value of ω in [-2.0, -0.5] should be used since both the μ_{e^2} and r' maxima are quite broad in this interval (Kay 2001). However, if $\Delta\mu_{e^2}$ and $\Delta r'$ are examined instead (Figure 1), it is evident that ω be set at –1.0 (harmonic mean) since this value results in the minimum mean squared error and the maximum rank correlation coefficient. A series of such analyses for sets of drillholes with different dips suggested that an appropriate value of ω would be –1.0 (Kay, 2001). As a result all of the conductivity logs acquired in the study area were power average-composited using this value of ω and a sample interval of 1.5m. These composited logs are used for the simulation of the deposit.

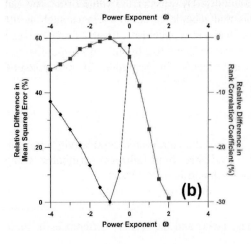

Figure 1. Relative difference in mean square error and rank correlation coefficient as a function of ω for composites generated from a cross-validation analysis of a set of representative drillholes (1.5m composite length).

4.2. SIMULATION PARAMETERS

Having generated conductivity composites, copper grades are simulated with ScoIS and the Markov-Bayes approximation, which are conditional to the copper assays and conductivity logs. In this section, the practical aspects of the simulation method used and its assumptions are examined. ScoIS requires the selection of a set of cutoffs for the hard data (copper). The cutoffs were chosen to adequately characterise the overall copper data histogram as well as the metal content (Dimitrakopoulos, 2004). The latter is important in representing the copper metal quantity in the deposit, where 10% of the higher-grade samples may represent 50% of the metal in the deposit. Details of the application at Kidd Creek are included in Kay (2001).

The inference and modelling of indicator variograms for each copper cutoff involved two steps, namely identification of the three principal axes of the variogram ellipsoid, and modelling of experimental variograms along these three directions. For cutoffs above the 85th percentile, a different approach was followed (e.g., Dimitrakopoulos, 2004) where variogram parameters were experimentally adjusted to minimise order-relation problems. All indicator variograms were modelled using different linear combinations of the same set of basic structures (in this instance a nugget effect and an exponential model). Also, variogram parameters were varied smoothly from one cutoff to the next (Dimitrakopoulos, 2004; Goovaerts, 1997).

As previously described, the Markov-Bayes assumption allows the conductivity and copper-conductivity covariance models to be deduced from the copper covariance model and the calibration parameter $B(z_k)$ for each of the copper cutoffs z_k. However, this requires a set of cutoffs for the conductivity data to be specified. Eleven conductivity cutoffs were selected to ensure adequate characterisation of the declustered conductivity cumulative distribution function. This was achieved by selecting cutoffs that divided the conductivity values into classes of approximately equal frequency.

The resulting calibration parameters are presented in Figure 2. It can be seen that the calibration parameters are low for copper grades less than about 3.0%. However, for higher copper cutoffs, the calibration parameter stayed reasonably constant or increased slightly. These results are consistent with the general observations from the copper-conductivity scatter-plots, which suggest that the conductivity logs were a poor predictor of the copper grade in low-grade areas due to low signal level.

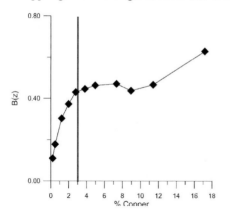

Figure 2. Markov-Bayes calibration parameter versus copper grade for the conductivity logs.

Also as previously discussed, conductivity and the copper-conductivity spatial models for each copper cutoff can be derived from the modelled copper-copper spatial model using a Markov-type hypothesis. Although there are no rigorous tests to verify the validity of this hypothesis, a useful check is to compare these derived variogram models with the corresponding experimental variograms. This check was performed on data from the test area for a range of copper cutoffs and is presented in Kay (2001). The results suggest that the derived models matched the spatial (auto) correlation associated with conductivity logs for low, median and high copper cutoffs.

In contrast, the cross correlation between the copper and conductivity was less well modelled using the Markov-Bayes assumption. Kay (2001) demonstrated that employing the Markov-Bayes assumption results in a variogram with a nugget to sill ratio of 0.80, whereas as the experimental data suggested that a value of 1.00 would be more appropriate. This effect was less marked for higher copper cutoffs, as the spatial correlation between copper and conductivity was quite well modelled for higher cutoffs (Kay, 2001). These results were further confirmation that conductivity logs were poor predictors of copper grade in areas with low copper concentrations. However, in spite of this shortcoming, it appeared that the Markov-Bayes hypothesis was valid.

The copper simulations generated in this study were produced using Zhu (1991)'s implementation of the Markov-Bayes simulation algorithm. In this paper the selection of the variogram parameters for cutoffs below the 20th and above the 85th percentile are explored. The reader is referred to Kay (2001) for an explanation of the other parameters. As discussed previously the variogram parameters for cutoffs between the 20th and 85th percentile can be inferred from manually fitting variograms. However, for cutoffs outside this interval, manually fitting variograms is unreliable since the associated experimental indicator variograms are very erratic. In the present study another approach was used to infer the variogram models for the extreme copper cutoffs. This approach consisted of the following steps:

1. Construct the experimental histogram (H^e) and indicator variograms (V^e) for all copper cutoffs using the declustered copper samples;
2. Infer variograms models (V^i) for the copper cutoffs above the 15th and below the 85th percentiles by manually fitting variogram models to V^e;
3. Assign starting values to the variogram models (V^h) for copper cutoffs below the 15th and above the 85th percentiles;
4. Generate copper simulations using the copper samples and the variogram models V^i and V^h;
5. Calculate the indicator variograms (V^{ml}) for cutoffs above the 15th and below the 85th percentile using the simulated copper values and compare these with the corresponding experimental indicator variograms V^e. If a visual inspection indicates that they are similar then proceed to Step 6. Otherwise adjust the manually fitted variograms V^i and return to Step 2;
6. Compare the histogram of simulated copper values (H^m) with that of the declustered copper assays for cutoffs that occur between the 15th and 85th percentiles. If they agree then proceed to Step 7, otherwise return to Step 2;
7. Compare the histogram of simulated copper values (H^m) with that of the declustered copper assays for the cutoffs below the 15th and those above the 85th percentiles. If they agree then proceed to Step 8, otherwise return to Step 3;
8. Examine the order relation corrections required during the generation of the simulations for cutoffs that occur between the 15th and 85th percentiles. If these are excessive, then return to Step 2, otherwise proceed to Step 9;
9. Examine the order relation correction required during the generation of the simulations for the cutoffs below the 15th and those above the 85th percentiles. If these are excessive, then return to Step 3, otherwise proceed to Step 10;
10. Stop - an acceptable set of copper indicator variograms has been generated.

The above procedure relies on determining how well the histogram of simulated copper values mimics the declustered histogram of copper assays. In the present study two measures were used. These were the *average difference* between the histogram of simulated values and the true experimental histogram, and the *average relative difference* between the two histograms. This latter quantity is important because, even though the high cutoffs only represent a very small proportion of the sample and simulated data, a small difference between the true and simulated histograms can be very economically significant. For example, an absolute difference of 1% between the two histograms may not be significant for the median class where 16% of the samples lie (i.e. 16±1%). However, the same absolute difference is highly significant for a high-grade class, where only 1% of the assay samples lie (i.e. 1±1%). The *average relative difference* reflects the relative importance of such differences. The procedure also requires that the number and magnitude of order relation corrections be examined in Steps 8 and 9 to determine whether an acceptable number of corrections were required, and this was selected to be an average magnitude of the probability corrections in the order of 0.01.

The iterative procedure described above was used to infer the variogram parameters for the high copper cutoffs in the study and resulted in a more finely tuned set of simulations (Kay 2001). This is shown in Figure 3 which presents the *average difference* and *average relative difference* for the set of simulations based on the final set of variogram parameters. The final set of variogram parameters resulted in realisations matching the experimental variograms very well for the high grade copper classes (Figure 4). With respect to the order relation corrections, the 'final' set of variogram parameters perform well (Kay, 2001).

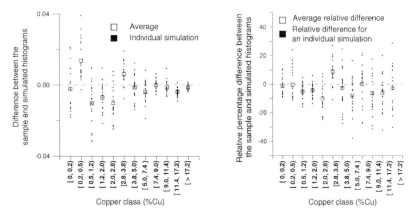

Figure 3. Difference (left) and relative difference (right) between histograms of the copper composites and the ensemble of copper simulations. Also shown is the average copper simulation.

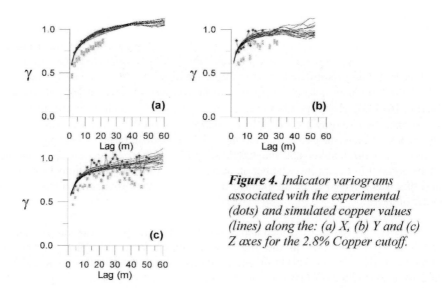

Figure 4. Indicator variograms associated with the experimental (dots) and simulated copper values (lines) along the: (a) X, (b) Y and (c) Z axes for the 2.8% Copper cutoff.

To show the effect of the contribution of conductivity data to the generation of copper simulations, three different groups of simulations are examined. Group A is the set of simulations generated using copper assays only. Group B was identical to A except that the copper assays associated with one drillhole fan are removed. The third set, Group C, has the same set of copper assays as Group B, but also includes the conductivity logs collected in the drillhole fan whose copper assay data had been excised from Group B.

Figure 5(a) presents a vertical section through a Group B simulation. The vertical section is in the plane where the excised drillhole fan was contained. With respect to the overall distribution of simulated copper grades, the Group A and B simulations appear to be similar (Kay, 2001). However, differences can be observed. Figure 5(b) displays some of the differences between a Group A and a Group B simulation that both use the same random seed. By removing the copper assays in the drill fan the resulting Group B simulation has considerably higher grades than the Group A simulation in the mineralised zone centred at X=275 (ie the zone outlined in red). However, as Figure 5(b) indicates, the opposite is true in the second mineralised zone (i.e. the zone outlined in black). The figure suggests that removing the copper assays in this zone results in much lower simulated copper grades.

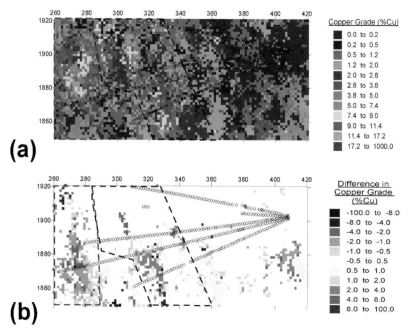

Figure 5. Vertical section through a representative Group B simulation (a) simulated Group B copper values; and (b) difference between the Group C and corresponding Group A simulated copper values. Also shown are the sample locations associated with the excised drillhole fan.

Figure 6 illustrates the effect of using conductivity logs in the simulation process. It is evident that the Group C simulations, using the conductivity logs collected in the drillfan, are qualitatively similar to the A and B simulations. For example, the Group C simulation contains the two mineralised zones discussed earlier. However, Figure 6(b) shows that using the conductivity logs results in simulations that are very similar to the Group A simulations. For instance in the plane of the vertical section, the difference between the equivalent Group A and C simulations is less than ±0.5% Cu for more than 90% of the vertical section. Moreover, in the remaining 10% of the section, the difference is less than ±1.0% Cu with only a few simulated points differing by more than ±5.0% Cu. This suggests that, in the plane of the vertical section using conductivity logs results in simulations that are very similar to those based purely on copper assays.

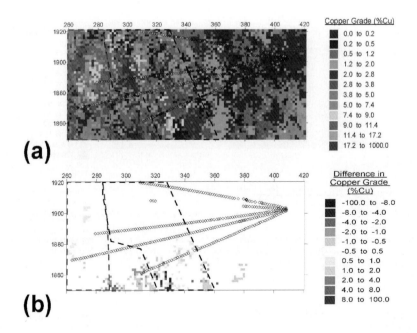

Figure 6. Vertical section through a representative Group C simulation (a) simulated Group C copper values; and (b) difference between the Group C and corresponding Group A simulated copper values. Also shown are the sample locations associated with the excised drillhole fan.

5. Conclusions

The ability to technically integrate geophysical data with conventional assays in metalliferous mines is important during the developmental and mining stages where orebody models are constructed. The present study described the conditional simulation of ore grades in an application that integrates drill core assay data and downhole geophysical logs. This application was used to assess the variability of copper grades at the Kidd Creek base metal mine, Canada. Composites were generated using generalised power averaging which aims to maximise extraction of information from the conductivity logs. Sequential co-indicator simulation was applied to 'hard' (copper assay) and 'soft' (conductivity logging) data using the Markov-Bayes approximation. The selection of the cutoff grade was based on the quantity of metal, and the iterative calibration of indicator variograms at very high cutoffs was performed to ensure convergence with declustered copper assay statistics. Validation of the simulations suggests the Markov-Bayes approximation works reasonably well. It was shown that simulated realisations of copper grades are of comparable quality, when some of the assay composites in selected drillholes are replaced by conductivity data. This suggests that replacing some assaying with logging can generate savings with little loss of information.

Acknowledgments

The support of the Kidd Creek Mine, Ontario, Canada and Falconbridge Ltd is gratefully acknowledged.

References

Basford, P., Kelso, I., Briggs, T., Clifford, M., Anderson, R., and Fullagar, P., 2001, Development of a Short-term Model using Petrophysical Logging at Century Mine, North Queensland: *Australian Society of Exploration Geophysicists Preview*, No. 92, 19-24.

Bryan, R.C. and Roghani, F., Application of Conventional and Advanced Methods to Uranium Ore Reserve Estimation and the Development of a Method to Adjust for Disequilibrium Problems. *17th APCOM Symposium*, 1985, p. 109-120.

Claerbout, J. and Muir. F., Robust Modeling With Erratic Data. *Geophysics*, vol. 38, 1973, p. 826-844.

David, M., *Handbook of Applied Advanced Geostatistical Ore Reserve Estimation*. Elseveir, New York, 1988.

Deutsch, C. and Journel, A., *GSLIB Geostatistical Software Library and Users Guide*. Oxford University Press, New York, 1992.

Dimitrakopoulos, R., *Risk Assessment In Orebody Modelling And Mine Planning: Decision-making with uncertainty*. BRC Notes, SME Annual Meeting & Exhibit, Denver Colorado, 2004, p. 300.

Dimitrakopoulos, R. and Desbarats, A., Geostatistical Modeling of Gridblock Permiabilities for 3D Reservoir Simulators, *SPE Reservoir Engineering*, 1993, p. 13-18.

Dimitrakopoulos, R. and Kaklis, K., Integration of Assay and Cross-hole Tomographic Data in Orebody Modelling: Joint Geostatistical Simulation and Application at Mount Isa Mine, *Queensland. Transactions, IMM*, 110 (Jan.-April), 2001, p. B33-B39.

Fallon, G., Fullagar, P. and Sheard, S., Application of Geophysics in Metalliferous Mines, *Australian Journal of Earth Sciences*, vol. 44, no. 4, 1997, p. 391-409.

Fullagar, P.K., and Fallon, G.N., 2001, Geophysical GradeEstimation at Mines, *Australian Society of Exploration Geophysicists Preview*, No. 90, 30-32.

Goovaerts, P., *Geostatistics for Natural Resources Evaluation*. Oxford University Press, New York, 1997.

Kay, M., Geostatistical Integration of Conventional and Downhole Geophysical Data in the Metalliferous Mine Environment. MSc thesis, University of Queensland, 2001.

Miller, S. and Luark, R., Spatial Simulation of Rock Strength Properties Using a Markov-Bayes Method, *International Journal of Rock Mechanics and Mining Science and Geomechanics Abstracts*, vol. 30, no 7, 1993, p. 1631-1637.

Swan, A. and Sandilands, M., *Introduction to Geological Data Analysis*. Blackwell Science, Oxford, 1985.

Zhu, H., *Modeling Mixture of Spatial Distributions with Integration of Soft Data*. Phd Thesis, Stanford University, 1991.

Zhu, H. and Journel, A., Formatting and Integrating Soft Data: Stochastic Imaging via the Markov-Bayes Algorithm: in Soares, A. (Ed.) *Geostatistics Troia '92*. Kluwer Academic Publishers, Dordrecht, 1992, p. 1-12.

THE KRIGING OXYMORON: A CONDITIONALLY UNBIASED AND ACCURATE PREDICTOR (2nd EDITION)

EDWARD ISAAKS, Ph.D.
ISAAKS & Co.

Abstract. An analysis of conditional bias and its impact on mineral resource estimation is presented. A simple method is proposed for building a long-term mineral resource block model that accounts for conditional bias, change-of-support, and the information effect at the time of mining.

1. Introduction

Accounting for change-of-support, the information effect, and conditional bias are problems well known to mineral resource modelers. Although the methods proposed for dealing with change-of-support and the information effect are little more than approximations, case studies suggest these methods are useful (David, 1977; Journel and Huijbregts, 1978; Matheron, 1984; Parker, 1980; Isaaks and Srivastava, 1991; Deraisme, 2000). However, the same cannot be said for conditional bias. A literature review reveals that conditional bias is poorly understood and that many of the claims are misleading.

Krige (1994; 1996; 1999) claims the preliminary prerequisite of all resource estimators is the elimination of conditional bias. Sinclair and Blackwell (2002) claim that conditional bias contributes to the discrepancies noted between the prediction of recoverable resources and production. David, Marcotte, and Soulie (1984) propose a correction for conditional bias and claim that this correction will reduce the discrepancy between predicted resources and production. Pan (1998) proposes a correction for conditional bias followed by a correction for the smoothing induced by the first correction. However, it can be shown that these two corrections are circular in the sense that the final smoothing correction re-introduces conditional bias. Guertin (1984) proposes a solution that she claims can be easily implemented as a correction factor for *any mineral resource estimation* or grade control system. Deutsch and McLennan (2003) argue that conditionally simulated block model values are both conditionally unbiased and accurate predictors of the tons and grade that will be recovered at the time of mining. However, as will be shown these claims are not correct despite their wide acceptance.

A conditionally unbiased and accurate predictor[1] is an oxymoron. The estimator for a long-term mine planning block model may be conditionally unbiased but then the

---

[1] Accuracy is defined as the ability of the long-term block model to predict the actual tonnage and average ore grade that will be recovered at the time of mining.

histogram of block estimates will be smoothed yielding inaccurate predictions of the recoverable tons and grade above cutoff grade. Conversely, if the histogram of block estimates provides accurate predictions, then the block estimator is necessarily conditionally biased. The estimator for a long-term mine planning block model cannot be conditionally unbiased and simultaneously accurate as claimed by Deutsch and McLennan (2003). David (1977) recognized the oxymoron by pointing out that one can accurately estimate the histogram of block grades but then one cannot localize the blocks. Alternatively, one can estimate as accurately as possible the grades of precisely located blocks (thereby minimizing conditional bias) but then the block histogram will be smoothed. The only exception to this apparent contradiction occurs when the block estimates are perfectly correlated with the true block grades. In this unlikely scenario the block model is both a conditionally unbiased and accurate predictor.

This paper provides an analysis of conditional bias and its impact on mineral resource estimation. Although it may not be possible to eliminate conditional bias from the grade control estimator it can be evaluated through conditional simulation. Block models for long-term mine planning can be built using simulation methods that not only quantitatively account for the conditional bias of future grade control estimators, but also for future change-of-support and information effects.

Section 2 defines two types of mineral resource block models on the basis of how the block estimates are used by the mine. These definitions provide the key to understanding the role of conditional bias in mineral resource modeling. Section 3 provides a formal definition of conditional bias and describes a simple check. Section 4 examines the impact of conditional bias on prediction when the block estimates are used for selection at the time of mining e.g., grade control. Section 5 examines the problem of predicting the tons and grade that will be recovered at the time of mining given that selection will be made using grade control estimates based on future blast hole data. Section 6 describes how to build a long-term mine planning block model by conditional simulation that accounts for a future conditionally biased grade control estimator, the information effect, and a change of support.

2. Two Types of Block Models

Mineral resource models can be classified into one of two types depending on how the block estimates are used by the mine operation.

Type 1: Models whose block grade estimates are used to predict the tons and average grade of ore material that will be recovered each annual, semi-annual, or quarterly period over the life-of-mine are classified as Type 1. Individual block estimates are typically derived from relatively sparse diamond drill hole (DDH) data. Predicted recoveries made from Type 1 estimates are useful for feasibility studies, long and short term mine planning, and the estimation of production schedules etc. Individual block estimates are not used for selection at the time of mining. Thus, it is not necessary to know the precise location or recoverable grade of each ore block. Knowledge of the distribution of recoverable[2] block grades to be mined in the future for each period is sufficient. Type 1 models are often referred to as long-term (mine planning) models.

Type 2: Models whose block grade estimates are used for selection at the time of mining are classified as Type 2. Individual block volumes are equivalent to a selective mining unit (SMU) with the grade of each block typically estimated from neighboring blast hole (BH) grades. The use of these estimates to distinguish between ore and waste is commonly known as *grade control*.

3. Definition of Conditional Bias

3.1 NOTATION

D : The deposit or domain of interest.

$[Z(\mathbf{u}), \mathbf{u} \in D]$: A stationary random function consisting of a set of point support random variables.

$Z_v(\mathbf{u}) = \frac{1}{|v|} \int_{v(\mathbf{u})} Z(\mathbf{u}')d\mathbf{u}'$: A random variable of support v centered at location $\mathbf{u}$.

$[Z_v(\mathbf{u}), \mathbf{u} \in D]$: A stationary random function consisting of a set of random variables of support v. The random function $[Z_v(\mathbf{u}), \mathbf{u} \in D]$ is written as Z_v to simplify notation.

$F_v(z; \mathbf{u} \mid (n)) = \text{prob}\{Z_v(\mathbf{u}) \leq z \mid (n)\}$: Non stationary cumulative conditional distribution function (ccdf) of the random variable $Z_v(\mathbf{u})$ at the location $\mathbf{u}$ conditioned by n data.

$F_D(z; v \mid (n)) = \frac{1}{|D|} \int_D F_v(z; \mathbf{u}' \mid (n))d\mathbf{u}'$: The probability that the grade of a randomly selected SMU within the domain D will be no greater than the cutoff z.

$[Z_{v^*}(\mathbf{u} \mid (n)), \mathbf{u} \in D]$: A non stationary random function consisting of a set of random variables where each RV $Z_{v^*}(\mathbf{u} \mid (n))$ is of the form $\sum_{i=1}^{n} w_i Z(\mathbf{u}_i)$ with $\sum w = 1$. The random function $Z_{v^*}(\mathbf{u} \mid (n))$, $\mathbf{u} \in D$ is written as Z_{v^*} to simplify notation.

$F_{v^*}(z; \mathbf{u} \mid (n)) = \text{prob}\{Z_{v^*}(\mathbf{u} \mid (n)) \leq z\}$: The non-stationary ccdf of the random variable $Z_{v^*}(\mathbf{u})$ at the location $\mathbf{u}$ conditioned by the (n) data.

$F_D(z; v^* \mid (n)) = \frac{1}{|D|} \int_D F_{v^*}(z; \mathbf{u}' \mid (n))d\mathbf{u}'$: The probability that the estimated grade $Z_{v^*}(\mathbf{u})$ of a SMU randomly selected within D will be no greater than z.

3.2 DEFINITION

The conditional expectation is given by:

$$E\{Z_v \mid Z_{v^*} = z\} = h(z) \qquad \forall z \qquad (1)$$

---

[2] The recoverable grade is the actual grade recovered given that selection is based on estimates typically made from blast hole data at the time of mining.

where the function $h(z)$ may be linear or non linear. However, if we impose the condition:

$$h(z) = z \qquad \forall z \qquad (2)$$

the function $h(z)$ will be linear through the origin with a slope of 1.0 and the estimator Z_{v*} is conditionally unbiased by definition (Journel and Huijbregts, 1978).

The conditionally unbiased relation (2) can be re-written as:

$$E\{Z_v - Z_{v*} \mid Z_{v*} = z\} = 0 \qquad \forall z \qquad (3)$$

Equations (2) and (3) imply that the average of the estimator Z_{v*} above cutoff is an unbiased estimate of the average of the corresponding true values Z_v:

$$E\{Z_v \mid Z_{v*} > z\} = E\{Z_{v*} \mid Z_{v*} > z\} \qquad \forall z \qquad (4)$$

Equation (4) can also be written as:

$$E\{Z_v - Z_{v*} \mid Z_{v*} > z\} = 0 \qquad \forall z \qquad (5)$$

3.3 A CHECK FOR CONDITIONAL BIAS

The linear regression of Z_v on Z_{v*} is given by $E\{Z_v \mid Z_{v*} = z\} = a*z + b$ where a is the slope and b the intercept. Thus, if the slope of the linear regression of Z_v on Z_{v*} is not equal to 1 or the intercept is not equal to 0, then the estimator Z_{v*} is conditionally biased, e.g.,

$$\begin{aligned} E\{Z_v \mid Z_{v*} = z\} &= h(z) \\ a*z + b &= z, \ \forall z \quad \text{iff } a = 1 \text{ and } b = 0 \end{aligned} \qquad (6)$$

The linear regression model also provides some insight on the relationship between the two random functions Z_v and Z_{v*} e.g., the slope a is given by:

$$a = \frac{\text{cov}(Z_v, Z_{v*})}{\text{var}(Z_{v*})} = \frac{\sigma_v}{\sigma_{v*}} \rho_{vv*} \qquad (7)$$

where σ_v^2 and σ_{v*}^2 are the variances of Z_v and Z_{v*}. Thus, for a conditionally unbiased estimator:

$$a = \frac{\sigma_v}{\sigma_{v*}} \rho_{vv*} = 1.0 \qquad (8)$$

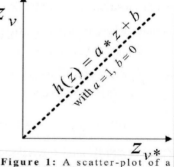

Figure 1: A scatter-plot of a conditionally unbiased estimator.

Two important observations can be made from Equation (8).

1. Since in practice, the correlation between the true and estimated values is: $\rho_{v\,v*} < 1$, then for a conditionally unbiased estimator, necessarily: $\sigma_v^2 > \sigma_{v*}^2$. In other words, the estimates of a conditionally unbiased estimator are smoothed

2. Conversely, if the two distributions $F_D(z;v|(n))$ and $F_D(z;v^*|(n))$ defined in section 3.1 have equal variances: $\sigma_v^2 = \sigma_{v^*}^2$, then necessarily $a < 1$. That is, the estimator Z_{v^*} is conditionally biased.

4. Type 2 Estimates and their Recovery Functions

Recall that the estimates of a Type 2 estimator are used by the mine operation for the selection of ore at the time of mining.

4.1 NOTATION

The type 2 estimator is denoted by a double asterisk, e.g., $Z_{v^{**}}(\mathbf{u})$.

$F_{v^{**}}(z;\mathbf{u}|(n))$: The non stationary ccdf of the RV $Z_{v^{**}}(\mathbf{u})$ at location $\mathbf{u}$ conditioned by the (n) data.

$F_D(z;v^{**}|(n)) = \dfrac{1}{|D|} \int_D F_{v^{**}}(z;\mathbf{u}'|(n))d\mathbf{u}'$: The non stationary conditional probability that the estimated grade $Z_{v^{**}}(\mathbf{u})$ of a randomly selected SMU within the domain D will be no greater than z. Note that this distribution is commonly estimated in practise.

$F_v(z;\mathbf{u}|v^{**},(n))$: The non stationary ccdf of the RV $Z_v(\mathbf{u})$ at location $\mathbf{u}$ given that $Z_{v^{**}}(\mathbf{u}) \leq z$ and the (n) conditioning data.

$F_D(z;v|v^{**},(n)) = \dfrac{1}{|D|} \int_D F_v(z;\mathbf{u}'|v^{**},(n))d\mathbf{u}'$: The non stationary conditional probability that the true grade $Z_v(\mathbf{u})$ of a randomly selected SMU within the domain D will be no greater than z given that its estimated grade $Z_{v^{**}}(\mathbf{u})$ is no greater than z. Note that this distribution is not known nor is it commonly estimated in practice.

4.2 ACTUAL RECOVERIES GIVEN THAT SELECTION IS MADE USING ESTIMATED GRADES.

The following recovery functions describe the *actual but unknown* quantities that will be recovered given that selection is made using the estimates $z_v^{**}(\mathbf{u})$. The recovered tonnage is given by:

$$T_D(z) = T_o[1 - F_D(z;v^{**}|(n))] \qquad \forall z \qquad (9)$$

The actual but unknown quantity of recovered metal is given by:

$$Q_D(z) = T_o \int_z^{\infty} z' \, dF_D(z';v|v^{**},(n)) \qquad \forall z \qquad (10)$$

The actual but unknown recovered grade is given by:

$$m_D(z) = \frac{Q_D(z)}{T_D(z)} \qquad (11)$$

4.3 ESTIMATED RECOVERIES GIVEN THAT SELECTION IS MADE USING ESTIMATED GRADES

The recovery equations provided by (10) and (11) are not useful since the distribution $F_D(z; v\,|\,v^{**},(n))$ is not known or commonly estimated in practice. However, by replacing the unknown distribution with the commonly estimated distribution $F_D(z; v^{**}\,|\,(n))$, one can estimate the recoveries as follows:

$$\hat{T}_D(z) = T_o[1 - F_D(z; v^{**}\,|\,(n))] \qquad \forall z \qquad (12)$$

The estimated recovered quantity of metal is given by:

$$\hat{Q}_D(z) = T_o \int_z^\infty z'\, dF_D(z'; v^{**}\,|\,(n)) \qquad \forall z \qquad (13)$$

and the estimated recovered grade is given by:

$$\hat{m}_D(z) = \frac{\hat{Q}_D(z)}{\hat{T}_D(z)} \qquad (14)$$

If the estimator $Z_{v^{**}}(\mathbf{u})$ is conditionally unbiased, then the estimated recoveries (13) (14) will be equal to the actual recoveries (10) (11) since conditional unbias implies the following:

$$E\{Z_v \,|\, Z_{v^{**}} > z\} = E\{Z_{v^{**}} \,|\, Z_{v^{**}} > z\}$$

$$\Rightarrow \frac{\int_z^\infty z'\, dF_D(z'; v\,|\,v^{**};(n))}{1 - F_D(z; v^{**}\,|\,(n))} = \frac{\int_z^\infty z'\, dF_D(z'; v^{**}\,|\,(n))}{1 - F_D(z; v^{**}\,|\,(n))} \qquad \forall z \qquad (15)$$

$$\Rightarrow \int_z^\infty z'\, dF_D(z'; v\,|\,v^{**};(n)) = \int_z^\infty z'\, dF_D(z'; v^{**}\,|\,(n))$$

$$\Rightarrow F_D(z'; v\,|\,v^{**};(n)) = F_D(z'; v^{**}\,|\,(n))$$

Thus, it appears[3] that the estimator $Z_{v^{**}}(\mathbf{u})$ must be conditionally unbiased in order to provide accurate predictions of the tons and grade that will be delivered to the mill. Ideally, the estimator $Z_{v^{**}}(\mathbf{u})$ will also minimize the conditional variance $E\{[Z_v - h(z)]^2\}$ (Journel and Huijbregts, 1978) so as to minimize ore loss and dilution or misclassification at the time of mining.

5. Type 1 Estimates and their Recovery Functions

Recall, that Type 1 estimates are used to predict the tons and grade of ore that will be recovered in the future at the time of mining. They are not used for selection at the time of mining.

---

[3] Section 6 describes how conditional simulation can be used to accurately predict the tons and grade that will be delivered to the mill in spite of a conditionally biased grade control estimator.

5.1 NOTATION

Type 1 estimates are denoted by a single asterisk, e.g., $z_{v^*}(\mathbf{u})$.

$F_{v^*}(z;\mathbf{u}\,|\,(n))$: The non stationary ccdf of the RV $Z_{v^*}(\mathbf{u})$ at location $\mathbf{u}$ conditioned by the (n) data.

$F_D(z;v^*\,|\,(n)) = \frac{1}{|D|}\int_D F_{v^*}(z;\mathbf{u}'\,|\,(n))d\mathbf{u}'$: The non stationary conditional probability that the estimated grade $Z_{v^*}(\mathbf{u})$ of a randomly selected SMU within the domain D will be no greater than z. Note that this distribution is commonly estimated in practice.

5.2 THE RECOVERY EQUATIONS

Recoverable tonnage:
$$\hat{T}_D(z) = T_o[1 - F_D(z;v^*\,|\,(n))] \qquad \forall z \qquad (16)$$

Recoverable quantity of metal:
$$\hat{Q}_D(z) = T_o \int_z^\infty z'\,dF_D(z';v^*\,|\,(n)) \qquad \forall z \qquad (17)$$

Recoverable grade:
$$\hat{m}_D(z) = \frac{\hat{Q}_D(z)}{\hat{T}_D(z)} \qquad (18)$$

Recall, that (9), (10), and (11) provide the actual recoveries given that selection will be made using the estimates $Z_{v^{**}}$ in the future. Thus, to be useful the recoveries predicted by (16), (17), and (18) must be equal to those given by (9), (10), and (11). However, this is a problem since there is nothing in (16), (17), and (18) that guarantees equivalence to (9), (10), and (11). This problem is recognized within the mining industry where a common solution is to impose additional constraints on the estimators $z_{v^*}(\mathbf{u})$ and $z_{v^{**}}(\mathbf{u})$, e.g.,

Condition 1. $F_D(z;v^*\,|\,(n)) = F_D(z;v^{**}\,|\,(n'))$ - This condition requires the histogram of the type 1 estimates to be equal to the histogram of the type 2 estimates within D. For example;

- In practice, the future distribution $F_D(z;v^{**}\,|\,(n))$ is estimated using smoothing relations and the change of support hypothesis (Journel and Huijbregts, 1978; Isaaks and Srivastava, 1989; Sinclair and Blackwell, 2002).
- The distribution $F_D(z;v^*\,|\,(n))$ is then made to match as close as possible to the estimated distribution $F_D(z;v^{**}\,|\,(n))$ by controlling the number of samples used to estimate z_{v^*} locally (Deutsch and McLennon, 2003).

Condition 2. $E\{Z_v - Z_v^{**}\,|\,Z_v^{**} > z\} = 0 \;\; \forall z$ - This condition requires the type 2 estimator to be conditionally unbiased.

The equivalence between the predicted recoveries (16), (17), and (18) given conditions (1) and (2) and the actual recoveries (9), (10), and (11) is easily confirmed.

However, condition (1) may not be that easy to impose on the estimator Z_{v^*}. The change of support and information effect may render the distributions $F_D(z;v^*|(n))$ and $F_D(z;v^{**}|(n))$ incomparable. Thus, at best this practice amounts to nothing more than an approximation.

Condition (2) may also be difficult if not impossible to impose on the future estimator $Z_{v^{**}}$. Although kriging is said to be a *conditionally unbiased estimator,* in reality it is conditionally unbiased if and only if the distribution of $Z(\mathbf{u})$ is normal and its mean $E\{Z(\mathbf{u})\}$ is known (David, 1977). The problem is that almost all distributions of $Z(\mathbf{u})$ in mining applications are non-normal with relatively large coefficients of variation and large coefficients of skew. Because of this, it is very difficult if not impossible for the mine operator to insure that the grade control estimator is conditionally unbiased.

5.3 THE OXYMORON

Note, that although the estimator Z_{v^*} is an accurate predictor of recoveries (9), (10), and (11) given conditions (1) and (2), Z_{v^*} is almost certain to be conditionally biased. For example, from condition (2),

$$\frac{\sigma_v}{\sigma_{v^{**}}} \rho_{vv^{**}} = 1.0 \tag{19}$$

and from condition (1),

$$\sigma_{v^*} = \sigma_{v^{**}} \tag{20}$$

and since $\rho_{vv^*} < \rho_{vv^{**}}$ with near certainty then,

$$a = \frac{\sigma_v}{\sigma_{v^*}} \rho_{vv^*} < 1.0 \tag{21}$$

that is, Z_{v^*} is almost certain to be conditionally biased. Thus, in spite of conditional bias, Z_{v^*} may be an accurate predictor of recoverable resources given conditions (1) and (2).

6 Conditional Simulation and Prediction

This section proposes a method for building the long-term block model using conditional simulation via the LU decomposition of the covariance matrix, (Davis, 1987).

6.1 NOTATION

$Z_{\tilde{v}}(\mathbf{u})$ - the tilde above a variable denotes a conditionally simulated value. Otherwise the notation for the simulated variables and their distributions is identical to the definitions provided in section 4.1

6.2 CONDITIONAL SIMULATION

Consider the following vectors of point support Gaussian random variables:

$\mathbf{Y}_1 = [Y(\mathbf{u}_i), i = 1, n]'$ - a vector of (n) $N(0,1)$ random variables located at DDH sample locations $\mathbf{u}_i$ $i = 1, n$,

$\mathbf{Y}_2 = [Y(\mathbf{u}_j), j = 1, s]'$ - a vector of (s) $N(0,1)$ random variables located at blast hole (BH) locations $\mathbf{u}_j$ $j = 1, s$, and

$\mathbf{Y}_3 = [Y(\mathbf{u}_k), k = 1, t]'$ - a vector of (t) $N(0,1)$ random variables located at the discretization point locations $\mathbf{u}_k$ $k = 1, t$ of the SMU.

Note that some of the locations may be co-located, e.g., $\mathbf{u}_i = \mathbf{u}_j$, $\mathbf{u}_i = \mathbf{u}_k$, $\mathbf{u}_j = \mathbf{u}_k$ for some i, j, k (see Figure 2).

The corresponding covariance matrices are given by:

$\mathbf{C}_{11} = \text{cov}(\mathbf{Y}_1 \mathbf{Y}_1')$ with dimension $n \times n$

$\mathbf{C}_{21} = \text{cov}([\mathbf{Y}_2', \mathbf{Y}_3']' \mathbf{Y}_1')$ with dimension $m \times n$ where $s + t = m$.

$\mathbf{C}_{22} = \text{cov}([\mathbf{Y}_2', \mathbf{Y}_3']'[\mathbf{Y}_2', \mathbf{Y}_3'])$ with dimension $m \times m$.

The covariance matrix between the random vectors $\mathbf{Y}_1, \mathbf{Y}_2$, and $\mathbf{Y}_3$ can be decomposed into the product of a lower and upper triangular matrix, e.g.,

$$\begin{bmatrix} \mathbf{C}_{11} & \mathbf{C}_{12} \\ \hline \mathbf{C}_{21} & \mathbf{C}_{22} \end{bmatrix} = \begin{bmatrix} \mathbf{L}_{11} & \mathbf{0} \\ \hline \mathbf{L}_{21} & \mathbf{L}_{22} \end{bmatrix} * \begin{bmatrix} \mathbf{U}_{11} & \mathbf{U}_{21} \\ \hline \mathbf{0} & \mathbf{U}_{22} \end{bmatrix} \qquad (22)$$

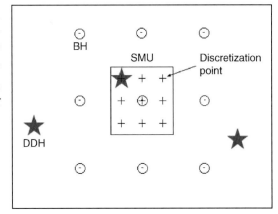

Figure2: Example locations of the random variables Y relative to a SMU. The stars represent $\mathbf{Y}_1$ at DDH locations, while the circles represent $\mathbf{Y}_2$ at BH locations and the plus signs symbolize $\mathbf{Y}_3$ at the discretization points of the SMU. Note the co-location of some of the variable locations.

Next, we interpret the relatively sparse DDH data $z(\mathbf{u}_i)$, $i = 1, n$ as a realization of the random vector $\mathbf{Y}_1$, e.g.,

$$y_1(\mathbf{u}_i) = \varphi(z(\mathbf{u}_i)), \quad i = 1, n \qquad (23)$$

where $\varphi(\cdot)$ is the normal score transform. Realizations of the random vectors $\widetilde{\mathbf{Y}}_2$ (at BH locations) and $\widetilde{\mathbf{Y}}_3$ (at SMU discretization point locations) can be simulated conditional to the transformed DDH data $\mathbf{Y}_1$ as follows:

$$\begin{bmatrix} \mathbf{Y}_1 \\ \widetilde{\mathbf{Y}}_2 \\ \widetilde{\mathbf{Y}}_3 \end{bmatrix} = \begin{bmatrix} \mathbf{L}_{11} & \mathbf{0} \\ \mathbf{L}_{21} & \mathbf{L}_{22} \end{bmatrix} * \begin{bmatrix} \mathbf{L}_{11}^{-1}\mathbf{Y}_1 \\ \mathbf{W} \end{bmatrix}$$

$$= \begin{bmatrix} \mathbf{Y}_1 \\ \mathbf{L}_{21}\mathbf{L}_{11}^{-1}\mathbf{Y}_1 + \mathbf{L}_{22}\mathbf{W} \end{bmatrix} \qquad (24)$$

where $\mathbf{W}$ is a random vector of (m) iid $N(0,1)$ random variables. Multiple realizations of the vectors $\widetilde{\mathbf{Y}}_2$ and $\widetilde{\mathbf{Y}}_3$ each conditional to $\mathbf{Y}_1$ (and to each other) are obtained by generating realizations of the *iid* random vector $\mathbf{W}$ and evaluating,

$$\begin{bmatrix} \widetilde{\mathbf{Y}}_2 \\ \widetilde{\mathbf{Y}}_3 \end{bmatrix} = \mathbf{L}_{21}\mathbf{L}_{11}^{-1}\mathbf{Y}_1 + \mathbf{L}_{22}\mathbf{W} \qquad (25)$$

for each realization of $\mathbf{W}$. A single conditional simulation of the SMU grade at location $\mathbf{u}_0$ is given by:

$$z_{\widetilde{v}}(\mathbf{u}_0) = \frac{1}{t}\sum_{k=1}^{t} \varphi^{-1}(\widetilde{y_3}(\mathbf{u}_k)) \qquad (26)$$

The corresponding estimated SMU grade made from conditionally simulated blast hole grades is given by:

$$z_{\widetilde{v}**}(\mathbf{u}_0) = \sum_{j}^{s} \lambda_j B[\varphi^{-1}(\widetilde{y_2}(\mathbf{u}_j))] \qquad (27)$$

where λ are ordinary kriging weights for example and $\varphi^{-1}[\widetilde{\mathbf{Y}}_2]$ are simulated DDH values at the blast hole locations. $B[\cdot]$ is a user defined function for transforming simulated DDH grades to simulated BH grades. For example, the function $B[\cdot]$ could be used to add noise or deviations to the vector of simulated DDH grades $\varphi^{-1}[\widetilde{\mathbf{Y}}_2]$ (Parker and Isaaks, 1992; Journel and Kyriakidis, 2004).

The distributions $F_{\tilde{v}}(z; \mathbf{u}_0 \mid (n))$ and $F_{\tilde{v}}(z; \mathbf{u}_0 \mid \tilde{v}^{**}, (n))$ of the conditionally simulated values $z_{\tilde{v}}(\mathbf{u}_0)$ and $z_{\tilde{v}^{**}}(\mathbf{u}_0)$ are generated by repeated applications of (25), (26), and (27). For example by using an efficient LU algorithm it may be practical to simulate as many as 500 equi-probable pairs of $z_{\tilde{v}}(\mathbf{u}_0)$ and $z_{\tilde{v}^{**}}(\mathbf{u}_0)$ for each SMU.

Thus, the simulated tonnage recovered over the domain D is given by:
$$\tilde{T}_D(z) = T_o[1 - F_D(z; \tilde{v}^{**} \mid (n))] \qquad \forall z \tag{28}$$
The simulated actual quantity of recovered metal is given by:
$$\tilde{Q}_D(z) = T_o \int_z^\infty z' \, dF_D(z'; \tilde{v} \mid \tilde{v}^{**}, (n)) \qquad \forall z \tag{29}$$
The simulated actual recovered grade is given by:
$$\tilde{m}_D(z) = \frac{\tilde{Q}_D(z)}{\tilde{T}_D(z)} \tag{30}$$

Equation (26) solves the change of support problem by computing a simple spatial average from a number of jointly simulated point values within the SMU. Note that each simulated point value is back-transformed before averaging.

Equation (27) provides a simulation of the grade control estimator using simulated blast hole grades. Note, that (27) includes a user-definable function enabling the user to simulate the relationship between the DDH and BH grades if known (Parker and Isaaks, 1992; Journel and Kyriakidis, 2004). Thus, the impact of poorer quality blast hole assays on the predicted recoveries can be put into the estimation of recoverable resources here. Equations (29) and (30) simulate the actual recovered quantity of metal and recovered grade given that the SMU are selected by their grade control estimate. The key is the simulated conditional distribution of the true SMU grades given their grade control estimates. This distribution quantitatively accounts for any conditional bias inherent in the grade control estimator as well as for any associated misclassification.

7 Conclusions

- If the block estimates are to be used for selection (grade control), then it is desirable to minimize conditional bias. Although conditional bias may be minimized, it likely cannot be eliminated.
- If the grade control estimator is conditionally biased, the predictions of the long-term mine planning model should quantitatively account for the bias. Such an accounting can be evaluated through conditional simulation.
- If the block estimates are not used for selection at the time of mining, but rather for the prediction of the tons and grade that will be recovered in the future, then whether or not the block estimator is conditionally biased is irrelevant to the accuracy of predicting the future recoveries.
- The predictions of the long-term model should quantitatively account for the ore loss and dilution (misclassification) that will occur at the time of mining. Again, such an accounting can be evaluated through conditional simulation.

- And finally, conditional simulation provides an easy solution to change of support. Long-term mine planning models with block sizes equivalent to the SMU are easily simulated. With good software, conditional simulation via the LU decomposition of the covariance matrix is as practical as ordinary kriging.

8 References

David, M., Marcotte, D. and Soulie, M., Conditional bias in kriging and a suggested correction, In Geostatistics for Natural Resource Characterization, G Verly, M David, AG Journel & A Maréchal, eds, Riedel Publishers, Dordrecht, pp 217-244, 1984.

David, M., *Geostatistical Ore Reserve Estimation*. Elsevier, Amsterdam, 1977.

Davis, M.W., Production of conditional simulations via the LU decomposition of the covariance matrix. Math Geology, 19(2):91-98, 1987.

Deraisme, J. and Roth, C., The information effect and estimating recoverable reserves. Geovariances. www.geovariances.fr/publications/article6/index.php3, 2000.

Deutsch, C.V. and McLennan, J.A., Conditional bias of geostatistical simulation for estimation of recoverable reserves. Can. Inst. Min. Metall. Bull., 2003.

Guertin, K., Correcting conditional bias, In Geostatistics for Natural Resource Characterization, G Verly, M David, AG Journel & A Maréchal, eds, Riedel Publishers, Dordrecht, pp 245-260, 1984.

Isaaks, E. H. and Srivastava, M. R., *An Introduction to Applied Geostatistics*. Oxford University Press, New York, 1989.

Journel, A.G. and Huijbregts, C.J., Mining Geostatistics. Academic Press, New York, 1978.

Krige, D.G., A basic perspective on the roles of classical statistics, data search routines, conditional biases, and information and smoothing effects in ore block valuations. Conference on Mining Geostatistics, Kruger National Park, 1994.

Krige, D.G., A practical analysis of the effects of spatial structure and of data available and accessed, on conditional biases in ordinary kriging. 5th International Geostatistics Congress, Wollongong, Australia, 1996.

Krige, D.G., Conditional Bias and Uncertainty of Estimation in Geostatistics. Keynote Address for APCOM'99 International Symposium, Colorado School of Mines, Golden, CO., 1999.

Matheron, G., The selectivity of the distributions and the "second principle of geostatistics", In *Geostatistics for Natural Resource Characterization, G Verly, M David, AG Journel & A Maréchal, eds*, Riedel Publishers, Dordrecht, pp 421 – 433, 1984.

Pan, G., Smoothing effect, conditional bias and recoverable reserves. Can. Inst. Min. Metall. Bull. V. 91, no. 1019, pp81-86, 1998.

Parker, H., The volume-variance relationship: a useful tool for mine planning. In P. Mousset-Jones, editor, *Geostatistics*, pages 61-69, New York, 1980. McGraw Hill, 1980.

Parker, H. and Isaaks, E.H, *The Assessment of Recoverable Reserves for the Minifie Deposit using Conditional Simulation*, Kennecott-Niugini Mining Joint Venture, Lihir Project Resource/Reserve Calculations, 1992.

Sinclair, A.J., and Blackwell, G.H., *Applied Mineral Inventory Estimation*. Cambridge University Press, Cambridge, 2002.

POST PROCESSING OF SK ESTIMATORS AND SIMULATIONS FOR ASSESSMENT OF RECOVERABLE RESOURCES AND RESERVES FOR SOUTH AFRICAN GOLD MINES.

D.G. KRIGE [1], W. ASSIBEY-BONSU [2] AND L. TOLMAY [2]
[1] Private Consultant, South Africa
[2] Gold Fields Limited, South Africa

ABSTRACT. This study is based on a comprehensive data base from a section of a large deep South African gold mine. The upper section of the area covered was accepted as providing the known data for purposes of estimating in a deeper extension of this section. Ore blocks in this extension were valued using the data from the upper 'known' area together with data in development raises typically 150m apart in the deeper extension area. Estimation techniques used were Simple Kriging (SK) with post-processing, and Simulations and the recoverable block estimates were compared with the known follow-up 'actual' values of these blocks. The study shows that the direct SK post-processing and repeated simulation approaches, if applied efficiently, can provide equally useful tools for computing global recoverable resources. However, the direct SK post-processed technique provided the only advanced practical estimates of individual ore blocks for short-term mine planning, grade control and ore resource/reserve classification.

1 Brief Historical Background to Ore Block Valuations

The main objectives of block valuations in South African gold mines have always been, and still are:
- To provide management and shareholders with a reliable inventory of the mine's basic asset, i.e. its ore resources and reserves classified into categories as required by the relevant codes.
- The estimation of tonnages and grades expected to be obtained from mining in short and medium term time categories e.g. monthly, quarterly and annually, *and from individual stopes and mine sections.*
- Where the average ore grade is not sufficiently high to warrant 100% mining of the ore body, proper advanced indications for the selection of blocks above the break even or cut off.
- The planning of grade control so as to produce a profile of acceptable production and financial targets.

The birth of Geostatistics and kriging in South Africa more than 50 years ago resulted from the statistical explanation of the presence of conditional biases in the orthodox valuation techniques (Krige 1951). Kriging, properly applied, eliminated these biases but provided smoothed estimates. To overcome the smoothing effect, some geostatisticians introduced the practice of Ordinary Kriging (OK), but with a limited data search, and without using the global mean or applying simple kriging (SK). This could rectify the problem of smoothing but had the effect of re-introducing the conditional biases, which in fact had led to the birth of geostatistics. The practice of using a limited search cannot be condoned (Krige, 1996, 1997, 2001; McLennan and Deutsch, 2004). It is also theoretically impossible to meet both objectives of conditional unbiasedness and the absence of smoothing *on the basis of specific fixed grade estimates for individual blocks*.

A logical advance towards a solution of the problem of smoothing was the substitution of probability estimates for fixed individual kriged estimates. This was effected by using uniform conditioning, direct or indirect conditioning (Assibey-Bonsu and Krige 1999b) and various other post processing procedures, e.g. spectral postprocessor (Journel, Kyriakidis and Mao 2000). Simulation techniques have also been proposed for producing unsmoothed and unbiased block recoverable estimates. However, single simulations could be unsmoothed but will be conditionally biased (Krige and Assibey-Bonsu, 1999a), and repeated simulations, when averaged, will produce smoothed values. Lately, McLennan and Deutsch (2004) have suggested "conditional non-biased simulation" based effectively on the introduction of the concept of probability estimates via repeated simulations, in substitution for specific block estimates. This is a form of post processing, or conditioning, as practiced in the direct processing of kriged estimates.

The authors (McLennan and Deutsch (2004)) compared such estimates with straight kriged estimates, but not with kriged estimates after post-processing. Their analyses covered only estimates on a global basis. The comparison of these techniques for estimates for individual blocks and local small production areas, and the overall effect on grade profiles over time, were not considered.

The above background calls for block estimates to be *globally unbiased as well as for individual blocks and mine sections*, and also properly processed to eliminate any 'smoothing' effects. The argument that final selection of ore blocks as ore or waste is done at the stage when the more intensive sampling data are available on a proper SMU (Selective Mining Unit) basis and that 'unbiasedness' and 'unsmoothing' are only necessary on a global basis, does not hold except possibly, but still to a more limited extent only, for open cast mines. A detailed examination and comparison of kriged estimates before and after post processing with the recent simulation approaches on both a global and more local short-term basis, including individual blocks, is therefore justified. This is the objective of this paper based on a massive set of 'actual' data from a large deep level South African gold mine.

2 Basic Principles of the Two Main Techniques

Conditioning of unbiased estimates is based on the principle of replacing these smoothed estimates with probability distributions representing the expected follow-up 'actual' grades with each kriged estimate as the mean of the distribution at a variance level equal to that expected for the 'actual' grades. For *direct conditioning* this is effected by super imposing on the kriged estimate for each block, a 'simulated' distribution of expected 'actual' values with a variance equal to the difference in variability between the smoothed and 'actual' grades (Assibey-Bonsu and Krige, 1999b). The end result is an estimated unsmoothed tonnage-grade curve to replace the smoothed kriged estimate. For *uniform conditioning* this is done, not for individual SMU blocks, but for groups of local SMU blocks within larger panels or blocks. In this study all references to SMU's and 'blocks' refer to 2D ore units of 20x20m in the plane of the ore body.

The Sequential Gaussian Simulation (SGS) technique was used for generating the simulation realizations. Simple Kriging was used to determine the parameters of the Gaussian conditional cumulative distribution function at respective locations. The SGS was generated using the GSLIB software (See Clayton and Journel, 1992). Post-processing of the simulation results adopted in this paper is similar to that proposed by McLennan and Deutsch (2004). It involves the distribution of the repeated simulated values, say 50, for each block as an estimate of the unsmoothed tonnage-grade curve for the block, and thus that it reflects the uncertainty of the mean of the relevant 50 realisations as an estimate of the 'actual' grade. The assumption is, thus, that the variance of the 50 simulated values straddling the mean of the 50 simulated grades for each block reflects the probability distribution of the 'actual' grades.

The objective of this analysis is to apply these two main techniques using a set of 'actual' data to provide estimates for comparison with 'actual' follow-up information on a practical basis, so as to determine the validities of the alternative techniques and their relative efficiencies.

3 The Data Base

The 'actual' data used should ideally represent the grades of SMU blocks as estimated on the basis of the more extensive data, which will be available at the time of the actual selection during production. Such data for a block can never be complete and *thus the presence of the inevitable information effect in the setting of the target of the SMU 'actuals' for the dispersion variance of block estimates and of the tonnage-grade curves is an essential requirement for both the SK post-processing and the repeated simulation techniques.*

The data used in this study comprised an area of 2Km x 2Km on the Ventersdorp Contact Reef (VCR) on a large deep mine with a total number of nearly 43000 underground sample values recorded as cm.g/t grades and reflecting a direct measure of the gold concentration per unit area of the ore body. The ore body strikes approximately in a north south direction at an average dip westwards of 28^0, and with an average mining width of 139 cm.

Fig. 1 shows the total area covered. The VCR in this area forms a geologically homogenous population with no significant grade trend with depth. The area was split into an upper section of 2Km x 1Km with about half of the samples from stope (or panel) faces and from development exposures, including a set of 8 raises 150m apart and extending into strips 1 and 2 in the lower follow-up area. These values are accepted as 'point' values from the 'known' database to be used to estimate ore resource blocks of 20m x 20m in the deeper 'follow-up' section. The valuation is largely an extrapolation exercise since no regular underground drilling is practiced (Assibey-Bonsu and Krige, 2003).

The 'known' point values in the deeper section were also regularized into 363 data blocks and used as the follow-up 'real' block values for judging the comparative efficiencies of the estimates for these blocks. Follow-up blocks with less than 5 samples per block were discarded because the information effect for such blocks will be abnormally high. The 'known' block values for both the upper and lower areas are shown on Fig. 1 in 4 shades of grey/black and the patterns for both areas confirm the absence of any significant trend.

The geostatistical details of the 'actual' data for points and blocks are recorded in Table 1, and the 3-parameter lognormal distribution model and variogram in Figs 2 and 2B respectively. The third parameter of 255 cm g/t provides acceptable fits for the lognormal distributions of points and of block grades. Table 1B shows the variogram parameters for normal score and relative models. Fig. 4 also shows the target tonnage grade curves for the 363 'actual' 20 x 20m blocks for the follow-up area.

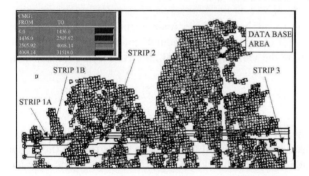

Figure 1: Showing study area and strips representing different production periods.

	No Units	Mean cm.g/t	Std Dev	3 Par Log n b=255	
				Mean	Variance
Upper Area					
Points	18684	3061	4660	7.541	1.15
Blocks 20 by 20m	1335	2918	2317	7.907	0.403
Follow up area					
Total Points	23372	3281	4210	7.667	1.064
Strips 1 to 3:					
Blocks 20 by 20m	363	3311	2420	7.961	0.466

Table 1: Showing details of the database of point and 20x20m block grades

	DIRECTION	NUGGET	C1	R 1	C2	R 2	C3	R 3
POINTS	120 DEGREES	0.414	0.372	30	0.116	70	0.098	500
	30 DEGREES			25		70		180
BLOCKS	120 DEGREES	0.26	0.4	65	0.3	85	0.03	702
20 BY 20	30 DEGREES			32		85		204
Relative	30 DEGREES	0.126	0.153	33	0.14	85	0.068	204
20 BY 20	120 DEGREES			43		63		700

Table 1B: Showing relative and normal score semi variogram parameters.

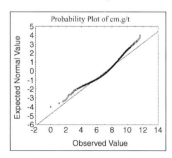
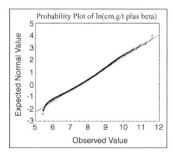

Figure 2: Showing 3-parameter lognormal distribution models with additive constants of zero and 255 for Data Base (Points).

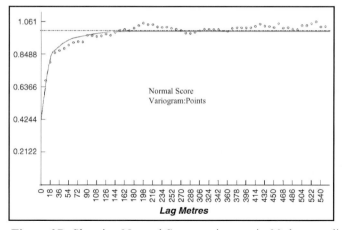

Figure 2B: Showing Normal Score variogram in 30 degrees direction for points.

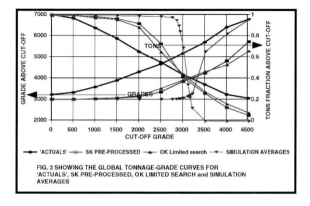

FIG. 3 SHOWING THE GLOBAL TONNAGE-GRADE CURVES FOR 'ACTUALS', SK PRE-PROCESSED, OK LIMITED SEARCH and SIMULATION AVERAGES

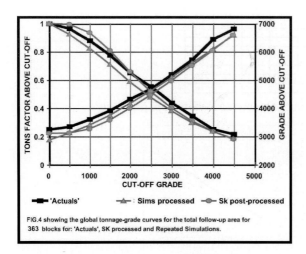

FIG.4 showing the global tonnage-grade curves for the total follow-up area for 363 blocks for: 'Actuals', SK processed and Repeated Simulations.

4 Estimation Techniques – Global Results

For all comparisons of dispersion variances, tonnage-grade curves, and correlations with slopes of regression, etc., the estimates and 'actuals' were normalized by transforming the grades to Logarithm (grade + 255) (see Fig. 2). This caters for the proportional effect, and ensures the approximation to Normal distributions, and thus provides for linear regression trends with slopes for measuring the presence and extent of conditional biases.

The following techniques have been used for the global follow-up area

- *Ordinary Kriging with a limited search.*

This technique is aimed at overcoming the 'smoothing' effect of an extensive search routine or of Simple Kriging (SK). The search parameters used were as follows:
Minimum number of point data……………………..2
Maximum number of point data…………………….8
The results and comparisons with the follow-up 'actual' data are summarized in Table 2 and shown in Fig.3. *The estimates show some elimination of the smoothing effect but serious conditional biases and cannot be recommended for detailed mine planning purposes.*

STRIP	No.Blks	Dispersion Variances		OK vs Act	
		Actuals	OK	Corr.Coef.	Regr.Slope
1/3 total	363	0.466	0.204	0.401	0.606

Table 2: Showing dispersion variances and correlation details for 'actuals' and OK estimates on Ln(x+b) basis.

- *Simple Kriging with no post processing.*

With no post-processing or conditioning, this technique largely overcomes the problem of conditional biases (slope of regression = 0.96) but the results are 'smoothed' with a Ln Variance of 0.087 compared to that for the 'actuals' of 0.47 (see Fig 3, Table 3 and

4). The results provide some provisional indications for the selection of blocks above cut-off but will be conservative for recoverable grades and optimistic for the corresponding tonnages. *Nevertheless, this technique provides a base for performing post-processing for which the effective absence of conditional biases is essential and it provides some advanced indication of which blocks are likely to be mined above cut-off.*

- *Simple Kriging with post-processing*

The above results were post-processed with direct conditioning (Assibey-Bonsu and Krige, 1999b) to provide the comparison with 'actuals' on a tonnage grade basis as shown in Fig. 4 and Table 3. For this purpose ten cut-off grades were used as shown in the figure. The dispersion variance of the estimates and the grade-tonnage curve are not ideal but approach those of the 'actuals'.

STRIP	No. Blks	Actuals	Ln Variances SK Pre.	Sim Avgs	Variances of Probability distributions SK proc.*	Sims.direct	Equiv. sims
1A total	59	0.318	0.111	0.041	0.310	0.597	0.441
1Btotal	58	0.379	0.099	0.022		0.556	0.435
1A+1B	117	0.348	0.105	0.032		0.577	0.438
2	96	0.479	0.115	0.021	0.360	0.563	0.422
3	150	0.536	0.054	0.012		0.556	0.433
1+2 total	213	0.414	0.110	0.027			0.431
1/3 total	363	0.466	0.087	0.021			0.432

Table 3: Showing dispersion variances for 'actual', and SK and Sims. Pre- and post-processed estimates on Ln(x+b) basis.
* Graphical

The differences between these estimates and the 'actuals' result from dispersion variances for the SK estimates of 0.31 to 0.36 compared to the follow-up variance of 0.35 (strip 1, i.e. 1A+1B) to about 0.5 (strips 2 and 3), i.e. an apparent remaining smoothing effect. However, the 'actual' dispersion variance is too high due to the presence of a low information effect resulting from an average of some 12 values inside each follow-up block. The SK estimates cover a higher information effect and a correspondingly lower dispersion variance effectively in line with the actual position during production when selections are restricted to values external to the blocks. *This stresses the importance of all post-processing procedures to take proper account of a realistic information effect.*

STRIP	Correlation Coef. with 'actuals'			Regr. Slope		
	S K pre	Sim Avgs	first simn	SK pre	Sim Avgs	first simn
1A total	0.645	-0.113	-0.147	1.108	-0.315	-0.100
1Btotal	0.436	-0.111	0.282	0.854	-0.458	0.242
1A+1B	0.548	-0.110	0.061	1.034	-0.364	0.046
2	0.309	-0.046	-0.048	0.644	-0.223	-0.039
3	0.380	-0.022	0.050	1.380	-0.143	0.046
1+2 total	0.431	-0.077	0.010	0.860	-0.302	0.008
1/3 total	0.392	-0.057	0.028	0.956	-0.269	0.023

Table 4: Showing correlations and regression slopes for SK and Sims vs. 'actuals' on Ln(x+b) basis.

- *Repeated simulations – block averages for 50 iterations*

The simulation approach recently proposed by McLennan and Deutsch (2004) was applied to the point data from the database and using the corresponding semi-variogram. The results from 50 iterations were first averaged for each block to provide the 'smoothed' block tonnage grade curve shown on Fig. 3 and the correlation results in Tables 3 and 4. *Unlike the original SK un-processed block estimates, these simulation averages do not meet the requirement for the effective absence of conditional biases.*

- *Repeated simulations on a probability basis.*

In this approach the 50 simulated grades for each set of iterations are accepted as a probability model for each block. The resultant global tonnage grade curve and results, shown in Fig 4 and in Table 3, serve the purpose of comparison with the 'actuals' and SK with post-processing above. In this case the dispersion variance for the estimates of 0.43 compares well with that of 0.47 for the 'actuals'. However, a study of the sets of simulation distributions show a departure from the 3-parameter model used for the other block estimates and 'actuals'. To overcome this problem, the untransformed mean and variance for each block were used to calculate the theoretical equivalent 3-parameter variances for the simulation distributions with the same untransformed means and variances. These averaged 0.432 for strips 1 to 3 which agrees reasonably well with the 'actuals'. Note that *the information effect did not feature in the simulation process, and effectively produces results with a variance close to that for perfect block valuations, i.e., the grades indicated can be too optimistic*

5. Main Conclusions For Global Estimates

The techniques covered in paragraph 4 above, other than SK processed and Simulations on a probability basis, cannot be recommended and leave only the latter two for further consideration.

Note that the repeated simulations do little to distinguish between individual blocks for the guiding of the selection process in mine planning in advance of actual selection during production. At the other extreme end the position for a single simulation for individual blocks shows virtually no correlation with the 'actuals' and serious conditional biases (coefficient = 0.028, and regression slope = 0.023, see Table 4, first simulation). *The latter highlights the danger of selection of any single simulation realization, e.g., median realization, for mine planning.*

5.1 RELATIVE ECONOMIC PROFITS

The two techniques have also been compared with the actuals on the basis of an elementary financial analysis of *'relative profits'* defined as:

(tons above cut-off) x (grade above cut-off – cut-off)

and the results are shown in Fig 5 for the global position for the follow-up area. There is a reasonable agreement of both approaches with the 'actual'. *The critical conclusion is that globally the two approaches can produce results very close to those for the follow-ups. The situation for individual blocks and local small areas will be discussed in the following paragraph.*

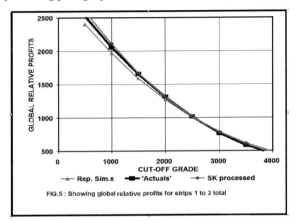

FIG.5 : Showing global relative profits for strips 1 to 3 total

6. Results for Subdivisons of the Global Area

Paragraphs 3 and 4 cover the global position for the whole follow-up area. In order to focus on the estimates and grade control problems specifically for short-term production, the follow-up area was split into 3 main strips as shown in Fig. 1. Strips 1A and 1B cover the first and second 20m extensions into the estimation area, with measured resources and mining periods of approximately 4 months each. Strips 2 and 3 cover further extensions of 40m and 80m respectively, both with indicated resources and with mining periods of 8 months to 16months and 16 to 32 months respectively.

The results for the 'actuals' and the 2 techniques remaining for further consideration have been further examined for the sub-divisions represented by the individual strips 1 to 3 (See Tables 3 and 4 and Fig. 4). The individual strips shows fairly stable dispersion variances for 'actuals', simulation averages and SK pre processed, but a decline for the latter two in their correlation levels with the 'actuals' (see Tables 3 and 4). This is to be expected as the distance of known data accessed for block valuations increases steadily from strip 1 to strip 3. *The regression slopes are heavily in favour of the SK pre-processed estimates* against negative slopes for Simulation averages (-0.3 to -0.46, i.e. serious conditional biases).

The post-processed versions of these two sets of estimates cannot be directly correlated with the 'actuals' in the light of their probability nature as distinct from the specific SK pre-processed and the simulation average figures for individual blocks and sections. However, the general tenor of the latter figures should carry through to the 'post-processed' versions.

For this reason the position for strip 1A was analysed in some detail and is summarized in Figs. 6A and 6B. The note under par. 5 above applies particularly to this strip when the two figures are compared for simulation averages and SK pre-processed estimates vs. 'actuals'. *Fig 6A demonstrates clearly that the simulation averages per block provide effectively no correlation with the 'actuals' and show maximum conditional biases; i.e. no contribution to the problem of doing selective planning on a block basis in advance of the final selection when more data will be available. In contrast to simulations the SK pre-processed results in Fig. 6B show a reasonable correlation level of 0.65 and virtually no conditional biases.* The general tenor of these figures in fact does carry through to the 'post-processed' versions, and is confirmed by the relative profits for these 2 techniques vs. 'actuals' as demonstrated in the correlation graphs in Fig 7A and 7B.

The results for the simulation averages and the corresponding probability estimates are evidently due to the fact that on the South African gold mines virtually no conditioning data are available for block estimates as these estimates are done essentially on an extrapolation basis. *The simulations are thus not fully conditional and should only be used for global patterns and tonnage/grade curves.*

7. Overall Conclusions

This paper stresses the main principles in geostatistical applications to mine resource valuation, which should be accepted and practiced by all concerned:

i) Where at all practical, alternative techniques and their detailed procedures followed, such as various kriging approaches with a choice of search routines, simulations etc., should be compared using an actual data base. This will provide actual follow-up data for correlations with the estimates, including measures of comparable efficiencies. Where such actual data are not available, a suitable simulation base could be used. It is disturbing that the geostatistical literature over many years of outstanding achievements, have provided very few such studies. They could have eliminated many misunderstanding between some practitioners.

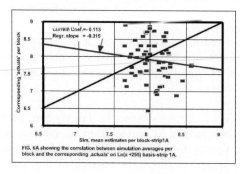

FIG. 6A showing the correlation between simulation averages per block and the corresponding 'actuals' on Ln(x +255) basis-strip 1A.

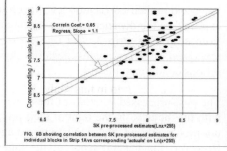

FIG. 6B showing correlation between SK pre-processed estimates for individual blocks in Strip 1A vs corresponding 'actuals' on Ln(x+255)

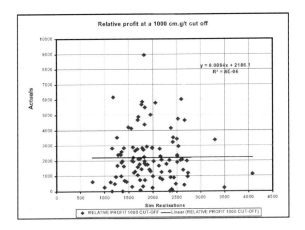

Figure 7A: Relative Profit at a 1,000cmg/t cut-off (simulation)

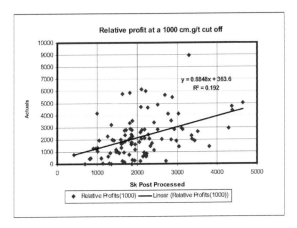

Figure 7B: Relative Profit at a 1,000cmg/t cut-off (SK post-processing)

ii) The principle of conditional unbiasedness, which gave rise to the birth of geostatistics more than fifty years ago, is still valid today. This principle cannot be reconciled with any unsmoothed estimates such as OK with limited search routine. Any new technique, however sophisticated, must be tested on a practical follow-up study, as mentioned above. The only solution at this stage is via some form of probability estimates, as used in this study.

This paper shows that, for global estimates, there is little to choose between SK post-processed and simulation probability estimates, particularly for deep level mining where blocks are valued largely on an extrapolation basis. For short-term individual block estimates, however, kriging with post-processing shows a distinctive advantage over repeated simulations. For any mine, where some advanced drilling is available at the resource valuation stage, a detailed practical study, similar to this, seems necessary to compare the alternative techniques at various levels of data densities, particularly for individual short term block estimates. However, an earlier pilot study (Assibey-Bonsu and Krige, 1999a), but not on the same scale and basis, indicated similar results as shown in this paper.

8. Acknowledgements

Acknowledgement is made to Gold Fields Limited, for permission to publish this paper.

9. References

Assibey-Bonsu, W. and Krige, D.G. (1999a) Practical *Problems in the estimation of recoverable reserves when using Kriging or Simulation Techniques.* International Symposium on Geostatistical Simulation in Mining, Perth Australia, October 1999

Assibey-Bonsu, W. and Krige, D.G. (1999b) Use *of Direct and Indirect Distributions of Selective Mining Units for Estimation of Recoverable Resource/Reserves for New Mining Projects,* APCOM'99 International Symposium, Colorado School of Mines, Golden, October 1999.

Assibey-Bonsu, W. and Krige, D.G.(2003). *An analysis of the practical and economic implications of systematic underground drilling in deep South African gold mines.* APCOM 2003 International Symposium, Cape Town, May 2003 (SAIMM).

Journel A.G., Kyriadkidis P.C., and Mao S. (2000). *Correcting the smoothing effect of estimators: a spectral postprocessor.* Mathematical geology, Vol. 32, No7, October 2000.

Deutsch C. V. and Journel A.G. (1992). *Geostatistical Software Library and User Guide.* Oxford University Press, 1992., pp340

Krige, D.G. (1951). *A statistical approach to some basic mine valuation problems on the Witwatersrand :* J. of the Chem. Metall. and Min. Soc. of S.A. December, 1951 - discussions and replies March, May, July and August 1952.

Krige, D.G. (1960). *On the departure of ore value distributions from the log-normal model in South African gold mines.* J.S.A.I.M.M., November 1960, January and August 1961.

Krige, D.G. (1962). *The application of correlation and regression techniques in the selective mining of gold ores.* 2nd APCOM Symposium, University of Arizona, April, 1962.

Krige, D.G. (1996). *A practical analysis of the effects of spatial structure and data available and used, on conditional biases in ordinary kriging* - 5th International Geostatistics Congress, Wollongong, Australia, 1996.

Krige, D.G. (1997). *Block Kriging and the fallacy of endeavouring to reduce or eliminate smoothing. Keynote address*, 2nd Regional APCOM Symposium, Moscow State Mining University, August 1997

Krige, D.G. (2001). *Comment on paper by Journel and others on a Spectral Postprocessor.* Mathematical Geology,Vol 33, No.6, 2001.

McLennan, J.A. and Deutsch, C.V. (2004*) Conditional Non-Bias of Geostatistical Simulation For Estimation of Recoverable Reserves.* Canadian Inst. of Min. and Met. (CIM) Bulletin, May 2004.

THE PRACTICE OF SEQUENTIAL GAUSSIAN SIMULATION

MAREK NOWAK[1] and GEORGES VERLY[2]
[1] *Nowak Consultants Inc. 1307 Brunette Ave, Coquitlam BC V3K 1G6*
[2] *Placer Dome Inc. 1055 Dunsmuir St, Vancouver BC V7X 1P1*

Abstract. The theory of simulation is relatively well documented but not its practice, which is a problem since simulation is not as robust as linear estimation. As a result, many costly mistakes probably go undetected. In this paper, a process for simulation is introduced with the objective of reducing the likelihood of such mistakes. The context is sequential Gaussian simulation within the mining industry. However, a significant part of the process can be applied in other simulation framework.

Four of the most important aspects of the process are discussed in detail. A gradual trend adjustment is suggested as a post-simulation step. A modified bootstrap approach is presented to deal with the grade uncertainty that accounts for spatial dependence between the samples. A number of pre- and post-simulation checks are also discussed. Some post-simulation adjustments of the simulated values are suggested to improve on the quality of the simulation.

All of the approaches, solutions and checks presented in this paper are simple, flexible, and can be easily implemented by a practitioner.

1 Introduction

Sequential Gaussian simulation starts by defining the univariate distribution of values, e.g., assay grade values, performing a normal score transform of the original values to a standard normal distribution, and assuming multi-normality of the normal scores. The multi-normal assumption ensures that the conditional distribution at a given location is normal with mean and variance provided by simple kriging (SK). Simulation of normal scores at grid node locations is done sequentially most often with SK using the normal score variogram and a zero mean (Isaaks, 1991, Deutsch and Journel, 1998, Goovaerts, 1997). Once all normal scores are simulated, they are back-transformed to original grade values.

Although the simulation methodology is well documented, a practical process leading to valid and representative realizations of in-situ grades is rarely a focus of attention within the geostatistical community. To a practitioner, this can lead to frustration in applying a methodology that may produce poor results. There is a need for a simulation procedure that is systematic, robust and easy to follow. This need led Placer Dome to design a process for sequential Gaussian simulation. Note that a significant portion of the process can be applied to other simulation algorithms (see Figure 1).

Figure 1 shows that the process is more complex than just normal score transformation, variogram modeling, simulation and back-transformation. A number of steps have been added to improve on the span of uncertainties, trend reproduction, reproduction of data distribution, reproduction of a variogram model, reproduction of correlated variables, and choice of optimistic and pessimistic scenarios. Although some of these steps have yet to be implemented, generally the process is closely followed by Placer and is described in Nowak and Verly (2004).

This process is based on specific difficulties encountered during real case studies, in particular:

- Simulated values may not adequately follow general trends, especially away from data locations.
- Bootstrapped distributions may be almost identical when created from large data sets.
- Average and/or variability of simulated data may be substantially different from average and/or variability of the conditioning data.
- Variograms of simulated values may be different from the variogram models.

This paper is a detailed discussion of the following portion of the process:

- Trend analysis.
- Bootstrap grades.
- Check/Adjust simulated normal scores (histograms and variograms).
- Check/Adjust distribution of simulated grades.

2 Trend analysis

Trends are not always well reproduced in sequential Gaussian simulation (Steps I.1.2 and I.1.8 in Figure 1a). This is because of the stationarity assumption necessary for the normal score transform and the assumption of a constant zero mean in the SK algorithm. One simple way to deal with this problem is to filter the trend and simulate the residuals of the original values (Deutsch, 2002). Unfortunately, this solution may produce simulated grade values that are negative. An obvious way out is to reset the negative values to zero, but this may result in a significant bias and poor reproduction of the trend.

A second solution consists in defining the local prior means to be used by SK with a correction factor for all kriging variances (Goovaerts, 1997; Deutsch, 1998). This solution was not tried by the authors but it is suspected that it may lead to difficulties in the reproduction of the original values distribution and it does not address the fact that the normal score transform is global within a geological domain.

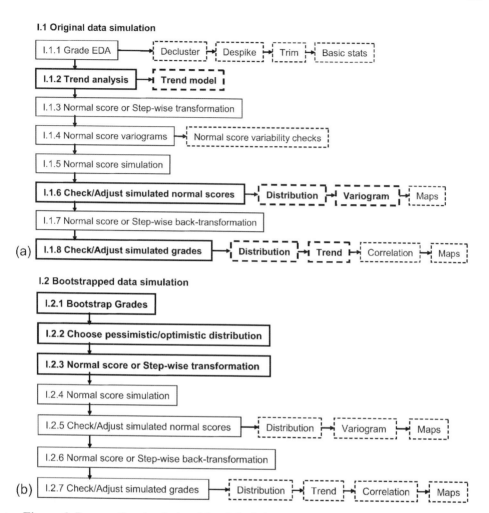

Figure 1. Process for simulating (a) original data and (b) bootstrapped data. The topics discussed in this paper are highlighted. The other process steps are discussed in Nowak and Verly (2004).

A third solution is given by Leuangthong and Deutsch (2004) who suggest a step-wise normal score transform. The method consists in defining the trend and residuals followed by a normal score transform of the residuals conditional to the trend. In practice, the residuals are classified according to a series of trend value intervals and there is one standard normal score transform of the residual per interval. This method is very promising because the normal score transform is conditional to the trend. The method ensures that there is no trend in the normal score space and that a proper normal score variogram is used. Finally, the method greatly reduces the number of negative grade values after the step-wise back-transform.

This method can be modified to a transformation of the original values conditional to the trend instead of residuals conditional to the trend, which would ensure that there are no negative grades after back-transformation.

Although the step-wise normal score transform is very promising, other solutions for trend reproduction have been tried by the authors. These solutions rely on a definition of a trend at all grid locations and on the average simulated model. It is assumed that the trend represents a relatively smooth surface and can be assessed by OK with a high nugget effect. An example of the trend values compared with the original data is presented in Figure 2a. The first attempt consisted of filtering the trend and simulating residuals. This approach, however, was abandoned because of a significant amount of simulated negative grades. Other attempts were made to correct for the trend of the simulated normal score values or the back-transformed simulated values. The best results have been obtained by adjusting back-transformed values according to:

$$Sim_{tr}(x) = Sim(x) \cdot w(x) \tag{1}$$

where $Sim(x)$ is the simulated value at location x before the trend adjustment, $Sim_{tr}(x)$ is the simulated value after trend adjustment, and $w(x)$ is a correction factor calculated as follows:

$$w(x) = (c(x) - 1) \cdot v(x) + 1$$
$$c(x) = Tr(x) / Avsim(x)$$
$$v(x) = \sigma_{kr}(x) / \sigma_{kr\,max}$$

where $Tr(x)$ is the trend value at location x, $Avsim(x)$ is the average simulated value, $\sigma_{kr}(x)$ is the kriging standard deviation and $\sigma_{k\,rmax}$ is the maximum kriging standard deviation at any given node.

The kriging standard deviation $\sigma_{kr}(x)$ affects the amount of the adjustment. If a simulated node is very close to conditioning data then $v(x) \cong 0$ and no adjustment is made. On the other hand, a maximum adjustment is made far from data locations. Note that a similar progressive correction, i.e., a correction dependent on the distance from the data, has been discussed by Xu (1997). The advantages of the approach are:

- The average of simulated values is similar to the trend, in particular away from data locations.
- The coefficients of variation of the simulated values before and after the correction have been observed to be quite similar in practice.
- The correction is simple and can be done on already simulated values.
- The correction is flexible in the sense that $\sigma_{k\,rmax}$ can be replaced by an arbitrary value.

The disadvantage of the approach is the difficulty to infer the trend everywhere, in particular far from data locations.

Figure 2b shows a comparison of the trend with the average simulation before the trend adjustment, and Figure 2c presents the comparison after the adjustment for the trend. Clearly, there is a substantial improvement in the reproduction of the trend when the adjustment is made.

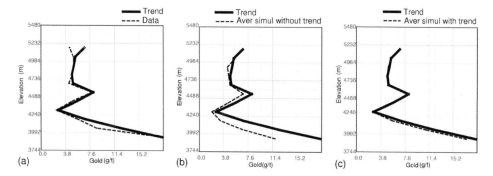

Figure 2. Comparison of the trend (solid line) along elevation with (a) conditioning data, (b) average simulation before trend adjustment, (c) average simulation after trend adjustment

3 Check/Adjust simulated normal scores

Post-simulation checks are necessary to ensure a reasonable reproduction of the distribution and spatial correlation (Step I.1.6 in Figure 1a). In a first step both the histogram and the variogram of the simulated normal scores are checked against the original normal score histogram and variogram. All the realizations should be considered at the same time for the checks to avoid natural fluctuations between the realizations. The verification of the results should take place within the same zone that has been used to get the simulation parameters, i.e., the declustered grade distribution, the normal score transform, and the normal score variograms (Figure 1a, Steps I.1.1, I.1.3, and I.1.4).

3.1 HISTOGRAMS

The simulated normal score histogram check may reveal that the simulated distribution is not standard normal. This section discusses (1) the case of the average of the simulated values different from 0.0, (2) the case of the variance of simulated values different from 1.0, and (3) a gradual adjustment of the simulated value to a standard normal distribution.

3.1.1 Simulated normal score average different from 0.0.

The difference may result from an improperly defined validation zone, i.e., the zone within which the simulation results are validated. Usually, this zone should be similar to the zone within which the simulation parameters (histogram, normal score transform, variogram) are calibrated. The difference may also result from an improper declustering of the original distribution. Two possible solutions are:

- Modification of the validation zone. If for example the non-zero average is due to a significant amount of simulated values at some distance from the conditioning data, and at the same time conditioned to low assays on the edges of the drilled out area, a modification of the validation zone that excludes areas far from conditioning data may reduce significantly the difference observed. Figure 3 illustrates the impact of such a modification of the validation zone. Here, in the original validation zone the average of simulated values is -0.11 but in the modified validation zone the average is -0.02 which is close to the 0.0 data average. The modified validation zone is limited to the area close to the conditioning data, extending not further than a search radius used for polygonal declustering.
- Adjustment to declustering weights. If a polygonal declustering is used, the search radius may be inappropriate. In other words, the original data distribution has not been properly defined.

If the source of the difference is unknown and there is reason to believe that the original distribution mean (= 0.0) is correct, the simulated values may have to be adjusted as per sub-section 3.1.3.

3.1.2 Variance of simulated values is different from 1.0

As for the average, the difference in variance may result from an improper validation zone or improper declustering and the solutions proposed earlier for correcting the average may be applied.

Another reason for a difference in variance is a possible inconsistency between the normal score transform and the normal score variogram. By construction, the normal score conditioning values are standard normal within the zone of interest Z (e.g. one geology domain within the validation zone), which means that the dispersion variance of the normal scores within Z is 1.0, i.e.:

$$D^2(0|Z) = \overline{\gamma}(Z,Z) = 1.0 \qquad (2)$$

where $\overline{\gamma}(Z,Z)$ is the average normal score variogram value within Z. The normal score variogram fit should be consistent with the above equality, which means that the variogram sill should be larger than one if the zone Z is not very large with respect to the variogram range, as it can be in the case of local grade control.

In practice, the variogram is often fitted first with a sill of one (Figure 1a, Step I.1.4). The value of $\overline{\gamma}(Z,Z)$ should then be computed. If the $\overline{\gamma}(Z,Z)$ value is within 5% of one, a simple rescaling of the variogram values is reasonable, otherwise a variogram model adjustment (sill and range) is suggested (Figure 4).

3.1.3 Gradual adjustment of simulated normal score average and variance.

If the source of the difference in mean (≠ 0.0) and/or variance (≠ 1.0) is unknown and there is reason to believe that the original N(0,1) distribution is correct, the simulated values may have to be adjusted. The following approach is a progressive correction that depends on the distance of the simulated node from the conditioning data.

First, a maximum possible adjustment at a given node $Sim_{tr\,max}(x)$ is defined by a simple standardization (mean = 0 and variance = 1):

$$Sim_{tr\,max}(x) = (Sim(x) - Av_{Gsim}) / \sigma_{Gsim}$$

where $Sim(x)$ is the original simulated value at location x, Av_{Gsim} is the global average of all simulated values and σ_{Gsim} is the global standard deviation of all simulated values.

The actual adjustment $Sim_{tr}(x)$ is defined as follows:

$$Sim_{tr}(x) = (Sim_{tr\,max}(x) - Sim(x)) \cdot ratio(x) + Sim(x)$$
$$ratio(x) = \sigma_{sim}(x) / \sigma_{max\,sim}$$
(3)

where $\sigma_{sim}(x)$ is the standard deviation of the simulated values at the selected node, and σ_{maxsim} is the maximum standard deviation of the simulated values from all nodes.

Note that for a node located on a conditioning data, $ratio(x)=0$ and $Sim_{tr}(x)=Sim(x)$, i.e., there is no correction. As the node gets further from the conditioning data, the value of $ratio(x)$ gradually increases from zero up to one, and the value of $Sim_{tr}(x)$ gradually varies from $Sim(x)$ to $Sim_{tr\,max}(x)$.

As shown in Figure 5, this adjustment results in a modification of both the average and the variance of the simulated values. Note that the adjustment does not result in an average and variance equal to 0 and 1 respectively, but there is a substantial improvement. Note also that the adjustment described in this section is a gradual affine correction that will not correct the shape of the distribution. If it is necessary to also correct the shape of the distribution (i.e., adjusting to a N(0,1) distribution), then a more sophisticated approach can be used (Xu, 1994).

3.2 VARIOGRAMS

The variograms of the simulated values can deviate from the modeled variograms. A deviation from the original model may adversely impact the simulation results, especially when the focus of the study is on variability of the mined blocks. The difference between the simulated and modeled continuities (variograms) may be caused by (1) poorly fitted variograms, (2) a modeler's decision to fit according to geological interpretation, and (3) unknown reason.

The first two sources of differences are counter-acted by data conditioning. Regardless of the original variogram model, continuities of the experimental data are to some extent imprinted on the simulated continuities, especially when there are lots of data as in mining. The impact of the data can be checked by comparing conditional and unconditional simulations. If deemed necessary, both variogram model range and sill may be adjusted to achieve the desired results. As shown in Figure 6 the adjustment to the variogram model results in improved, albeit not perfect, continuities of the simulated values.

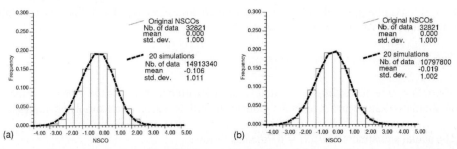

Figure 3. Comparison of simulated values with the data: (a) validation domain identical to simulation zone, (b) validation domain extending not further from the data than a search radius used for polygonal declustering

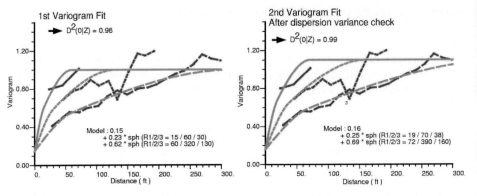

Figure 4. Example of variogram models before and after normal score variability check. a) Variogram model with total sill of 1.0 results in a dispersion variance within the validation zone of 0.96. b) Modified model with total sill of 1.10 results in a dispersion variance of 0.99.

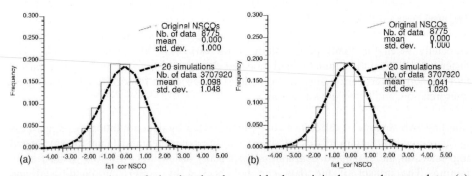

Figure 5. Comparison of simulated values with the original normal score data: (a) before the correction (b) after the correction of both average and variance

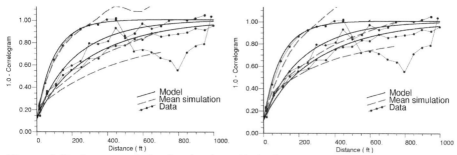

Figure 6. Variograms of simulated values (dashed curves) compared to the variogram models (solid curves) and to the experimental variograms (bulleted line) along different directions. Before correction for the sill and ranges (a), the simulated value variogram show more continuity than the experimental variograms. After correction (b), there is a better match between the two.

4 Check/Adjust simulated grades

All checks and sometimes the adjustments made in normal score space are necessary but not sufficient to ignore the checks on the simulated values after back-transformation (Step I.1.8 in Figure 1a). This is especially true because of a potential compounding effect of the corrections made. Although the writers are not aware of significant problems related to the series of corrections, their effect on the final simulated grades should be studied. Comparisons should be made with the original data within the validation envelope. Histograms, probability plots, scatterplots and visual checks of maps of simulated values are useful tools. Care should be given to ensure that the simulated mean grade in a geological domain is similar to the average estimated grade in that domain. If they are different, the simulated grades may have to be adjusted either by modifying some pre-simulation parameters, such as a trimming value, and re-simulating, or by a simple adjustment of the simulated values to the required average.

5 Bootstrap grades

Two main levels of uncertainty can be identified: geological (rock types) and grade uncertainty. Only the grade uncertainty is discussed in this section, but the same discussion applies to the geological uncertainty.

Current simulation practice often relies on the assumption that the distribution of in-situ grade values is known from the declustered grade histogram. The additional risk associated with an imperfect knowledge of the actual grade distribution should be addressed, resulting in better reproduction of the space of uncertainty.

Using a bootstrapping methodology, statistical fluctuations can be investigated by sampling from the original distribution (Steps I.2.1, I.2.2, and I.2.3 in Figure 1b). A typical procedure consists in creating a series of possible datasets by drawing randomly with replacement as many values, with the attached declustering weights, as there are in the original distribution. The fluctuations between the various datasets are then investigated.

When there are many sample values, such as in mining, the classical bootstrap approach results in datasets that are very similar to each other. This similarity would be perfectly correct if the sample values were uncorrelated, but this is not the case in a typical mining situation.

Spatial correlation can be addressed by drawing fewer values from the original distribution (Srivastava, pers. comm.). Indeed, the variance of the mean grade is:

$$Var1(\text{Mean}) = \frac{1}{N^2} \sum \sum C_{ij}$$

where C_{ij} is the covariance for the distance between sample i and j, and can be deduced from the variogram.

If P values are drawn randomly from the original dataset, the variance of the mean is:

$$Var2(\text{Mean}) = \frac{1}{P} Var(\text{Data})$$

where $Var(\text{Data})$ is the variance of the original data set.

The required fluctuation for the mean is achieved if P is chosen such that $Var2(\text{Mean}) = Var1(\text{Mean})$, then:

$$P = \frac{Var(\text{Data})}{Var1(\text{Mean})}$$

Note that this formula could be refined to account for declustering weights.

Figure 7 illustrates the impact of bootstrap on the possible means of the original distribution. If no bootstrap is applied, the standard deviation of the mean is zero, i.e., the mean is fixed (Figure 7a). If the classical bootstrap is applied, the standard deviation of the means is 0.09 (Figure 7b). If the spatial bootstrap is applied, the standard deviation of the means increases to 0.036 (Figure 7c).

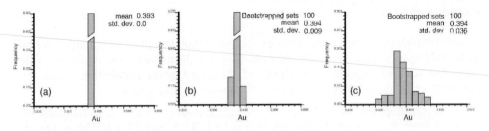

Figure 7. (a) Data mean - no bootstrap. (b) Typical bootstrap - mean distribution. (c) Spatial dependence bootstrap - mean distribution.

The bootstrapping may be done on data from all geological domains or on data from one domain at a time. If the former is used, the choice of optimistic (high average) and pessimistic (low average) distributions is more difficult, because the distributions from one or two domains may influence the results. The authors feel that bootstrapping per domain is a better solution. Under those circumstances, a pessimistic/optimistic distribution is truly pessimistic/optimistic in all domains. Of course, care should be given when choosing the bootstrapped distributions for simulating the grades. The distributions should not be overly pessimistic or optimistic. The choice of pessimistic/optimistic distributions can be limited to a specific area, or can be based on low/high metal content or NPV.

Prior to the final choice of the optimistic and pessimistic scenarios, it may be useful to have some insight on the potential impact of that choice on simulated values. Applying a cut-off grade on the bootstrapped distribution corrected for change of support may provide such insight.

Once a bootstrapped distribution is chosen, it is used first to generate a standard normal score transform (Figure 8a). The bootstrapped distribution and its transform are then used to convert the original grade values to normal score values (Figure 8b). The cumulative frequencies of the original sample grades are deduced from the bootstrapped distribution, then used to get the corresponding normal score values. Note that the resulting normal score values are not standard normal. For example, in the case of an optimistic bootstrapped distribution, the average of the normal score values of the original grade values is less than zero.

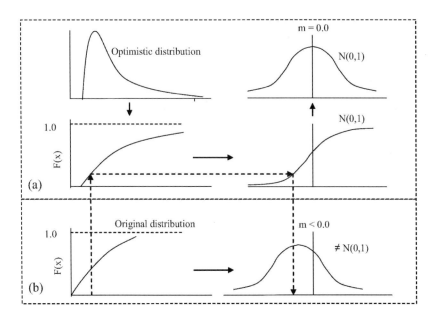

Figure 8. (a) Standard normal score transform based on a bootstrapped distribution. (b) Original grade distribution converted to normal scores using the standard normal score transform of the bootstrapped distribution.

6 Conclusions

A process for sequential Gaussian simulation is presented, which contains more steps than the usual normal score transformation, variogram modeling, simulation and back-transformation. A significant portion of the process may be used for other simulation methods, such as sequential indicator simulation. The authors believe that using similar processes in the mineral industry would avoid many costly mistakes.

Four of the most important aspects of the process are discussed in detail: trends, bootstrapping, checks, and adjustment of the simulated values.

Sequential Gaussian simulation often fails to correctly reproduce trends because of its strong stationarity requirement. A simple, albeit approximate, solution consists of adjusting for the trend after the simulation, via a gradual correction that depends on the distance to the conditioning data.

It is important that the simulation correctly reproduces the space of uncertainty. A modified bootstrap approach is presented to deal with the grade uncertainty. The modification is made to account for the spatial dependence between the samples. A similar approach can also be used to deal with the geological uncertainty.

To ensure high quality of the simulated values, a number of validation checks at different stages of the simulation are necessary. The checks start at the pre-simulation stage when experimental dispersion variances are compared against their theoretical values. Next, a series of checks followed by possible adjustments are done in the normal score space, and later similar checks and the adjustments are completed after back-transformation and trend addition. Most of the checks simply consist of comparing simulation and conditioning data statistics within a validation envelope. The potential adjustments are progressive, depending on the distance to the conditioning data.

References

Deutsch, C.V. and A.G. Journel, 1998, *GSLIB: Geostatistical Software Library and User's Guide*, Oxford University Press, New York, 380 pp.
Deutsch C.V., 2002, *Geostatistical Reservoir Modeling*, Oxford University Press, New York, 376 pp.
Goovaerts, P., 1997, *Geostatistics for Natural Resources Evaluation*, Oxford University Press, New York, 467 pp.
Isaaks, E.H., 1991. *Application of Monte Carlo methods to the analysis of spatially correlated data*, Unpublished PhD thesis, Stanford University.
Leuangthong O. and Deutsch, C.V., 2004. Transformation of Residuals to avoid Artifacts in Geostatistical Modelling with a Trend, *Mathematical Geology*, Vol 36, No 3, p. 287-305.
Nowak M., and Verly G., 2004. A Practical Process for Simulation, with Emphasis on Gaussian Simulation, Submitted to Orebody Modelling and Strategic Planning 2004 Symposium, Perth, Australia.
Xu, W., and Journel, A.G., 1994. Posterior identification of histograms conditional to local data. In *Stanford Center for Reservoir Forecasting Report* (SCRF) 7. Stanford Center for Reservoir Forecasting, School of Earth Sciences, Stanford, CA, USA

SPATIAL CHARACTERIZATION OF LIMESTONE AND MARL QUALITY IN A QUARRY FOR CEMENT MANUFACTURING

J. ALMEIDA[1], M. ROCHA[1] & A. TEIXEIRA[2]
[1] CIGA, Centro de Investigação em Geociências Aplicadas, FCT/UNL, Monte de Caparica, 2829-516 Caparica, Portugal. ja@fct.unl.pt
[2] SECIL, Outão, Portugal

Abstract. The aim of this study is to characterize the quality of the limestone and marl raw material exploited in a quarry for cement manufacturing by the SECIL Company (southern Portugal) based on the spatial distribution and variability of the chemical components (SiO_2, Al_2O_3, Fe_2O_3, CaO and MgO).
The first step of this study consists of the construction of sets of simulated images of these chemical components, using the Direct Sequential Simulation and Co-simulation algorithms. In the second step, the simulated images are combined on the quality indices LSF (lime saturation factor), SIM (silica modulus), ALM (alumina modulus) and CS (lime and silica ratio) in order to estimate local distribution laws of these indices. The local uncertainty and the probability of occurrence of extreme values are a tool of prime importance for the planning of temporal exploitation, regarding the proportioning optimisation mixture of raw materials coming from different quarry stopes.

1 Introduction

A set of parameters is currently used in cement manufacturing to characterize the quality of the raw material and to ensure the attendance of the quality of the produced cement. In Portugal, the SECIL Company uses four quality parameters (LSF – lime saturation factor; SIM - silica modulus; ALM – alumina modulus and CS – lime and silica ratio) and the magnesium grade (IPQ, 2001).
The LSF represents the relationship between the amount of calcium in the cement and the maximum amount theoretically possible for combining with other elements. It has a major influence in the manufacturing process and on the quality of the final product. An optimal LSF ranges between 1 and 1,02. It is calculated through the following relationship of grades, when expressed in weight percentage:

$$LSF = \frac{CaO}{1,8 SiO_2 + 1,18 Al_2O_3 + 0,65 Fe_2O_3} \quad (1)$$

The SIM is the second most important parameter to control the final product and it is calculated through the relationship between the grade of silica and the sum of the alumina and iron grades:

$$SIM = \frac{SiO_2}{Al_2O_3 + Fe_2O_3} \tag{2}$$

A high SIM has the advantage of producing cement with high content of silicates, consequently with high mechanical resistance. Optimal values range between 2,4 and 2,6.

The ALM represents the relationship between the alumina and iron in the raw material; values should range between 1,5 and 1,7.

$$ALM = \frac{Al_2O_3}{Fe_2O_3} \tag{3}$$

Also the relationship between the calcium and the silica (CS) should be higher than 2:

$$CS = \frac{CaO}{SiO_2} \tag{4}$$

Finally, the magnesium grade (MgO) should be below than 5% in weight.

Raw materials exploited in marl and limestone quarries are combined amongst themselves to obtain optimal mixtures and if necessary with additives so that the final product presents quality parameters within adequate ranges.

Control of quality is done as soon as possible, starting from the quarry stopes. Samples collected from regular meshes of holes are chemically analysed by fluorescence of X rays on five chemical components: SiO_2, Al_2O_3, Fe_2O_3, CaO and MgO. Based on this regular but scarce information the main objective of this study is to provide images of the most probable values of these indices on each stope as well as the global and local uncertainty.

2 Methodology

In this work a stochastic simulation methodology is presented to characterize the quality of raw material within each stope according to the described quality parameters. The characterization of each chemical component and calculation of indices *a posteriori* instead of a direct characterization of the indices is preferable once the spatial variability in the quarry is mainly related to the deposition of each component and not on the quality indices themselves.

The main goal of the proposed methodology is to produce sets of images of five medium to highly correlated components, following the steps:

1. Exploratory data analysis of the chemical components in study: SiO_2, Al_2O_3, Fe_2O_3, CaO and MgO;
2. Application of Principal Component Analysis (PCA) and selection of the Principal Components (PC) that explain most of the initial variance;
3. Calculation of experimental variograms for the chemical components and PC selected and fitting of theoretical models;
4. Correlation analysis between chemical components and PC selected (calculation of correlation indexes);
5. Stochastic simulation of N_s images of the PC, using the Direct Sequential Simulation (DSS). These simulated images are used as secondary variables in the following steps of the proposed methodology.

6. Stochastic simulation of N_s images for each of the chemical components using the Direct Sequential Co-simulation. These images are conditioned to the experimental measurements (primary variable) and to the simulated images of PC (secondary variables).
7. Making use of the N_s simulated images of the chemical components, calculation of N_s correspondent images for each quality index (SIM, ALM, LSF and CS), following formulas (1) through (4) node by node. At each node location, the set of N_s values constitutes an estimation of the local histogram of each quality index giving the most probable value and the uncertainty.
8. Construction of probability maps showing areas where the quality indices (SIM, ALM, LSF, CS and the MgO grade) exhibit values in the optimal intervals;
9. Upscaling of the values defined at a small-scale block (step 8) to the stope boundary block size.
10. Construction of final indicator maps delimiting areas where the quality indices exhibit values in the optimal intervals.

2.1 BACKGROUND OF DIRECT SEQUENTIAL CO-SIMULATION WITH A SET OF SECONDARY VARIABLES

The DSS algorithm (Soares, 2001) was applied to produce simulated images of the k selected PC, respectively $Z_{PC1}(x)$, $Z_{PC2}(x)$,... $Z_{PCk}(x)$. Next step consists of the co-simulation of the chemical components conditioned to the previous simulated images of $PC_1, PC_2, \ldots PC_k$ as secondary images.

Direct Sequential Co-simulation with a set of secondary variables constitutes an extension of the initial algorithm proposed by Soares, 2001, and can be summarized as follows (Almeida et al, 2002):
1. Define a random path visiting each node of a regular grid of nodes.
2. At each node x_u, simulate the value $z^s(x_u)$ using the DSS algorithm:
   a) Identify the local mean and variance of $z(x)$ in x_u location, $z(x_u)^*$ and $\sigma^2_{sk}(x_u)$, using the simple co-located kriging estimator with a multiple set of secondary variables:

$$z(x_u) = \sum_{\alpha=1}^{n} \lambda_\alpha z(x_\alpha) + \sum_{i=1}^{k} \lambda_{PCi} Z_{PCi}(x_u)$$

Using the matrix formalism, the simple co-located kriging system with two secondary variables collocated in x_u and n neighbourhood samples is defined as follows (for sake of simplicity using two PC, PC_1 and PC_2):

$$\begin{bmatrix} 1 & C_{12} & \cdots & C_{1n} & C_{1u}^{PC1} & C_{1u}^{PC2} \\ C_{21} & 1 & \cdots & C_{2n} & C_{2u}^{PC1} & C_{2u}^{PC2} \\ \vdots & & \ddots & \vdots & \vdots & \vdots \\ C_{n1} & C_{n2} & \cdots & 1 & C_{nu}^{PC1} & C_{nu}^{PC2} \\ C_{u1}^{PC1} & C_{u2}^{PC1} & \cdots & C_{un}^{PC1} & 1 & C_u^{PC1PC2} \\ C_{u1}^{PC2} & C_{u2}^{PC2} & \cdots & C_{un}^{PC2} & C_u^{PC2PC1} & 1 \end{bmatrix} \cdot \begin{bmatrix} \lambda_1 \\ \lambda_2 \\ \vdots \\ \lambda_n \\ \lambda_{PC1} \\ \lambda_{PC2} \end{bmatrix} = \begin{bmatrix} C_{1u} \\ C_{2u} \\ \vdots \\ C_{nu} \\ C_u^{PC1} \\ C_u^{PC2} \end{bmatrix}$$

Where:

$C_{\alpha\beta}$ - Covariance of the primary variable between samples at locations x_α and x_β

$C_{\alpha u}^{PC1}$ - Cross-covariance between primary variable at location x_α and PC_1 at location to estimate x_u

$C_{\alpha u}^{PC2}$ - Cross-covariance between primary variable at location x_α and PC_2 at location to estimate x_u

C_u^{PC1PC2} - Cross-covariance between secondary variables PC_1 and PC_2 at location to estimate x_u; equals zero.

λ_α - Weights of primary information

λ_{PC1} and λ_{PC2} - Weights of secondary variables

$C_{\alpha u}$ - Covariance of the primary variable between samples at locations x_α and location to estimate x_u

C_u^{PC1} - Cross-covariance between primary variable and secondary variable PC_1 at location to estimate x_u

and $\alpha=1...n$; $\beta=1...n$ (number of neighbouring samples of x_u).

b) Locally resample the histogram of $z(x_u)$, for instance using a normal score transform (φ) of the primary variable $z(x)$, and calculate $y(x_u)^* = \varphi(z(x_u)^*)$;
c) Draw a value p from a uniform distribution $U(0,1)$;
d) Generate a value y^s from $G(y(x_u)^*, \sigma^2_{sk}(x_u))$: $y^s = G^{-1}(y(x_u)^*, \sigma^2_{sk}(x_u), p)$;
e) Return the simulated value $z^s(x_u) = \varphi^{-1}(y^s)$ of the primary variable.

3. Loop until all nodes are simulated.

Assuming Markov-type approximation, the cross-covariance function can be calculated using the following relation in terms of covariance or correlograms (Almeida and Journel, 1994, Goovaerts, 1997):

$$C_{12}(h) \approx \frac{C_{12}(0)}{C_{11}(0)} C_{11}(h)$$

$$\rho_{12}(h) \approx \rho_{12}(0)\rho_{11}(h)$$

This approximation enables the inference of the primary variable is performed taking into account the spatial covariance of the primary variable and the correlation index between each secondary variable PC_1, PC_2, ... and the primary variable (ρ_{PC1} and ρ_{PC2}):

$$C_{\alpha u}^{PC1} = \rho_{PC1}.C_{\alpha u}(h) \text{ and } C_{\alpha u}^{PC2} = \rho_{PC2}.C_{\alpha u}(h)$$

2.2 ZONATION OF RAW MATERIAL IN THE QUARRY STOPES

Direct Sequential Co-simulation produces simulated images of the five chemical components on a small-scale block. The set of N_s simulated values constitutes an estimation of the local cumulative distribution function for each variable within each node.

For each small block centred on location x_u quality parameters values were computed using formulas (1) through (4) obtaining Ns local values. For each parameter p, the N_s locally calculated values could be classified according to an indicator variable $I^p(x_u; N_s)$:

$$I^p(x_u; N_s) = \begin{cases} 1 & \text{if } p(x_u, N_s) \in \text{optimal range for parameter } p \\ 0 & \text{otherwise} \end{cases}$$

where $p(x_u, N_s)$ is the value of parameter p in x_u calculated taking the N_s realization.

The average of the indicator values represent the proportion that small blocks belonging to the optimal range:

$$\frac{\sum_{N_s} I^p(x_u, N_s)}{N_s} = prob\ p(x_u, N_s)$$

Two situations remain to solve: a) upscaling the small-scale block calculations to large-scale blocks at the same size of the stope (50m x 4m x 20m height) and; b) transform the probability values $prob\ p(x_u, N_s)$ into indicator values delimiting good and poor quality zones.

For a large block $v(u)$ constituted by N_v small blocks, it is estimated the following distribution law, which represents the proportion in volume of the large block $v(u)$ belonging to the optimal range for parameter p:

$$F^*(v_u, p) = \frac{1}{N_s N_v} \sum_{i=1}^{N_s} \sum_{j=1}^{N_v} I^p(x_j, i)$$

The final step consists on the calculation of probability thresholds p_c to transform the proportion maps $F^*(v(u), p)$ into binary maps:

$$I(v_u, p) = \begin{cases} 1 & \text{if } F^*(v_u, p) > p_c \\ 0 & \text{otherwise} \end{cases}$$

The calculation of the threshold for parameter p, was based on the local and global probabilities of each large block to belong to each one of the categories (Soares, 1992, Almeida et al, 1993, Pereira et al, 1997).

3 Case study

The target area for exploitation is mainly constituted of grey and yellow limestone to marl (Kullberg et al, 2000). Orientation of the layers changes among N(80° to 90°)W in the most west area, N(65° to 70°)W in the central area and, approximately, N70°W in the east area. Dip varies with more sloping at west (50° to 60°) than at east (≈ 45°).

A set of samples was extracted from a regular mesh of 131 vertical holes (Figure 1) for an area of 8 hectares. The total length of each hole is 20 meters (height of the steps of the exploitation), with a spacing of 10 meters in the N-S direction and 50 meters in the E-W direction. Each sample is a mixture of powder rock representing an average of grades for about 20 meters height, that validates the sample values in the characterization of three-dimensional blocks with the same vertical height.

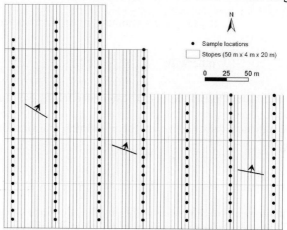

Figure 1. Spatial location of the samples in the studied area, orientation of the layers and design of stopes of 50 x 4 x 20 m³ height.

Univariate statistics and correlation indexes were calculated for all initial data considered. The results are summarised in Tables 1 and 2.
It is observed a high positive correlation among SiO_2, Al_2O_3 and Fe_2O_3. The correlation between CaO/SiO_2, CaO/Al_2O_3 and CaO/Fe_2O_3 is also high, although negative. The MgO is not correlated with the remaining chemical components, meaning that its deposition is independent from the remaining components. This evidence is premonitory that two main PC will be necessary to synthesize all the initial information.

	N° samples	Mean	Median	Variance	Skewness index
SiO_2		10,44	9,65	13,67	1,08
Al_2O_3		4,29	3,70	3,78	1,57
Fe_2O_3	131	2,11	1,90	0,49	1,48
CaO		41,60	41,47	13,11	-0,60
MgO		3,85	3,83	2,91	0,14

Table 1. Basic statistics of the initial dataset.

	SiO_2	Al_2O_3	Fe_2O_3	CaO	MgO
SiO_2	1,0000	0,9781	0,9468	-0,8631	-0,2403
Al_2O_3		1,0000	0,9586	-0,8161	-0,3103
Fe_2O_3			1,0000	-0,7653	-0,3566
CaO				1,0000	-0,2781
MgO					1,0000

Table 2. Correlation indexes between chemical components.

PCA algorithm was applied to synthesize the initial dataset to a reduced number of PC. Eigenvalues of the five axes and their explanation percentages are presented in Table 3. Figure 2 shows the graphical representation of the correlation indices between each initial variable and the PC and the projection of the 131 samples. According to Table 3 it is verified that first two PC transport more than 98% of the initial variance and that is enough to be used as secondary variables in the simulation of the images synthesizing the initial dataset.

Axis	Eigenvalues	Proportion of population variance (%)	Accumulated proportion (%)
1	3,711034	74,221	74,221
2	1,216618	24,332	98,553
3	0,052199	1,044	99,597
4	0,019663	0,393	99,990
5	0,000486	0,010	100,000

Table 3. Eigenvalues of each axis and proportion of population variance.

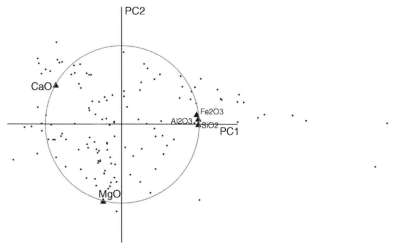

Figure 2. Graphical representation of the correlation indexes between initial variables and PC and sample projections on the two first PC.

	Model	C_1	a (N80°W)	a (N10°E)	Anisotropy
PC_1	Sph	0,74	200	30	6,67
PC_2		0,24	230	40	5,75
SiO_2		13,67	175	30	5,83
Al_2O_3		3,78	175	30	5,83
Fe_2O_3	Sph	0,49	250	35	3,33
CaO		13,11	125	20	6,25
MgO		2,91	200	60	5,83

Table 4. Models of variograms fitted to PC_1, PC_2 and chemical components.

Experimental variograms and fitting of theoretical models of spherical type (*Sph*) were made for the two selected PC and the five initial variables (see examples in Figure 3 and the parameters list in Table 4). Both PC_1 and PC_2 and all variables exhibit strongly anisotropic variograms (relationships between 3,33 and 6,67) where the main direction is related with the geological orientation of the layers.

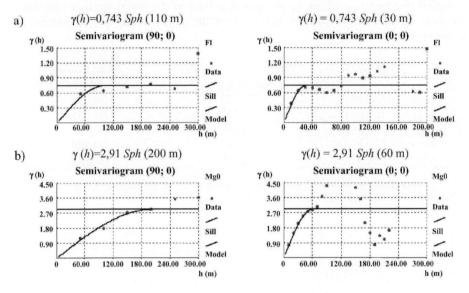

Figure 3. Experimental variograms and theoretical models fitted: a) PC_1; b) MgO.

The area in study was subdivided in a regular grid of 320 x 250 = 80000 small blocks with 1m by 1m length by 20 m height. Fifty images of PC_1 and PC_2 (to use as secondary variables) were simulated using DSS algorithm and fifty correspondent images of the initial variables were simulated using the proposed Direct Sequential Co-simulation algorithm conditioned to the PC_1 and PC_2 images.
The Figure 4 illustrates and example of a set of simulated images of PC_1, PC_2 and co-simulated images for each one of the initial variables. Each set of simulated images of the initial variables allows the calculation of a simulated image of the quality parameters as described in formulas (1) through (4). In order to validate the proposed method Table 5 shows the basic statistics for a set of simulated images.

	Number of blocks	Mean	Median	Variance	Skewness index
SiO_2		10,45	9,59	14.20	1,29
Al_2O_3		4,29	3,81	4,22	1,30
Fe_2O_3	80000	2,11	1,93	0,61	1,29
CaO		41,60	41,76	7,37	-0,71
MgO		3,86	3,82	3,30	0,90

Table 5. Basic statistics of one simulated set of images.

CHARACTERIZATION OF LIMESTONE AND MARL QUALITY IN QUARRY 407

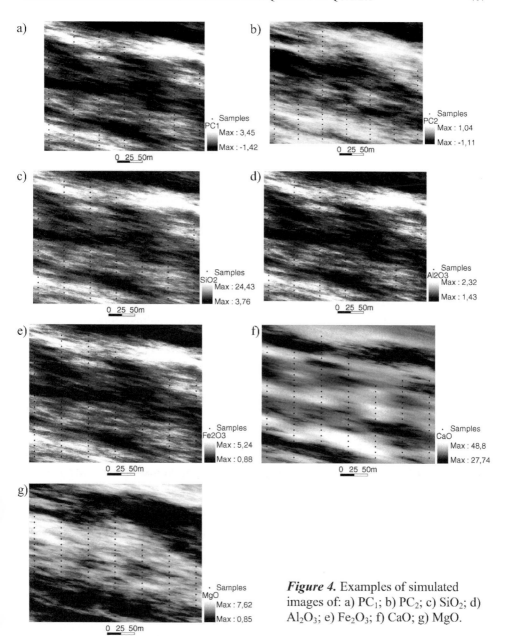

Figure 4. Examples of simulated images of: a) PC_1; b) PC_2; c) SiO_2; d) Al_2O_3; e) Fe_2O_3; f) CaO; g) MgO.

In order to upscaling all small blocks to a set of stope boundary blocks, 310 large blocks with 50 x 4 x 20 m³ each were digitalized in the studied level of the quary (Figure 1). In the sequence of the proposed methodology local probabilities of each block to present parameters in the class of adequate quality were calculated and final probability maps were classified as indicator maps. For illustrative purposes, local probabilities and the limits of best areas are presented for the LSF and SIM parameters in Figure 5.

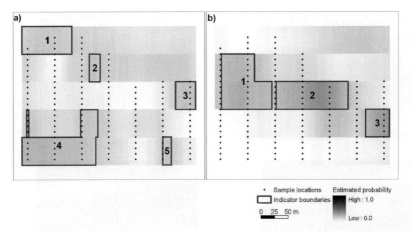

Figure 5. a) Probability of LSF ∈ [0,66 ; 1,02] and identified areas; b) probability of SIM ∈ [1,8 ; 3,0] and identified areas.

4 Conclusions

The presented case study shows a successful application of a multiple corregionalization simulation methodology, through the use of the main components of PCA as secondary variables and co-simulation of the main variables using these components as secondary variables. This methodology has the advantage of avoiding the problem of modelling multiple corregionalizations when a set of dependent variables is taken into account.

The final maps constitute an essential tool in the short-medium term planning of the exploration, allowing with a certain spatial resolution of the quality of the raw material exploited in each stope. The knowledge *a priori* of the most probable values and correspondent uncertainty of these chemical components and quality indexes in each exploitation step (local cumulative distribution functions), allow the optimal proportioning of raw materials, giving rise to a minimization of the costs namely addition of additives and stabilization of grades.

References

Almeida, A. and A. Journel 1994. Joint simulation of multiple variables with a Markov-type corregionalization model. Mathematical Geology 26(5), 565-588.
Almeida, J.; Soares, A. & Reynaud, R., 1993, Modelling the Shape of Several Marble Types in a Quarry, Proceedings of the XXIV International Symposium APCOM, Montreal, 3: 452-459.
Almeida, J.; Santos, E. Bio, A., 2002, Use of geostatistical methods to characterize population and recovery of Iberian hare in Portugal, Submitted to the Fourth European Conference on Geostatistics for Environmental Applications, Barcelona
Goovarets, P., 1997, Geostatistics for natural resources evaluation, Oxford University Press, New York, 483 p.
IPQ, 2001, Norma Portuguesa NP EN 197- 2001: Cimento: Composição, Especificações e Critérios de Conformidade para Cimentos Correntes, Instituto Português da Qualidade.
Kullberg M.C.; Kullberg J.C. & Terrinha P., 2000, Tectónica da Cadeia da Arrábida, in Tectónica das regiões de Sintra e Arrábida, Mem. Geociências, Museu Nac. Hist. Nat. Univ. Lisboa, 2, 35-84.
Pereria, M. J., Almeida, J, BritoG., Soares, A. and Zungailia, E. 1997, Stochastic Simulation of Sediment Quality for a Dredging Project, V. Pawlowsky-Glahn (ed.), Proceedings of IAMG'97, CIMNE, Barcelona, 2: 899-904.
Soares, A., 1992, Geostatistical estimation of multi-phase structures. Mathematical Geology, 24(2): 149-160.
Soares, A., 2001, Direct Sequential Simulation and Cosimulation. Mathematical Geology, 33(8), 911-926.

A NON-LINEAR GPR TOMOGRAPHIC INVERSION ALGORITHM BASED ON ITERATED COKRIGING AND CONDITIONAL SIMULATIONS

E. GLOAGUEN, D. MARCOTTE and M. CHOUTEAU
Department C.G.M., C.P. 6070 succursale centre-ville, Montréal, Québec, Canada

Abstract. A new constrained velocity tomography algorithm based on ray approximation is presented. This algorithm is based on slowness covariance modeling using experimental travel time covariance. The computed covariances, the measured travel times and additional slowness values allow cokriging and conditional simulation. Among several realizations, the one that minimized the L1 norm is chosen as the best velocity field. In the proposed method the raypaths must be known. Starting with a homogeneous velocity field, an iterated solution is computed updating the raypaths applying Snell-Descartes' law on the best velocity field after each iteration. First, the advantage of an iterated solution is presented. Then, the proposed approach is compared to a classical LSQR algorithm using a synthetic model and real data collected for geotechnical evaluation in a karstic area. The tomographies on synthetic models show that geostatistical methods provide comparable to or better results than LSQR. For both methods, additional velocity constraints reduce uncertainty and improve spatial resolution of the inverted velocity field. Also, the simulation on synthetic models increases the spatial resolution compared to LSQR. The real data analysis shows that the proposed method gives very consistent results with respect to the drilling log information.

1 Introduction

Ground Penetrating Radar (GPR) is a non-destructive geophysical technique which uses radio waves (10 to 2000 MHz) to investigate electrical properties of the ground. A popular method of GPR data acquisition is cross-hole tomography. The transmitter, located in one hole, emits an electromagnetic wave impulse. The travel time to a receiver located in a coplanar hole is recorded. The goal is to determine the spatial distribution of slowness from the different travel times, a fundamentally non-linear problem. Common approaches discretize the plane between the holes in a series of cells in each of which the slowness is considered constant (Holliger et al., 2001). Commonly used tomographic algorithms (LSQR (Paige and Saunders, 1982)) use the ray approximation for wave propagation. These algorithms require the user to specify critical parameters often obtained only by trial and error. We propose an iterative method based on a stochastic model for the cell slowness. The first step consists of identifying the slowness stationary covariance structure from the non-stationary travel times. This is accomplished by using the ray approximation for wave propagation. The covariances and cross-covariances are linearly related through a geometric matrix describing the paths. The second step consists of simple cokriging and conditional simulation of the slowness field from the arrival times and the ray approximation. Travel times modeling is performed using all the simulated velocity field. The simulation that produces the travel time vector that minimizes the L1 norm compared to the measured travel times is used to compute new propagation paths applying Snell-Descartes' law. The system is updated and a new solution is obtained. Usually, after a few iterations the solution obtained is stable. Moreover, a substantial reduction in travel time Mean Square Error is observed with these final simulations compared to the classical cokriging solution or to alternative inversion algorithms.

It is possible to have velocity information along the holes, for example, using borehole reflection surveys. LSQR and geostatistical methods allow including these additional data. This allows a dramatic increase in the spatial resolution and also decreases the uncertainty on velocity estimates. First, the GPR technique is presented and a classical tomography method is briefly described. Then, the theory of the proposed method is presented. The proposed method and classical tomography are compared using a synthetic stochastic model. Finally, LSQR and the geostatistical method are used to image a karstic zone in a geotechnical study.

2 Ray based tomography

An easy way to approximate a wave path in propagation mode is to use the ray. A ray is defined as the curve that connects a transmitter to a receiver, and lies perpendicular to the wave front (Berryman, 2000). For ElectroMagnetic propagation, the ray geometry depends on the electric property contrasts, and, thus, on the velocity contrasts as described by Snell-Descarte's law.

In ray-based tomography, the field is discretized as a series of cells. For each transmitter-receiver pair, the length of each segment of ray path that crosses a cell is computed. All the segment lengths are organized in a (sparse) matrix L, called the parameter matrix, which describes the geometry of the rays. L is of size nt observed times by np cells (of constant slowness). Equation 1 represents the linear relation between travel time vector t and the slowness vector s.

$$Ls = t \qquad (1)$$

This equation represents the forward modeling of the travel time. The slowness field "the unknown" must be estimated by inversion of Equation 1.

2.1 REGULARIZATION AND EQUALITY CONSTRAINTS IN CLASSICAL INVERSION

Generally, in Equation 1, L cannot be inverted directly. In most cases, the problem is ill-posed. The linear system is modified to include a regularization term (Menke, 1989).

$$\begin{bmatrix} L \\ kD \end{bmatrix} [s] = \begin{bmatrix} t \\ 0 \end{bmatrix} \tag{2}$$

where k is a scalar and D is typically the discrete first derivative (flatness) of the slowness field. D can also be taken as the identity matrix (smoothness). The solution can be both smoothed and flattened by taking a weighted sum of the identity matrix and the derivative matrix.

When slowness values are known within the field that is to be inverted, it is suitable to force solution to fit the known values. The implementation of such equality constraints is easy in linear systems (Menke, 1989). Equation 1 is modified to take into account the velocity constraints:

$$\begin{bmatrix} L \\ M \end{bmatrix} [s] = \begin{bmatrix} t \\ sc \end{bmatrix} \tag{3}$$

where M is a matrix of size $sc \times np$, sc is the vector of known cell values. In each row, M is equal to one in the column corresponding to a known value and zero elsewhere.

In this study, the LSQR algorithm (Paige and Saunders, 1982), a classical tomography algorithm is used. This is a conjugate gradient type algorithm with Golub-Kahan bidiagonalisation (Berryman, 2000). The algorithm converges quickly and is particularly effective for sparse matrices. However, the convergence criteria must be carefully chosen to avoid the algorithm iterating on noise. Here, the correlation from one iteration to the next, and the derivative of the sum of the residuals were used as convergence criteria. Flatness and smoothness regularizations were combined.

2.2 PROPOSED METHOD

The stochastic approach for linear system inversion was first presented in Franklin (1970). Being linearly related, slowness and travel time covariance matrices are also linearly related. The linear relation between slowness and travel time covariances is:

$$cov(t,t) = L cov(s,s) L^T + C_0 \tag{4}$$

where $cov(t,t)$ is the $nt \times nt$ travel time covariance matrix, $cov(s,s)$ is the $np \times np$ slowness parameter covariance matrix and C_0 is the travel time nugget effect.

The covariance matrix for the slowness is specified by choosing the model function and its parameters (nugget, sill, and range). Once the model function is selected, the slowness covariance parameters are estimated by an iterative search in the low-dimension covariance parameter space.

2.3 COKRIGING

Cokriging (Chilès and Delfiner, 1999) is a mathematical interpolation and extrapolation tool that uses the spatial correlation between a secondary variable (here, the measured travel times) and a primary variable (here, the slowness) to improve estimation of the primary variable at unsampled locations. When an acceptable slowness covariance model is obtained, the slowness field is cokriged using the arrival times and any available slowness data. It is also easy to go further and to impose slowness gradients or even any kind of linear constraint to the solution. The simple dual cokriging weights Γ are given by:

$$\Gamma = \begin{bmatrix} cov(t,t) & cov(t,sc) \\ cov(sc,t) & cov(sc,sc) \end{bmatrix}^{-1} \begin{bmatrix} t \\ sc \end{bmatrix} \quad (5)$$

where cov signifies covariance, sc are the known slowness cells and s are the slowness cells that are to be inverted.

The cokriging estimator Z_g^* for the slowness is given by:

$$Z_g^* = \Gamma^T \begin{bmatrix} cov(t,s) \\ cov(sc,s) \end{bmatrix} \quad (6)$$

2.4 SIMULATION

By construction, cokriging gives a smooth estimate of the slowness field. It may be desirable and informative to obtain various reasonable solutions showing the kind of variability that can be expected from the slowness covariance model adopted. This is obtained by using geostatistical simulation algorithms rather than cokriging. There exist many efficient simulation algorithms (Chilès and Delfiner, 1999). The Fast Fourier Transform Moving Average simulation (FFT-MA) is a fast simulation algorithm for generating regular grid non conditional Gaussian stationary processes (Le Ravaloc et al, 2000). Conditioning of FFT-MA simulation is performed by cokriging, using the same weights as in Equation 5. For each tomography, several simulations are computed. For each simulation, travel times are computed using Equation 1. The best simulation is defined as the one that minimizes the sum of the absolute difference between the computed and the measured times (L1 norm). Contrary to LSQR, both cokriging and conditional simulation retrieve exactly the slowness data.

3 Results

3.1 ADVANTAGES OF CURVED RAYS

Because the true velocity field is not known, the first iteration is performed using straight ray approximation. Of course, the straigth ray is not a satisfying approximation. Figure 1 and Figure 2 show the influence of low and high velocity anomalies along the raypath, respectively. Intuitively, a low velocity anomaly appears smaller in the straight ray reconstructed images because the ray convergence toward the high velocity zone. Conversely, a high velocity anomaly appears larger in the reconstructed images than in reality. A well known technique is to update

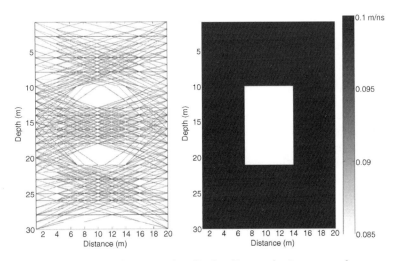

Figure 1. Left: raypaths. Right: Low velocity anomaly.

the raypath after each iteration taking into account the velocity cell constrasts (Berryman, 2000).

Figure 3 shows 6 iterations of the proposed method on a synthetic model. The model consists of a rectangular anomaly (represented by white dashed line) of 0.125 m/ns in a medium of 0.1 m/ns. The optimized covariance model is an isotropic spherical model. The range is 6 m, the slowness sill is 3 $(ns/m)^2$ and the travel time nugget effect is 1 ns^2. The covariance model stays the same for all the iterations. For each iteration the best of 100 simulations is chosen. Figure 3 shows that curved ray tomography allows an increase in the spatial resolution and reduces numerical artifacts. After only one curved ray iteration, the reconstructed anomaly is well recovered. At the fourth iteration, there remains only few artifacts. Figure 4 shows the L1 norm of 20 iterations. This figure illustrates that after the fourth iteration there is no improvement in the reconstructed image.

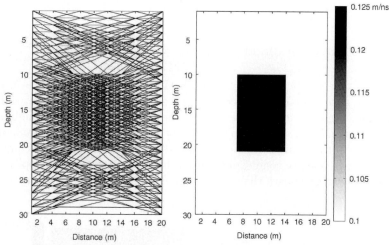

Figure 2. Left: raypaths. Right: High velocity anomaly.

3.2 TOMOGRAPHY ON SYNTHETIC DATA

In this section, the results of constrained LSQR and simulation tomographies on a synthetic velocity model are presented. The synthetic velocity model is presented in Figure 5a. Transmitter and receiver positions are also plotted in Figure 5a. A curved ray tracing algorithm based on graph theory was used to compute synthetic travel times (Berrymann, 1991; Moser, 1991). The modeled slowness covariance function is Gaussian with ranges 6 m along the horizontal axis and 3 m along the depth axis. The travel time nugget effect is 1 ns^2 and the slowness sill is 0.2 $(ns/m)^2$. The velocity in every cell intersected by the two boreholes is fixed as a constraint. Velocity constraints are implemented as presented in Equations 3, 5 and 6.

Figures 5b and 5c present the LSQR and the best simulation tomographies. For both methods, the main features of the velocity field are recovered. But, iterative simulation allows an improvement in the spatial resolution. The correlations between the velocity model and tomography images are 0.73 and 0.89 for the constrained LSQR and the iterated simulation, respectively.

4 Geotechnical evaluation in a karstic area

Borehole GPR measurements were performed to complement the site characterization of a planned expansion of a cement plant, including a mill and a reclaim facility adjacent to existing buildings. The whole site is located in a karstic environment. The overburden is an irregular residual clay layer overlying a limestone bedrock. Sixteen holes were visited during the survey (Figure 6). A RAMAC system with 100 MHz borehole antennas was used for the survey. Single-hole reflection mea-

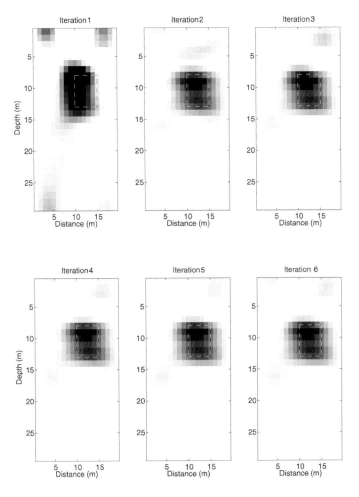

Figure 3. Curved ray iteration for the proposed method on block synthetic model

surements were performed in each hole and nineteen tomographic panels were acquired. In this article, four holes have been used to perform velocity tomography (AR13, AR08, AR12 and AR18 in Figure 6). Because they are nearly coplanar, they were included in the same 2D tomography. Slowness constraints were obtained by inversion of single-hole radar profiles (Giroux et al., 2004). Figure 7a shows the result of constrained LSQR tomography and the stratigraphy obtained from drilling logs. Figure 7b shows the constrained simulation that minimizes the L1 norm. The modeled slowness covariance function is Gaussian. The ranges are 15 m along the horizontal axis and 6 m along the depth axis. The travel time nugget effect is 10^{-6} ns^2 and the slowness sill is 1.5 x 10^{-6} (ns/m)^2. The stratigraphy is also shown. It is clear, that the proposed method offers a better match with drilling log information. Also, the conditional simulation provides an image easier

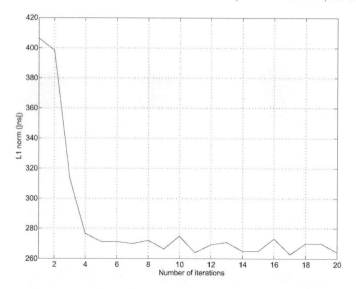

Figure 4. Objective function for block synthetic model

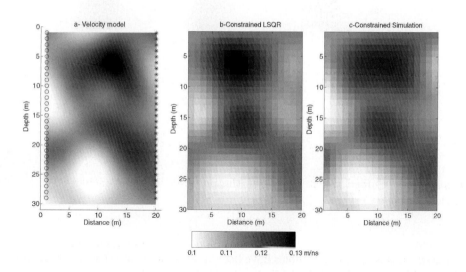

Figure 5. a- velocity model (o: transmitter, ⋆: receiver). b- constrained LSQR. c- constrained simulation

to interpret geologically. As expected, the velocity of the clay is lower (about 0.07 m/ns) than the velocity of the limestone (about 0.12 m/ns). These values compare well with the theoretical ones (Dubois, 1995; Feschner et al., 1998). Moreover, strong artifacts are generated by LSQR that render the interpretation difficult.

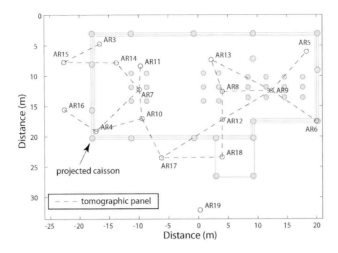

Figure 6. Borehole locations in the survey site

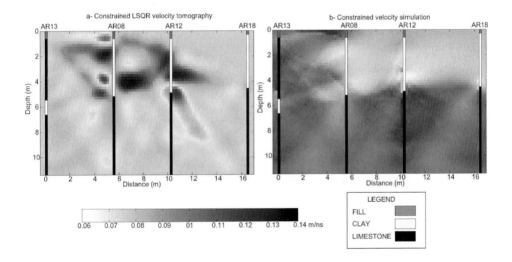

Figure 7. a-Constrained LSQR b-Conditional simulation

5 Conclusions

It has been demonstrated that the geostatistical tomography gives similar to or better results than LSQR. Moreover, the proposed method allows to take into account the non linearity of raypaths. During the real data analysis, simulation allows finding a velocity field in excellent agreement with the drilling log information.

Acknowledgements

We would like to thank Dr. Bernard Giroux for providing his ray path modeling code. We also thank Maria Annecchione for editing this manuscript.

References

Berryman, J.G., *Analysis of approximate inverses tomography*, Optimization and Engineering, vol. 1, 2000, p. 437-473.

Chilès, J. P. and Delfiner, P., *Geostatistics: Modeling spatial uncertainty*, Wiley Series on Probability and Statistics, 1999.

Dubois, J-C., *Borehole radar experiments in limestone: analysis and data processing*, First Break, vol.13 no. 2, 1995, p. 57-67.

Fechner, T., Pippig, U., Richter, T., Corin, T., Halleux, L. and Westermann, R., *Borehole radar surveys for limestone investigation*, in Proc. GPR1998, Lawrence, Kansas, USA, 1998, p. 127-132.

Franklin, J. N., *Well-posed stochastic extensions of ill-posed linear problems*, J. Math. Anal. Apll., vol. 31, no. 1, 1970, p. 682-716.

Giroux, B., Gloaguen, E. and Chouteau, M., *Geotechnical application of borehole GPR - a case history*, In Proc. GPR2004, Delft, The Netherland, 2004, p. 249-252.

Holliger, K., Musil, M. and Maurer, H.R., *Ray-based amplitude tomography for crosshole georadar data*, Journal of Applied Geophysics, vol. 47, 2001, p. 285-298.

Le Ravalec, M., Noetinger, B. and Hu, L. Y., *The FFT Moving Average generator: an efficient numerical method for generating conditioning gaussian simulation*, Mathematical Geology, vol. 32, 2000, p. 701-723.

Menke, W., *Geophysical data analysis*, Academic Press, 1989.

Moser, T. J., *Shortest path calculation of seismic rays*, Geophysics, vol. 56, 1991, p. 59-67.

Paige, C.C. and Saunders, M.A., *LSQR: an algorithm for sparse linear equations and least-squares*, ACM Trans. Math. Soft., vol. 8, no. 1, 1982, p. 43-71.

APPLICATION OF CONDITIONAL SIMULATION TO QUANTIFY UNCERTAINTY AND TO CLASSIFY A DIAMOND DEFLATION DEPOSIT

SEAN DUGGAN [1] AND ROUSSOS DIMITRAKOPOULOS [2]

[1] *MRM-Placers, De Beers Group Services, Aon House, 117 Hertzog Blvd, Cape Town, South Africa.*
[2] *WH Bryan Mining and Geology Research Centre, University of Queensland, Brisbane, Australia*

Abstract. Since the early 1900's diamonds have been known to occur in aeolian placers in south western Namibia. At Namdeb's Elizabeth Bay Mine diamonds are extracted from the fine to coarse grit layers in a sequence of stratigraphic horizons formed during periods of vigorous wind action. Significant capital expenditure is required to extend the life of mine at Elizabeth Bay and, as this is an inherently high-risk deposit, a sound understanding of the risks associated with the resource estimates is required. Various methods were evaluated to quantify the uncertainty of the thickness estimates and to facilitate classification according to the SAMREC guidelines. The thickness of the resource has a significant impact on the mining method as well as volume calculations. This investigation involves the use of conditional simulation of thickness to derive a method for classifying the resource. The simulations were used to construct block conditional distribution functions and evaluate a number of uncertainty measures, including conditional variance, conditional coefficient of variation, interquartile range and probability interval. A method employing conditional simulation to assess the efficiency of sample spacing is briefly presented. The approaches using coefficient of variation calculations provide promising results that enable classification of uncertainty related to the thickness of the Elizabeth Bay resource.

1. Introduction

Diamonds were discovered in a contemporary aeolian placer in the vicinity of Lüderitz in 1908. Recent mining, from 1990 to the present, has focussed on the Quaternary to Recent aeolian placers at Elizabeth Bay, c. 40km south of Lüderitz where the economic horizons comprise siliceous grits to small pebble size beds (mostly 2 - 8mm clast size). In order to continue mining into the future an additional capital expenditure is required to extend the life of mine and in addition to mining and treatment difficulties there is a risk associated with the uncertainty of the estimate of grade, average diamond size and resource thickness. This study addresses uncertainty related to the thickness of the economic part of the deposit.

Good spatial characteristics make the use of geostatistical estimation methods possible for most variables including resource thickness. The diamonds are found in a sequence of fine to coarse grit horizons formed during periods of vigorous wind action and, in parts, fluvial reworking. The mineralised component of the deposit comprises a thin upper deflation grit known locally as Grey Beds, overlying a thicker sequence called

Brown Beds. In the north-east of the deposit the Brown Beds are underlain by the Red Beds (or Fiskus Sandstone), but because of limited representative sampling Red Bed data was excluded from this study.

Although Namibia does not have a specific code for classification of resources the guidelines outlined in the South African Code for Reporting of Mineral Resources and Mineral Reserves (SAMREC) are accepted by most mining companies. However, this code, like other international codes, provides only broad guidelines and is in no way quantitative.

2. Resource Thickness

The thickness of the ore body determines the mining method used (and therefore the carats recovered) and is used to obtain a local estimate of volume. Ordinary kriging was used to estimate grade, average diamond size and resource thickness in the three main stratigraphic horizons into 100m x 100m blocks.

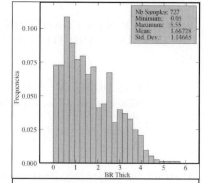

Figure 1: Histogram of Brown Beds thickness; sample data

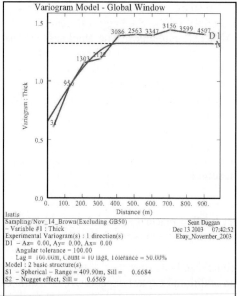

Figure 2 : Brown Bed thickness variography

The average thickness of the Grey Beds at Elizabeth Bay is 0.60m and the underlying Brown Beds are, on average, 1.67m thick. The histogram derived from sample data for the Brown Bed horizon is shown in Figure 1.

The Brown Bed thickness semi-variogram (Figure 2) shows an isotropic, Spherical model with a range of 410m and a nugget effect of about 50%. A similar model was obtained for the Grey Beds with a double structure with ranges of 125m and 500m. Ordinary kriging was used to estimate the average thickness of both horizons using 100m x 100m blocks.

Figure 3 illustrates the estimated resource thickness for the Grey Beds and the estimation standard deviation (Figure 3, right). The latter reflects the sample density (black dots), as expected.

3. Method

A summary of the method used to find a measure of uncertainty is outlined in Figure 4. The technique requires successful conditional simulation realizations of the variable under study on a dense grid of nodes (data support). Reproduction of the data histogram and variogram model are validated for each realization. The change of support of the realizations into the required block size is performed or alternatively, simulated directly on blocks (e.g., Godoy, 2003). The second part of the method is to derive the conditional distribution functions (cdf's) from the set of simulated resource models, describing the uncertainty about the unknown thickness values for each block. The third step involves computing the uncertainty measure(s) from the cdf of each block and finally classifying the blocks into a specific category of resource by selecting thresholds using the set of available, simulated ore body models.

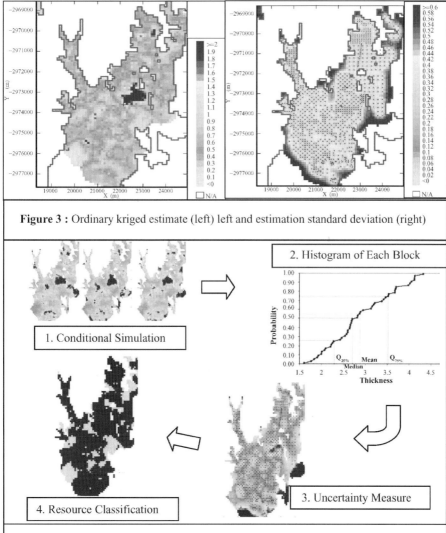

Figure 3 : Ordinary kriged estimate (left) left and estimation standard deviation (right)

Figure 4 : Illustration of methodology

4. Conditional Simulation

The conditional simulation of resource thickness for the Grey Beds and Brown Beds at Elizabeth Bay was carried out using a Turning Bands simulation algorithm (Lantuéjoul, 2002). One hundred realisations were generated with a regular discretisation of 10 x 10 x 1. A total of 920 samples were used for simulating the Grey Beds and 727 for the Brown Beds. The simulation domain was defined by a polygon delineating the edge of the resource and includes 2882 blocks for the Grey Beds and 2910 blocks for the Brown Beds. Statistics of the back transformed data compared favourably with the raw data and variograms obtained from point conditional simulation confirmed the accuracy of the simulations. The simulation mean (e-type) is plotted on the upper left of Figure 5 and is compared to five realizations. The e-type estimate shows a greater amount of smoothing than OK (Figure 3) and the individual realizations show high variability.

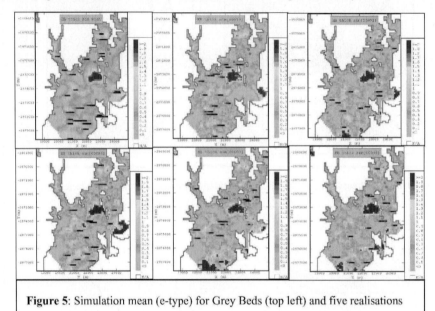

Figure 5: Simulation mean (e-type) for Grey Beds (top left) and five realisations

5. Measures of Uncertainty

The steps of conditional simulation and change of support are followed by the derivation of a set of conditional distribution functions (cdf) related to the simulated variable denoted by Z. Each of these functions provides a measure of local uncertainty in that it relates to a block attribute Z(u) at a specific location u within the deposit. From the distribution it is possible to read the probability that Z(u) is valued above any given cut-off z_k:

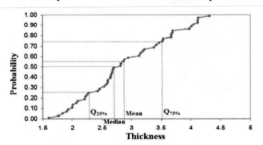

$$F(u;z) = P\{Z(u) \le z_k\}, \quad \forall z$$

Figure 6: Example of a local conditional distribution function (cdf) for thickness at a given block

Figure 6 shows a cumulative conditional distribution function modelling the possible uncertainty about the thickness for a given block at location u. Each discrete point in the cdf corresponds to a simulated value z'(u). The graph is a cumulative histogram containing all simulated values assigned to the block by each one of the realisations. A continuous function is interpolated between the discrete points to enable the assessment of probabilities for any cdf value.

A variate of summary statistics and uncertainty measures can be derived from the cdf and used to support decision making. In this study a number of basic measures are investigated and these include a conditional variance, a conditional coefficient of variation (relative standard deviation), the interquartile range, and a probability interval.

5.1 CONDITIONAL VARIANCE
The conditional variance measures the spread of the cdf around its mean value z_E^*:

$$CVar(u) = \sum_{k=1}^{k+1} [\bar{z}_k - z_E(u)]^2 \cdot [F(u;z_k) - F(u;z_{k-1})]$$

where:
- z_k, $k=1,\ldots K$, are K threshold values discretising the range of variation of z values
- $\bar{z}_k$ is the mean of the class z_{k-1}, (z_{k-1},z_k) which in case of a within class linear interpolation model corresponds to $\bar{z}_k = (z_{k-1} + z_k)/2$

- $z_E^*(u)$ is the expected value of the cdf approximated by the discrete sum:

$$z_E^*(u) = \sum_{k=1}^{K+1} \bar{z}_k \cdot [F(u;z_k) - F(u;z_{k-1})]$$

5.2 CONDITIONAL COEFFICIENT OF VARIATION
The conditional coefficient of variation (CCV) or relative conditional standard deviation corresponds to the conditional standard deviation divided by the mean. The CCV expresses variability as a percentage of the mean, and is calculated as follows:

$$CCV(u) = \frac{\sqrt{\sum_{k=1}^{k+1} [\bar{z}_k - z_E(u)]^2 \cdot [F(u;z_k) - F(u;z_{k-1})]}}{z_E^*(u)}$$

5.3 CONDITIONAL INTERQUARTILE RANGE
The conditional interquartile range (IQR) is defined as the difference between the upper and the lower quartiles of the distribution:

$$IQR(u) = q_{0.75}(u) - q_{0.25}(u) = F^{-1}(u;0.75) - F^{-1}(u;0.25)$$

5.4 PROBABILITY INTERVAL

The probability that the unknown is valued within an interval (a, b), termed probability interval, is calculated as the difference between cdf values for thresholds b and a:

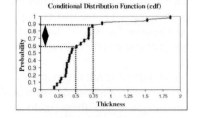

$$\Pr ob\{Z(u) \in (a,b)\} = [F(u;b) - F(u;a)]$$

6. Results

The conditional variance (Figure 7) shows a poor correlation with sample density because there are high and low values related to both high and low sample density. However, the thicker part of the Grey Beds corresponds to a higher conditional variance value. Visual examination suggests there is a proportional effect between the conditional variance and thickness. The implication is that the conditional variance may not be an appropriate measure of local uncertainty if the objective includes the comparison between zones with different magnitudes of thickness.

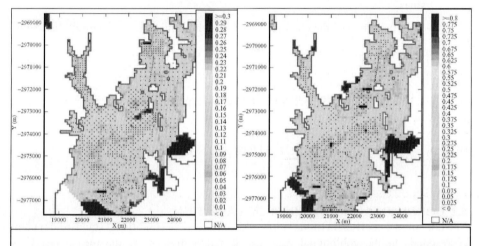

Figure 7: Conditional variance for thickness; Grey Beds (left) and Brown Beds (right)

Figure 8 (left and middle) illustrates CCV values calculated for each block for the Grey and Brown Beds. There is a good correlation between CCV and conditioned sample points (points in Figure 8) and the identification of similar zones is better than with the conditional variance. Comparison with the estimation standard deviation (Figure 3) shows a difference. A further advantage of the CCV is that it is expressed as a percentage of the mean. The CCV values for the two horizons differ and that will make the selection of common threshold values for resource categories difficult.

The interquartile range (IQR) is a relative measure that does not use the mean as the centre of the distribution. This measure ignores the internal distribution of probability densities, leading to over representing uncertainty. Visual examination of IQR values for both horizons suggests higher uncertainty about the higher thickness values. The effectiveness of the IQR as an aid in classification may be improved by dividing by a median or mean to make the measure dimensionless (Lantuéjoul, 2003). The IQR divided by the mean for each block (Figure 8, right) shows an irregular change from low to high values making identification and selection of thresholds difficult. The major

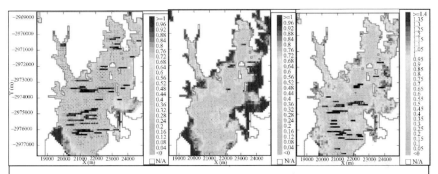

Figure 8 : Conditional coefficient of variation for Grey Beds (left), Brown Beds (middle) and IQR (right)

drawback of using IQR is that it ignores the distribution of probability densities, leading to an over representation of uncertainty.

Probability intervals for a range of thicknesses at Elizabeth Bay could be selected between, for example, 0.5m and 0.75m. Like some of the previous measures, there is a good correlation between regions of dense sampling and high probability values. Where the resource thickness is less than 0.5m a less costly mining method can be used and this measure is ideal for establishing probability of finding a thickness of 0.5m or less in 100m x 100m blocks. However the measure is less useful for determining thresholds related to uncertainty.

7. Resource Classification

The resource classification criterion is based on the uncertainty measures derived from the cdf of each block in the resource model. Each criterion requires the selection of a threshold value that reflects the error tolerance that is acceptable for the block estimate.

Consider the CCV as a classification criterion. Given a threshold, λ, the 100m block will be classified

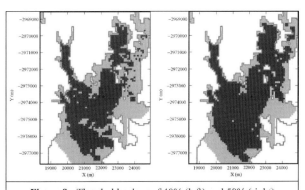

Figure 9 : Threshold values of 40% (left) and 50% (right) for the Grey Beds

at an Indicated or Inferred level of confidence depending on whether the CCV is less than or greater than the threshold; If CCV< λ then Indicated and if CCV $\geq \lambda$ then the block is Inferred. Figure 9 shows threshold values of λ =40% and 50% for the Grey Bed horizon where the central darker coloured blocks will be at an Indicated resource category.

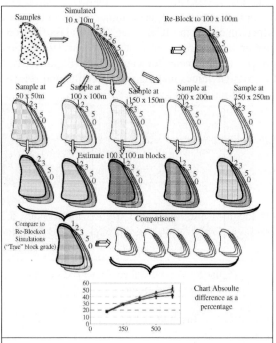

8. Assessment of Sample Spacing Efficiency

The method, outlined in Figure 10 uses conditional simulation to quantify the expected error (% average difference between estimated and "actual"). The steps include initial analysis of the data, generation of a suitable conditional simulation, sampling each realisation using a given spacing, estimation of the

Figure 10: Illustration of method for finding optimum sample spacing

attribute, calculating the expected difference error (%), repeat the process to generate expected errors and finally calculate the mean coefficient of variation of expected differences per block. The method was applied using data from a part of the Grey Beds at Elizabeth Bay. The deposit was simulated on a 10m x10m grid and node values,

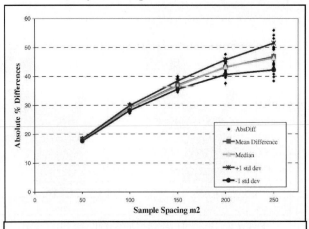

Figure 11: Expected differences for each sample spacing

selected at a specific spacing were used to estimate thickness in 100m x 100m blocks. Fifty realisations using the sequential Gaussian simulation method (e.g., Dimitrakopoulos and Luo, 2004) were used to assess sample spacing at 50m, 100m, 150m, 200m and 250m intervals.

The results of sampling the Grey Bed simulations are shown in Figure 11. The spread of differences found with 50 realisations for

50m x 50m sampling is very small increasing marginally to 100m x 100m sampling. The percentage absolute difference is sensitive to low "actual" thickness values. The expected difference for 50m x 50m sampling is about 18%, increasing to 55% for 250m x 250m sampling.

9. Discussion of Results and Conclusions

The methods using conditional coefficient of variation calculations provide promising results for quantifying uncertainty related to resource thickness estimates at Elizabeth Bay. The application of specific threshold values requires further work but these methods provide a useful tool for resource classification. Although not strictly quantitative when the results are combined with similar values calculated for grade, stone size and revenue they will enable a good overall classification of the resource to be made.

Of the measures evaluated the CCV provides the most reliable measure to use for resource classification and although the method remains semi-quantitative it is possible to use a CCV value to assign a resource category for the Elizabeth Bay Beds. Table 1 shows resource categories assigned by using CCV values where the Inferred category is assumed to be larger than Indicated or Measured and has been subdivided into "upper" and "lower" units. The classes have been determined predominantly by visual means but taking cognizance of the method referred to above and knowledge of the deposit. Figure 12 illustrates the application of CCV categories (similar to Table 1) on the Grey Beds (left) and the Brown Beds (right).

Table 1: Proposed Resource Classification categories derived from CCV values of thickness at Elizabeth Bay

CCV		Classification
0	0.2	Measured
0.2	0.4	Indicated
0.4	0.6	High level Inferred
0.6	0.8	Low level Inferred
0.8	2	Not in resource (Deposit)

The decision to allocate blocks to either an Inferred or Indicated resource category at Elizabeth Bay remains the responsibility of the "competent person" but using the CCV as a measure provides a good, quantitative method to use as an aid, which with similar measures on other variables, will make quantitative classification possible.

The method of assessing sample spacing efficiency can be used to determine the distribution of % errors for each of the 100m x 100m estimation block, and the mean errors mapped for each of the nominated sample spacings. The absolute average of expected % error in estimated thickness ranged from 18% at a spacing of 50 x 50m up to 55% for samples spaced at 250 x 250m.

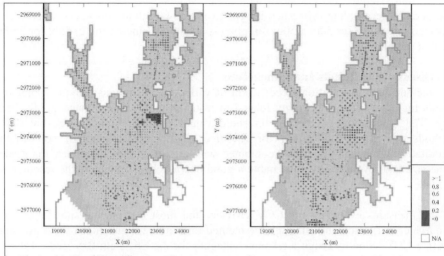

Figure 12: Final Classification of resources according to figures shown in Table 1 for the Grey Beds (left) and the Brown Beds (right)

Acknowledgements

Namdeb Diamond Corporation is acknowledged for allowing the presentation at Geostats 2004 and publication of this information. Dr. Marcello Godoy is acknowledged for his contribution to this study and Dr. Christian Lantuéjoul is thanked for his informative comments and discussion.

References

Dimitrakopoulos, R. and Luo, X., 2004. *Generalized sequential Gaussian simulation on group size v and screen effect approximations for large field simulations.* Mathematical Geology, v. 36, no. 5, pp. 567-591.

Duggan, S. P. 2002. *Mineral Resource Estimate for two portions of Elizabeth Bay.* Internal De Beers report written for Namdeb, November.

Godoy, M., 2003. *The effective management of geological risk in long-term production scheduling of open pit mines.* PhD Thesis, The University of Queensland, Brisbane, 252p.

Lantuéjoul, C. *Geostatistical Simulation.* Berlin: Springer, 2002, 192p.

Lantuéjoul, C. *Comparison of various criteria for resource classification.* Report written for De Beers, October, 2003.

GEOSTATISTICAL SIMULATION TECHNIQUES APPLIED TO KIMBERLITE OREBODIES AND RISK ASSESSMENT OF SAMPLING STRATEGIES

JACQUES DERAISME[1] and DAVID FARROW[2]
[1]*Geovariances, 49 Ave Franklin Roosevelt, Avon, 77212 France*
Deraisme@geovariances.com
[2]*MRM - TSS, De Beers Consolidated Mines Ltd, P Bag X01, Southdale 2135 South Africa david.farrow@debeersgroup.com*

Abstract. Typically a kimberlite diatreme has several different geological zones. The upper portion is generally filled with the sedimentary crater facies, the central zone is more typically an in situ massive series of volcanic breccias and the lower regions comprise a complex root zone. Depending on the local degree of erosion, not all zones remain at any particular kimberlite occurrence.

A method of simulating the simpler internal geologies seen in the central region had previously been developed using a geometrical technique. In the upper reaches of the diatreme zone, the geologies have more complicated geometries and the approach adopted for the central regions needs to incorporate a more sophisticated method of simulating the internal geologies.

The similarity between the sedimentary facies that comprise the crater zone infill and the sequences that the oil industry targets as oil reservoirs suggest a similar technique could be applied to the simulation of internal geology of crater zone of kimberlite pipes.

Previous work has shown that a truncated gaussian approach can be useful, but the restrictions on facies relationships have limited its implementation. Plurigaussian simulation allows more complex interrelationships to exist between the simulated zones.

In conjunction with other geometric simulations, plurigaussian simulation can be used to guide sampling programs to optimise sampling layouts and sample size and ensure that the goals of the sampling programs are attainable. This paper focuses on the application of the combination of these simulation techniques and will be illustrated by a case study.

1 Introduction
The Orapa Kimberlite Mine forms the basis for this study. The mine, located 240 km west of Francistown in the western portion of the Botswana Central district, produces approximately 6 Million Carats per year with a value of almost US$ 500 million.

Typically kimberlite pipes have a number of differing zones or facies. Three facies, namely "crater", "diatreme" and "hypabyssal" have been recognised in the Orapa orebody. The Orapa orebody comprises two volcanic pipes, which coalesce to form a single crater. The focus of this study is the crater facies rocks of the upper portion of the southern pipe.

The crater facies comprise a predominantly sedimentary type sequence of volcanic materials that have been re-deposited into the volcanic crater. Application of the Plurigaussian Simulation technique to the crater facies has been investigated for assisting with the creation of geological block models and determining an optimal sampling configuration.

Numerous boreholes and pit exposures have been combined into a digital geological model using GEMCOM™ software for analysis and visualization, that is typically a time consuming process. During mining, and as additional drilling is undertaken, more data becomes available. Incorporating additional data into the digital model requires that the models be regenerated, to ensure that the spatial distributions are maintained, once again a time consuming process. As a consequence, the digital models are not updated with any regularity and the mining models and the geological models are frequently out of phase. This results in sub-optimal Resource Management. An algorithmic method of more rapidly generating an overall geological model which can be updated on a regular basis is a significant advantage.

The initial search for a suitable algorithm explored the truncated gaussian approach. Despite showing promise, it did not prove very effective. The plurigaussian simulation methodology implemented in the latest release of the geostatistical software, ISATIS™, was another option which offers enhanced capabilities for geological modelling and has been successfully applied to the geological simulation of oil reservoirs. The application of this method is the subject of this paper.

2 The Geology of Orapa

A review of the Orapa geology is given in Field et al. (1997) and readers are referred to this paper for a more detailed introduction. For the purposes of this study only the major rock types of the crater facies are summarised.

The Orapa pipes intrude into the Archean basement granite-gneiss and tonalities and the sedimentary rocks and lavas of the Karoo Supergroup. They were covered by extensive thicknesses of Cenozoic and Mesozoic deposits. The deposit comprises two pipes, named the southern and the northern lobes. Rocks belonging to the crater, diatreme and hypabyssal facies, as described by Hawthorne (1975), have been recognised.

The Crater Facies deposits are well preserved and divisible into epiclastic, volcaniclastic and pyroclastic varieties. The epiclastic deposits are those in which sedimentary processes can be identified and comprise a wide variety of types including talus deposits, debris flow material, boulder beds, grits and lacustrine shales. Those deposits with no obvious mechanism of deposition are termed volcaniclastic. They are highly variable in character, with well sorted bedded horizons but dominated by coarse massive, matrix supported types. No convincing directional sedimentary structures have been found within the deposit. Basal Hetrolithic Breccias which apparently mark the base of the crater zone deposits occur intermixed with the volcaniclastic deposits. Pyroclastic deposits show evidence of direct deposition by volcanic mechanisms. The Pyroclastic deposits are present only in the northern pipe and comprise materials that exhibit evidence of pyroclastic fall, flow or surge.

3 The Plurigaussian Simulation Methodology

Plurigaussian simulation (PGS) aims to simulate categorical variables, such as geological facies, by the intermediate simulation of two continuous Gaussian variables. Facies are obtained by applying thresholds to the Gaussian simulated values. A detailed review of the PGS method is given in Armstrong et al. (2003).

The basic idea is to start out by simulating at grid locations one (Truncated Gaussian Simulation or TGS) or two (PGS) gaussian variables with a variogram characterizing the spatial continuity of the lithotypes indicators. Then a "rock type rule" is used to convert these values into lithotypes. The conversion is using the bijection between the gaussian values and the cumulated distribution function (cdf.). Therefore the application of that method requires to inform each grid node by the an estimate of the cdf. This step is carried out by calculating the so-called vertical proportion curves. By interpolating these proportions on the 3D grid, we get a 3D matrix of proportions.

PGS is an extension of TGS, the latter implying a rather strict stratigraphic sequence: because the simulated Gaussian values are continuous, the application of a threshold practically means, in the simple case of 3 facies, that for going from facies 1 to facies 3, it is likely to have a transition through facies 2. By using 2 gaussian functions, all transitions facies 1 to 2 or 1 to 3 or 2 to 3 are authorized.

The concept of non stationary proportion curves is central in PGS/TGS, where the so-called rock type rule plays an essential role in producing realistic models that represent the transitions between the different facies. The key point is that the Gaussian variables and the indicators are linked by means of thresholds but, even if the indicators are not stationary, they can be obtained by truncation of stationary Gaussian variables, which can be easily simulated. Initial applications were made in the petroleum industry where this approach seems natural due to the sedimentary origin of the reservoirs. The analogies with orebodies where mineralization occurs in layers forming a consistent stratigraphy justifies the application of the same conceptual model in this case study.

The process of PGS has three steps:
- o determination of the vertical proportion curves from statistics on the drill-hole data. A vertical proportion curve represents the profile along the vertical of the proportions of each facies level by level. These statistics are highly dependent on the choice of a particular surface, the reference surface, which can be interpreted as a guide to the system of deposition of the different lithological facies. The drillhole data will then be transformed into a "flattened" space where the reference surface represents the horizontal surface at zero elevation. The simulations of the Gaussian variables will be

processed in the flat space before being transferred to the real stratigraphic space.
- choice of a model describing the relationships between the different facies. This includes the definition of the lithotype rule, the correlation between the two Gaussian variables and their variogram models.
- generation of gaussian values at data locations. This is the most difficult and original part of the method, because at the data locations only facies are known but this does not tell the corresponding gaussian values. A special statistical method called a Gibbs sampler is used to generate these values.
- simulation of the two Gaussian variables followed by truncation to obtain the facies indicators. Finally the simulated facies are transferred to the structural grid.

4 Data Sources

o

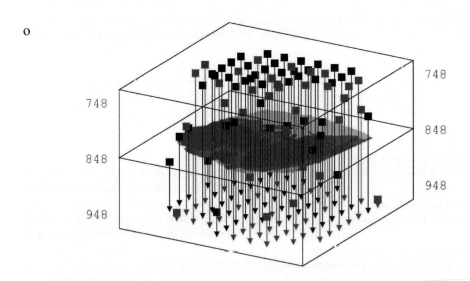

Figure 1: Vertical boreholes spaced on a 100m square grid, sub-sampled 200m grid boreholes in red.

The most recent geological model was transferred to a 5m * 5m * 5m block model. This was taken as the starting point of the work. Simulated boreholes were generated from the geological block model on 100 m, 150m and 200 m square grids (Figure 1). The facies observed on the simulated boreholes from the geological model are considered to be "reality" and are used to condition the facies simulations. The aim is to determine how much drilling is required to produce an accurate model of the pipe geology and associated volumes.

5 The Simulation

5.1 CHOICE OF A REFERENCE SURFACE

This is a crucial decision that has consequences on all stages of the process, data analysis and simulated images. In a sedimentary context the reference surface is meant to represent the direction perpendicular to the deposition of the different facies. When comparing facies parallel to that surface, more similarity is expected and consequently more correlation between boreholes than along parallel plane surfaces will be observed. The consequence on the simulated images will be to force the facies to be stacked in parallel to the reference surface. In the present case, a bowl shaped surface showing the angle of dip of the bedded horizons in accordance with the proximity to the pipe boundaries was used.

5.2 VERTICAL PROPORTION CURVES

The boreholes were discretized by cores of 5m in length and repositioned relative to the reference surface. For each case corresponding to the different horizontal spacing the vertical proportion curves have been calculated and averaged within polygons designed in order to take account of the lateral facies change (Figure 2).

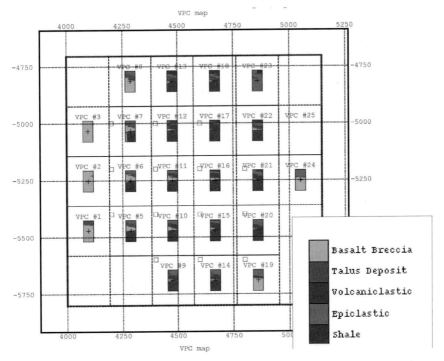

Figure 2: Vertical proportion curves calculated from boreholes within 2D polygons.

5.3 3D PROPORTIONS

The vertical proportion curves have then been interpolated on each grid cell by a kriging procedure with a rather long range (2 km) variogram, expressing the very gradual change of the lithotype proportions. This gives a 3D matrix of proportions that will be used to calculate local thresholds on the Gaussian random functions (Figure 3).

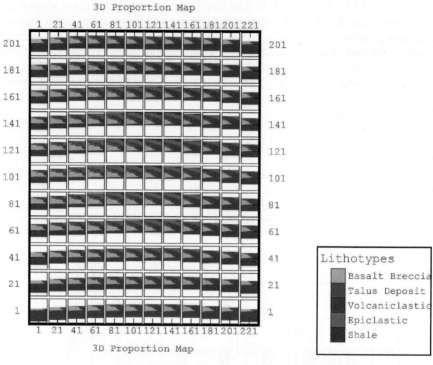

Figure 3: 2D representation of the 3D proportions interpolated on the grid in the flat space.

5.4 LITHOTYPE RULE

The knowledge of the lithotype proportions is not sufficient to derive the values of the thresholds to be applied on the simulated Gaussian values. Additional information on a partition of the 2D gaussian space is required. Depending on the number of facies, there is a finite number of rectangular partitions that may represent the possible relationships between the Gaussian random functions and the lithotypes. From these we chose the most sensible from a geological point of view, considering the probable transitions between the facies. For instance, since the shale occurs on the top of the diatreme adjacent to the epiclastic deposit, while the basalt breccia mark the base of the crater, it is appropriate to differentiate the corresponding lithotypes on the first Gaussian function. The representation (Figure 5) is schematic: the thresholds will be calculated at the simulation stage. In this example the lithotype rule implies that Basalt, Epiclastic and Shale are dependent only on thresholds applied on the first Gaussian function, while the Talus and Volcaniclastic also depend on the second Gaussian function.

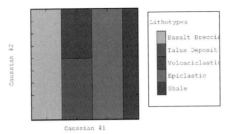

Figure 4: Rectangle lithotype rule.

Once the lithotype rule is defined, the variograms of the two Gaussian functions can be modelled. The prevailing role played by the proportion curves does not mean that the choice of the variogram has no consequence. This is illustrated in the Figure 5, where two ranges of the variogram of the first Gaussian function were tried as an example. In the lower picture the 3 lithotypes (orange, blue and purple), that are only discriminated by the first Gaussian function look much less continuous than on the upper picture. The second Gaussian function was simulated in correlation with the first (coefficient of correlation of 0.7) in order to make the Talus facies (green) preferentially conformable to the Breccia facies (orange).

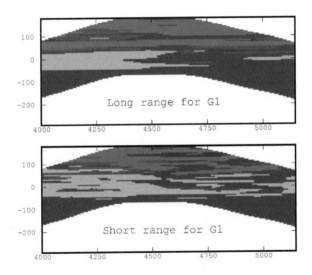

Figure 5: Cross section (in the flat space system) of two simulations changing the horizontal range of the variogram associated to the first Gaussian function.

5.5 CONDITIONAL SIMULATIONS

The simulations, achieved by means of the turning bands method, are performed in flat space, then transferred to the real "stratigraphic" space. Figure 6 compares the original geological model to 3 different realizations obtained from plurigaussian simulations using either no boreholes (just the average proportions) and called "non conditional" or boreholes (BH) spaced every 200m or every 100m. It is observed that with an increasing availability of data, the simulations converge towards the supposed reality.

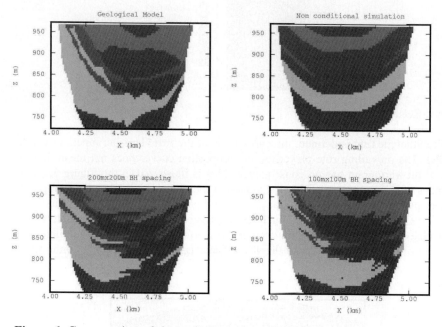

Figure 6: Cross section of three plurigaussian simulations with increasing information rates compared to the original block model.

6 Results

In the scope of evaluating the level of uncertainty in the volume of each lithotype, statistics have been calculated on the difference between the original block model and the simulations based on different levels of information. By comparing different borehole spacings (100m, 150m and 200m) it appears that 150m provides a satisfactory global estimation of the different facies (Figure 7 where the density of boreholes has been transformed into metres drilled).

The detailed analysis of the volumes of the different lithotypes has been made by levels 25m high (Figures 8 and 9). Compared to the geological model, it appears that the sampling using boreholes on 100m spacing guarantees maximum reduction in uncertainty. Use of boreholes on 200m spacing boreholes leads to an average uncertainty of about 10%, rising to 20% for some levels.

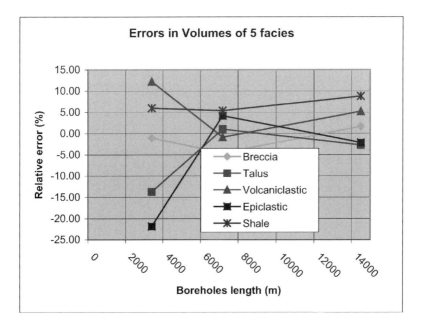

Figure 7: Relative errors on the global volumes from simulations preformed with increasing sampling by boreholes.

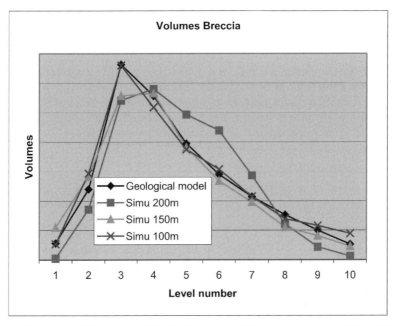

Figure 8: Volumes of Breccia on 25m high levels, of the geological model and the simulations with different sampling

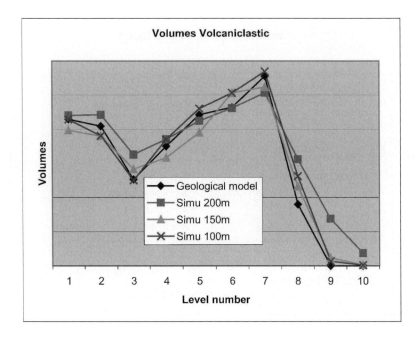

Figure 9: Volumes of Volcaniclastic, on 25m high levels, of the geological model and the simulations with different sampling.

7 Conclusions

The plurigaussian simulation process has proved to be very efficient in providing images reproducing the main features of the geology encountered in kimberlite crater deposits. It appears this may be a useful addition to the process of geological modelling. This will be explored in an operational context. Besides, the quantification of the confidence as a function of the number of holes will aid in economic decision making.

References
Hawthorne, J.B.. Model of a kimberlite pipe. Physics and chemistry of the earth, 1975, Vol 9, p.1-15.
Armstrong A., Galli A., Le Loc'h G., Geffroy G. and Eschard R., *Plurigaussian Simulations in Geosciences*, Springer, 2003.
Bleines. B., Deraisme J., Geffroy F . et al., *Isatis software Manual*, Géovariances and Ecole des Mines de Paris, 2004.
Chilès J. P. and Delfiner P., *Geostatistics Modeling Spatial Uncertainty*, Wiley & Sons, 1999, p. 531-535.
Deraisme J. and Farrow D., *Quantification of uncertainties in geological modelling of kimberlite pipes*, APCOM 2003, The South African Institute of Mining and Metallurgy, 2003.
Field, M., Gibson, J.G., Wilkes, T.A., Gababotse, J. and Khutjwe, P. (1997). The geology of the Orapa A/K1 kimberlite Botswana: Further insight into the emplacement of kimberlite pipes. Russian Geology and Geophysics, Vol.38, No.1. Proceedings of the Sixth International Kimberlite Conference, Vol. 1, p.24-41.

MODELLING 3D GRADE DISTRIBUTIONS ON THE TARKWA PALEOPLACER GOLD DEPOSIT, GHANA, AFRICA

THOMAS R. FISHER*, KADRI DAGDELEN**, and A. KEITH TURNER*
*Department of Geology and Geological Engineering
**Department of Mining Engineering
Colorado School of Mines
Golden, Colorado 80401-1887 USA

Abstract. In the Precambrian Tarkwaian Group of Ghana, gold is preferentially located as paleoplacers within quartz-pebble conglomerates. Gold distributions are intimately associated with sedimentologic and stratigraphic features of the host rocks. In this situation, traditional geostatistical methods have not provided accurate predictions of ore grades and reserves, due to difficulties in properly incorporating geologic information in the geostatistical estimation.

Application of Transition Probability/Markov geostatistical techniques allowed us to combine geologic concepts and domain knowledge with indicator and Gaussian-based estimation techniques. Vertical variability relationships within stratigraphic sequences, as measured by borehole data, were used to predict lateral distributions of lithologic facies. The result was a set of 3-D spatial relationships that reflect an integration of geologic concepts and readily observable geologic attributes.

This approach provides an alternative to more traditional geostatistical ore deposit modelling. It provides a statistically sound, lithofaces-based prediction of gold grades and uncertainty of the predictions, constrained by geology and the 3D geological framework.

1 Introduction

Mine profits are largely determined by accurate estimation of ore reserves and correct classification of material as either ore or waste during mining operations. Accurate prediction of ore grades and reserves requires incorporation of geological data, knowledge, expertise, and concepts. In the Paleoproterozoic Tarkwaian Banket Formation of Ghana, West Africa, gold is preferentially located in a succession of paleoplacer within quartz-pebble conglomerates. Thus, gold distributions are associated with sedimentologic and stratigraphic features of the host rocks.

At the Tarkwa mine, simple kriging based on 100m x 100m diamond drill (DD) boreholes underestimates gold values by as much as 20 percent below reported mill head grades (Gold Fields Ghana, Ltd, 2003). It is believed that this inaccurate prediction of ore grades may result from lack of characterization of depositional environments of the host rocks in estimation of ore reserves. Underestimation may not cause great difficulties during current opencast mining operations, but the success of projected future underground operations with less densely spaced borehole control will depend on much more precise predictions of both ore grades and reserves based on appropriate geologic models.

2 Regional Geologic Setting

The Tarkwa region (Figure 1) is contained within the Man-Leo Shield in southwestern Ghana. The geology is dominated by the Paleoproterozoic Birimian Supergroup, a series of meta-volcanic belts and intervening meta-sedimentary basins that formed as primitive island arcs accreted to the Archean craton (Sylvester and Attoh, 1992). The Birimian terrane consists of five NE-SW trending volcanic belts, named from east to west (and youngest to oldest): the Kibi-Winneba Belt, the Ashanti Belt, the Sefwi Belt, the Bui Belt, and the Bole-Navrongo Belt.

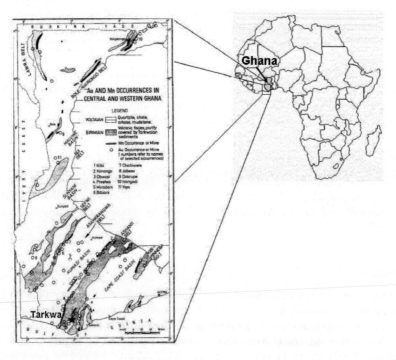

Figure 1. Index map and simplified geology of the Tarkwa Region (after Fisher and Turner, 2002).

The Tarkwa mine is located at the southwest end of the Ashanti Belt or "Tarkwa Syncline" (Figure 1). At least five episodes of deformation have affected the syncline. The Tarkwa depositional basin was formed during the first episode of deformation and is filled by a fining-upward sequence of Proterozoic clastic sedimentary rocks, known as the Tarkwaian Group, that is indicative of an extensional half-graben geometry as described by Frostick and Reid (1987). The Tarkwa Syncline contains the largest accumulation of Tarkwaian-type sediments and the highest paleoplacer gold concentrations of the Ashanti Belt. The underlying Birimian is considered a possible source of the gold in the Tarkwaian placers (Boadi, et al, 1991), although there is much disagreement amongst researchers regarding this theory.

The Tarkwaian Group consists of four formations, the Kawere, the Banket, the Tarkwa Phyllite, and Huni Quartzite. The Banket, main gold-bearing unit of the Tarkwaian, consists of up to 160m of relatively mature quartzites and conglomerates. Gold is found in paleoplacers within a succession of stacked, tabular units of alluvial fan, braided-stream, and valley-fill deposits derived from a southeastern source. At the Tarkwa Mine, the primary Banket gold concentrations are located in the A1, A3, and C zones. Gold is extracted from five areas of the mine; Pepe Anticline, Mantraim, Akontansi East, Akontansi Ridge, and Kottraverchy.

3 Modelling Procedure

An overview of the adopted modelling procedure (Fisher, 2004) is given in Figure 2. Geostatistical methods used incorporated site-specific geological information, knowledge, and experience to translate raw observations into a 3-D probabilistic model of gold distribution in the Pepe Anticline area of the mine. The top of Figure 2 shows the three major types of information incorporated within the process – exploration drilling data, field observations, and geologic maps and cross-sections. Gold Fields Ghana Ltd. provided much of these data in digital formats, but these historical data sources were supplemented by personal observations and discussions with mine personnel during an extensive site visit in early 2002. These data consisted of two major types – data associated with exploratory drilling that were used to create a borehole database (upper left of Figure 2), and stratigraphic and structural information (Box 4, Figure 2) that formed the basis of a 3-D geologic framework model of the Pepe area of the mine.

The modelling procedure has two major components. The first component, shown on the left side of Figure 2 (Boxes 1-3), involved statistical assessment of the observations contained in the borehole database to define statistically and geologically meaningful sedimentological units, herein called "statistical facies", and subsequently the development of probabilistic distributions of gold-assay values for each "statistical facies". The second component, shown on the right side of Figure 2 (Boxes 4-7), involved several steps to develop a 3-D probabilistic model of the spatial distribution of the "statistical facies". The final step in the modelling process (shown as Box 8, Figure 2) produces a 3-D probabilistic model of the gold distribution by combining the 3-D probabilistic distribution of the "statistical facies" (produced from the second component and shown as Box 7, Figure 2) with the probabilistic distributions of gold-assay values for the "statistical facies" (produced from the first component and shown as Box 3, Figure 2). The production of a 3-D probabilistic model of the gold distribution permits several useful applications to mine planning and operation (Box 9, Figure 2). Details of this modelling procedure are provided in the following sections.

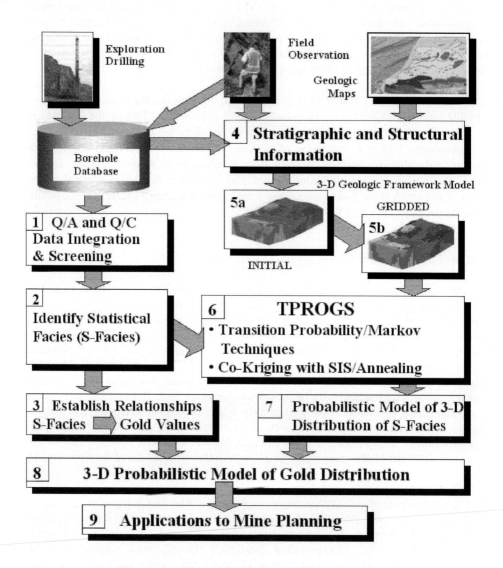

Figure 2. Flowchart of the Analysis Process (from Fisher, 2004).

3.1 AVAILABLE DATA

Data from 53 DD, continuously-cored exploration boreholes located on a nominal 100m by 100m grid within the Tarkwa mine were extracted from the main Gold Fields Ghana, Ltd. Mine databases. The critical observations were contained in several data sources. These had to be merged to form a single consistent borehole database for this study.

When completed, this database contained for each borehole: 3-D coordinates, stratigraphic and lithologic observations (including 19 categorical and continuous geologic variables) developed by geologists logging the cores, plus gold assay values obtained from core samples. An extensive QA/QC data-screening process was undertaken to ensure this master database was error-free and appropriately formatted for use in the study (Box 1, Figure 2).

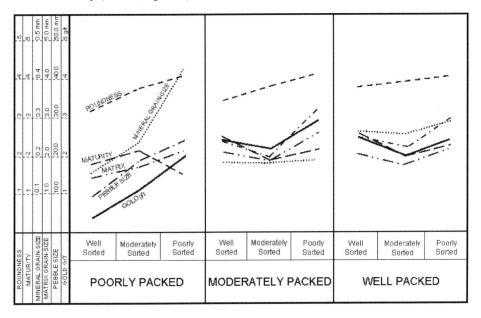

Figure 3. Relationships between gold values and selected sedimentological parameters for "sorting-packing" (pebble and larger-size material) pair groups. Values for each parameter are averaged from 528 core samples. Roundness refers to pebble or cobble roundness, mineral grain-size pertains to heavy mineral assemblage (primarily magnetite and hematite). Maturity and matrix refer to coarse sand-sized and smaller material.

3.2 IDENTIFICATION AND CREATION OF STATISTICAL FACIES

Studies of the Tarkwaian Banket by Sestini (1976) and Hirdes and Nunoo (1994) demonstrate that gold distributions reflect stratigraphic features, such as channel geometry and lithofacies, and are related to sedimentological parameters. Such relationships are geologically reasonable in placer deposits (Burton and Fralick, 2003). The master borehole database contained significantly greater numbers of observations than had been available in the earlier studies, so a series of correlations was computed between gold assay values and stratigraphic and sedimentological parameters. Although the data proved to be somewhat noisy, the conclusions reached by the earlier studies remained valid. Figure 3 shows gold values increasing as grain/pebble packing and sorting improves, and varying systematically with other selected sedimentological parameters. A k-means clustering method (Wishart, 2001) from ClustanGraphics® using a modified version of the Gower general similarity coefficient (Gower, 1971), permitting simultaneous handling of both categorical and continuous variables, was used to identify geologically meaningful groups from the parameters (Fisher, 2004). Six clusters were selected and designated as "statistical facies" (Box 2, Figure 2). The "statistical facies" can be related to real world conditions in a manner similar to the "electro-facies" concept of Doveton (1994a).

3.3 ESTABLISHMENT OF GOLD DISTRIBUTIONS FOR STATISTICAL FACIES

By coding the membership of each sample in the borehole database according to its "statistical facies," and associating the samples with their proper assay values, it was possible to compute distinct gold distributions for each "statistical facies" (Box 3, Figure 2). These distributions form the basis of gold assay **cdf**'s for each "statistical facies".

3.4 DEVELOPMENT OF 3D GEOLOGICAL FRAMEWORK

A 3D geologic framework was constructed to constrain the geostatistical simulations (Fisher, 2004). A series of correlated and tied cross-section panels was developed from the available 100m x 100m spaced DD borehole data. Tops and bases of principal gold-bearing Banket members A1, A3, and C were "picked" along with positions of the main thrust fault planes (Fisher, 2004). This interpretation provided opportunity to properly locate and correlate the main thrust faults and distinguish relationships within the three main fault blocks identified within the study area.

Stratigraphic horizons and fault planes were individually hand contoured. These maps were then digitized and cartographically registered in ESRI's ArcMap® GIS software. The results were then exported to Golden Software's SURFER® 8.0 contour and volume modeling software, where the digitized contours of each horizon were gridded to produce a series of individual 3D surfaces. Within SURFER® it was possible then to stack the constructed surfaces, including interpreted fault planes, into a 3D framework volume model (Box 5a, Figure 2 and Figure 4). This resultant model was then discretized into 5m by 5m (in the X-Y plane) by 0.5m (in the Z-direction or vertical direction) 3D cells (Box 5b, Figure 2) that were used by the T-PROGS transition probability software.

3.5 TRANSITION PROBABILITY MODELLING

Markov chain analysis forms the basis of the transition probability approach (Box 6, Figure 2). It has been successfully used in stratigraphy and sedimentology to discover statistically significant and fundamental patterns of lithological repetition (e.g., Doveton, 1994b). An initial 1D Markov model can be extended to a 3D Markov model to predict lateral distributions of sedimentary facies, where lateral variation has been under- sampled, by using Walther's Law (Middleton, 1973) and changing the associated mean lengths in accordance with geological expectations and observed field data (Weissmann, et al, 1999). The T-PROGS software (Carle, 1999) was selected for this phase of the modeling because it permits incorporation of categorical variables and subjective observations into the model. The reader is referred to Carle (1996), Carle and Fogg (1996) and Weissmann, et al, (1999) for more in-depth explanation and details.

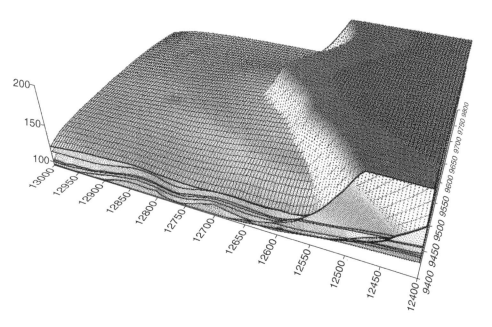

Figure 4. Display of initial 3D geologic framework model (Fisher, 2004).

3.6 SEQUENTIAL INDICATOR SIMULATION AND SIMULATED ANNEALING

Sequential Indicator Simulation, or "SIS", (Journel and Kyriakidis, 2004) was iteratively applied to the 3D Markov model transition probabilities defined in the previous section and substituted for the indicator cross-variogram in a co-kriging step (where the 3D Markov model conditioned the simulation) to produce 3D realizations of the modeled facies (Weissmann, et al, 1999). The SIS method was used because no assumptions are necessary about the shape of the gold distributions. The Markov chain model controlled factors such as global lithofacies proportions calculated from the conditioning borehole data and juxtapositional patterns. The co-kriging equations are solved using a basis function approach (Carle, 1996) - a more computationally efficient method. Subsequently, Sequential Quenching, the "zero-temperature" case of Simulated Annealing (van Laarhoven and Aarts, 1987), was applied to improve the geological reality of the SIS realization and reflect additional constraints (Fisher, 2004). The Sequential Annealing step improves the match of the final realization with the originally computed transition probabilities. The SIS realization was gradually perturbed so as to match defined characteristic lengths and facies continuities. No changes of facies were allowed at the boreholes, because these locations were "known", but at other locations a series of stochastic variations in facies assignment was possible.

Repetition of this process provided for the production of an uncertainty-based 3D distribution of "statistical facies" (Box 7, Figure 2).

3.7 DEVELOPING A 3D PROBABILISTIC MODEL OF GOLD DISTRIBUTION

Our ultimate goal was to produce a lithofacies-based gold grade uncertainty distribution (Box 8, Figure 2). This was accomplished, on a cell-by-cell basis, combining the uncertainty-based distribution of statistical facies (Box 7, Figure 2) with the previously computed **cdf** for the gold distribution of each "statistical facies" (Box 3, Figure 2). The computational process involves several steps. Because the probabilities of facies occurrences were altered during the quenching phase, the probabilities from the quenching step were used to compute a "global" **cdf** for all facies. Each model cell was examined in turn, and the most probable facies for each cell selected from this **cdf**. A facies having been selected, the appropriate **cdf** of gold distribution was selected and used to assign a gold grade to the cell. This process was repeated multiple times, producing several different gold grade estimates for each cell (Fisher, 2004).

All of the assigned grades were then used to compute a cumulative grade distribution, and uncertainties in grade, for the entire realization. This information was accumulated and presented as **pdf** plots. Values were assigned to appropriate selective mining units (SMUs). Mean grades were computed for the SMUs and the material in the SMU classified as either ore or waste according to pre-selected cutoff values. Thus, the process provides a statistically sound, lithofacies-based gold grade uncertainty prediction constrained by geology and the 3D geological framework.

4 Conclusions and Applications to Mine Planning and Operations

Construction of a statistically sound, lithofacies-based 3D gold-grade distribution model, which incorporates uncertainty in the predictions, supports several important mine planning and optimization applications. Efficiency of mine planning can be improved by using 3D geological models during exploration, using new borehole data "on-the-fly" as it becomes available.

By substituting differing geological concepts prior to the Sequential Indicator/Simulated Quenching steps, we can apply the process and methodology defined herein to orebodies other than sedimentologically controlled paleoplacers. For instance, Carlin-type deposits, layered mafic/ultramafic intrusions, massive sulfide, or other ore deposits may also be evaluated by applying appropriate 3D geological model(s).

Lastly, traditional ore control techniques classify material into ore or waste categories based on estimated average grade assuming *no uncertainty* exists on the estimated grades (Coşkun, 1997). Adding uncertainty to the estimates of grade, based on knowledge of geologic conditions, allows methodologies such as the loss function concept (Coşkun, 1997; Dagdelen and Coşkun, 1998) to be applied with greater confidence and accuracy.

Acknowledgements

We would like to acknowledge and thank Gold Fields Exploration, Inc., Denver, Colorado; Gold Fields Ghana, Ltd., Accra, Ghana, Africa; Gold Fields Ltd., South Africa, and the personnel of the Tarkwa Mine, Tarkwa, Ghana, for providing access to the Tarkwa Mine data, and for their generous funding of this research. We also thank the U.S.G.S. Water Resources Branch, Tucson, Arizona for funding modification and research of the T-PROGS Transition Probability Geostatistical Software.

SURFER® is a registered trademark of Golden Software, Inc., Golden, Colorado. ArcMap® is a registered trademark of ESRI, Redlands, California. ClustanGraphics® is a registered trademark of Clustan Limited, Edinburgh, Scotland.

References

Boadi, I. O., Norman, D. I., and Appiah, H., Source Terrane for Tarkwa Paleoplacer Deposit, Ghana, *in*, Pagel, M., and Leroy, J. L., (eds.), *Source, Transport, and Deposition of Metals*, Proceedings of the 25th Ann. SGA, Nancy France, Balkema Publishers, 1991, p. 641-648.

Burton, J. P., and Fralick, P. W., Depositional Placer Accumulations in Coarse-Grained Alluvial Braided River Systems, *Economic Geology*, vol. 98, no. 5, 2003, p. 985-1001.

Carle, S. F., *A Transition Probability-Based Approach to Geostatistical Characterization of Hydrostratigraphic Architecture*, Unpublished Ph.D. Dissertation, University of California, Davis, 1996, 233p.

Carle, S. F., *T-PROGS: Transition Probability Software, Version 2.1*, University of California, Davis – Hydrologic Sciences Graduate Group, 1999, 78p.

Carle, S. F., and Fogg, G. F., Transition Probability-Based Indicator Geostatistics, *Mathematical Geology*, vol. 28, no. 4, 1996, p. 453-476.

Coşkun, B., *Risk Quantified Ore Control*, Unpublished Master's Thesis, Colorado School of Mines, 1997, 119p.

Dagdelen, K., and Coşkun, B., Risk Quantified Ore Control in Open Pit Gold Mining, *in*, Basu, A. J., (ed.), *Computer Applications in the Minerals Industries International Symposium*, Australian Institute of Mining and Metallurgy, Publication Series No. 5/98, 1998, p. 9-12.

Doveton, J. H., *Geological Log Analysis Using Computer Methods*, AAPG Computer Methods In Geology, No. 2, 1994a, p. 169.

Doveton, J. H., Theory and Applications of Vertical Variability Measures from Markov Chain Analysis, *in*, Yarus, J. M., and Chambers, R. L., (eds.), *Stochastic Modeling and Geostatistics – Principles, Methods, and Case Studies*, AAPG Computer Applications in Geology, No. 3, 1994b, p. 55-64.

Fisher, T. R., *Three-Dimensional Sedimentological and Geostatistical Modelling in Precambrian Paleoplacers of the Tarkwa District, Ghana*, Unpublished Ph.D. Dissertation, Colorado School of Mines, 2004, (in preparation).

Fisher, T. R., and Turner, A. K., Application of hybrid 3D modelling methods to prediction of ore grades in stratabound deposits, *in, Proceedings of IAMG Berlin*, September 2002.

Frostick, L. E., and Reid, I., Tectonic Control of Desert Sediments in Rift Basins Ancient and Modern, *in*, Frostick, L. and Reid, I., *Desert Sediments: Ancient and Modern*, Geological Society of London Special Publication No. 35, 1987, p. 53-68.

Gold Fields Ghana Limited, *A Technical Report of the Tarkwa Gold Mine, Ghana*, Gold Fields Ghana Ltd, and Iamgold Corporation, http://www.iamgold.com/reports/etc/Tarkwa-TechRpt.Final.pdf, 2003, 50p.

Gower, J. C., A General Coefficient of Similarity and Some of Its Properties, *Biometrics*, vol. 27, p. 857-874.

Hirdes, W., and Nunoo, B., The Proterozoic Paleoplacers at Tarkwa Gold Mine, SW Ghana: Sedimentology, Mineralogy, and Precise Age Dating of the Main Reef and West Reef, and Bearing of the Investigations on Source Area Aspects, *in*, Oberthur, T., (ed.), *Metallogenesis of Selected Gold Deposits in Africa*, Geologisches Jahrbuch, Reihe D, Heft 100, 1994, p. 247-311.

Journel, A. G., and Kyriakidis, P. C., *Evaluation of Mineral Reserves: A Simulation Approach*, Oxford University Press, New York, 2004, 216p.

Middleton, G. V., Johannes Walther's Law of the Correlation of Facies, *Geological Society of America Bulletin*, vol. 84, p. 979-988.

Sestini, G., Sedimentology of a Paleoplacer, The Gold-bearing Tarkwaian of Ghana, *in* Amstutz, G. C., and Bernard, A. J., (eds.), *Ores in Sediments*, IUGS, Series A, No. 3, Springer-Verlag, 1976, p. 275-305.

Sylvester, P. J., and Attoh, K., Lithostratigraphy and Composition of 2.1 Ga Greenstone Belts of the West African Craton and Their Bearing on Crustal Evolution and the Archean-Proterozoic Boundary, *Journal of Geology*, vol. 100, p. 337-393.

van Laarhoven, P. J. M., and Aarts, E. H. L., *Simulated Annealing: Theory and Application*, D. Reidel Publishing Company, Dordrecht, 1987, 186p.

Ward, J., Hierarchical Grouping to Optimize an Objective Function, *Journal of the American Statistical Association*, vol. 58, 1963, p. 236-244.

Weissmann, G. S., Carle, S. F., and Fogg, G. E., Three-Dimensional Hydrofacies Modeling Based on Soil Surveys and Transition Probability Geostatistics, *Water Resources Research*, vol. 35, no. 6, 1999, p. 1761-1770.

Wishart, D., k-Means Clustering with Outlier Detection, Mixed Variables, and Missing Values, *Proceedings of GfKl 2001 (25th Annual Conference of the German Classification Society, The University of Munich, 14-16 March, 2001, Munich)*.

CONDITIONAL SIMULATION OF GRADE IN A MULTI-ELEMENT MASSIVE SULPHIDE DEPOSIT

N.A. SCHOFIELD
Hellman and Schofield, P.O. Box 599, Beecroft, NSW 2119, Australia

Abstract. In the past decade, conditional simulation methods have been used widely to model the distribution of grades in precious and base metal deposits. Often a number of simulations are used as a basis for evaluating the risk of ore and waste misclassification and improving ore selection practices. The complexity of the application of simulation methods can depend on the nature of the mineralization being modelled. A single element deposit such as gold in a disseminated style of mineralization typical of epithermal gold deposits may be an example of a less complex application. Multi-element deposits with multiple geostatistical sample populations in complex structural settings such as the Cannington silver-lead-zinc deposit represent more challenging applications. This paper discusses a method to generate relatively large scale conditional simulations of mineralization geometry and multiple elements in such deposits.

1 The Mineral Deposit

The Cannington base metal deposit was discovered in 1990 by BHP Minerals (Bailey, 1998). The silver-lead-zinc mineralization is associated with a diverse package of siliceous and mafic rocks with extensive retrogression and alteration. A zoning of base metals is evident within the Southern Zone which is consistent with the interpreted isoclinal fold structure. The lode horizons are defined by the spatial distribution of the base metals. The mineralization types totalling 10 to date, describe the geometry, economic, geochemical and textural relationships within the deposit. Locally, the mineralised sequence around the fold shown in Figure 1 is commonly referred to as the Footwall (CW, NS and CK mineralization types), Hanging-wall (BM, BL and KH) and Hinge (GH, GHB) areas. Mining began in the rich silver-lead-zinc concentrations hosted mainly within the Glenholme mineralization (GH, GHB) within the Hinge area.

2 Modelling using Conditional Simulation

Since early 2000, the Cannington mine geologists have been experimenting with modelling of the distribution of mineralisation types and mineralisation grade using a combination of Probability Field conditional simulation (PF) (Froidevaux, 1992, Srivastava, 1990) and Sequential Gaussian Simulation (SGS) (Gomez-Hernandez and Journel, 1992). The process is ongoing. The focus of this paper is the distribution of lead, zinc and silver within a suite of mineralised units in the Hinge area marked GHB in Figure 1.

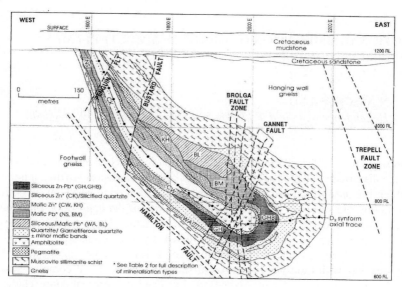

Figure 1: Geologic section through the Cannington Deposit (after Bailey, 1998)

3 The Drill-Hole Data and Statistics

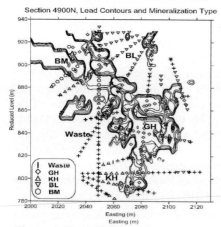

Figure 2: Typical drill-hole cross

Figure 2 presents a typical drill-hole cross-section through the mineralization in the Hinge area. The spatial distribution of the logged mineralisation types is shown by different symbols while the contours map the lead grade concentration in the drill holes. The higher grades of lead and silver occur in the Glenholme (GH) mineralization which forms the central body in the Hinge area. The geometry of the mineralization types varies along the northerly extension of the fold nose, pinching and swelling in response to structural influences. The overall trend of the mineralised structure is around N12E and inclined at around 10 degrees. The lower grade KH mineralization occurs to the west of the GH while the lower grade BL mineralization occurs to the east of and above the GH. The hanging wall lead mineralization (BM) occurs in the upper left quadrant of the section.

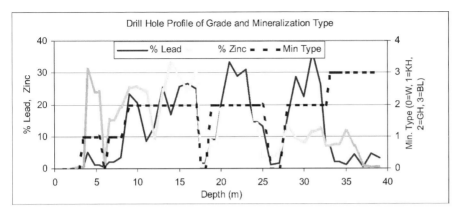

Figure 3: Typical drill-hole profile of lead-zinc and mineralization type.

	Lead %				Silver ppm				Zinc %			
	KH	GH	BL	BM	KH	GH	BL	BM	KH	GH	BL	BM
Mean	2.07	14.28	2.15	8.38	64	573	92	283	4.48	8.58	1.70	5.88
Std. Dev.	4.99	11.83	4.40	10.56	129	563	496	370	6.33	6.08	2.83	5.78
Minimum	0.0	0.0	0.0	0.0	0	0.0	0.0	0.0	0.0	0.0	0.0	0.0
Median	0.13	11.70	0.28	2.43	22	410	15	96	1.15	7.89	1.15	4.72
Maximum	40.80	60.00	44.25	43.10	1550	10700	21900	2600	33.40	38.50	33.40	35.50
Number	2779	7120	2779	1234	2779	7120	2779	1234	2779	7120	2779	1234

Table 1: Summary statistics of lead, silver and zinc in the mineralization types.

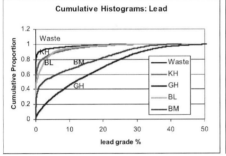 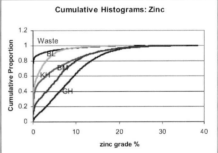

Figure 4: Cumulative histograms of lead and zinc in the mineralization types.

Figure 3 shows a typical drill-hole profile across the mineralization emphasising the sharp changes in lead grade that occur at the transition of one mineralization type to another. Silver tends to follow lead closely but not so zinc as the figure shows.

Summary statistics of lead, silver and zinc in all mineralization types are shown in Table 1. Figure 4 presents cumulative histograms of lead and zinc in all mineralization types. The economic dominance of the GH mineralization is obvious from average concentrations of lead, silver and zinc in this mineralization. Bivariate correlation matrices shown in Figure 5 indicate lead and silver are strongly correlated in all mineralization types. Linear correlations are shown in the upper triangle and the rank order correlations are shown in the lower triangle.

GH Mineralization			
Elements	Lead	Silver	Zinc
Lead	1.000	0.820	0.515
Silver	0.910	1.000	0.296
Zinc	0.662	0.532	1.000
KH Mineralization			
Lead	1.000	0.904	0.516
Silver	0.827	1.000	0.434
Zinc	0.720	0.681	1.000
BM Mineralization			
Lead	1.000	0.901	0.456
Silver	0.942	1.000	0.365
Zinc	0.701	0.672	1.000

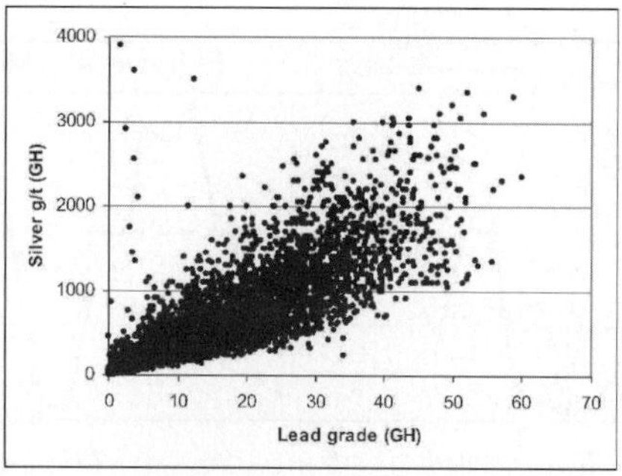

Figure 5: Element correlations and silver-lead scatter-plot in GH.

4 Spatial Continuity of Mineralization Geometry

A numerical sequence is used to describe the mineralization types. Where possible, the sequence should correspond to the physical geological sequence such as the stratigraphic sequence or a nested pattern of alteration. The numerical sequence used in the present case is shown in Table 2 below.

Mineralization	Waste	KH	GH	BL	BM
Sequence no.	0	1	2	3	4

Table 2: Numerical coding of the sequence of mineralization types

The numerical sequence allows continuity of the mineralization types to be described in terms of indicator continuity functions. These functions describe the transition from one set of mineralization types to another e.g. the indicator continuity function for the threshold 0.5 describes the transition from the Waste type to all other mineralization types. Figure 6 shows the geometry indicator continuity maps for the four transition thresholds in the horizontal plane.

5 Spatial Continuity of Lead, Silver and Zinc Grades

The normal scores transforms (Deutsch and Journel, 1992) of these elements are used to describe their spatial continuity within each mineralization type. Inadequate data or the complexity of the mineralization geometry can cause difficulties in the description of these continuities for some mineralization types. In such cases, the element continuity properties of the dominant mineralization type may be adopted or the data from two or more mineralization types may be combined to provide a more appropriate description of the element continuities for those types.

6 Conditional Simulation of Mineralization Types and Grades

The spatial distribution of the mineralization types can be simulated using the method of Probability Field (PF) simulation proposed by Srivastava 1990. In the present case, the non-conditional field of probabilities is generated with the Sequential Gaussian Simulation method (SGS). The conditioning of the probability field to the local mineralization type data is based on the estimate of cumulative probability for each threshold in the numerical sequence of mineralization types shown above. The approach is a very fast way to generate large scale conditional simulations of categorical variables which are both geologically and statistically acceptable. Table 3 below compares the proportions of simulated mineralization types in two conditional simulations compared to the data coding. The differences between the simulations and the data are mainly due to data clustering.

Volume Proportion of Mineralization Types					
	Waste	KH	GH	BL	BM
Samples	0.321	0.140	0.358	0.119	0.062
Simulation 1	0.387	0.099	0.306	0.140	0.068
Simulation 2	0.376	0.099	0.305	0.145	0.074

Table 3: Volume proportions of mineralization types in samples and two simulations.

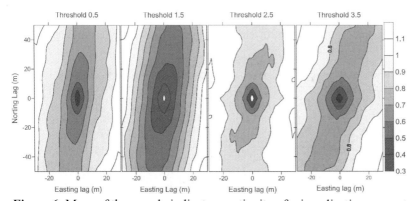

Figure 6: Maps of the sample indicator continuity of mineralization geometry

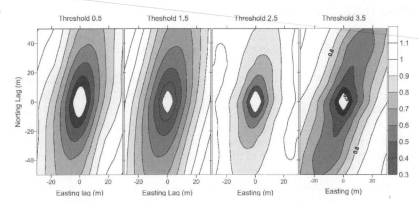

Figure 7: Maps of simulation indicator continuity of mineralization geometry

The continuity maps of the simulated mineralization types shown in Figure 7 show greater continuity at short scale than that of the data shown in Figure 6. This is a deliberate choice to make the maps of mineralization types more realistic, generating relatively smooth contacts between the mineralization types. Simulations of the metal grade distributions for each mineralization type using SGS are found to honour reasonably the univariate, bivariate and spatial properties of the metals (not shown).

Within each mineralization type, the metal distributions are simulated to be independent of the mineralization type boundary as suggested in Figure 3. Figure 8 presents section maps showing the spatial distribution of simulated mineralization types and simulated lead grades.

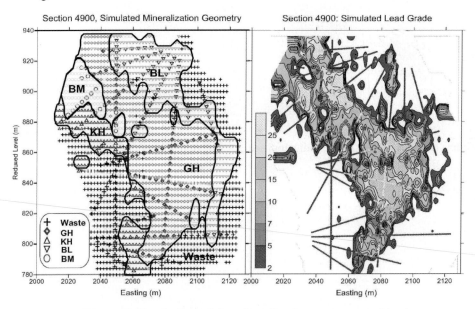

Figure 8: Simulations of the mineralization geometry and lead

7 Conclusions

The Cannington lead-zinc-silver mineralization presents a significant challenge to resource modelling using conditional simulation. The spatial distribution of the mineralization types is complex and dominates the problem of ore definition.

The approach described in this paper generates spatial conditional simulations which reasonably reproduce the statistical and spatial properties of the input data and provide plausible models of the distribution of mineralization types and metal grades.

Results of using the simulations for stope planning and grade prediction indicate the simulations provide an improved basis for locating stopes and reliable predictions of overall stope grades compared to the manual modelling of mineralization type distributions and ordinary kriging estimation of grades. Simulations also provide a better appreciation of the variability in the ore grade that will be realized over time in the mining of stopes.

Acknowledgements

The author acknowledges the cooperation of BHP-Billiton, in particular Alan Edwards in undertaking the work that forms the basis of this paper.

References

Bailey, A., Cannington silver-lead-zinc deposit. in *Geology of Australian and Papua New Guinea Mineral Deposits*, (Eds: A. A. Beckman and D.H. Mackenzie), pp 782-792, 1998. The Australian Institute of Mining and Metallurgy, Melbourne, Australia.

Deutsch, C.V. and Journel, A.G., GSLIB Geostatistical Software Library and User's Guide. Oxford University Press, New York, 1992.

Gomez-Hernandez, J.J. and Journel, A.G., Joint Sequential Simulation of MultiGaussian Fields. In *Geostatistics Troia '92* (Ed. A. Soares), Volume 1, pp85-94, 1992. Kluwer Academic Publishers, London.

Froidevaux, R. Probability Field Simulation. In *Geostatistics Troia '92* (Ed. A. Soares), Volume 1, pp 73-84, 1992. Kluwer Academic Publishers, London.

Srivastava, R.M., An Application of Geostatistical Methods for Risk Analysis in Reservoir Management. 1990. *SPE Paper No 20608.*

THE ULTIMATE TEST – USING PRODUCTION REALITY. A GOLD CASE STUDY

PAUL BLACKNEY, CHRISTINE STANDING and VIVIENNE SNOWDEN
Snowden Mining Industry Consultants
PO Box 77, West Perth, Western Australia, AUSTRALIA, 6872

Abstract. The McKinnons gold mine is owned by Burdekin Resources NL ('Burdekin') and is located within the Cobar Basin in New South Wales, Australia. Open pit mining began in February 1995 and was completed in December 1996. High grade, low grade and mineralised waste stockpiles were processed in separate campaigns up to April 2000 and this production data provides a unique opportunity to test the accuracy of industry standard estimation techniques.

A multi-support data set is available from the mine, comprising exploration reverse circulation (RC) and diamond core (DD) drillholes (25 m sections), RC grade control drilling (12 mE by 12 mN and 6 mE by 6 mN patterns) and some blasthole data. This information was used to create comparative feasibility-stage and grade control resource estimates which were compared with each other and with the actual tonnes and grades processed.

Using the wide-spaced data, feasibility-stage resources were estimated using ordinary kriging, multiple indicator kriging, uniform conditioning, conditional simulation and a global change of support method. Kriging neighbourhood analysis and conditional bias tests were used to determine appropriate panel sizes for estimation. Some models were deliberately constructed using panel sizes that led to conditional bias to test the effect of bias in mine reconciliation. Comparative grade control estimates were then created from the close-spaced data to demonstrate the effects of this additional information.

Estimations were found to be relatively insensitive to estimation method and the band of uncertainty for models based on exploration data was within ± 10% in terms of tonnes and grade. Although confidence limits improved significantly to within ± 3% in tonnes and grade for grade control estimates, it was found that estimates were sensitive to whether hard or soft boundaries were used to control the estimation.

The study illustrates that the bounds of uncertainty reflected by conditional simulations are indicative of the confidence limits for a given choice of geological model and the associated geostatistical parameters and estimation technique. Any changes in these inputs will all have some affect on the accuracy of the estimate. Even production estimates may be affected by perception, particularly related to the effective cut-off applied.

Practitioners should be aware that resource risk will include cumulative errors arising from a complex process. Once projects are in production it is important to check and validate the key assumptions made during the project evaluation. The only way to achieve this satisfactorily is through the process of reconciliation, that is the comparison of actual production (tonnes, grade and metal) with predictions (resources, reserves and mine plans). This is an often-ignored but vital aspect of the mine value chain, and allows a reality check on the feasibility process which can trigger remedial action if necessary.

1 Introduction

During closure of the McKinnons gold mine, Burdekin compiled a comprehensive data package containing feasibility-stage data, mining grade control data and gold production results for high and low grade ore types. Elliot et al (2001) described the reconciliation of production data compared with the feasibility study estimates prepared before mining began in 1995. The input data has now been used in a study to quantify the accuracy of current day industry standard estimation techniques used to estimate resources in the Australian gold mining sector.

2 Outline of study

There are a number of nodes of uncertainty which can impact on resource estimation, including data integrity, geological interpretation, grade estimation error and ore mining control. The uncertainty may be demonstrated by comparing the tonnes, grade and metal estimates of planning models with production reality. This is currently topical in Australia because of the proposed revision to the JORC Code (JORC 1999) which encourages the Competent Person to quantify risk/uncertainty associated with resource/reserve estimates.

In this study, resource models were created using the wide-spaced drillhole and mineralisation domains as interpreted during the McKinnons feasibility study. Comparative grade control models were created from the close-spaced mining data, more detailed mineralisation domains and, for some estimates, the production grade control polygons as mined. Estimation methods included ordinary kriging (OK), multiple indicator kriging (MIK), sequential indicator simulation (SISIM), uniform conditioning (UC) and global change of support (GCOS) methods. Model block sizes were determined from kriging neighbourhood analysis and conditional bias tests. The actual tonnages and grades processed during the life of the mine were used to test the accuracy of the exploration and grade control models using the various estimation methods.

3 Background

Between 1990 and 1994, RC and diamond drilling on 25 m sections was used to define the resource for feasibility studies. Following the decision to mine in 1994, the entire deposit was drilled on a 12 mE by 12 mN pattern with some infill on a 6 mE by 6 mN pattern (*Figure 1*) to provide pre-production grade control modelling.

All high grade ore (>1.3 g/t Au) was processed by November 1997 and low grade stockpiles (0.7 to 1.3 g/t Au) were processed by October 1998. Processing of mineralised waste (0.3 to 0.7 g/t Au) was completed in April 2000. The actual production for each ore type is listed in *Table 1*

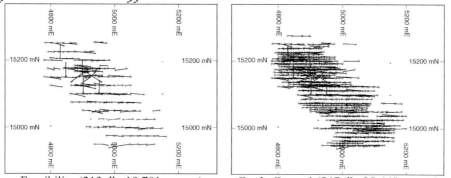

Figure 1 McKinnons drillhole location trace plans.

Ore	K tonnes	Grade
High Grade (>1.3 g/t)	1,204	2.64 g/t
Low Grade (0.7 to 1.3 g/t)	399	1.08 g/t
Mineralised Waste (0.3 to 0.7 g/t)	1,067	0.65 g/t
Total	**2,670**	**1.61 g/t**

Table 1 McKinnons production data (Elliot et al 2001).

4 Estimation methods

Kriging neighbourhood analysis and conditional bias tests were employed prior to estimation to determine a suitable block size for unbiased estimation (Krige, 1996). OK is optimal for near normal distributions (Goovaerts, 1997) but is often used in estimation of skewed data, together with capping or top cutting of sample grades. MIK and UC are non-linear techniques which accommodate highly-skewed or mixed data distributions (Glacken and Snowden, 2001) and deliver recoverable resources which represent the appropriate selective mining unit (SMU).

SISIM (Deutsch and Journel, 1998) can be used to produce a number of equally-probable realisations which can be re-blocked to represent the dimensions of a target SMU and reported to determine the probability and grade above cut-off for each SMU. The tonnage and grade spread of the simulation results reflect the confidence limits in tonnage and grade and can be used to derive the confidence limits associated with a given mining area or period.

5 Feasibility data analysis and modelling

For this study, the available datasets and subsequent models were clipped to the limits of the end-of-mine open pit survey to define a consistent space for comparison. The statistical and spatial character of the feasibility data was investigated within sub-domains of a global envelope defined at a grade threshold of ~0.1 g/t Au. Within this envelope three main statistical domains were defined, comprising a central high grade domain separating two lower grade domains. The geology is complex and mineralisation is believed to be structurally controlled and so domains were essentially defined by grade. The gold grade distributions within all domains displayed a high positive skewness, with coefficients of variation of approximately 2.5.

Gold grade continuity was primarily investigated using standardised indicator variograms. The indicator analysis revealed nugget effects representing 10% to 43% of the total sill, with most thresholds displaying a nugget effect of 20% to 35% of the total sill. All three domains displayed patterns of rotating anisotropy associated with increasing grade. The variogram model parameters from these analyses were used to control MIK estimation and SISIM computations. OK estimation was based on back-transformed variogram model parameters.

A kriging neighbourhood analysis, using the median indicator variogram, was employed to determine the optimal kriging block size for the high grade central domain (Krige, 1996). From a range of block sizes and locations tested, a 12.5 mE by 12.5 mN block on a 2.5 m bench was identified as optimal, with this block size returning regression slopes of up to 0.92 and kriging efficiencies of up to 68%. OK and MIK models were estimated using this block size and conditional bias sensitivity models were created using block sizes of 5 mE by 5 mN on 2.5 m benches and 25 mE by 25 mN on 5 m benches.

Simulation models were based on 100 realisations generated on a node spacing of 2.5 mE by 2.5 mN by 1.25 mRL. The realisations were re-blocked to supports of 5 mE by 5 mN and 12.5 mE by 12.5 mN blocks on 2.5 m benches, and 25 mE by 25 mN blocks on 5 m benches, and presented in terms of the median and 90% confidence limits.

6 Grade control data analysis and modelling

The grade control (GC) data comprised all the original wide-spaced exploration drillholes plus the additional close-spaced RC drilling completed for grade control (*Figure 1*). Categorical indicator kriging using a 0.1 g/t Au indicator grade was applied within three large scale domains to define the grade control mineralised envelope. The categorical model was visually compared to the input drillhole data to determine a probability threshold that closely reproduced the spatial patterns evident in the drillholes (*Figure 2*, left). The mineralised envelope was then further domained into seven regions that reflected a weakly developed depletion zone overprinting primary grade continuity trends (*Figure 2*, right).

The domaining was successful in producing singular distributions for the lower grade domains but mixed distributions were still evident in the higher grade domains. An indicator method was therefore selected for grade estimation. The indicator variography analysis revealed nugget effects ranging from 40% to 75% of the total sill and ranges of a few metres to 30 m vertically and from 15 m to 190 m in the horizontal plane depending on domain and indicator grade. Two of the seven domains exhibited rotating anisotropy associated with increasing indicator grade.

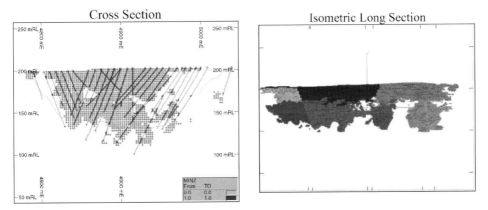

Figure 2 GC sections, 0.1 g/t Au indicator section (left) and structural domains (right).

Block grades were estimated using both MIK and OK. The grade control estimation block size was set to 5 mE by 5 mN on 2.5 mRL benches to reflect the expected mining selectivity. Two MIK estimates were computed, one with a hard boundary (HB) between structural domains and a second using a soft boundary (SB) condition. The OK estimate was computed using the hard boundary and grade top-cuts, which varied by domain. *Figure 3* illustrates the presentation of these three models on a typical bench.

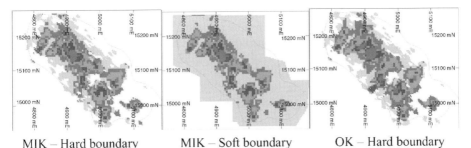

MIK – Hard boundary   MIK – Soft boundary   OK – Hard boundary
Figure 3 Example bench plans grade control block models.

A kriging neighbourhood analysis using the OK variogram later revealed significant variation in regression slope and kriging efficiency by domain. For example, in a well drilled domain, 96% of the model blocks have regression slopes better than 0.9 and 83% of the blocks have kriging efficiencies better than 70%. In contrast, in a more poorly drilled domain, only 66% of the blocks have regression slopes that are better than 0.9 and only 21% of the blocks have kriging efficiencies better than 70%.

One hundred realisations of the grade control data were generated using SISIM. The domain controls and parameters for the simulation were the same as those applied to the HB MIK. A node spacing of 2.5 mE by 2.5 mN by 1.25 mRL was employed

Simulations were re-blocked to represent the behaviour of 5 mE by 5 mN by 2.5 mRL and 10 mE by 10 mN by 2.5 mRL SMU's for comparison with the kriged models (*Figure 4*).

Polygons representing the production grade control interpretation were available for the upper part of the pit. This interpretation was used to create a traditional polygonal model by calculating the top-cut average grade of the grade control samples located within each polygon.

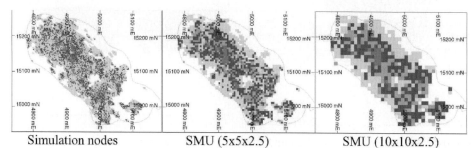

Simulation nodes　　　　SMU (5x5x2.5)　　　　SMU (10x10x2.5)
Figure 4 Example median grade control SISIM bench plans

7 Comparative Results

7.1 EXPLORATION MODELS

7.1.1 OK versus MIK
The OK and MIK exploration models demonstrated very little global sensitivity to estimation method. Estimation selectivity showed only marginal changes as the block size was increased from 5 mE by 5 mN by 2.5 mRL to 25 mE by 25 mN by 5 mRL due to the diffuse nature of the mineralisation and the relatively high nugget effect.

7.1.2 Kriging versus UC versus GCOS
The closest comparison between the theoretical GCOS and UC estimates of the 6.25 mE by 6.25 mN by 2.5 mRL SMU grade-tonnage relationship occurred for the 25 mE by 25 mN by 5 mRL OK model. This is despite the kriging neighbourhood analysis supporting the smaller 12.5 mE by 12.5 mN by 2.5 mRL panel size for kriging. This result probably reflects the limitations of the simple kriging neighbourhood analysis completed for this study and demonstrates that overall, estimation into the larger panel size suffers less conditional bias.

7.1.3 Kriging versus simulation
The comparisons between the exploration OK, MIK and SISIM models are illustrated in *Figure 5*. The models are based on a block size of 12.5 mE by 12.5 mN by 2.5 mRL. There is a bias noted whereby the average grades of the OK and MIK models are located towards the upper confidence limits defined by the SISIM models which is believed to be due to differences in treatment of the high grade tail of the grade distribution.

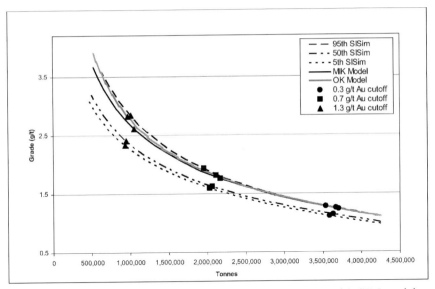

Figure 5 Grade-tonnage curves of the exploration OK, MIK and SISIM models.

The simulations validate the optimal model block size determined by kriging neighbourhood analysis as the kriged block models reflect a similar grade/tonnage profile to the simulations. The OK model appears to be slightly oversmoothed compared with the MIK model.

Using SISIM to quantify confidence limits, and making the assumption that the range of simulations between the 5th and 95th percentiles give a reasonable representation of the space of uncertainty, the 90% confidence limit at a cut-off of 0.3 g/t is $\pm 1\%$ on tonnage and $\pm 7\%$ on grade. At a 0.7 g/t cut-off, the confidence limit is $\pm 2\%$ on tonnage and $\pm 10\%$ on grade. The tonnage uncertainty again increases at a 1.3 g/t cut-off, where the confidence limit is $\pm 4\%$ on tonnage but the confidence limit remains at $\pm 10\%$ in grade.

7.2 GRADE CONTROL MODELS

7.2.1 OK versus HB MIK versus SB MIK

Grade-tonnage reporting from the HB MIK, SB MIK and HB OK models within pit is presented in *Figure 6*. The highest metal profile is presented by the HB OK model, followed by the HB MIK and then the SB MIK models. Tonnage and grade reporting is similar between all estimates at the 0.7 g/t and 1.3 g/t cut-offs, however the influence of the soft boundary assumption is readily apparent at the 0.3 g/t cut-off. Tonnage and grade predictions at the 0.7 g/t and 1.3 g/t cut-offs are within 0% to 8% of each other for all models. At the 0.3 g/t cut-off, the SB MIK model predicts 20% more tonnage at 16% less grade.

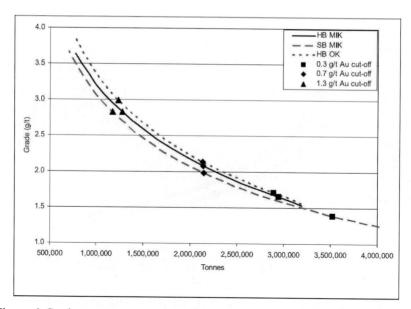

Figure 6 Grade-tonnage comparisons between HB MIK, SB MIK and HB OK grade control estimates.

7.2.2 HB MIK versus SISIM

The comparison of the HB MIK estimate to the SMU predictions provided by SISIM indicated that the grade-tonnage profile of the kriged estimate was closer to that predicted for a 10 mE by 10 mN by 2.5 mRL SMU rather than the 5 mE by 5 mN by 2.5 mRL block size used during estimation (*Figure 7*).

This outcome is somewhat at odds with the results of the kriging neighbourhood analysis which suggested there was minimal conditional bias at a 5 mE by 5 mN by 2.5 mRL block size within the majority of the pit. Some of the mineralisation domains located deeper within the pit may be oversmoothed but it is surprising that the simulations show that the effective resolution of the kriging overall is 10 mE by 10 mN by 2.5 mRL. This outcome may be a function of the relatively high nugget effect shown by the variograms.

Using the 10 mE by 10 mN by 2.5 mRL re-blocked simulations to define the limits of uncertainty for grade control, the 90% confidence limit is $\pm 0\%$ for tonnage and $\pm 3\%$ for grade at a cut-off of 0.3 g/t. At a 0.7 g/t cut-off, the confidence limits are $\pm 1\%$ for tonnage and $\pm 2\%$ for grade. The tonnage uncertainty increases slightly to $\pm 3\%$ at a 1.3 g/t cut-off, but remains at $\pm 2\%$ for grade.

8 Reconciliation of models with production

Actual production is shown, together with the comparable exploration UC and grade control MIK estimates in *Figure 8*. There is minimal uncertainty due to estimation block size, with the UC change of support from the 12 mE by 12 mN by 2.5 mRL model being marginally lower grade than from the 25 mE by 25 mN by 5 mRL model.

The HB MIK grade control model is at the upper limit of the estimates while the SB MIK grade control model presents a grade-tonnage profile closer to that of the production reporting, except that this model suggests the presence of considerably more tonnage than realised at a cut-off of 0.3 g/t. If the HB MIK model is considered to be appropriate, there may have been potential to improve the metal recovery (*Figure 9*).

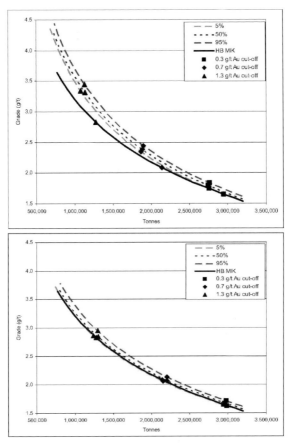

Figure 7 Grade-tonnage of HB MIK (5 mE by 5 mN by 2.5 mRL) with 90% confidence limits of SISIM at 5 mE by 5 mN by 2.5 mRL support (top) and 10 mE by 10 mN by 2.5 mRL support (bottom).

At the high grade cut-off of 1.3 g/t, the HB MIK model suggests slightly more tonnes at a higher grade might have been achieved. At the cut-off of 0.7 g/t, the model suggests considerably more tonnage could have been recovered at a similar head grade and at the 0.3 g/t cut-off more tonnage could have been recovered.

The 0.7 g/t production estimate represents less tonnes at higher grade than any of the models, perhaps suggesting the actual SMU cut-off applied during mining was higher than 0.7 g/t. This could well be due to the polygonal grade control approach applied to unsmoothed sample grades and would be in line with the volume-variance relationship.

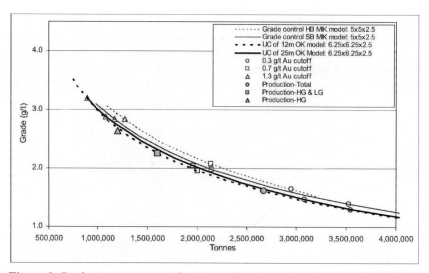

Figure 8 Grade-tonnage curves for the exploration and grade control models compared with production results.

The average grades of the HB MIK and SB MIK models within the production outlines as dug agree with those determined by the polygonal modelling (*Figure 10*). However, it is possible some metal was misclassified and the effective lower cut-off of the production model is about 0.5 g/t (as per the block cut-off for both the hard and soft boundary models) rather than 0.3 g/t as reported by production records.

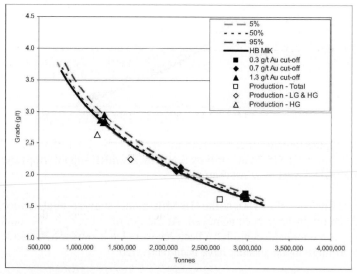

Figure 9 Actual production results and grade-tonnage curves for HB MIK (5 mE by 5 mN by 2.5 mRL) with 90% confidence limits of SISIM at 10 mE by 10 mN by 2.5 mRL support.

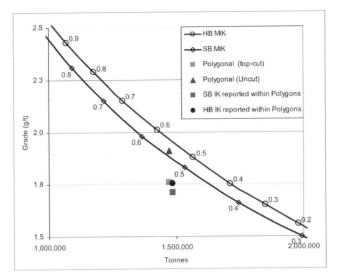

Figure 10 Grade-tonnage reporting above 175 mRL within production interpretation compared to HB MIK and SB MIK.

9 Conclusions

Exploration models show very little sensitivity to estimation method and change of support models using UC are comparable with theoretical GCOS results. The quality of the estimate using 25 mE by 25 mN by 5 mRL blocks with change of support to 6.25 mE by 6.25 mN by 2.5 mRL SMUs is marginally improved compared with estimating into the optimal smaller blocks identified by kriging neighbourhood analysis.

Global 90% confidence limits for 12.5 mE by 12.5 mN by 2.5 mRL blocks based on SISIM of exploration data show tonnage can be estimated within ±1% and grade within ±7% at a 0.3 g/t cut-off. At 0.7 g/t the tonnage uncertainty increases to ±2% and grade to ±10%. At 1.3 g/t, the tonnage uncertainty increases to ±4% and grade uncertainty does not change.

A surprising degree of sensitivity is evident depending on whether grade control models are based on hard or soft domain boundaries and, indeed, whether based on a polygonal estimation method.

Global 90% confidence limits for 10 mE by 10 mN by 2.5 mRL blocks based on SISIM of grade control data show tonnage can theoretically be estimated within ±1% and grade within ±3% at cut-offs of 0.3 g/t and 0.7 g/t cut-off. At 1.3 g/t, the tonnage uncertainty increases to ±3%.

Optimal block sizes defined for exploration models by kriging neighbourhood analysis are confirmed by re-blocking of conditional simulations. However, the effective resolution of the grade control kriging is 10 mE by 10 mN by 2.5 mRL, which is somewhat at odds with the results of the kriging neighbourhood analysis which supports blocks of 5 mE by 5 mN by 2.5 mRL.

The authors have observed complementary behaviours, for example the close alignment of the volume-variance relationship using different techniques, but have also noted unexpected biases, for example at the grade control stage, where the uncertainty should be minimised, yet there remains a sensitivity related to the treatment of domain boundaries.

The learning from this study is that no one approach can be guaranteed fool-proof and, although theoretical tests are partially successful, there will always remain a degree of uncertainty due to inherent variability, geological interpretation and boundary control, estimation technique, scale of mining and human perception. The use of contingencies during feasibility assessments is recommended to determine the impact of both high and low scenarios on the value of the project under consideration.

Acknowledgements

We would like to acknowledge the considerable efforts of M. Franks, S. Hackett and I. Glacken in assisting with this study, Burdekin Resources N.L. for permitting us to use the data from the McKinnons gold deposit for this paper and Snowden Mining Industry Consultants Pty Ltd for providing time and resources for this study to be undertaken.

References

Deutsch, C.V., and Journel, A.G., 1998. GSLIB Geostatistical Software Library and User's Guide. Second Edition. (Oxford University Press: New York).

Elliott, S.M., Snowden, D.V., Bywater, A., Standing, C.A. and Ryba, A., Reconciliation of the McKinnons Gold Deposit, Cobar, New South Wales, in *Mineral Resource and Ore Reserve Estimation – The AusIMM Guide to Good Practice* (Ed: A.C. Edwards), 2001, p. 257-268.

Glacken, I M. and Snowden, D V, Mineral Resource Estimation in Mineral Resource and Ore Reserve Estimation – The AusIMM Guide to Good Practice (Ed: A.C. Edwards), 2001, p. 189-197.

Goovaerts, P., Geostatistics for Natural Resources Evaluation. Oxford University Press. 1997, 483 pp

Joint Committee of the Australasian Institute of Mining and Metallurgy, Australasian Institute of Geoscientists, and Minerals Council of Australia, 1999. Australasian Code for reporting of identified Mineral Resources and Ore Reserves, 1999 edition.

Khosrowshahi, S, and Shaw, W J 1997. Conditional simulation for resource characterisation and grade control – principles and practice, in Proceedings World Gold '97 conference, pp 275-282 (The AusIMM, Singapore, 1997).

Krige, D G, 1996. A practical analysis of the effects of spatial structure and of data available and accessed, on conditional biases in ordinary kriging, in Geostatistics Wollongong '96 (Eds: Baafi E Y, and Schofield, N A) pp 799-810 (Kluwer, The Netherlands, 1997).

ORE-THICKNESS AND NICKEL GRADE RESOURCE CONFIDENCE AT THE KONIAMBO NICKEL LATERITE (A CONDITIONAL SIMULATION VOYAGE OF DISCOVERY)

MARK MURPHY[1], HARRY PARKER[2], ANDREW ROSS[1] and MARC-ANTOINE AUDET[3]
[1]Snowden Mining Industry Consultants, [2]AMEC Americas Limited,
[3]Falconbridge Nouvelle Caledonie SAS

Abstract. Tropical weathering on the ridges of the Koniambo massif in New Caledonia has produced nickel mineralisation of variable thickness. Conditional simulation studies of nickel grade and ore-thickness (a proxy for ore tonnage) were used to quantify the resource risk and to generate constraining envelopes for resource classification.

Ore-thickness intercepts were created from vertical drilling and converted to 2D point data. Many drillholes that did not meet the selection criteria were included as barren, and these holes imposed a strong positive skewness on the data histograms. Indicator variography revealed that both grade and ore-thickness continuity is quasi-isotropic. One-hundred 2D sequential indicator conditional simulations were generated for each attribute on a 10 m by 10 m grid for the three deposit areas. This paper focuses on results from one area, the Centre sector.

The 2D simulation realisations were reblocked to generate a panel mean for each simulation, and the distribution of the 100 panel means were found to be near normal. Tonnages were computed for each panel from the mean simulation thickness and deposit-average bulk density. Relative 90% confidence limits were then computed for each panel using normal distribution assumptions, the panel distribution standard deviations, and the panel means. However, because confidence limits also depend on production rate, the panel relative 90% confidence limits were scaled to the production increment (quarterly, annual) of interest by incorporating assumptions from the standard error of the mean. The rescaled values revealed the risk on an annual production basis was low for both attributes. On a quarterly production basis, the nickel grade risk was low in all areas, but only areas of close-spaced drilling achieved target levels of tonnage risk. The relative 90% confidence risk maps were then used as a guide for resource classification of the deposit and also to focus an infill drilling programme to support a feasibility study to be used by the sponsors to finance the project.

The simulation method takes the local variability of the data into account, a clear advance over traditional estimation variance techniques. The requisite drillhole spacing required to achieve a desired level of confidence varies within the Koniambo deposit, being tighter in high-variability areas (mixture of ore and pinnacles of waste rock) and broader in low variability (more homogenous) areas.

1 Introduction

Falconbridge Nouvelle Caledonie SAS (Falconbridge) required a risk assessment of the resources at the Koniambo nickel laterite project in New Caledonia. The main aim of the assessment was to determine if additional drilling was required to achieve acceptable levels of confidence in ore grade and tonnage for the project feasibility study. The levels of confidence were to be established through 2D conditional simulations of ore-thickness (a proxy for tonnage) and nickel grade.

Arik (1999) and Yamamoto (2001) have considered that confidence measures should reflect the local data configuration and the local variability of data. Their approach is to calculate these components separately. Conditional simulation avoids this, and allows tractable assessment of risk at any anticipated production scale from a single set of data.

2 Geology and mineralisation

The nickel deposits on the ridges and elevated plateaus of the Koniambo massif are typical of the laterites that have developed under tropical weathering conditions on ultramafic bodies throughout New Caledonia (Figure 1, left). Figure 1 (right) is a generalised Koniambo weathering profile that consists of variable thicknesses of limonite and saprolite that reach a combined maximum thickness of approximately 40 m. High-grade ores are characterised by boxworks of garnierite in the saprolite-limonite transition and the upper levels of the saprolite.

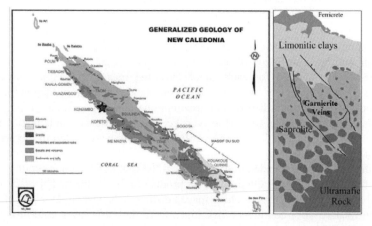

Figure 1 New Caledonia geology (left) and generalised laterite profile (right)

The structural controls and continuity of high-grade nickel mineralisation are not readily identified from vertical drilling. Close-spaced drilling is required to determine the variability of the bedrock topography and the distribution of low-grade boulders and waste pinnacles. Drilling on a 56 m pattern (80 m ' quincunx' pattern) is considered adequate for global resource estimates in the limonitic horizon. However a 28 m pattern (40 m ' quincunx' pattern) is considered necessary to provide confidence in geological interpretations in the more complex saprolite horizons.

3 Input data

The levels of confidence in ore-thickness and nickel grade at Koniambo were evaluated from 2D data because the mineralisation has a very large lateral extent (10 km x 4 km) relative to the depth of the deposit profile.

The study was conducted in two phases. Preliminary runs were made in 2002, and confidence limits obtained indicated the need for 30,000 m of further drilling. The work was updated after the drilling was completed in 2003, and that work is described herein.

Falconbridge divided the deposit into three sectors. This paper focuses on results from the largest area, the Centre sector. The 2D data for both attributes were created from drilling intercepts that met minimum grade and ore-thickness criteria determined by Falconbridge. Figure 2 below shows example locations of the ore thickness data used in the risk assessment study at the Centre sector. The study was constrained to a boundary interpreted to be the limit of mineralisation in the sector.

Figure 2 Centre sector 2D data, ore-thickness (left) and nickel grade (right)

Statistics for each attribute were computed using an 80 mE by 80 mN declustering window to account for the variable spacing of drilling (Figure 3). The summary statistics reveal that both attributes have skewed distributions, with over 35% of the data being less than Falconbridge's minimum grade criterion of 2% Ni.

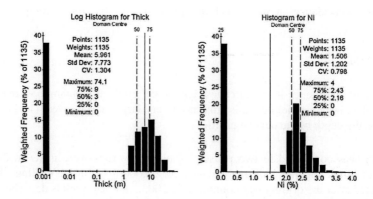

Figure 3 Ore intercept declustered histograms, ore-thickness (left) and nickel grade (right)

4 Variography

Indicator semivariograms were computed for multiple indicator thresholds of the two attributes. The ore-thickness indicator semivariograms are well structured up to the 10 m ore-thickness threshold, although the experimental structure is poor above this threshold. The nickel grade indicator semivariograms reveal a very short-range structure, with the longer range structure declining for the higher thresholds.

Figure 4 and Figure 5 summarise the sill and range values interpreted to fit the experimental indicator semivariograms of ore-thickness and nickel grade of the Centre data. Both attributes show a pattern of decreasing range with increasing indicator threshold. Indicator nugget effects of nickel grade increase with threshold. However the indicator nugget effects of ore thickness are consistent. There is also a pattern of moderate anisotropy for the lower thresholds, but higher thresholds are isotropic.

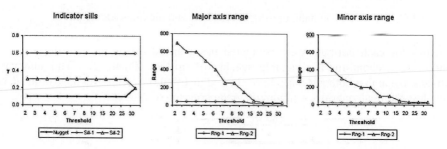

Figure 4 Ore-thickness standardised indicator variography sills (left) and ranges (right)

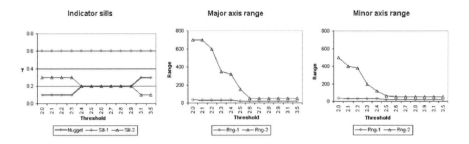

Figure 5 Nickel grade standardised indicator variography sills (left) and ranges (right)

The complex changes in variography with changes in threshold also support the choice of multiple indicator simulation instead of sequential gaussian simulation, which would have been an easier method to implement.

5 Simulation

Due to the low-grade spike of zero values associated with each attribute, sequential indicator simulation was selected to provide point-support realisations of ore-thickness and nickel grade. Independent simulation of each attribute was justified because nickel grade is independent of ore-thickness where ore-thickness is greater than zero. Twelve indicator thresholds (as listed on the x-axes Figure 4 and Figure 5) were selected for sequential indicator simulation of the ore-thickness and nickel grade. The first threshold for each attribute was set to partition the barren data, and higher thresholds were set at key attribute values of interest.

Using the input data and indicator variography models, 100 conditional simulations were computed for ore-thickness and nickel grade using the GSLIB, SISIM program for sequential indicator simulation (Deutsch and J ournel, 1998). The simulations were validated by comparing the input statistics and variography with the simulation outputs.

The simulations reproduced the input means of each attribute (Figure 6 and Figure 7), although with marginally lower E-type averages for ore-thickness.

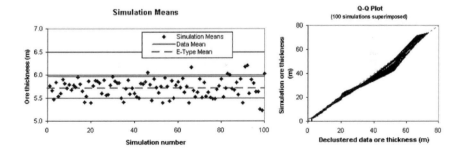

Figure 6 Ore-thickness simulation means (left) and Q-Q plot (right)

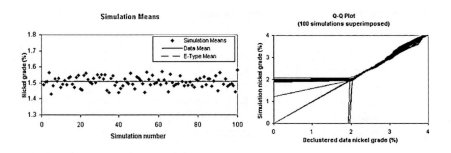

Figure 7 Nickel grade simulation means (left) and Q-Q plot (right)

The data histogram and variogram reproduction was also acceptable for both attributes. However, the E-type estimate of ore-thickness was marginally lower than that of the input data, but the two means were coincident for nickel grade. Images of one realistisation and the E-type averages for ore-thickness and nickel grade are shown in Figure 8 and Figure 9.

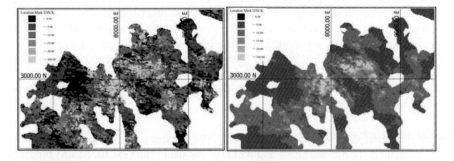

Figure 8 Ore-thickness simulation 001 (left) and E-type average (right)

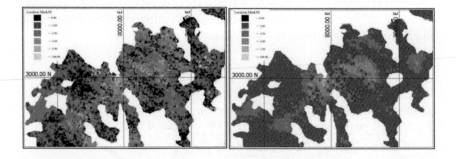

Figure 9 Nickel grade simulation 001 (left) and E-type average (right)

The simulation and E-type images show some spatial correspondence of zones of thicker ore and higher nickel grade. The ore-thickness simulation shows higher relative variability than the nickel grade results.

6 Reblocking and confidence limits

The simulation results of each attribute were averaged or reblocked into 100 m square panels (nominally 100, 10 m by 10 m spaced nodes) to derive a mean value for each panel of each simulation (Figure 10). For ore thickness, zero values were retained to reflect the estimated ore tonnage. However, for nodes having simulated thickness of zero, nickel grade values were set to null prior to reblocking so that the reblocked average would reflect the grade of the panel ore tonnage. Note that this is required because the input data only has an associated grade when the thickness is greater than zero.

The resulting distributions of the panel means were then interrogated to derive key statistics of the distributions. Due to the boundary constraint imposed on the study area, the peripheral 100 m square panels contained less than 100 simulation nodes. For these edge panels, the number of contained nodes captured was used to compute the panel proportion. Resource tonnages for each panel were calculated as the product of panel area, proportion, average reblocked thickness, and a density value of 1.5 t/m^3.

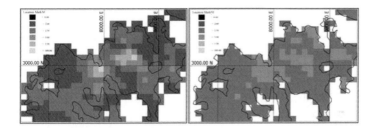

Figure 10 Panel averages for ore-thickness (left) and nickel grade (right)

The shape of the reblocked mean distributions for each panel is near normal, which is a feature consistent with the Central Limit Theorem of statistics. This theorem states that a distribution of means tends towards a normal distribution, as the number of samples used to compute the mean becomes large.

Confidence limits for the mean can be calculated using normal distribution theory. For example, the mean ± 1.645 standard deviations, contains 90% of the area under the standard normal distribution curve. The 90% confidence limits can be expressed relative to the mean of each panel distribution to give the relative 90% confidence limits (1.645 x panel standard deviation / panel mean). Specifically, for the reblock means of ore-thickness or nickel grade, the relative 90% confidence limits quantify the variability that can be expected (9 times out of 10) during production. For this study, the target acceptance threshold for Measured and Indicated resource classification was set to 90% confidence limits within ± 15% of the mean for a given production period, as discussed further below.

The relative 90% confidence limits were computed for each panel within the study area as shown in Figure 11. These plots confirm the intuitive conclusion that the lowest risk for both attributes occurs where the data spacing is closest (see Figure 2 for data locations). Additionally, the risk for ore-thickness (tonnage) is significantly higher than the risk for nickel grade, with the nickel grade meeting the benchmark of 90% confidence limits within ±15% of the mean for most of the Centre area on a panel-by-panel basis.

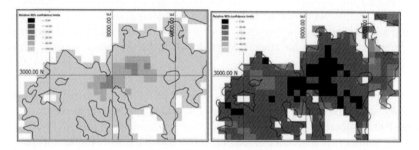

Figure 11 Panel relative 90% confidence limits for ore-thickness (left) and nickel grade (right)

One of the objectives of the simulation approach is to take into account the local variability of the data in assessing risk. Figure 12 shows two panels, A and B, with near identical data configurations in the centre of the study area, along with the ore thickness input data and the relative 90% confidence limits of panels within this area. The histograms of the simulation panel means for panels A and B are shown to the right of the data map. A kriging estimation variance approach to risk assessment would have given the same kriging variance and confidence limits for both panels. However, the relative 90% confidence limits using simulation are ± 30% for panel A and ± 20% for panel B.

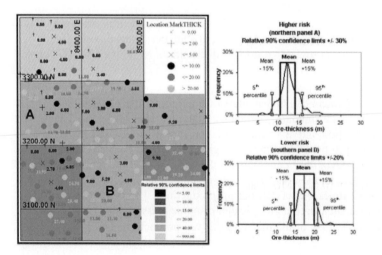

Figure 12 High (A) and low (B) risk panels and distributions of panels means for each panels; 5th and 95th percentiles are compared to interval ± 15% of the panel mean.

7 Production scaling of confidence limits

The relative errors computed on a panel-by-panel basis do not accommodate the fact that multiple panels will be mined during any mine production period. Because a production schedule is not yet available to allow reblocking or aggregation of panels to reflect actual production periods, it was assumed that a number of panels, n, of similar character (grade and/or depth and/or thickness) would be mined in a given production period. Further, the panels were assumed to be large enough to assume independence between panels. This assumption permits adoption of an approximation of the standard error formula ($\sigma/\sqrt{n}$) to compute the standard deviation for panels that are aggregated to represent a larger volume of mine production. For samples of size n from a large population, relative 90% confidence limits can be estimated by the product of the standard error of the mean and the standard normal deviate or z-value (1.645) of the confidence limit of interest.

For the purposes of this study, the size of the sample n is derived from the ratio of the production period tonnage to the tonnage within each reblocked panel. This formula assumes independence of realisations for each of the panels constituting a production period. The semivariograms show first ranges of less than a panel width (100 m) and secondary ranges of 200 m or less. Given these features and the fact that the majority of the study area is defined by 80 m spaced sampling or less, the assumption of independence was taken to be reasonable.[1] The example below shows this calculation for one panel in the study area using the assumption of a production rate of 2.5 Mt per annum and shows the relative 90% confidence limits are ± 11.8% per annum when multiple panels of the same risk character are scheduled to the meet the production requirement.

$$\text{Sample size (n)} = \frac{\text{Tonnage for production period}}{\text{Tonnage of reblocked panel}} = \frac{2,500,000}{120,259} = 20.8 \text{ panels}$$

$$\text{Relative standard deviation } (\sigma_R) = \frac{\text{Panel standard deviation}}{\text{Panel mean}} = \frac{3.85}{11.79} = 32.6\%$$

$$\text{Relative 90\% confidence limit} = \pm \frac{z_{90\% \text{ confidence}} \cdot \sigma_R}{\sqrt{n}} = \pm \frac{1.645 \cdot 32.6}{\sqrt{20.8}} = \pm 11.8\%$$

Using this method the relative 90% confidence limits were computed for annual (2.5 Mt) and quarterly (0.625 Mt) production periods for ore-thickness.

The quarterly production (0.625 Mt) map of ore-thickness shows that most of the deposit area has relative 90% confidence limits exceeding ± 15% of the panel mean. Only areas of dense drilling and panels with low ore tonnage (generally thin or partially filled panels) are below this threshold. The thin ore areas have smaller confidence intervals, as a large number of "equivalent" panels are mined in the

---

[1] A data check of the semivariogram of residuals: [simulated reblocked values – the mean reblocked value for panels], showed slight dependencies at 100 m, and were the study to be repeated, a larger panel size would be chosen to reduce/remove these dependencies.

production period. This result is an undesirable artefact of the method; however the tonnage involved is small. On an annual basis (2.5 Mt), most of the study area has relative 90% confidence limits below the ± 15% ore-thickness target value. However, on a quarterly basis most of the study area has relative 90% confidence limits above the ± 15% ore-thickness target value.

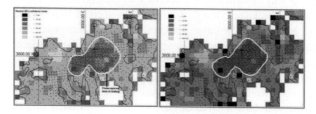

Figure 13 Ore-thickness relative 90% confidence limits, scaled to production rates of 0.625 Mt (left) and 2.5 Mt (right); point markers are drillhole collar locations

8 Discussion

From this study, Falconbridge concluded that the variability of nickel grade presented a low-risk. Wide spaced drilling (80 m spacing) adequately defined the nickel grade with relative 90% confidence limits within ± 15% of the mean for 100 m square panels. However, tonnage risk was considered to be high on a panel-by-panel basis. Rescaling the risk to quarterly and annual production periods revealed that the annual risk was acceptable, but that close-spaced drilling would be required to increase the confidence in tonnage to the target of relative 90% confidence limits within ±15% of the mean on a quarterly basis.

The 20% relative accuracy confidence limits are shown in Figure 13 (left panel) and enclose an area of dense drilling. The required 90% confidence limits for Measured Resources were eased to ± 20%, provided steps were taken to ameliorate the additional risk. The risk is ameliorated by implementing detailed drilling on a 10 m spacing a year in advance of production and increasing the number of mining faces (exposed ore) available for production.

This was a pioneering study for the authors and organisations involved. It was successful in that the initial study was completed rapidly, focused infill drilling requirements to reduce risk and assisted Falconbridge in managing the risk profile for a billion-dollar project.

9 References

Arik, A. (1999). An Alternative Approach to Resource Classification. 1999 APCOM Proceedings, SME. Denver. p 45-53.
Deutsch, C.V. and J ournel (1998). GSLIB GeostatisticalSoftware and User's Guide. Second Edition. Oxford Press. New York.
Yamamoto, Y.K (2001). Computation of global estimation variance in mineral deposits. In: Computer Application in the Minerals Industries, Xie, Wang & J iang (eds). Swets & Zietlinger. Lisse. p 61-65.

MINERAL RESOURCE CLASSIFICATION THROUGH CONDITIONAL SIMULATION

TOMASZ M. WAWRUCH and JORGE F. BETZHOLD
Vice-Presidency Mineral Resources,
Anglo American Chile, Santiago

Abstract. Through the Mineral Resource Classification the quantification of uncertainty/confidence on modelled geometry/estimates is to be addressed and established. Estimates represent different levels of reliability. The classification role is to ensure that the characteristics, quantity and quality of mineralised material is adequate for the proposed project or mining program, assuring the use of full plant capacity and optimisation of the mining and metallurgical performance down stream to the final products.

1 Introduction

The purpose of the paper is to discuss a method of Mineral Resource classification based on conditional simulation applicable at production scale.

Evaluation, classification and reconciliation are integral parts of the Mineral Resource management process. Drilling and sample collection together with QA/QC validation periodically update this process. The aforementioned system is "factor-dependent". Any sophisticated method applied to the modelling and evaluation is worthless if sample collection, preparation and chemical assays amongst other "factors" are not properly controlled.

Evaluation on its own tries to predict short, medium and long term scheduled mining output. It is worthwhile to look for the right methods to provide good tonnage and grade predictions to benefit the mineral resource management and mining program. As a result of the imprecision of evaluation, Mineral Resource modelling undergoes different magnitudes of discrepancies in comparing to the actual results. This is one reason for which Mineral Resource classification is required. It gives to competent persons, who are aware of the risks and consequences involved in inadequate mineral resource prediction, the opportunity to express a confidence concerning the estimates assigned to rocks to be mined. Monitoring the production results through the reconciliation process gives an important feedback in order to measure discrepancies, quantify errors and tune back the system if performance is unsatisfactory.

Some characteristics the mineral resource classification should fulfil are as follows:
- Classification must be transparent and objective to what is to be achieved through the set of parameters defined integrally and accordingly to the geometry/grade modelling and interpolation strategy.
- The classification method must be reproducible and auditable based on quantifiable principles, applicable at the production scale.
- Mineral Resource classification should consider the production scale over a given period of time to be reconciled. This refers to the support size/estimation error relationship.
- Acceptability concerning the discrepancy between prediction and actual results should be explicitly established according to the nature of mineralization and the operational requirements.
- Through the Mineral Resource classification output, it should be possible to define areas where the confidence concerning geometry/estimates requires improvement. The increased confidence should be quantified as a function of money spent on drilling and sampling collection campaigns.

2 Classification methodology

The proposed Mineral Resource classification approach quantifies the confidence in the evaluation result. The reliability of the estimated tonnage/grade at location "x" is established and measured as a function of variance calculated through a given number of conditional simulations.

This Mineral Resource classification methodology is based on:

1. Multiple realisations of the sequential Gaussian conditional simulation. Other simulation routines and multi-realisation engines can also be considered (Deutsch, 2002; Goovaerts, 1997).
2. Calculation of the coefficient of variability for local mining unit
3. Change of support related to 1 to 3-monthly and yearly production panels
4. Estimation error according to established requirements in terms of % of error and % of confidence limit.

As an example for the Base Metals industry, widely accepted rules of Mineral Resources classification are as follows:

- Mineral resources are classified as **Measured** when the local estimate, whose variability is corrected to monthly to quarterly production units, is estimated within 15% error at a 90% confidence limit.

- Mineral resources are classified as **Indicated** when the local estimate, whose variability is corrected to yearly production units, is estimated within 15% error at a 90% confidence limit.

- The mineral resources that do not fulfil the aforementioned criteria are classified as ***Inferred.***

2.1 MULTIPLE REALISATION OF SEQUENTIAL GAUSSIAN SIMULATION

The objective of conditional simulation is to generate an equally probable set of realizations that account for proportion/distribution and spatial geometry/grade variability inferred from conditioning data at global/local scale. During the simulation, simulated nodes are visited in a random fashion. The conditioning is extended to all of the data available within a neighbourhood of the location being simulated, including the original data and all previously simulated values. Given that the estimated model is inferred from sample statistics that are uncertain because of limited number of samples, *the purpose is to provide the measure of uncertainty given by the differences between **N** alternative simulated values at location **x**.* Different simulations impart different global statistics and spatial features on each realisation. In this way, it is possible to establish the spectrum of possible values at any location. (Deutsch, Journel-1998)

The number of realisations needed depends on how many are judged sufficient to model the uncertainty being addressed. A cumulative coefficient of variability (COV) from a low number of simulations has a variable behaviour, which stabilises as the number of simulations from which it is calculated, increases. After certain number of simulations, the oscillation of COV between succeeding simulated realisations stabilises. Taking this fact into account the number of simulations to define the uncertainty model is established.

The reliability of the simulations is checked by comparative analysis done in three ways:
- Reproduction of global statistical distribution
- Spatial grade variability/continuity model reproduction
- Local grade reproduction between conditioning values and sets of "N" simulations

To ensure that simulated realizations correctly reproduced the spatial theoretical grade variability/continuity model, simulated realizations are submitted to the spatial variability analysis of each given realization. Every single simulated realization is then checked using the same spatial variability formula and the same parameters as those used to get the variogram/correlogram experimental points of the raw variable.

The spectrum of spatial geometry/grade variability models along selected directions covers the interval of possible solutions. Considering the uncertainty of the conditioning database and spatial grade variability model, the family of "N" simulations is accepted as fairly representing the imposed spatial variability/continuity model.

2.2 CALCULATION OF COEFFICIENT OF VARIABILITY FOR LOCAL BLOCK

The coefficient of variability - COV, a dimensionless measure, defines the magnitude of deviation relative to the average. The whole mineralised domain is analysed. The measure of relative dispersion for each local block, based on "N" simulated values is calculated. The distribution of COV varies over the orebody as a function of the amount of conditioning data, local grade variability and the spatial grade variability/continuity model. In a densely sampled area, the expected variability among the multiple simulations of the variable is expected to be lower than in a poorly sampled part of the orebody. Nonetheless, even over the densely sampled areas, where the local grade variability is high, high values of the coefficient of variability can be expected.

2.3 CHANGE OF SUPPORT TO PANEL PRODUCTION

A meaningful way to classify Mineral Resources is to take account of the variability of local blocks within a bigger production volume. The change from local volume/grade variability to the variability of the production panel tends to ensure that the expected metal contained within monthly and annual production units is estimated with an error not greater than the established tolerance at given confidence limit.

The way to express local geometry/grade variability as a variability of bigger mining units is through the variability reduction factor (Isaaks, Srivastava-1989). The calculation takes into account averages of COV computed for local mining blocks and production panels.

A series of author's exercises with different kinds of deposits have shown that global monthly or annual variability reduction factors are not the best values to apply for Mineral Resource Classification purposes. Spatial COV maps usually show different local configurations for different geological situations. Using one global correction value increases the uncertainty for low variability areas and gives more confidence to high variability areas. It is considered that a more objective way to do this is by introducing local variance reduction factors.

It is proposed to divide the block population into classes of COV. The number of classes depends on what even-frequency per class is targeted. The classification is based on calculations computed within each group. Assuming that "n" local COV classes are analysed, "n" variability reduction factors are computed.

A randomly positioned production scale panel allows to accumulate a number of observations considered sufficient to calculate the COV for a given production scale. Following this, the variance reduction factor for monthly and annual production units is computed and applied to the local COV. In this way the local variability is corrected to production scale variability as if it was a fraction of bigger support.

$$f_R = COV_{panel\ production} / COV_{Local}$$

f_R - variability reduction factor
COV_{Local} - local coefficient of variability

Confidence limits are addressed by the mean of formula where the local coefficient of variability COV is corrected by variance reduction factor.

$$f_R * COV_{Local} * Z_C \leq 15\%$$

Where,
Z_C -confidence limit coefficient; accepted Normal distribution (0,1) if local COV is reasonably Gaussian

2.4 CLASSIFICATION CURVES

Having obtained the following:

- Percentile proportion of each class
- Local average of COV for each class
- Two COV for monthly and annual production panels corrected by variability
- reduction factor,

it is possible to visualise three curves called, for Mineral Resource Classification purposes, **Classification Curves**.

Pairs obtained from the class central point and average local COV are used to construct the local COV curve that increases as a function of increasing percentile population.

The measured and indicated classification curves are based on calculations computed within each group. The common feature of them is their decreasing nature, throughout the increasing percentile population.

The three curves create two intersections if Measured, Indicated and Inferred Mineral Resources are present. The curve at lower local relative variability represents the partition between Measured and Indicated Mineral Resources. The second decreasing curve at higher local variability establishes the limit between Indicated and Inferred Mineral Resources. At each intersection the proportion of Mineral Resources belonging to one of three Mineral Resources confidence classes can be read. Their Y-axis equivalents establish the separation thresholds between Measured, Indicated and Inferred Mineral Resources in terms of local COV.

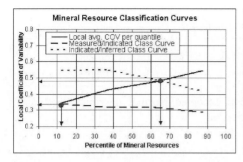 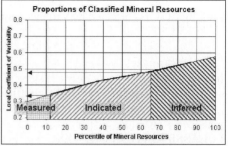

Figure 1. Graphic representation of Mineral Resource classification results

Based on the local COV at production support (monthly and annual) it is possible that one or both intersection points are not present. If the spatial continuity, sampling density and local variability are unfavourable, the proportion of Measured or Indicated Mineral Resources will not be present or will show a low proportion of the total.

3 Sensitivity of the proposed methodology

3.1 HISTOGRAM GRADE REPRODUCTION

Global/local grade distribution – since the coefficient of variability is used as a measure of local grade dispersion, any oversight of global or local grade distribution distorts the local relative variability distribution. As a consequence might be, underestimated local grades at assumed correctly reproduced spatial variability/continuity model increase values of coefficient of variability. This increases the amount of Mineral Resources at lower confidence class, which correctly classified would have been assigned a better category.

3.2 SPATIAL GRADE VARIABILITY REPRODUCTION

Spatial grade variability/continuity model – if incorrectly reproduced and accepted, it could significantly change the classified Mineral Resource proportions. If the spatial variability range of simulated realizations is too long, it produces a false effect of better continuity. The improved spatial grade continuity implies relatively higher dispersion variance distribution for the given production reference unit. As a result, more Mineral Resources can be classified with higher tonnage/grade uncertainty. The effect is predominantly pronounced over areas with scarce conditioning data (Figure 2, Table 1).

Short continuity ranges promote a faster decrease in variability whilst changing support from local increments to panel production units. Small values of the variability reduction factor increase the proportions of Indicated Mineral Resources.

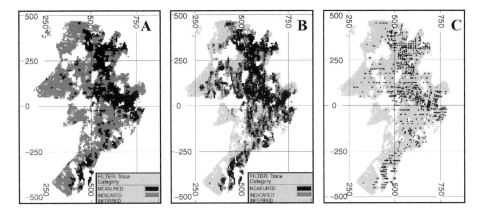

Figure 2. Distributions of Classified Mineral Resources as a function of spatial variability range reproduction: A: short-range, B: long-range, C: spatial data location

	Classified Mineral Resources – Proportions [%]		
Case	Measured	Indicated	Inferred
A: Short range	29.4	65.8	4.8
B: Long-range	29.7	28.6	41.7

Table 1. Proportions of Classified Mineral Resources as a function of spatial variability range reproduction. Stable % proportion of Measured Resources shows the data-driven effect and variable % proportion between Indicated/Inferred Resources reveals the model-driven results.

3.3 SIZE OF PRODUCTION PANEL

The concept of production panels may have a real or theoretical aspect. In the case of having a mining program for a given period of time, programmed areas/volumes can be used to calculate the COV of panel production as increments of local COV. This is a case to verify and quantify the uncertainty on tonnage/grade for the existing extraction program.

Without a mining program (project, pre-feasibility study) a theoretical approach to calculate variability for a production period can be used. Vertical dimensions can be taken from a possible number of benches envisaged in the production program. This can be estimated by comparing to other mines/projects of a similar nature. To scale up the horizontal dimensions, the character of spatial variability/continuity model (isotropy, anisotropy) and distribution of existing or assumed opened mineralised faces/stopes are decisive factors. Once established, the size of production panel can be an object for sensitivity study.

3.4 NUMBER OF LOCAL BLOCKS WITHIN PRODUCTION PANEL

Different orebodies present their proper geometric particularity. Whilst computing the statistics on local COV within monthly or annual production panels, some of the panels over the borders of the domain gather only a small number of local COV's. To ensure the robustness of statistics, panels with small numbers of local blocks should be discarded.

3.5 FREQUENCY OF LOCAL BLOCKS PER CLASS

The suggested number of classes for the local COV should be between 4 and 10. This means that a small orebody could not be divided into an elevated number of classes having too few local blocks per class.

Sensitivity analysis on the aforementioned issues should be carried out. The output for the exercise is a family of classification curves for Measured/Indicated and Indicated/Inferred Mineral Resources. Through them the uncertainty of the Category proportions for the Mineral Resource is assessed. The average of computed answers is accepted to break-up the classified Mineral Resource proportions.

4 Quantified confidence improvement - example

This section shows the application of the discussed methodology. Together with the classification method, the reconciliation between predicted and actual proportions of classified Mineral Resources is presented.

The classification method applied Sequential Gaussian Conditional Simulation as engine to create conditioned, equally probable grade distributions. The local variability was expressed through the change of support for monthly and annual production panels. The criterion of an error within 15% at 90% confidence limit was used.

The orebody had been intercepted by 98 boreholes. The average distance between them was greater than 50m. Ranges of the spatial grade variability model were less than 50m and represented only subtle geometrical anisotropy. Following the procedure discussed in this paper, an initial classification of Mineral Resources has shown no Mineral Resources classified as Measured and only 12% of Indicated Mineral Resources (Figure 2). The quantity of Measured/Indicated resources indicated a need for new information to improve the confidence in geological resources.

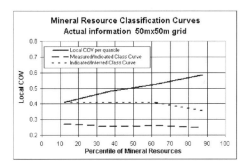

Figure 2. Graphic assessment of Mineral Resource classification

An exercise was carried out to quantify "a priori" a possible amount of Mineral Resources to be upgraded as a function of new drilling information. This was done by simulating virtual drilling campaigns. To optimise the drilling program, different sets of drilling grids were analysed.

It was assumed that in spite of only 346 samples from 98 boreholes regularly distributed over the orebody, the average grade and variance would not change drastically as a result of new data collection. This assumption has been assessed using the set of 51 simulations generated for the purpose of classification. At a 90% confidence limit the expected discrepancy concerning average grade was defined as 5.4% and variance as 6.8%.

Following this, the discussed classification methodology was applied. It was concluded that among multiple exploration strategies, a sampling grid of approximately 30m x 20m would allow to have 13% of Measured and 52% of Indicated Mineral Resources (Figure 3).

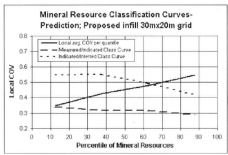

Figure 3. Prediction of Mineral Resource classification proportions based on virtual exercise

The actual exploration drilling program contains 72 new boreholes. In total 729 samples were conditioning the geometry and the grade distribution within the orebody. The uncertainty on geometry was assessed through probabilistic models and grade estimates within the orebody were reproduced using sequential conditional Gaussian simulation.

The classification methodology applied to Mineral Resources assigned the confidence level in the following (Figure 4) proportions: Measured Resources 8%, Indicated Resources 60% and Inferred Resources 32%.

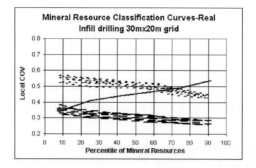

Figure 4. Mineral Resource classification based on the updated database

The targeted proportion of classified Mineral Resources had been reached. The use of a family of classification curves allowed to assess the uncertainty concerning the confidence on Mineral Resource classified proportions.

5 Conclusions

Mineral Resource classification is an integral part of Mineral Resource evaluation and reconciliation. It constitutes an important strategic tool allowing to assess tonnage/grade uncertainty for the mining program.

A golden formula to classify mineral deposits does not exist. Different methods to express confidence in Mineral Resource evaluation are employed. Although the common practices have been developed, the robust approach toward the Mineral Resource classification method through the uncertainty quantification is not always exercised.

The method presented in this paper proposes to quantify confidence through equally probable, spatially conditioned multi-realizations. As an engine to create the conditional spatial grade distribution the sequential conditional Gaussian simulation was used. The classification approach integrates transparency, objectivity and geostatistical tools commonly used in modelling and evaluation. The relevant issue is to express the uncertainty of Mineral Resources as a function of production panels that is to be reconciled for a determined production period. The reproducibility of the classification method is achieved through parameters defined numerically. This makes the classification method easily auditable. Classification curves allow visualising the classification output.

This classification approach includes quantification of confidence in estimated Mineral Resources. It can be applied to geological projects and mining operations. The results are submitted to a continuous monitoring and validation process through reconciliation figures on a monthly basis.

Acknowledgements

We would like to acknowledge Anglo American Chile for supporting this research.

References

Deutsch, C.V., *Geostatistical Reservoir Modelling*, Oxford University Press, 2002.
Deutsch, C.V. and Journel, A.G., *GSLIB: Geostatistical Software Library and User's Guide*, 2nd Edition, Oxford University Press, 1998.
Goovaerts, P., *Geostatistics for Natural Resources Evaluation*, Oxford University Press, 1997.
Isaaks, E.H. and Srivastava, M.R., *An Introduction to Applied Geostatistics*, Oxford University Press, 1989.

GEOSTATISTICAL INVESTIGATION OF ELEMENTAL ENRICHMENT IN HYDROTHERMAL MINERAL DEPOSITS

ABANI R. SAMAL[1]* AND RICHARD H. FIFAREK[1] AND RAJA R. SENGUPTA[2]
[1] Dept. of Geology, Southern Illinois University, Carbondale, IL 62901-4324, USA
[2] Dept. of Geography, McGill University, 805 Sherbrooke St. W.
Montreal, Quebec, Canada H3A 2K6
*Email: arsamal@yahoo.com

Abstract. Hydrothermal hypogene processes enrich or deplete rocks in specific suites of elements to form mineral deposits. Subsequent geochemical processes, such as near surface oxidation, commonly remobilize previously developed element associations. Late stage oxidizing fluids and elemental enrichment/depletion are commonly guided by permeable geologic structures in the deposit. An investigation of element redistribution is possible using a cross-covariance analysis between pairs of elements. The maximum positive cross-covariance of a pair of variables yields a vector, known as the lag vector. This lag vector may indicate the direction and distance of element displacement from their original loci.

This paper discusses the application of cross-covariance in modeling the anisotropy of metal redistribution as a function of late oxidation in the Pierina hydrothermal Au-Ag deposit. Cross-covariance analyses of assay values from drill-hole samples in both the oxidized and unoxidized zones are calculated and lag vectors [$l_{xy(O)}$, where x and y are elements in a zone O] are derived to infer a preferred path of metal remobilization. The azimuths of lag vectors for the element pairs Ag-Au, Cu-Au and Cu-Ag in the oxidized zone correspond to the orientations of recognized faults and fractures in the deposit. This implies that the remobilization of Au, Ag and Cu by oxidizing fluids was strongly controlled by specific fault or joint sets. The data for all three element pairs from the unoxidized zone suggested structural controls different from those of the oxidized zone.

These results imply that a cross-covariance analysis for pairs of elements may be used to infer structural controls on fluid flow which might be responsible for element remobilization and possible enrichment. Such an analysis may be useful in mineral exploration for predicting metal enrichment and the location of exotic (transported) deposits.

1. INTRODUCTION AND BACKGROUND INFORMATION:

The spatial pattern of element distribution in hydrothermal deposits results from the overprinting of multiple alteration events. The late stage, near-surface oxidation of deposits is related to vertical movements of the groundwater table, and typically results in the downward transportation of elements, possibly with some lateral component of movement. Fluid flow and element mobilization are guided by major structural features of high permeability (faults, joints, etc.). Oxidation is important to the economics of mining large, low-grade, disseminated metal deposits in that it releases metals encapsulated in sulfides, thereby making the ore amenable to low-cost extraction technologies, and in the enrichment of metal grades.

Element distributions in mineral deposits can be modeled numerically through the application of geostatistics. A multivariate geostatistical analysis using maximum cross-covariance values can be applied to model the spatial dependency of two metals and to relate the results to orientations of mineralized faults and joints (Samal and Fifarek, 2003). In this study, we explore the application of cross-covariance analysis to assay data from the Pierina (Peru) Au-Ag deposit where late stage oxidation has clearly remobilized metals.

1.1 Cross-covariance and lag effect:

With the assumption of second-order stationarity and ergodicity, the covariance ($C_{i(h)}$) of any variable i measures spatial dependency of the *same* variable (or values of the same property of material) at two locations, where h is the distance of separation (a vector) between the two locations. Similarly, under assumption of joint second-order stationarity, the cross-covariance function $C_{ij(h)}$ measures the spatial dependency between *two* variables i and j, here concentrations of two elements, separated by vector h. The cross-covariance between two elemental concentrations i and j is expressed as (Equation 1):

$$C_{ij\,(h)} = \frac{1}{n}\sum (i_h - m_i)(j_{-h} - m_j) \tag{1}$$

where m_i is the mean of the variable i and m_j is the mean of the variable j

The cross-covariance analysis is not an even function (Wackernagel 1998). The asymmetric behavior of the cross-covariance function between two variables in isotopic, heterotopic or partially heterotopic datasets is seen in the assay values of gold deposits. The dataset used in our cross-covariance analysis is partially heterotopic, i.e. data for all variables are not available for all sample locations. The asymmetry can be defined as $C_{ij(h)} \ne C_{ij(-h)}$, where $C_{ij(-h)}$ is the cross-covariance of i and j separated by a distance h but in the opposite direction. But if both the sequence of variables and the sign of the lag (h) are changed, the value of the cross-covariance function $C_{ij(h)} = C_{ji(-h)}$ (Wackernagel, 1998; Isaaks and Srivastava, 1989). Cross-covariance values can be positive when the variables at the end points of the h vector are on the same side of their means, i.e, $i_h > m_i$ and $j_{-h} > m_j$ or, $i_h < m_i$ and $j_{-h} < m_j$; where i_h is the value of i at the head of the vector h, and j_{-h} is the value of j at the tail of the vector h. Depending on how far they are from their respective means, the value of a positive cross-covariance will be high or low. So, if i and j are extremely high values (enrichment of both elements) or extremely low values (depletion of both elements), both cross-covariance values will be positive and high (not low). But if one element is enriched and the other element is depleted, then the cross-covariance is negative.

The lag effect is the vector (l_{ij}) separating the locations of extreme values of two variables, which in some geological environments may be due to the delay in enrichment of one element with respect to the other at two different locations (Isaaks and Srivastava, 1989, Goovaerts, 1997). This offset distance is also termed the delay effect when time-series data is considered, such as in most environmental applications (Wackernagel, 1998). In the oxidized zone of a hydrothermal mineral deposit, the offset between the concentrations of two elements is due to differences in their mobility resulting in the enrichment and depletion of different elements at different locations. In a preliminary exploratory study of the Pierina deposit (Samal and Fifarek, 2003, Samal, Fifarek, Sengupta 2003 and Samal, Fifarek, Mohanty 2004), lag vectors were derived that generally corresponded to the orientation of specific major fault and joint systems. Deriving the lag-vectors from maximum positive cross-covariance values (maximum values in any direction) ignores other higher cross-covariance values. In this paper other high cross-covariance values are taken into consideration for three pairs of elements (Ag-Au, Cu-Au, and Cu-Ag) in the Pierina deposit.

1.2   Deposit geology:

The Pierina deposit is located in the Ancash Province of Peru. It is a world class, high sulfidation, epithermal Au-Ag deposit with anomalous but uneconomic concentrations of Cu, Zn, As and Hg. The geology and genesis of the deposit are presented by Fifarek and Rye (in press), from which the following summary is taken.

The Au-Ag ore-body is sub horizontal, elongates N-S, and almost entirely hosted by rhyolite ash flow tuffs that overlie porphyritic andesite and dacite lavas and are adjacent to a crosscutting and interfingering dacite flow dome complex. Alteration and mineralization occurred 14.5 Ma ago as a result of the expulsion of fluids, gasses and metals from an underlying magma. Highly acidic fluids formed at the level of the deposit where slowly rising magmatic vapors condensed and mixed with cool meteoric groundwater. The progressive neutralization of these migrating acid-sulfate fluids led to zoned alteration assemblages from proximal vuggy quartz through quartz-alunite ± clay and intermediate argillic to distal propylitic. Copper-gold-silver mineralization largely followed alteration and is marked by the deposition of enargite (Cu_3AsS_4), electrum (Au-Ag), acanthite (Ag_2S) and related minerals. The primary elemental associations and concentrations in the sulfide deposit were established at this time.

A late oxidizing event related to a near-surface, steam-heated process was superimposed on the deposit during the waning stage of hydrothermal activity that was accompanied by a drop in the water table. These oxidizing fluids led to the destruction of sulfides and the formation of barite, hematite, goethite and minor jarosite. Consequently, the previously established elemental concentrations and associations were substantially modified due to the remobilization of most elements. Late oxidizing fluids pervaded rocks of the upper 200 to 300 m of the deposit and particularly followed open faults and joint sets.

Exploration drilling on mostly 50 m centers and assays of 1 m intervals of drill core or cuttings provided an extensive database of Cu, Au and Ag values. Additional datasets were generated from information on the distribution of alteration and fracture-filling minerals in the exploration drill holes. Together, the datasets constitute the basis for evaluating the remobilization of metals (Au, Ag and Cu) by oxidizing fluids. Approximately 24,500 data points were selected from the oxidized zone and approximately 14400 samples were selected from the unoxidized zone, by excluding widely spaced data points.

2. DATA ANALYSIS:

2.1   Data Preparation:

The drill-hole data were visually examined in a 3D environment using GEMCOM and Arc-GIS software-systems. For analytical purposes, a single table with records of Au, Ag and Cu and alteration details was created within GEMCOM. The oxidized zone is characterized by the presence of iron-oxides (FeOx) whereas the unoxidized part of the deposit is marked by the absence of FeOx. A solid model was then created for the alteration.

Using GEMCOM, two tables of data formatted for the geostatistical software ISATIS geostatistical software were created: one for the oxidized zone and the other for the unoxidized zone. A selected portion of the data was chosen from the area of regularly spaced drill-holes for analysis.

2.2   Geostatistical Analysis:

A univariate variography (covariance) analysis was used to model the anisotropy of individual elements in this mineral deposit. For pairs of elements, a cross-covariance analysis was used to derive lag vectors. The cross-variogram is an even function (Wackernagel, 1998, p 147) that fails to detect anisotropy and therefore is not relevant to this study.

ISATIS was utilized to analyze for the cross-covariance of the three variables, Au, Ag and Cu. Each set of data, oxidized or unoxidized, was analysed in 62 directions to cover all possible directions in 3D space with a $30°$ angular tolerance for each direction. Out of these 62 directions, 12 directions were on the horizontal plane (reference plane) and two in the vertical plane (up and down). The remaining 48 directions are defined in 3D space as 4 directions in 12 vertical planes whereby each plane includes one horizontal direction. On each plane, these 4 directions are separated by $30°$ between the horizontal and vertical directions. A 50m lag was chosen for all directions except the vertical directions where the lag distance was set as 10m.

The cross-covariance for Ag-Au, Cu-Au, and Cu-Ag pairs was calculated using the exploratory data analysis tool of ISATIS. It is noteworthy that ISATIS calculates the cross-covariance in each specified direction and its reverse direction, in other words, the ISATIS software calculates $C_{ij}(h)$ and $C_{ij}(-h)$. A cross-correlation analysis of the same pair of variables taken in the same sequence was performed in order to cross-check the results of the cross-covariance analysis. The cross-correlation (CC_{ij}) function (Equation 2) is:

$$CC_{ij(h)} = \frac{1}{n} \sum \frac{(i_h - m_i)(j_{-h} - m_j)}{\sigma_i \sigma_j}. \tag{2}$$

Experimental cross-covariograms for each direction were plotted for a comparative analysis made in two ways: 1) for each pair of variables, high positive cross-covariance values were ranked and the corresponding directions were compared with the results from the previous study (Samal & Fifarek, 2003), and 2) the directions were compared with recognized structural trends. For reasons of clarity only those directions with the top three cross-covariogram values are shown in the following figures.

3. RESULTS AND DISCUSSIONS OXIDIZED ZONE:

Among the three variables, Cu in the oxidized zone has the highest variance followed by Ag and Au (Table 1). This reflects the relatively wide range of Cu values and suggests that the leaching of Cu was more extensive and the element more mobile than Au and Ag.

Table 1: General statistics of variables: oxidized zone

	Mean	Variance		Covariance
Au	1.4 ppm	14.3 ppm^2	Ag & Au	61.4 ppm^2
Ag	12.7 ppm	1493.3 ppm^2	Cu & Au	128.7 ppm^2
Cu	128.7 ppm	139047.4 ppm^2	Cu & Ag	980.7 ppm^2

From a comparison of covariogram plots (not shown) of the three variables it is clear that the spatial dependency of Au and Ag is very similar. The covariance of Au and that of Ag at shorter distances of separation exhibit very high values along ENE to ESE directions (azimuths of 60°, 90°, 120°) at a lag-interval of less than 50m. The ranges of 150m (approximate) are higher along these directions than in other directions (e.g., azimuths of 0°). Additionally, the covariance values fall rapidly to low values along generally North-South directions.

The cross-covariograms of Ag and Au (Fig. 1) indicate the lag vector ($l_{AgAu(O)}$) is oriented along an East-West to ENE-SSE (Azimuth 90° & 60°) directions with a shallow dip of 30° (±5°) toward East. Experimental cross-correlogram plots of Ag-Au produce a similar pattern as that of the cross-covariograms (Fig. 2). The sequence of maximum to lower cross-correlogram values is along the same directions (Azimuth 90° & 60° and dip of 30° (±5°) toward East) as seen in the cross-covariograms.

The cross-covariograms of Ag and Cu yield a preferred lag vector ($l_{AgCu(O)}$) of azimuth 60°, dip 30°, followed by vectors with azimuth 270°, dip 60° and azimuth 240°, dip 60° (Figure 3). It can be inferred that, with respect to Cu enrichment (or depletion), a significant lateral movement of Ag has occurred in ENE-WSW to E-W directions. With the angular tolerance (30°) used in the analysis, it is likely that elemental remobilization is controlled by joints and faults aligned approximately ENE-WSW to E-W directions.

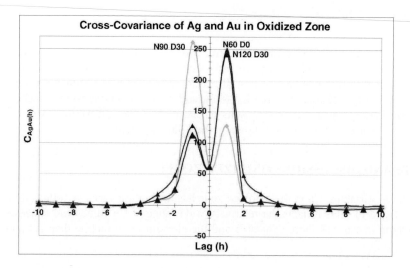

Figure 1. Cross-covariogram plots of Ag-Au in the oxidized zone (azimuths and dips shown for three highest values).

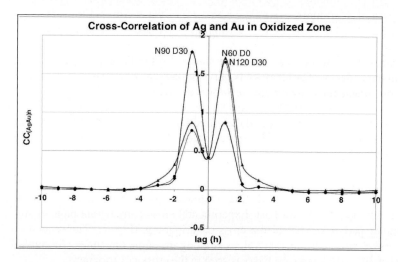

Figure 2. Cross-correlation plots of Ag-Au in the oxidized zone (azimuths and dips shown for three highest values).

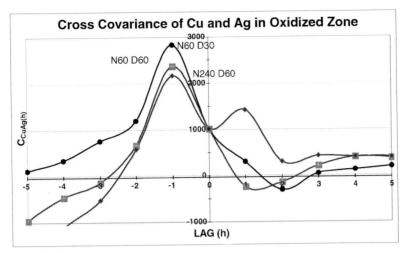

Figure 3. Cross-covariance of Cu – Ag in Oxidized zone (azimuths and dips shown for three highest values).

The cross-covariance plots of Cu and Au indicate a $l_{CuAu(O)}$ of azimuth 150°, dip 60°. Other prominent cross-covariance values imply vectors with azimuths of 270° dip 60° and 240°, dip 60° (Figure 4). From these observations, it is evident that, with respect to copper, the fluid transport and enrichment/depletion of gold and silver was in a general ENE-WSW to East-West direction.

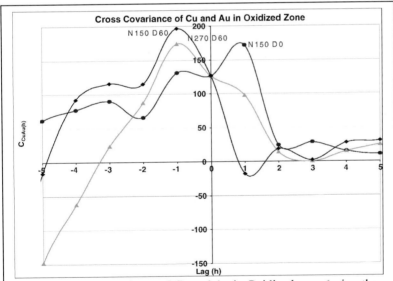

Figure 4. Cross-covariance of Cu and Au in Oxidized zone (azimuths and dips shown for three highest values).

In summation, Ag and Au have a similar spatial dependency pattern (as documented by their covariance values). The cross-covariance patterns of pairs of metallic elements (Ag-Au, Cu-Ag and Cu-Au) are useful in deriving vectors, which implies the shallow dipping transport of Au with respect to Ag along East, SSE and SSW directions. The cross-covariogram plots of Au-Cu and Ag-Cu pairs also suggest general West, East to ENE directions of fluid flow. Orientations of the major structural trends (joints and faults) that guided oxidizing fluids, as identified in this study, are summarized in Table 2. These are ENE to ESE and WSW to WNW directions, which are common in the cross-covariance analysis of all three pairs.

Table 2. Summary of prominent cross-covariance values

ANALYSIS	Rank	Representative directions	Comments
Cross-Covariance (Au & Ag)	1	Azimuth 90° and Dip 30°	Mostly ENE-ESE to WSW-WNW directions and shallow dips of 30° to 60° (±30°)
	2	Azimuth 240°	
	3	Azimuth 120° and Dip 30°(-h)	
Cross-Covariance (Ag & Cu)	1	Azimuth 60° and Dip 30°	
	2	Azimuth 60° and Dip 60°	
	3	Azimuth 240° and Dip 60°	
Cross-Covariance (Au & Cu)	1	Azimuth 150° and Dip 60°	
	2	Azimuth 270° and Dip 60°	
	3	Azimuth 150°	

3.1 Unoxidized Zone:

Data for the unoxidized zone were treated in the same manner as data for the oxidized zone. The unoxidized zone lies below the oxidized zone and is represented by fewer assays relative to the oxidized zone. For Ag - Au pairs, the lag vector ($l_{AgAu(U)}$) has an azimuth of 120° and dip of 30° (Fig 5). Other prominent directions of possible Au and

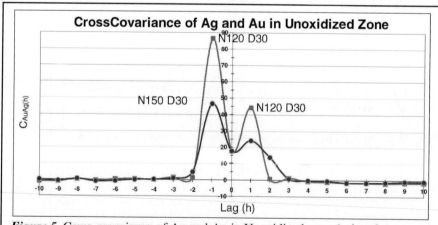

Figure 5. Cross-covariance of Ag and Au in Unoxidized zone (azimuths and dips shown for three highest values).

Ag movement during hydrothermal activity are along azimuth 300°, dip 30° and 150°, dip 30°. For Cu and Au pairs, the lag vector $l_{(AuCu(U))}$ has an azimuth of 90°, dip 60°, and a lag distance of separation of less than 50m followed by azimuths of 120°, dip 60°, and 270°, dip 60° (Fig 6). The analysis for Ag and Cu pairs suggests no preferred orientation of metal separation and fluid flow.

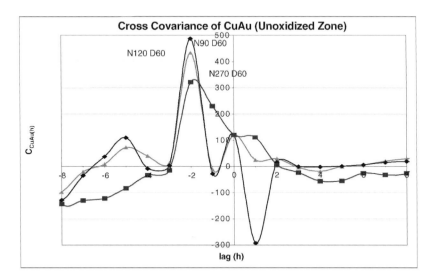

Figure 6. Cross-covariance of Cu and Au in Unoxidized zone (azimuths and dips shown for three highest values).

The azimuth 120° is common to Ag-Au and Au-Cu pairs and along which cross-covariance values are sufficiently high to suggest a prominent geologic trend. With an angular tolerance of 30°, the major directions of elemental remobilization are along ESE to east. This direction may represent a set of vectors along a set of fault or joint planes that are of pre-oxidation age. A structural study of mine exposures revealed a prominent set of faults and joints along this trend, as well as the other directions inferred (azimuths 120°, 90°, 240°, 270° & 300°) from this cross-covariance study.

4. CONCLUSIONS:

Based on the above observations, the following conclusions are possible.

i. A covariogram analysis suggests that Ag and Au have similar spatial patterns of distribution that differ from that of Cu in the oxidized zone of the Pierina deposit. Both elements show high covariance (a measure of spatial dependency) values at distances of separation less than 50m, whereas Cu shows no preferred lateral orientation of spatial dependency. Cross-covariance analysis of Ag & Au, and these two elements paired with Cu suggest downward and lateral mobilization of elements.

ii. In both oxidized and unoxidized zones, the general orientation of elemental mobilization is inferred to be along ENE, East, ESE, WSW and West directions with an angular tolerance of 30°. The major mineralized joints and faults in the Pierina gold deposit are oriented along these directions.

iii. Overall, the multivariate cross-covariogram analysis derived vectors of metal separation that coincide with recognized trends of major faults and joint sets in the Pierina deposit. Consequently, this type of analysis may be generally applied to hydrothermal mineral deposits as a means of identifying the structural features that guided hydrothermal and particularly oxidizing fluids resulting in the deposition and subsequent mobilization of metals.

Further research directions are suggested to refine and verify the validity of these results. These directions include the following:
- The odd parts of the cross-covariance (Goovaerts, 1997, p 73; Wackernagel, 1998, p 147; Webster and Oliver, 2000, p196) add to the anisotropic behavior of the results, whereas even parts of the cross-covariance are isotropic. It may be useful to model the odd parts of the cross-covariance and derive the lag-vectors for different pairs of the variables from the maximum and other high values.
- Using a geochemically identified immobile element in the pairs to better quantify distances of element separation and the location of enrichment zones.

5. ACKNOWLEDGEMENTS

The authors thankfully acknowledge Barrick Gold Company for providing the assay data used in this study and their support of geologic studies of the Pierina deposit. We thank Dr. Pierre Goovaerts for his valuable suggestions during the early stage of this research work.

6. REFERENCES:

Fifarek, R.H., and Rye, R.O., 2004, "Stable Isotope Geochemistry of the Pierina High Sulfidation Au- Ag Deposit, Peru: Influence of Hydrodynamics on SO_4^{2-}-H_2S Isotopic Exchange in Magmatic Steam and Steam-Heated Environments. Chemical Geology"

Goovaerts, P., "Geostatistics for natural resources evaluation", Oxford University press, 1997

Isaaks, E.H. and Srivastava, R.M. "An Introduction to Applied Geostatistics". Oxford University Press, New York. 561 p. 1989.

Samal. A. R, and Fifarek. R. H., 2003, "Application of Cross-Covariance in Geostatistical Modeling of Elemental Remobilization in Hydrothermal Mineral Deposits" in IAMG 2003 Conference Proceedings, Portsmouth, England, 6 pp.

Samal. A. R, and Fifarek. R. H., Sengupta. R, 2003, "Geostatistical modeling of elemental remobilization in hydrothermal deposits" in Abstratcts with programs, Vol 35, No 6, GSA annual meeting, Seattle, 2003

Samal. A. R., Fifarek. R. H. Mohanty. M. K., 2004, "Spatial modeling of elemental mobilizationin a hydrothermal gold deposit" in Abstratcts with programs, GSA North-Central Section, Annual meeting, St. Louis, 2004

Wackernagel Multivariate Geostatistics, Second Edition, Springer-Verlag Heidelberg, p. 145, 1998

Webster. R, Oliver. M., "Geostatistics for Environmental Sientists, J ohn Eiley & Sons, Ltd, 2000

VALUING A MINE AS A PORTFOLIO OF EUROPEAN CALL OPTIONS: THE EFFECT OF GEOLOGICAL UNCERTAINTY AND IMPLICATIONS FOR STRATEGIC PLANNING

EMMANUEL HENRY[*], DENIS MARCOTTE[**], and MICHAEL SAMIS[*]
*AMEC, 2020 Winston Park Drive, Suite 700,
Oakville, Ontario, L6H 6X7
**Department of Mineral Engineering, Ecole Polytechnique de Montréal, Quebec, H3T 1J4

Abstract. Mine valuation under market and geological uncertainty is an active research area. Twenty years ago, a seminal paper by Brennan and Schwartz described the application of Real Option Theory to the valuation of mines where metal prices are volatile. The study focused mainly on the impact of metal price uncertainty on the value of a mine. Geological uncertainty was not considered. For a simple mine model, this paper describes the close analogy between the decision to process a mining block at a given date and the European call financial option. The value of the European call depends primarily on the share price model, the present share price, the price volatility and the time to expiry. A mining block is either processed when the metal price covers the processing costs or otherwise stockpiled as waste. Metal prices and technical variables like grades, recovery, and costs are uncertain. Using geostatistical simulations, the study shows that grade uncertainty may introduce asymmetries in the block value greater than metal price uncertainty. The asymmetries are more pronounced for blocks with larger uncertainty. Greater value is given presently to these blocks assuming the block grades are perfectly known at the time of mining. The extension of this concept from individual blocks to the mine scale is done by considering a mine panel as equivalent to a portfolio of European call options. Implications for strategic planning are illustrated with a gold mine panel-scheduling example. Gold price was modelled with a Geometric Brownian Motion process. The case study shows that the value of the panel and its development strategy depend on the level of geological uncertainty and price volatility. However, the example shows that the benefits of optimising the panel under geological uncertainty is an order of magnitude below the benefits of resolving the geological uncertainty.

1 Introduction

Mine valuation under market and geological uncertainty is an active research area. Twenty years ago, a seminal paper by Brennan and Schwartz (1985) described the application of Real Option Theory to the valuation of mines where metal prices are volatile. This theory relies on former developments in Finance Theory on the evaluation of financial derivatives (financial options).

In the financial markets, an option is a contract giving the right, without the obligation, to buy or sell a share, or any financial instrument, at an agreed price and either at or before an agreed time in the future. An option that gives the right to buy a share at an agreed price (**E**) and a specific time in the future (**T**) is commonly called a European call option. **E** and **T** are called the exercise price and the time to expiry of the option, respectively. The share price (**S**) is usually volatile and at a time **T**, the option value (v_{opt}) will be:

$$v_{opt} = \max(S_T - E, 0) \quad [1]$$

If at time **T**, the share price is higher than the exercise price, then the owner of the option will be able to buy a share at price **E** and sell it immediately at price S_T, thus realizing a gain of S_T-**E**. If at time **T**, the share price is lower than the exercise price, the owner of the option will simply not exercise her option.

The value of an option, i.e. the price someone is ready to pay now to acquire the option (contract), depends on the price model and volatility, the current price, the exercise price, and the time to expiry. It is directly linked to the probability of the price being higher than the exercise price at expiry. Intuitively, the larger the time to expiry, the larger the price volatility, or the higher the current price, the higher the option value. Black and Scholes (1973), and Merton (1973), developed the first quantitative model for valuing European-like options with a share price following a Geometric Brownian Motion (Random Walk). Fig. 1 illustrates the value of a European call option as a function of the present price, a Geometric Brownian Motion price model, a time to expiry **T** of 1 year, and four annual volatilities (σ): 0%, 10%, 20%, and 30%. The exercise price **E** is $1. Fig. 1 shows that the value of the option increases with price volatility, and that the option can have a positive value even if the present price is below the exercise price. Practical valuation methods and algorithms for financial derivatives (a broader name for options) are found in Wilmott *et al.* (1995).

The Real Option Theory is the use of the Financial Option Theory to value real investments (see for details: Dixit and Pindyck, 1994, Amram and Kulatilaka, 1999, or Trigeorgis, 2000). Until recently, real option applications to mineral investments have considered mineral price volatility as the main source of uncertainty, and attempts to integrate geological uncertainty (as early as in Brennan and Schwartz, 1985, but see also Cortazar *et al.*, 2001, e.g.) were far from realistic. Carvalho *et al.* (2000) introduced a geostatistical simulation- and option pricing-based methodology to integrate geological models in the mine evaluation process. This paper focuses on the application of the Option Theory at the smallest scale in mines: the mining blocks. It illustrates the strong analogy between a mining block and a European call option, and shows how to use this analogy to value geological uncertainty and how geological and mineral price uncertainty interact. Implications for strategic planning are illustrated on a gold mine panel.

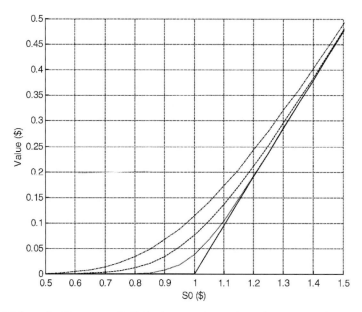

Figure 1. Value of a European Call Option as a Function of Current Price, S_0, and Price Volatility, σ; $T = 1$ year, $E = \$1$, and $\sigma = 0\%$ (continuous line), 10% (dotted line), 20% (dashed line), and 30% (dash-dotted line).

2 Analogy between a Mining Block and a European Call Option

A mining block contains **ton** tonnes of material at grade **g** of a mineral commodity sold at a price per unit **S**. The block may be developed or not. If the block is developed, it will be mined at a cost per tonne **m**, and then, either stockpiled as waste at a cost per tonne **stkp**, or processed at a cost per tonne **h** and marketed at a cost per unit **k**, depending on the benefit made by processing the block. Developing the block requires making the investment **dev** now, in order to be able to mine the block at time **T** from now. Recovery is denoted by ρ. Time discounting is ignored in this simple analogy, and the block is studied in isolation of the other blocks.

Traditionally, mine planners will make the decision to develop this block based on the block value v_{bl}:

$$v_{bl} = \text{ton} \cdot \text{g} \cdot \rho \cdot (S - k) - \text{ton} \cdot (m + h) \qquad [2]$$

and assuming grade and price are perfectly certain. If $v_{bl} > \text{dev}$, the investment is worth making and the block will be mined, otherwise, the block will be left un-mined.

However, if **g** and/or $S = S_T$ are uncertain at the time of decision, but if the true grade is known with certainty at the time of mining (for example, assuming that selection made on blast-hole sampling is exact, which is only a gross approximation), payoff from the block is similar to that of a European call option, with value:

$$v_{bl} = \max(\text{ton} \cdot g \cdot \rho \cdot (S - k) - \text{ton} \cdot (m + h), -\text{ton} \cdot (m + stkp)) \quad [3]$$

Equation 3 has the same form as Equation 1, and v_{bl} should therefore exhibit the same characteristics as v_{opt}:

- Grade uncertainty, like the mineral price, tends to increase the block value.
- Mineral price volatility tends to increase the block value.
- Grade uncertainty amplifies the effect of price volatility.
- The value of a block under price (and grade) uncertainty may increase with time T, if the volatility is large enough to compensate for the time discounting.

Table 1 summarizes the analogy between the mining block and a European call option.

Parameter	European Call Option	Mining Block
Time to Expiry	T	T
Exercise Price	E	ton (m + h)
Price	S	ton g ρ (S − k)
Cost of Not Exercising the Option	0	ton (m + stkp)

Table 1. Comparison between European Call Option and Mining Block.

3 Effect of Grade Uncertainty on the Value of a Mining Block

The European call option analogy of a mining block was investigated on a large panel in a gold mine, which is comparable to a portfolio of European call options.

The panel is made of 75 x 75 blocks, each of size 10 m x 10 m x 10 m. Geological information is provided by 50 m-spaced exploration drill holes. The gold distribution is lognormal with an average of 0.018 ounces per tonne (opt) and a coefficient of variation (CV) of 4. Gold grades are spatially variable with a strong nugget effect (40 % of the grade normal score variance, and ranges of 100 m north-south and 60 m east-west).

The mine planner is asked to determine which blocks should be developed now for production in two years. Production and economic parameters are shown in Table 2.

The gold price is modelled with a Geometric Brownian Motion process following the equation:

$$\frac{dS}{S} = \mu \cdot dT + \sigma \cdot dz \quad [4]$$

where μ is a trend, and dz is an increment to a standard Gauss-Wiener process. This model assumes no reversion to a long-term average, which is reasonable for gold, but not for most mineral commodities (see for example Schwartz, 1997, for more sophisticated commodity price models).

The focus of the first step is grade uncertainty while the price volatility and the effect of time are ignored. The panel is simulated 100 times on a fine grid and re-blocked to the nominal block size of 10 m x 10 m x 10 m. The simulations are then averaged to provide a map, illustrated in Fig. 2-a, similar to a kriged map. Mine planners are usually well aware that estimated (kriged) grade maps are uncertain, but for practical reasons, handle them as if they were certain, i.e. assuming they are an exact representation of the true grades, and overlooking local uncertainty associated with each block. Estimated grade maps are also generally smoother than in reality. Fig. 2-b shows a simulation outcome more representative of the true grade continuity.

Equation 3 is applied directly to the "certain" averaged model at time **T** = 0, assuming price is also certain and constant at S_0 = $350 per ounce. The blocks with value greater than the development cost **dev** should be developed and mined. Fig. 3-a shows the development outlines.

Recognizing that block grade-estimates are uncertain, and applying Equation 4 directly to each grade simulation outcome before averaging, gives the outlines in Fig. 3-b. The candidate area for development in Fig. 3-b is significantly larger than in Fig. 3-a, indicating the uncertain model generates more marginal blocks, i.e. blocks slightly higher than the economic break-even.

Parameter	Value	Variable Name in Text
Mining Production		
Block Tonnage	3,000 t	**ton**
Recovery	100%	ρ
Development Cost	$1 /t	**dev**
Mining Cost	$1 /t	**m**
Stockpiling Cost	$0 /t	**stkp**
Processing Cost	$3 /t	**h**
Marketing Cost	$0 /oz	**k**
Price Model		
Model	Geometric Brownian Motion	
Present Price	$350 /oz	S_0
Trend	0%	μ
Annual Volatility	12%	σ
Risk-Free Discount Rate	5%	
Planning Periods		
Time to Mining	2 years	**T**

Table 2. Production and Economic Parameters.

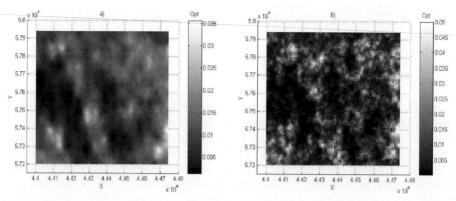

Figure 2. **a)** Average Grade of 100 Simulations; Reference Grade Model. **b)** Simulation Outcome Showing True Grade Variability.

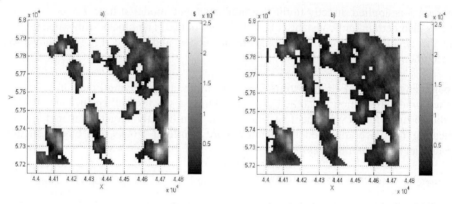

Figure 3. **a)** Value of the Blocks under Grade Certainty. **b)** Value of the Blocks under Grade Uncertainty. Only Positive Value Blocks are Shown.

Block values are plotted versus grades in Fig. 4. The black points correspond to the certain model, following the two linear equations:

$$v = -\text{ton} \cdot (m + stk) \qquad [5]$$

when $g < (m + h) / \rho (S - k)$, and otherwise:

$$v = \text{ton} \cdot g \cdot \rho \cdot (S - k) - \text{ton} \cdot (m + h) \qquad [6]$$

Light- and dark-grey points correspond to the uncertain model; light-grey points are for blocks with a CV greater than the average CV, 0.74, and dark-grey points, for blocks with a CV lower than 0.74. The figure shows that:

- Lower-grade blocks have more relative uncertainty (higher CVs) than higher-grade blocks. This is particular to this example and cannot be generalized.
- The higher grade-uncertainty, the higher the option value, as demonstrated by the light-grey points being above the dark-grey points.
- Grade uncertainty decreases the effective break-even cut-off grade (by approximately 20%, from the theoretical 0.011 opt down to 0.009 opt).

This result is in complete agreement with the findings made on financial options.

The impact of grade uncertainty is relatively large on the cut-off grade. However, this impact may be dampened, by the stockpiling cost for example. If **stkp** = $0.5 per tonne instead of $0 per tonne, grade uncertainty decreases the break-even cut-off grade by 10% only.

4 Cumulated Effect of Grade and Price Uncertainty on the Value of a Mining Block

Mineral price is usually considered as being volatile, i.e. uncertain, rather than certain. Both grade and price are then random variables that multiply each other in Equation 3. Price variance only is a function of time, with $\sigma_s(0)^2 = 0$, and $\sigma_s(T)^2$ increasing infinitely with time **T** for a Geometric Brownian Motion process.

In order to evaluate the impact of combining grade and price uncertainties, prices were simulated 1,000 times. No attempt was made to best fit the parameters to historic gold prices. However, the set of parameters in Table 2 is considered fair for the sake of the demonstration. Block values were calculated using Equation 3 and then averaged for each grade simulation and price simulation.

Fig. 5 shows block values as function of grade at **T** = 2 years (σ_s = 17%). In this example, grade uncertainty is far more important than price uncertainty.

5 Optimisation under Uncertainty

Optimisation of development outlines was performed on the maps shown in Fig. 3-a and 3-b using the assumption of uncertain grades. As commented earlier, uncertainty broadens somewhat the optimal design suggested by the certain value model. The broadened design increases the chances of capturing high-grade. The different designs suggested by the certain and uncertain models were applied successively to each of the 100 grade simulation scenarios. The average net value realized was then calculated. The procedure was also applied for the 100 optimal designs based on simulations, one for each simulation outcome. Each design was applied to all realizations and the average taken over the 100 x 100 possible combinations.

The results reported in Table 3 suggest that the most valuable mine design is the one which recognises the grade uncertainty (The comparison alone does not constitute a proof but provides useful indications).

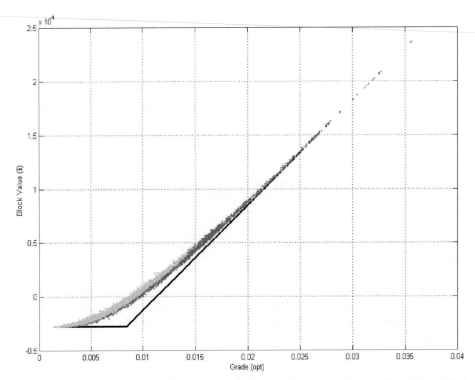

Figure 4. Block Value as a Function of Grades; Mining and Stockpiling Cost: $1 /t; Aligned Black Points: Grade Certainty; Dark-Grey Points: Grade Uncertainty with Block CV < 0.74; Light-Grey Points: Grade Uncertainty with Block CV > 0.74.

Design Based on	Base Scenario		Halved Variogram Range Scenario	
	Average Value	Relative Difference*	Average Value	Relative Difference*
Certain Grades	$4.5 M	0%	$2.4 M	0%
Uncertain Grades	$4.8 M	+6%	$2.7 M	+14%
Individual Simulations (Average)	$1.7 M	-61%	-$0.1 M	-103%

* To Certain Grade Model

Table 3. Value of Design Alternatives if Panel Mined at **T** = 2 years.

The design obtained by recognising grade uncertainty improves the certainty-based design value by 6% only. This is a small improvement and other biases in the mine optimisation parameters, such as assay results or cost estimates, would likely affect the design in the same order of magnitude as grade uncertainty. Table 3 also highlights that simulations are not useful in isolation: The best result achieved on a single simulation is about one-third of the result achieved using the option-based approach.

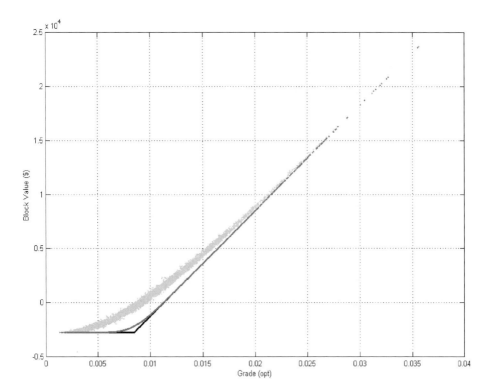

Figure 5. Block Value as a Function of Grade at **T** = 2 years; Mining and Stockpiling Cost: $1 /t; Aligned Black Points: Value with Grade and Price Certainty; Dark-Grey Points: Value with Grade Certainty and Price Uncertainty (σ_s = 17%); Light-Grey Points: Value with Grade and Price Uncertainty.

The study was repeated assuming a nugget effect of the normal scores of the grades equal to 10% of the sill, and halved ranges (50 m north-south and 30 m east-west). Lower panel values were obtained, as shown in Table 3, but the uncertainty-based optimisation is now 14% higher than the certainty-based optimisation.

6 Discussion and Conclusion

The methodology described in this paper provides a framework for integrating all sources of uncertainty, technical and/or financial. The complexity of Equation 3 can (and will most probably) be increased to include other important aspects of project evaluation, such as multiple minerals or foreign exchange uncertainty.

Grade uncertainty, as any other technical uncertainty, is project-specific and may or may not be discounted for risk, depending on the project analyst's application of Finance Theory. This is not the same as ignoring project uncertainty. Some analysts may view geological uncertainty as project specific and diversifiable, so that a risk adjustment is not necessary. Others may be of the opinion that geological uncertainty cannot be mitigated through diversification and would consider applying an appropriate risk adjustment. Systemic uncertainty in mineral prices, however, is not diversifiable, so prices are risk-adjusted. This was realised practically by performing risk-neutral price simulations instead of "real" price simulations.

The methodology is especially interesting for economically marginal projects (or project areas) only. It may be useful for evaluating near end-of-life investments, or capital-intensive push-backs in large open-pits. Uncertainty, technical or financial, is not that relevant for clearly uneconomic or clearly economic projects.

The impact of recovery was not studied in this paper. Recovery less than 1 will decrease the value of the block in Equation 3. It will also decrease the option value generated by grade uncertainty, by dampening the grade standard deviation.

The interest of the mining industry for uncertainty-based optimisation is encouraging. However, it is important to stress that the value added by an uncertainty-based optimisation may be an order of magnitude less than the value lost by not resolving the uncertainty. In the panel example illustrated here, optimisation under grade uncertainty improves the project value by 6%, for a value of $4.8 M. In comparison, the panel value if the true grade was known with certainty would be $15.6 M in average. In other words, geological uncertainty adds value to individual blocks, but destroys two-third of the true potential project value.

References

Amram, M. and Kulatilaka, N., *Real Options, Managing Strategic Investments in an Uncertain World*, Harvard Business School Press, 1999.
Black, F. and Scholes, M., The Pricing of Options and Corporate Liabilities, *Journal of Political Economics* vol. 81, 1973, p. 637-659.
Brennan, M. and Schwartz, E., Evaluating Natural Resource Investments, *Journal of Business*, vol. 58, no. 2, 1985, p. 135-157.
Carvalho, R., Remacre, A., and Suslick, S., Geostatistical Simulation and Option Pricing Techniques: A Methodology to Integrate Geological Models in the Mining Evaluation Projects, 6th *International Geostatistical Congress*, vol. 1, 2000, p. 1-10.
Cortazar, G., Schwartz, E. and Cassasus, J., Optimal Exploration Investments under Price and Geological-Technical Uncertainty : A Real Options Model, R&D Management, vol. 31, no. 2, 2001, p. 181-189.
Dixit, A. and Pindyck, R., *Investment under Uncertainty*, Princeton University Press, 1994.
Merton, R., Theory of Rational Option Pricing, *Bell Journal of Economics and Management Science*, vol. 4, no. 1, 1973, p. 141-183.
Schwartz, E., The Stochastic Behavior of Commodity Prices: Implications for Valuation and Hedging, *Journal of Finance*, vol. LII, no. 3, 1997, p. 923-973.
Trigeorgis, L., *Real Options, Managerial Flexibility and Strategy in Resource Allocation*, MIT Press, 2000.
Wilmott, P., Howison, S. and Dewyne, J., *The Mathematics of Financial Derivatives*, Cambridge University Press, 1995.

CLASSIFICATION OF MINING RESERVES USING DIRECT SEQUENTIAL SIMULATION

AMILCAR SOARES [1]
[1] *Environmental Group of the Centre for Modelling Petroleum Reservoirs, CMRP/IST, Av. Rovisco Pais, 1049-001 Lisbon, Portugal.*
e-mail: ncmrp@alfa.ist.utl.pt

Abstract

In mining operations, ore types are usually defined on the basis of technological criteria such as mining costs, processing plant performance and commercial costs, among others. Ore-type classification based on cut-off grades of estimated feed grades, tend to be biased when metal values or costs of ore type treatment are not linearly dependent on the feed grades. This paper presents an ore-type classification methodology based on jointly simulated grades. Direct sequential simulation (dss) and co-simulations (dscs) are the simulation techniques proposed to generate equiprobable images of different metal grades. Metal values and operating costs are then computed with several simulated grades of a block, in order to *a priori* classify the block, assigning it to the ore type which maximizes the profit or minimizes the costs of misclassification.
   A case study on the Neves Corvo mine illustrates the proposed methodology.

1 Introduction

In most mining operations, ore types are defined on the basis of technological criteria such as mining costs, processing plant performance and commercial costs, among others. Cut-off values of feed grades, together with geological criteria, are normally used for *a priori* classification of mining reserves into different ore types. However this classification can be severely biased when metal values or costs of ore type treatment are not linearly dependent on the feed grades and the *a priori* classification of mining reserves is performed on the estimated grades of blocks and stopes.
Suppose the value of a given stope is not a linear function of its grades, for example if metal recovery is highly non-linearly related with the feed grade, or the commercial costs have a non-linear dependence on penalty grades. Then, the decision of sending that stope to a given ore type stockpile based on the estimated feed grades cannot be the one that maximizes the profit of the stope.
If one knows not only the mean grade of a stope or block but also the local cumulative distribution function (cdf), the idea of the proposed methodology is to apply known non-linear functions of metal recovery, values, costs etc., to the cdf of a given stope, rather than to its estimated mean grade, in order to choose the best ore type stockpile.

Direct sequential simulation (dss) and co-simulations (dscs) are the simulation techniques proposed to generate equiprobable images of different metal grades. Metal values and operating costs are then computed with several simulated grades of a block in order to *a priori* classify the block in the ore type, which maximizes the profit or minimizes the costs of misclassification.

A case study on the Neves Corvo mine will illustrate the proposed methodology.

2 Case study: Neves Corvo mine

Neves Corvo is an underground tin-copper mine, which has been producing since the end of 1988. Considering the existing three orebodies, this study focuses on the Graça orebody that has been mined by a highly selective mining method (drift & fill) to maximize ore-type classification.

The main economic metal present in the ore is copper and the mineralisation can be described as being of the fissural or stockwork type. It is composed of veinlets and strings of sulphides and quartz, which cut mainly acid volcanic rocks, concordantly or not with the schistosity. The sulphides are mainly pyrite and chalcopyrite and the thickness of the veinlets may vary from a few millimetres to a few decimetres. The spatial distribution of the veinlets is highly irregular – as well as that of the grades – and does not show, in most situations, to be controlled by any particular geological feature. Cassiterite and stannite (tin and copper sulphide) are the main tin ores.

Two main ore types are defined in the Graça orebody: cupriferous ore (MC) and tin ore (MS), which are treated in different plants. The MS plant recovers copper and tin while the MC plant recovers only copper.

Data of Cu and Sn coming from drill-hole samples are available for this study.

3 Direct Sequential Simulation and Co-simulation

The principle of direct sequential simulation (dss) can be summarized as follows:
If the local cdfs are centred at the simple kriging estimate

$$z(x_u)^* - m = \sum_\alpha \lambda_\alpha(x_u)(z(x_\alpha) - m)$$

with a conditional variance identified by the simple kriging variance $\sigma^2_{sk}(x_u)$, the spatial covariance model or semivariogram is reproduced in the final simulated maps (Journel, 1994). The problem is that this simulation approach does not reproduce the histograms of the original variables (the local cdf cannot be fully characterized by only the local mean and variance).

The idea of direct sequential simulation (Soares, 2001) is to use the estimated local mean and variance, not to define the local cdf but to sample the constant global cdf $F_Z(z)$. Intervals of z are chosen from $F_Z(z)$, and simulated values $z^s(x_u)$ are subsequently sampled from them. These intervals are "centred" at the simple kriging estimate $z(x_u)^*$, being the interval range dependent on the simple kriging estimation variance $\sigma^2_{sk}(x_u)$ (Soares, 2001).

One of the main advantages of the proposed dss algorithm over traditional sequential indicator simulation (sis) and sequential Gaussian simulation (sGs) to simulate continuous variables is that it accommodates joint simulation of original variables without any prior indicator or Gaussian transformation.

In this case, the joint simulation of both metals Cu and Zn follows the Bayes rule. That means, that the simulation of a pair of values from a bi-variate distribution, say $F(Z_1,Z_2)$, is equivalent to generating the first value z_1 from the marginal distributions $F(Z_1)$ and the second from the conditional distribution, $F(Z_2|Z_1=z_1)$. In a spatial process with two correlated variables, $Z_1(x)$ and $Z_2(x)$, the first value z_1 is simulated from $F_{Z1}(x_u; z) = \text{prob}(Z_1(x_u)<z)$ at the location x_u and, afterwards, z_2 is generated from the conditional distribution $\text{prob}(Z_2(x_u)<z \mid Z_1(x_u) = z_1)$ (Almeida and Journel, 1994). The first variable is simulated with direct sequential simulation and the second variable using direct sequential co-simulation (Soares, 2001).

The same algorithm is then applied to simulate $Z_2(x)$ assuming the previously simulated $Z_1(x)$ as the abundant (known at every node) secondary variable. Co-located simple co-kriging is used to calculate $z_2(x_u)^*$ and to estimate $\sigma^2_{sk}(x_u)$ conditioned to neighbourhood data $z_2(x_\alpha)$ and the co-located datum $z_1(x_u)$ (Goovaerts, 1997).

One crucial issue of this sequential approach regards the choice of the variable to be simulated first. In sequential simulation algorithms local conditional distributions are estimated with some approximations, for example, the conditioning data is limited to a subset of samples (Gomez-Hernandez and Journel, 1993); hence, for variables with different spatial continuity patterns, the result is not independent of the order of the chosen sequence of variables to be simulated. Hence, practical criteria regarding the spatial pattern of the variables and its relative importance in the physical phenomenon are normally applied. In this case, Cu grades are simulated first – through direct sequential simulation – since Cu is the main metal, the most valuable one and, on top of that, reveals a more continuous spatial pattern.

4 Classification of mining reserves in ore types.

The usual procedure of classification of mining reserves in ore types consists in using the estimated average grades of each block as a threshold criterion to classify it. If a block value is a non-linear function of its grades, classification can be severely biased when performed with estimated average grades.

The idea of the proposed classification can be summarized in two basic points:

The classification is based on simulated grades, rather than estimated ones, which allow preserving the histograms of different metals, spatial pattern continuity and the spatial relationship between them.

The criteria to classify one given block in ore types will be based on the maximization of a profit function, or minimization of a cost function, applied to the joint simulated values of the block.

In the case study of the Neves Corvo mine, the processing plants have different metal recoveries and costs. The commercial costs, which include transport, shipping, insurance, treatment and refinement charges are also different for both metals, copper and tin.

Hence, the value of a given stope can be viewed as the difference between the metal value minus the treatment and the commercial costs. Suppose a block is located at x_u

with the feed grades of Cu and Sn: $z(x_u)$ and $y(x_u)$. The value of one tonnage of a block of MS ore type can be summarized as:

$$v_{MS}(x_u) = [s_{Cu} - c_{Cu}/z_c].z(x_u). \eta_{Cu} + [s_{Sn} - c_{Sn}/y_c].y(x_u). \eta_{Sn} - mc - plc_{MS} \quad [1]$$

that is, the sum of the copper value (net value minus costs) plus the tin value minus the mining costs and the MS plant costs. The tonnage of a block value of MC:

$$v_{MC}(x_u) = [s_{Sn} - c_{Sn}/y_c].y(x_u). \eta'_{Cu} - mc - plc_{MC} \quad [2]$$

which is the sum of the copper value minus the mining costs and the MC plant costs.
with: s – metals price of Sn and Cu ; η_{Cu}, η_{Sn} – Cu and Sn recovery at MC plant; η'_{Cu} – Cu recovery at MS plant; z_c, y_c – concentration grades; c_{Cu}, c_{Sn} – commercial costs; mc – mining costs; plc – plant costs

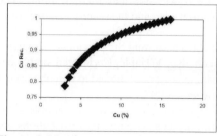
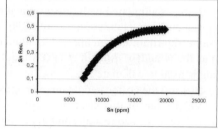

Figure 1. Metal recovery of Cu vs Cu (%) at MC plant and metal recovery of Sn vs Sn at MS plant.

Metal recovery of Cu and Sn are non-linear functions of the feed grades. Figure 1 shows the metal recovery of Cu vs Cu (%) at the MC plant and metal recovery of Sn vs Sn (%) at the MS plant.

The criterion to classify x_u as MC or MS is the maximization of the profit, $v_{MS}^l(x_u)$ or $v_{MC}^l(x_u)$, for the entire set of realizations $l = 1, Ns$, or, in other words, the minimization of the costs of misclassification. That is, x_u will be classified as the ore type that maximizes the value for the entire set of realizations. x_u is classified as cupriferous ore if:

$$\frac{1}{N_s}\sum_{l=1}^{Ns} v_{MC}^l(x_u) > \frac{1}{N_s}\sum_{l=1}^{Ns} v_{MS}^l(x_u) \quad [3]$$

x_u is considered as tin ore otherwise.

Note that as the value v is a non-linear function φ of the feeding grades z, $v^l(x_u) = \varphi[z^l(x_u)]$, a different result is achieved when this criterion is applied to an average grade of Cu or Sn at x_u:

$$\sum_{l=1}^{Ns} \varphi(z^l(x_u)) \neq \varphi\left(\sum_{l=1}^{Ns} z^l(x_u)\right)$$

An alternative criterion to [3] could be chosen in terms of costs rather than profits: the minimization of the costs of misclassification (Goovaerts, 1997). Suppose the following loss functions: the loss associated with classifying x_u as MC

$$L_1^l(x_u) = \begin{cases} 0 & \text{if } v_{MC}^l(x_u) > v_{MS}^l(x_u) \\ v_{MS}^l(x_u) - v_{MC}^l(x_u) & \text{otherwise} \end{cases}$$

and the equivalent loss associated with classifying x_u as MS:

$$L_2^l(x_u) = \begin{cases} 0 & \text{if } v_{MS}^l(x_u) > v_{MC}^l(x_u) \\ v_{MC}^l(x_u) - v_{MS}^l(x_u) & \text{otherwise} \end{cases}$$

The N simulated images allow calculating the average loss attached to the two types of classification:

$$\varphi_1(x_u) = \sum_{l=1}^{Ns} L_1^l(x_u) \text{ and } \varphi_2(x_u) = \sum_{l=1}^{Ns} L_2^l(x_u) \qquad [4]$$

the location x_u is declared to belong to MC or MS if it minimizes the corresponding average losses:

$$\varphi_1(x_u) > \varphi_2(x_u)$$

meaning that the costs of classifying x_u as MC are greater than the costs of classifying x_u as MS, hence x_u is classified as MS;

$$\varphi_2(x_u) > \varphi_1(x_u)$$

meaning that the costs of classifying x_u as MS are greater than the costs of classifying x_u as MC, hence x_u is classified as MC.

Both approaches [3] and [4] are equivalent and give the same results which are presented and compared with results following the more traditional routine of ore-type classification based on estimated grades.

5 Results

5.1 DATA ANALYSIS

Histograms of Cu and Sn were calculated from 524 samples from boreholes of the Graça orebody (Figures 2a and b). The Cu/Sn bi-plot shows the relationship between both elements (Figure 3). Spatial continuity main patterns of Cu and Sn can be summarized in the following: both Cu and Sn present a similar isotropic behaviour, modelled by an exponential model. Sn variogram presents a clear "nugget effect", representing 20% of the total variance, which is probably linked to the spatial dispersion of main tin mineralisations: cassiterite and stanite.

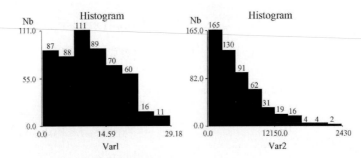

Figure 2. Histograms of Cu (a) and Sn (b).

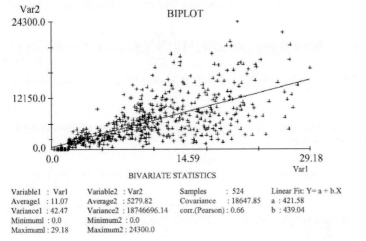

Figure 3. Bi-plot of Cu/Sn

From the bi-plot of Figure 3 one can visualize two populations with different behaviours regarding the correlation between Cu and Sn. If the total set of samples is split by a Cu threshold of 10%, the values with a Cu content lower than 10% show a higher correlation coefficient (r = .71) (Figure 4a) than the values with a Cu content higher than 10%, which do not present a significant correlation with Sn (r = .37) (Figure 4b). Cu/Sn cross variograms computed for those populations confirm the distinct spatial co-regionalisation behaviours.

5.2 JOINT SIMULATION OF CU AND SN

Cu and Sn grades were simulated in a regular grid of points (200x 110 x 20 nodes of 1x1x1m.). Sn values were simulated using direct sequential co-simulation assuming previously simulated maps of Cu as a secondary variable.
Simple collocated co-kriging was used for the estimation of Sn at each node of the regular grid visited during the sequential procedure:

$$z_1(x_u)^* = \sum_{\alpha=1}^{N} \lambda_\alpha(x_u)(z_1(x_\alpha) - m_1) + \lambda_u(x_u)(z_2(x_u) - m_2) + m_1 \quad [5]$$

CLASSIFICATION OF MINING RESERVES USING DIRECT SEQUENTIAL SIMULATION 517

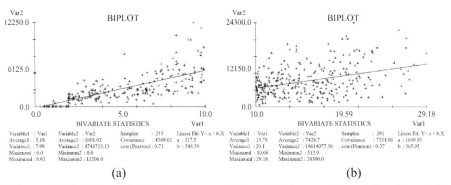

Figure 4. Bi-plot of Cu/Sn for the population with low grades of Cu a); and for the population with high Cu grades.

Collocated co-kriging is implemented with the Markov-type approximation of co-regionalization models: cross correlograms between $z1$ and $z2$, $\rho_{z1,z2}(h)$ are determined by the correlation coefficient $\rho_{z1,z2}(0)$ and the correlogram of $_{z1}$ $\rho z1(h)$: $\rho_{z1,z2}(h)=\rho_{z1,z2}(0) \cdot \rho_{z1}(h)$ (Goovaerts, 1997). Two local co-regionalisation models between Cu and Sn (described in 5.1) were adopted with the Markov-type approximation: "low" grades of Cu (<10%) with a correlation coefficient r = .71 and "high" grades of Cu (≥10%) with r = .37. An example of level 1, with "low" and "high" grades of Cu is shown in Figure 5. Note that in this case, under the Markov-type approximation, to estimate a local Sn value at the location x_u, the co-regionalisation model is dictated by the correlation coefficient of x_u.

Figure 5. Estimated maps of "low" (red) and "high" grades (blue) of Cu

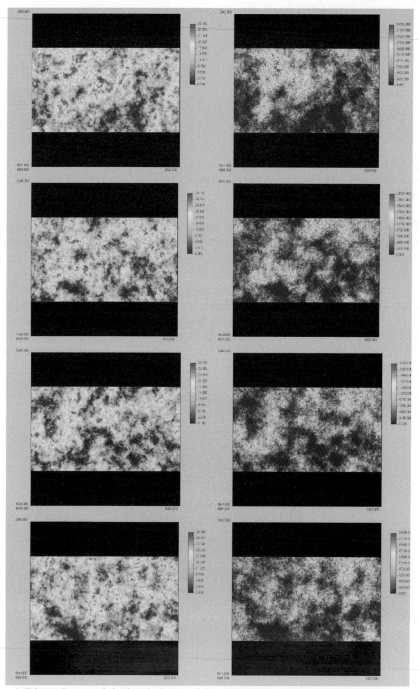

Figure 6. Direct Sequential Simulation and Co-Simulation : Four co-simulated pairs of Cu (left column) and Sn.

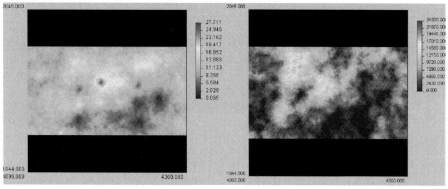

Figure 7. Average of 20 simulated maps of Cu (left) and Sn (right).

At the Cu "high" grades population (blue area of Figure 5), with a correlation coefficient r = .37, there is practically no influence of the secondary variable (simulated Cu grades). A set of 20 realizations of Cu and Sn were simulated for the entire area. Examples of four pairs of Cu and Sn images are presented in Figure 6. One example (level 1) of the average of 20 simulated maps of Cu and Sn are presented in Figures 7a and 7b, respectively. Notice that the influence of Cu at the Sn simulations is significant only at the "low" Cu grades area, where the correlation coefficient is high.

Marginal histograms and correlation coefficients of simulated Cu and Sn show a quite good match with the equivalent sample statistics.

Fig. 8 shows the variograms of the same two realizations of first level for Cu (left column) and Sn (right column). There is a very satisfactory match between the theoretical model (imposed to the simulations and co-simulations) and the experimental variograms of simulated values.

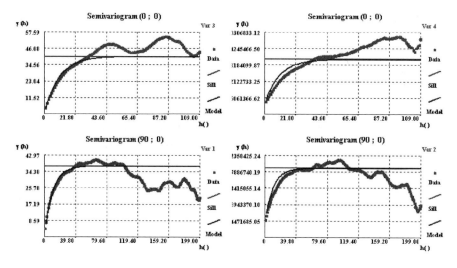

Figure 8. Variograms (experimental and model) of simulated values of Cu (left column) and Sn (right column) for two realizations.

5.3 CLASSIFICATION OF THE ORE TYPES

Each pair of simulated Cu and Sn values, at a given spatial location x_u, will feed the two profit functions [1] and [2] corresponding to the two different plant treatments and transport. Averaging out the profit of the 20 realizations for the two plant treatments will determine which ore type should be allocated to the spatial location x_u [3].

Sensitivity analysis has shown a high dependency of the profit of [1] on the tin metal prices. Four different tin metal prices were used, corresponding to those occurring during the period from the beginning of the mine's exploration until now. Figure 9 shows, in the right column, the two ore types classified on the basis of the simulated images for the four metal prices, from 4 US$/Lb (top) practiced in the end of the eighties, up to 2003 price of 2.4 US$/Lb (bottom). In the left column of Figure 9 the equivalent classification based on the average maps of Figures 7a and 7b is presented for comparison.

It is obvious that in both classifications the MS ore type decreases with the tin metal price. However, the classification of the average grade gives systematically higher proportions of MS than the simulations. This is quite expectable since the non-linear functions of Figures 1a and b, applied to a mean of a positively skewed histogram of Sn values (Figure 1b), tend to be greater than the mean of the non-linear transformation of each one of Sn grades. These highest proportions of MS reflect biased average-grades based classification: a systematic overestimation of MS proportion.

As a matter of fact, the continuous decreasing of tin metal price determined the very recent decision (taken in 2002) of the mine board to discommission the tin plant.

6. Final remarks

i)   This paper presents the use of stochastic simulation images of different metal grades to classify mining reserves in ore types. When costs and values can be allocated to the main mining operations, classification of ore types based on joint simulated metals are a much more accurate and unbiased alternative than the classical procedure of classification based on estimated grades.

ii)   This paper also shows that ore-type classification is a dynamic exercise of optimisation of future strategies, balancing historical decisions, the knowledge of reserves, and the near future of metal market prices, contracts, etc..

Considering the presented test case, when the decision of building a tin plant was taken, it was fully justified by the tin prices of that time. Once the tin plant was working, any classification should have been conditioned to its fixed and operational costs. According to the criteria followed in 5, most of the blocks are classified as MC. But that implies that a significant number of those blocks should remain unmined, given that the production capacity of the copper plant is limited. In this case, although we know that those blocks give, theoretically, more profit in the Cu plant, they should be sent to the tin plant as that will optimise the production capacity of both plants.

CLASSIFICATION OF MINING RESERVES USING DIRECT SEQUENTIAL SIMULATION 521

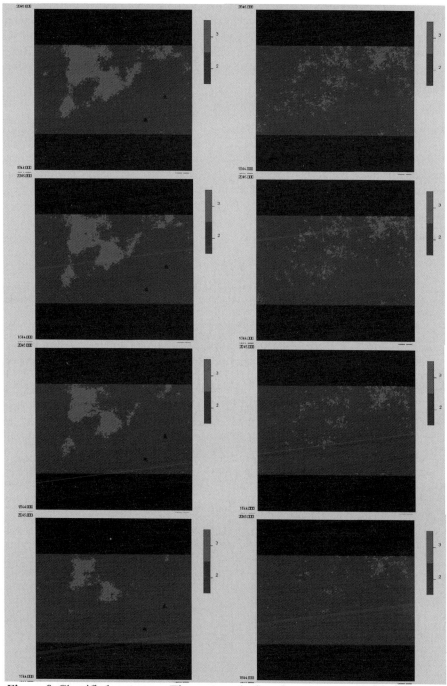

Figure 9. Classified oretypes – Tin oretype (red) and copper oretype (blue) based on 20 co-simulated pairs, and four different metal value of Tin (right hand side column); based on the average of the simulations (left hand side column).

iii) Finally, it is demonstrated that the combination of direct simulation and co-simulation is a very appropriate technique for the joint simulation of continuous variables. Recent applications of the dss can be found in environmental field , in soil pollution characterization (Franco C. et al, 2002), satellite image classification (Bio et al, 2002), ecological resources (Almeida et al, 2002) and in petroleum applications (Soares et al, 2001).

Acknowledgments

Somincor – Sociedade Mineira de Neves Corvo – is gratefully acknowledged for the permission to publish the data.
I would like to thank the valuable comments of two anonymous reviewers.

References

Almeida A., Journel A., 1994. Joint simulation of multiple variables with a Markov-type corregionalization model. Mathematical Geology 26(5): 565-588.
Almeida J., Bio A., Santos E., 2002. Use of Geostatistical Methods to Characterize Population and Recovery of Iberian Hare in Portugal. Proceedings of geoENV2002- Geostatistics for Environmental Applications. Barcelona.
Bio A., Carvalho J., Rosário L., 2002. Improving Satellite Image Forest Cover Classification with Field Data Using Direct Sequential Co –Simulation. Proceedings of geoENV2002- Geostatistics for Environmental Applications. Barcelona.
Caers J., 2000, Direct sequential indicator simulation. Proceedings of 6th International Geostatistics Congress. Cape Town. S.A..
Gomez-Hernandez, J., Journel A.G., 1993 - Joint Sequencial Simulation of MultiGaussian Fields. *Geostatistics TROIA'92*, Ed. Soares, A., Kluwer Pub., pp. 85-94.
Franco C.,Soares A. , Delgado J., 2002. Characterization of Environmental Hazard Maps of Metal Contamination In Guadiamar River Margins. Proceedings of geoENV2002- Geostatistics for Environmental Applications. Barcelona.
Goovaerts P., 1997. Geostatistics for Natural Resources characterization. Oxford University Press.pp 483.
Journel A.G., 1994. Modeling Uncertainty: Some Conceptual Thoughts. Geostatistics for the Next Century. ED Dimitrakopoulos R..kluwer Academic Pub.pp 30-43.
Soares A., 1998. Sequential Indicator Simulation with Correction for Local probabilities. Mathematical Geology, vol 30, N 6,, pp 761-765.
Soares A., 2001. Direct Sequential Simulation and Co-simulation. Mathematical Geology. 33-8. pp 911-926.
Soares A, Almeida J., Guerreiro L.. 2002. Incorporating Secondary Information using Direct Sequential Co-Simulation. To be published in Stochastic Modelling Vol II, American Association of Petroleum Geologists Pub.

USING UNFOLDING TO OBTAIN IMPROVED ESTIMATES IN THE MURRIN MURRIN NICKEL-COBALT LATERITE DEPOSIT IN WESTERN AUSTRALIA

MARK MURPHY[1], LYN BLOOM[2] AND UTE MUELLER[2]

[1] *Snowden Mining Industry Consultants, West Perth, Western Australia*
[2] *Edith Cowan University, Joondalup, Western Australia*

Abstract. Nickel and cobalt are key additives to modern alloys. The largest worldwide nickel-cobalt resources occur in surface laterite deposits that have formed during chemical weathering of ultramafic rocks at the Earth's surface. Geologically young deposits have formed by rapid weathering processes in tropical environments while older deposits that have formed in drier climates. At the Murrin Murrin mine in Western Australia the dry climate laterite deposits occur as laterally extensive, undulating blankets of mineralisation with strong vertical anisotropy and near normal nickel distributions. This deposit structure presents an estimation challenge for both classical and geostatistical resource estimation methods. In this paper, ordinary kriging and multiple indicator kriging estimation methods are applied to both the in situ and unfolded structural cases to obtain estimates for nickel and cobalt. Improvement in point grade estimation following the unfolding of the laterite blanket by vertical data translation prior to grade estimation is assessed in the light of close spaced grade control data. The results indicate that unfolding, particularly when combined with indicator kriging, improves both the nickel and cobalt estimates albeit only slightly in the case of cobalt.

1 Introduction

Nickel and cobalt are key metal additives in modern industry. Nickel and cobalt are primarily sourced from deep underground mines but the largest worldwide deposits where both metals occur are the near surface laterite deposits that have formed by weathering of ultramafic rocks in tropical or semiarid environments (Golightly 1981; Brand et al 1998). At Murrin Murrin in central Western Australia, surface weathering of ultramafic rocks in a semiarid environment has enriched nickel and cobalt to economically attractive concentrations approaching 2%Ni and 0.5%Co within smectite clay horizons. The nickel cobalt deposits at Murrin Murrin are flat lying, undulating blankets of 10 to 50 m thickness and lateral extents ranging from a few to tens of kilometres (Fazakerley and Monti, 1998).

2 The MM2 Dataset

One deposit area at Murrin Murrin, known as MM2 is the focus of this study. The data comprises samples collected from vertical drillholes during exploration and subsequent mining of the deposit. Exploration drilling was completed on a nominal 50 m square pattern and contains a local cluster of 12.5 m spaced holes (Figure 1, left). Grade control sampling was carried out on a 12.5 m square pattern to approximately 30 m below surface (Figure 1, right). For this study the drilling samples were accumulated into a composite length that matches the mining bench height of two metres. The exploration sampling was flagged as a subset of the grade control data and, both data sets were clipped to a boundary 30 m below surface and to a marginal ore processing threshold of combined nickel cobalt grade.

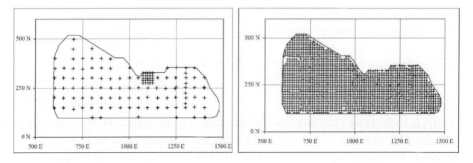

Figure 1. Exploration (left) and grade control (right) collar locations in the MM2 pit

For the purposes of this study the sampling from the grade control pattern is considered reality. Figure 3 shows cross sections through 250N (2:1 vertical exaggeration) with the 12.5 m spaced, bench height composites from grade control coded by nickel and cobalt grades within the ore envelope. These sections reveal that nickel forms a relatively continuous blanket of mineralisation with higher grades (>1.0 Ni%) defining an undulation in the nickel mineralisation across the area. In contrast, the high grade cobalt mineralisation (>0.06 Co%) is more pod-like but generally follows the blanket of nickel mineralisation.

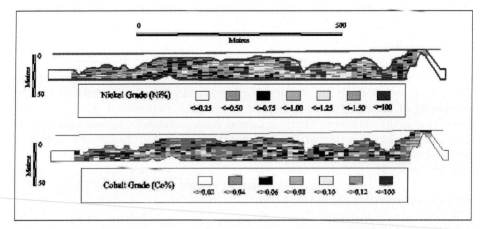

Figure 2. Cross section 250 N showing ore envelope and bench height composites

In Table 1 the summary statistics of both nickel and cobalt composites within the ore are compared for both the grade control and exploration sampling patterns. Declustered statistics were calculated using cell declustering to account for the clustered sampling in the exploration pattern.

Statistic	Nickel grade (%)			Cobalt grade (%)		
	Grade control	Exploration		Grade control	Exploration	
		Clustered	Declustered		Clustered	Declustered
Composites	13,414	1,046	1,046	13,411	1,046	1,046
Minimum	0.07	0.13	0.13	0.001	0.001	0.001
Maximum	2.67	2.23	2.23	0.887	0.887	0.887
Mean	0.85	0.84	0.80	0.058	0.054	0.053
Median	0.81	0.80	0.75	0.040	0.038	0.036
Standard deviation	0.38	0.39	0.38	0.054	0.054	0.056
CV	0.44	0.46	0.47	0.944	0.993	1.053

Table 1. MM2 grade summary statistics for grade control and exploration composites

The summary statistics show that the exploration sampling contains 1,046 samples compared to the 13,414 available from the final grade control pattern and that the nickel distribution is near normal while the cobalt distribution is highly skewed. Declustering produces in a minor reduction in the distribution means and a minor increase in data skewness. Accepting the grade control results as reality for this study, the exploration sampling statistics show that the exploration sampling pattern has been successful in determining the underlying mean and variability of both nickel and cobalt.

3 Unfolding

The large lateral extent and blanket geometry of nickel laterite deposits, combined with a strong vertical anisotropy, presents several problems for grade estimation from the exploration data. Of particular interest to mine planning is the correct reproduction of the lateral connectivity of higher grade zones as depicted in Figure 3. In Figure 3, a schematic cross section of a nickel laterite resource envelope and vertical drill holes is depicted against a backdrop of an estimation grid. The search neighbourhood used for estimation of the model nodes is shown as a flat lying ellipsoid with a shape dictated by the strong vertical anisotropy the deposit. A dashed line represents a surface of expected grade connectivity for this idealised deposit. It is assumed waste samples have been excluded from the estimation method.

In Figure 3, where drill holes are close together (near block A) or where the lies ore horizontally (near block B), the grade zones in the drill holes are reflected in the estimation model. However, where drilling is widely spaced and/or there is undulation of the surface of grade connectivity counter intuitive estimation results may occur (such as block C and block D). Problems of geometric controls arrecting grade estimation in situations of folded or undulated geometry have been recognised by prior authors (Wellmer & Giroux 1980, Dowd et al 1988, Lambert 2000, Sahin et al 1998, Sides and da Silva 1996). These authors have proposed several methods to remove estimation artefacts including domaining areas of similar geometry, data translation and application of local coordinate systems.

In this study, the vertical translation method was used to improve the grade connectivity of nickel grades within the study area (Murphy et.al. 2002).

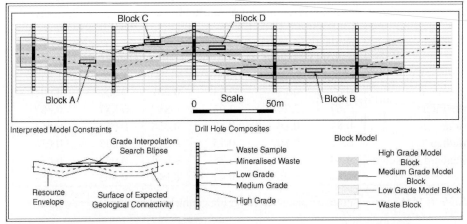

Figure 3. Schematic estimation model from vertical drilling in a finite domain

4 Variography

Traditional and indicator semivariograms were computed for the exploration data in both the in situ and unfolded data configurations. Twelve indicator thresholds were applied to both elements in 0.1%Ni increments ranging from 0.5 to 1.6%Ni, and 0.01%Co increments from 0.03 to 0.14%Co. The blanket geometry of the nickel mineralisation dictates that the minor axis of continuity is the downhole direction. Therefore, horizontal-plane semivariogram maps were used to test for the direction of maximum continuity in the horizontal plane and direction variograms were then computed for the axes of anisotropy.

For nickel, the traditional variography exhibits geometric anisotropy in the study area with a major axis of continuity as azimuth 70°. The variogram has a low nugget effect (0.02 of a sill of 1.00) and three nested spherical structures were fitted to the experimental data (0.45 sill, 7m x 30 m x 30m; 0.30 sill, 12 m x 50 m x 50 m; 0.23 sill, 15 m x 75 m x 100m). The variography of unfolded data has slightly longer ranges (0.45 sill, 7m x 30 m x 30m; 0.30 sill, 12 m x 60 m x 70 m; 0.23 sill, 15 m x 75 m x 200m). The nickel indicator semivariogram surfaces revealed patterns of rotational anisotropy with where the lower nickel thresholds having greater continuity east-west and higher thresholds having longer NE-SW continuity. There is a pattern of decreasing ranges and increasing nugget effect with increasing indicator nickel threshold and slightly longer ranges interpreted for the unfolded case.

For cobalt, the traditional semivariogram has a major axis azimuth of 100° and a nugget effect of 0.25. Again three nested structures were modelled for the in situ data (0.40 sill, 6 m x 20 m x 60m; 0.25 sill, 8 m x 50 m x 70 m; 0.23 sill, 10 m x 150 m x 200 m) and unfolded cases (0.40 sill, 8 m x 20 m x 20 m; 0.25 sill, 9 m x 30 m x 30 m; 0.23 sill, 10 m x 40 m x 40 m) with unfolding the data resulting in interpretation of much shorter ranges. Similar to nickel, the cobalt indicator semivariograms display a pattern of rotation anisotropy, increasing nugget effect, and decreasing indicator ranges with increasing indicator threshold. However, as a general comment the horizontal plane experimental variograms were poorly structured and the interpretations were based largely on the behaviour of the vertical, downhole results.

5 Estimation

The grade control sample locations were estimated from the exploration data with means of ordinary point kriging (OK) and multiple indicator kriging (IK) E-type estimates using the indicator thresholds discussed above (Journel, A.G., Huijbregts, C.J. 1978). Table 2 compares the grade control data statistics to the estimate made at each grade control location using the combinations of estimation method, data configuration and metal.

Stat	Nickel grade (%)					Cobalt grade (%)				
	Grade control	OK		IK		Grade control	OK		IK	
		In situ	Unfold	In situ	Unfold		In situ	Unfold	In situ	Unfold
Min.	0.07	0.16	0.15	0.32	0.32	0.001	0.001	0.001	0.001	0.001
Max.	2.67	2.06	2.00	2.07	2.12	0.887	0.887	0.887	0.887	0.887
Mean	0.85	0.85	0.85	0.87	0.87	0.058	0.055	0.060	0.056	0.061
Med.	0.81	0.83	0.83	0.84	0.85	0.040	0.052	0.055	0.052	0.056
S.D.	0.38	0.23	0.26	0.23	0.28	0.054	0.025	0.027	0.026	0.031
C.V.	0.44	0.26	0.31	0.26	0.32	0.944	0.452	0.459	0.464	0.503

Table 2. Summary statistics of exploration grade estimates compared to grade control

In terms of mine planning and the need to repay start-up capital expenditure in the early years of mining and processing, the corrected estimation of the amount of high-grade material at the exploration stage is critical to project feasibility. Despite the fact that the estimation results and input data are point values, pseudo grade volume curves have been generated for each estimate by assuming that each node represents an ore parcel of dimension 12.5 m E by 12.5 m N by 2 m RL. These curves give an insight into accuracy of high-grade volume estimates that can be expected from each estimation method and are shown in Figure 4.

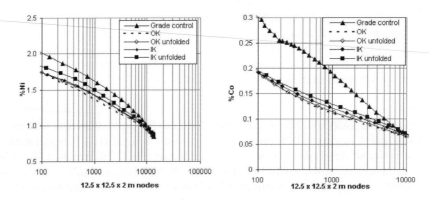

Figure 4. Pseudo grade tonnage curves for nickel (left) and cobalt (right)

6 Conclusions

The statistics in Table 2 reveal that for nickel, OK gives the most accurate estimate of the mean but the results plotted in Figure 4 show that the unfolded IK method is more accurate in estimation of the high grade ore. For cobalt, the best accuracy of both mean and high grade is also achieved for the combination of unfolding and IK estimation method albeit all methods poorly predict the amount of high grade cobalt.

References

Brand, N.W., Butt, C.R.M., Elias, M. (1998): Classification and features of Nickel Laterites. Cooperative Research Centre for Landscape Evolution and Mineral Exploration Report 74, CSIRO, Perth.
Dowd, P. A., Johnstone, S.A.W., Bower, J. (1988): The application of structurally controlled geostatistics to the Hilton Orebodies, Mt Isa, Australia. In: 21st Application of Computers and Operations Research In the Mineral (APCOM) Industry. A. Weiss (Ed), Society of Mining Engineers. (Vol 2), p 275-285.
Fazakerley, V.W., Monti R. (1998): Murrin Murrin nickel-cobalt deposits. In: Geology of Australian and Papua New Guinean Mineral Deposits, D. A. Beriman and D. H. Mackenzie (Eds), The Australasian Institute of Mining and Metallurgy. Melbourne. p 329-334.
Golightly, J. P. (1981). Nickeliferous Laterite Deposits. Econ. Geol., 75th Anniversary Volume, p 710-713.
Journel, A.G., Huijbregts, C.J. (1978). Mining Geostatistics. Academic. London.
Lambert, S. (2000): Geostatistical estimation and generalised spatial coordinate transformations. In: Geostats 2000, WJ Kleingeld and DG Krige (Eds), p 864-883, Cape Town.
McArthur, G.J. (1998): Using geology to Control geostatistics in the Hellyer deposit. Mathematical Geology, 20, 5. p. 342-366, Dordrecht.
Murphy, M. (1998): Murrin Murrin East resource estimation, Geostatistical Association of Australia Newsletter Sept 1998. [On-line] WWW: http://www.confed.com.au/gaa/.
Murphy, M, Bloom L and Mueller U (2002): Geostatistical optimisation of mineral resource sampling cost for a Western Australian nickel deposit. In Bayer U, et.al. (eds) IAMG 2002 Proceedings of the 8th Annual Conference of the International Association of Mathematical Geology, Terra Nostra Berlin, p.209-214.
Sahin, A Ghori, S.G., Ali, A.Z., El-Sahn, H.F., Наззan, H M. and Al-Sanounah, A. (1998): Geological controls of variograms in a complex carbonate reservoir, Eastern Province, Saudi Arabia. Mathematical Geology, 30, 3. p 309-322, Dordrecht.
Sides, E.J., da Silva, F.J. (1996): Application of variable search prism orientation for improved geological control on grade estimation at the Neves-Corvo copper-tin mine, Proceedings of the Conference on Mining Geostatistics, Geostatistical Association of South Africa, 182-199, South Africa.
Wellmer F.W., Giroux, G.H. (1980): Statistical and geostatistical methods applied to the exploration work of the Nanisivk Zn-Pb Mine, Baffin Island, Canada. Mathematical Geology, 12, 4, p 321-337, Dordrecht.

MEASURES OF UNCERTAINTY FOR RESOURCE CLASSIFICATION

LUIS EDUARDO DE SOUZA, JOÃO FELIPE C.L. COSTA and JAIR C. KOPPE
Mining Engineering Department, Federal University of Rio Grande do Sul
Av. Osvaldo Aranha, 99/504, 90035-190, Porto Alegre/RS, Brazil

Abstract. For many decades the mining industry regarded resource estimation and classification as a mere calculation requiring basic mathematical and geological knowledge. Often uncertainty associated with tonnages and grades were either ignored or mishandled. With initiatives to establish international standards for classifying mineral resources and reserves, it is important to establish the level of confidence in the results and correctly assess the error. Among geostatistical methods, Ordinary Kriging (OK) is probably the one most used for mineral resource estimation. It is known that OK variance is unable to recognize local data variability, which is an important issue when heterogeneous mineral deposits with higher and poorer grade zones are being evaluated. This study investigates alternatives for computing estimation variance from ordinary kriging weights that account for both the data configuration and the data values. These estimation variances are then used to classify resources based on confidence levels and their results are compared with those obtained by OK variance. The methods are illustrated using an exploration drill hole data set from a large Brazilian coal deposit. The results show the differences in tonnages within each class of resources when different measures of uncertainty are used.

1 Introduction

The mining industry has already recognized and established standards for resource evaluation and classification but now, with the increasing internationalization of mining companies, the development of internationally acceptable standards for this classification has become relevant.

Since 1994 the Council of Institutions of Mining and Metallurgy (CMMI), an international entity that congregates institutions from the United States (SME), Australia (AusIMM), Canada (CIM), United Kingdom (IMM) and South Africa (SAIMM), has proposed a set of definitions for the reporting and classification of mineral resources and reserves. These definitions were adopted later by a committee established in 1998 by the United Nations thus granting it truly international recognition.

The main mineral resource classification systems adopted worldwide are essentially based on sampling spacing, geological confidence and economical viability. These systems define classes of resources based on a degree of certainty associated with estimated tonnages and grades. Classes of in situ coal (measured, indicated and inferred) are defined based on the spatial distribution of the samples and the uncertainty associated with tonnages calculated for a given deposit or part of it. Thus, classifying in situ coal or coal resources requires the definition of the uncertainty associated with the estimate. However, what is not stated in the classification systems is how uncertainty should be measured, and even the JORC rules, such as in Table 1, provides a minimum necessary data density, but it does not specify or advise any estimation algorithm or how uncertainty should be assessed.

Classes of resources	Maximum extrapolation distance	Maximum spacing between points of observation[1]	Degree of uncertainty
Measured	500 m	+ 1 km; < 500 m	0 - 10%
Indicated	1,000 m	+ 2 km; < 1 km	10 - 20%
Inferred	2,000 m	+ 4 km	> 20%

Table 1. Classes of resources based on sampling spacing defined by the JORC system.
[1] The first distance is the acceptable limit and the second is the normally used distance.

Since classification codes are not prescriptive regarding the estimation method used, several geostatistical approaches have been suggested, mainly because these techniques provide a short, unambiguous identification of resources/reserves categories. Geostatistical estimate methods are suggested in most codes and these methods have become the accepted standard models for mineral resource estimates. Several geostatistical methods can be used to estimate and assess uncertainty. Among them ordinary kriging is probably the most widely used mainly due its specific features related simplicity and reliable estimates (Matheron, 1963; David, 1977; Journel, 1983; Isaaks and Srivastava, 1989). However, the geostatistical literature has discussed the misuse of ordinary kriging variance as an accurate measure of uncertainty, mainly because it is only variogram dependent and not data-value dependent, taking into account only the spatial arrangement of the samples, and consequently ignoring the local variability (Journel, 1986).

This study investigates some of the proposed alternatives that have been proposed to the kriging variance. Two distinct approaches to calculate estimate variance via ordinary kriging weights were used: (i) the interpolation variance (Froidevaux, 1993; Yamamoto, 1999) and (ii) the combined variance (Arik, 1999). The obtained results are compared with those derived from OK variance. All the methodologies were repeated to four different block sizes, trying to identify its influence on the estimates, as it is known the larger the block size the smaller the associated variance (Krige, 1996).

The estimate and the subsequent classification of resources into different classes or categories, according to the possible variations of these resources must provide a model that quantifies the risk on each category. A comparative study was carried on using an exploration data set from a large Brazilian coal deposit and the results show the impact in both tonnages and error in each resource category when the alternatives to uncertainty assessment are used.

2 Case study

The deposit object of this study is located in the southern Santa Catarina coal basin and it has been exploited since the early 1900's. The depositional environment imposed a particular geometry to this deposit with a longer continuity for coal thickness along the major axis of deposition and a short range along the orthogonal direction.

Since the samples used for resource assessment should be representative and present a high degree of confidence, all drill holes with poor reliability in terms of core yield or logging criteria were omitted from the deposit modeling. Thus, from the original 471 ddh, 340 were kept for thickness estimate purpose and 236 for the specific gravity. As the collars were not regularly spaced, a declustering procedure (Deutsch and Journel, 1998) was used to obtain a representative statistic for the entire area. In the sequence, spatial continuity analysis was carried out modeling the major and minor directions of anisotropy. A two-structure spherical variogram (Sph) model [γ(h)] was estimated from the experimental variogram points for the two variables as:

$$\gamma(h) = 0.045 + \left[0.057 Sph_{(1)} \left[\frac{hN-S}{467m}, \frac{hE-W}{233m} \right] + 0.233 Sph_{(2)} \left[\frac{hN-S}{6131m}, \frac{hE-W}{1897m} \right] \right] \quad (1)$$

for coal thickness, and:

$$\gamma(h) = 0.0015 + \left[0.0033 Sph_{(1)} \left[\frac{hN135E}{676m}, \frac{hN45E}{478m} \right] + 0.0029 Sph_{(2)} \left[\frac{hN135E}{3190m}, \frac{hN45E}{2440m} \right] \right] \quad (2)$$

for specific gravity, where $Sph = \begin{cases} \frac{3}{2}\frac{h}{a} - \frac{1}{2}\left(\frac{h}{a}\right)^3 & \text{if } h \leq a \\ \text{sill} & \text{if } h > a \end{cases}$, and a is the range of the variogram.

Coal thickness anisotropy coincides with the main axis of the coal basin. Specific gravity anisotropy directions are oriented according to the shoreline which used to divide the lacustrine/marine environment where this coal deposit was formed. Pyrite concretions presence and consequently the increase in the specific gravity is controlled by this shoreline orientation. Therefore, these two geological attributes not necessarily should have their major axis of anisotropy coincident.

The main parameters used for modelling the deposit into mineable blocks using ordinary kriging were: (i) minimum of 4 and maximum of 24 data located in the local neighbourhood of a given block being interpolated, (ii) 64 points used to discretise the block and obtain an average estimate of it, (iii) searching for samples around the block within the variogram ranges defining an ellipsoidal search, (iv) four different block sizes (175 x 175 m, 350 x 350 m, 525 x 525 m, and 700 x 700 m), (v) searching for samples in the local neighbourhood of a block dividing the search ellipsoid into octants. The variograms and the parameters used for kriging were cross validated (Isaaks and Srivastava, 1989).

Based on the available data and the JORC code standards for extrapolation distance and distance between samples (Table 1), the boundaries defined by geometric definitions were established. Adopting the usually recommended values, the areas covered by a single sample were disregarded for framing in measured or indicated coal in situ categories.

3 Assessing uncertainty

3.1 KRIGING VARIANCE

Ordinary kriging produces a set of estimates for which the variance of the errors is minimized through the use of the Lagrange multipliers and is usually referred to as the ordinary kriging variance:

$$\sigma_{OK}^2(u) = C(0) - \sum_{\alpha=1}^{n(u)} \lambda_\alpha^{OK}(u) C(u_\alpha - u) - \mu_{OK}(u) \qquad (3)$$

where $C(0)$ is the a priori variance of the data, λ_α^{OK} is the weight calculated for each datum in the neighbourhood of u, $C(u_\alpha - u)$ is the covariance from each datum and the location u and μ_{OK} is the Lagrange multiplier (Matheron, 1963). The kriging variance computed for a given point or block being estimated is essentially independent of the data values used in the estimation and it does not measure uncertainty, but just the spatial configuration of local data used to make the estimate. The link between kriging variance and data values is just through the variogram, which is global rather than local in its definition (Arik, 1999; Journel, 1986; Isaaks and Srivastava, 1989; Yamamoto, 1999).

3.2 INTERPOLATION VARIANCE

Yamamoto (1999) proposes the interpolation variance as the weighted average of the squared differences between data values and the OK estimate according the following expression:

$$s_0^2 = \sum_{i=1}^{n} \lambda_i \left[z(x_i) - z^*(x_0) \right]^2 \qquad (4)$$

where λ_i are the ordinary kriging weights, $z(x_i)$ are the n neighbor data close to the unsampled location (x_0), and $z^*(x_0)$ is the block estimate. This expression is exactly the same that proposed by Froidevaux (1993). It is data-value dependent and this definition requires all weights be positive since any negative weight could lead to a negative interpolation variance. There are several available solutions for avoiding negative weights and they can be basically divided into two types: (i) ordinary kriging weights can be constrained to be positive before solution of the ordinary kriging system (Barnes and Johnson, 1984; Herzfeld, 1989), or (ii) correct the negatives after kriging (Froidevaux, 1993; Journel and Rao, 1996; Deutsch, 1996). This study adopted the procedure proposed by Deutsch (1996), and his solution was implemented in the kriging routine *kt3d* of GSLIB (Deutsch and Journel, 1992).

3.3 COMBINED VARIANCE

Arik (1999) suggests an alternative measure to assess the uncertainty that is basically a combination of the kriging variance and the variance of the weighted average block value based on the data values used. The second is defined as:

$$\sigma_W^2 = \sum_{i=1}^{n} w_i^2 [Z_0 - z_i]^2 \quad i=1,\ldots,n \ (n>1) \tag{5}$$

where n is the number of data used, w_i are the ordinary kriging weights corresponding to each datum, Z_0 is the block estimate, and z_i are the data values. If there is only one datum, σ_W^2 is set to σ_{OK}^2. This component, called by Arik (1999) the local variance of the weighted average, is then used to calculate the combined variance (σ_{CV}^2) as follows:

$$\sigma_{CV}^2 = \sqrt{(\sigma_{OK}^2 \times \sigma_W^2)} \tag{6}$$

Suppose the example in Figure 1. For sake of simplicity, there are only seven data points surrounding the point 0 to be interpolated. Using an exponential variogram model with an isotropic range of 10, sill of 10, and nugget of 0, ordinary kriging was used to calculate the value at location 0, resulting in 592.7 with a kriging variance of 8.96.

If one changes the data set according to Figure 1b, keeping everything else the same, the new estimate would be 550.0. The kriging variance remains the same at 8.96, since the variogram parameters and data configuration were the same as the first run. Table 2 summarizes the results. One can observe the alternative variances reflect local variability.

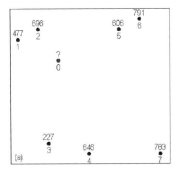
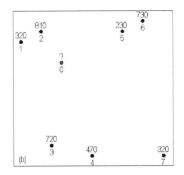

Figure 1. Sample data and location 0 (a) extracted from Isaaks and Srivastava (1989). The same data configuration with a different set of values (b).

Samples	Kriging Variance	Weighted Average	Combined Variance	Interpolation Variance
(a)	8.96	4114.6729	191.9667	28551.7148
(b)	8.96	11769.0791	324.6606	56772.0156

Table 2. Variances for the two data sets presented in Figure 1.

3.4 ERROR DEFINITION

Block models of four different sizes were defined for both variables using ordinary kriging. These models were validated and the results, including the estimated variance, used to define confidence intervals. For each block, the coal accumulation (t/m^2), expressed as a product of the thickness by density has its variance evaluated and expressed using (David, 1977):

$$\frac{\sigma_{xy}^2}{(xy)^2} = \frac{\sigma_x^2}{x^2} + \frac{\sigma_y^2}{y^2} + 2\rho_{xy}\frac{\sigma_x}{x}\frac{\sigma_y}{y} \qquad (7)$$

where σ_{xy}^2 is the variance of the product, ρ_{xy} is the coefficient of correlation, σ_x and σ_y are the standard deviations, x and y are the estimated block values to thickness and specific gravity. The third term in Equation (7) is null since the correlation between density and thickness is insignificant (Figure 2).

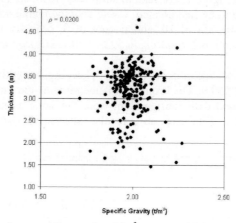

Figure 2. Scatterplot for specific gravity (t/m^3) versus thickness (m). Note the absence of correlation.

Assuming a Gaussian distribution to the error, the confidence interval with 95% probability of containing the mean can be approximated using (David, 1977):

$$\overline{t/m^2}(n) \pm t_{n-1,1-\alpha/2}\sqrt{\frac{\sigma_{xy}^2}{n}} \qquad (8)$$

where $\overline{t/m^2}$ is the inferred mean to n data, $t_{n-1,1-\alpha/2}$ is the $1-\alpha/2$ superior critical point for the t distribution with n - 1 degrees of freedom.

Thus, the global error for each coal in situ category was obtained using each accumulation block value as a weight to the block error, according to the theory of errors presented by Caputo (1969).

4 Discussion and conclusions

The approach presented was repeated for each one of the alternative variance discussed and for each block size tested. The results on either tonnages or error for each category are showed in Tables 3 and 4.

Block size	Tonnages of coal in situ (t x 10^6)		
	Measured	Indicated	Inferred
175 x 175 m	237.62	126.62	188.11
350 x 350 m	239.54	124.70	189.86
525 x 525 m	241.51	132.54	187.47
700 x 700 m	234.74	116.07	197.08

Table 3. Calculated tonnages of coal in situ for different block sizes.

Block size	Variance	Error (%)		
		Measured	Indicated	Inferred
175 x 175 m	σ_{OK}^2	4.43	7.61	11.33
	s_0^2	7.54	13.37	17.77
	σ_{CV}^2	3.12	5.61	8.34
350 x 350 m	σ_{OK}^2	3.64	6.92	10.53
	s_0^2	7.58	13.48	17.61
	σ_{CV}^2	2.81	5.38	8.01
525 x 525 m	σ_{OK}^2	3.19	6.33	10.46
	s_0^2	8.01	13.49	17.79
	σ_{CV}^2	2.68	5.14	7.96
700 x 700 m	σ_{OK}^2	2.76	5.84	9.47
	s_0^2	7.98	12.30	17.28
	σ_{CV}^2	2.44	4.58	7.49

Table 4. Confidence limits for the coal in situ calculated tonnages obtained via kriging variance (σ_{OK}^2), interpolation variance (s_0^2), and combined variance (σ_{CV}^2).

In Table 3, it is observed that the estimated tonnages have changed for the different block sizes tested, these variations were generally small, and only in the category of indicated in situ coal was the variation about 12%, with the increase of the block size. This seems to be related with a more complex geometry for this class as well as a different adherence that each size has regarding the geometric boundaries that define the resources categories. Table 4 shows that the calculated values of error using the methodology proposed by Yamamoto (1999) are substantially higher than the calculated ones with kriging variance as well as combined variance. Several blocks were classified as measured resources according to the geometric criteria, but could not be classified as indicated or even inferred due to uncertainty criteria. In this study, these blocks were not removed from the resources inventory or re-arranged into different categories, however if this variance was used for classifying a reduction on the resources would occur.

The fact that the case study consists of a tabular orebody, extremely continuous spatially and with abundance of information may have contributed to attenuate the differences between the results obtained via kriging variance and combined variance, and these factors may explain the small differences in terms of tonnages and error with the increment of the block size. Even so, there are significant differences in the calculated error using each one of the alternative variances.

Ordinary kriging (OK) variance (or its square root, the standard error) has been largely used as a measure for spread of the estimates, but since this parameter depends only on (i) the spatial continuity of the data and (ii) the spatial configuration of the observations, the error calculated using OK variance will be independent from the data values imposing severe limitation on its use. Therefore, the use of alternative measures of uncertainty allow a more accurate and coherent response. These measures for the uncertainty eliminate the subjectivity of using a fixed or empirical range of influence as discriminating factor among the categories of resources that do not respect the singularity of each mineral deposit.

References

Arik, A. An Alternative Approach to Resource Classification, *in Proceedings of the 28th International Symposium on Computer Applications in the Mineral Industries (APCOM'99)*, Colorado School of Mines, Golden, Colorado USA, 1999, p. 45-53.

Barnes, R.J. and Johnson, T.B. Positive Kriging, *Geostatistics for Natural Resources Characterization*, Reidel, Dordrecht, 1984, p. 231-240.

Caputo, H.P. *Matemática para a Engenharia*, Ed. Ao Livro Técnico S.A., Rio de Janeiro, 1969, 416 p.

David, M. *Geostatistical Ore Reserve Estimation, Developments in Geomathematics 2*, Elsevier Scientific Publishing Company, Amsterdam, 1977, 364 p.

Deutsch C.V. and Journel, A.G. *GSLIB: Geostatistical Software Library and User's Guide*, Oxford University Press, New York, USA, 1998, 335 p.

Deutsch, C.V. Correcting for Negative Weights in Ordinary Kriging, *Computers & Geosciences*, vol. 22, no. 7, 1996, p. 765-773.

Froidevaux, R. Constrained Kriging as an Estimator of Local Distribution Functions, in Capasso, V., Girone, G., and Posa, D., eds., *in Proceedings of the International Workshop on Statistics of Spatial Processes: Theory and Applications*, Bari, Italy, 1993, p. 106-118.

Herzfeld, U.C. A Note on Programs Performing Kriging with Nonnegative Weights, *Mathematical Geology*, vol. 21, no. 3, 1989, p. 391-393.

Isaaks, E.H. and Srivastava, M.R. *An Introduction to Applied Geostatistics*, Oxford University Press, New York, USA, 1989, 561 p.

Journel, A.G. Non-Parametric Estimation of Spatial Distributions, *Mathematical Geology*, vol. 15, no. 3, 1983, p. 445-468.

Journel, A.G. Geostatistics: Models and Tools for the Earth Sciences, *Mathematical Geology*, vol. 18, no. 1, 1986, p. 119-140.

Journel, A.G. and Rao, S.E. Deriving Conditional Distributions from Ordinary Kriging, *Stanford Center for Reservoir Forecasting (Report No. 9)*, Stanford University, 1996, 25 p.

Krige, D.G. A Practical Analysis of the Effects of Spatial Structure and of the Data Available and Accessed, on Conditional Biases in Ordinary Kriging, *in Proceedings of Geostatistics Wollongong '96*, vol. 2, 1996, p. 799-810.

Matheron, G. Principles of Geostatistics, *Economic Geology*, no. 58, 1963, p. 1246-1266.

Yamamoto, J.K. Quantification of Uncertainty in Ore-Reserve Estimation: Applications to Chapada Copper Deposit, State of Goiás, Brazil, *Natural Resources Research*, vol. 8, no. 2, 1999, p. 153-163.

INCORPORATING UNCERTAINTY IN COAL SEAM DEPTH DETERMINATION VIA SEISMIC REFLECTION AND GEOSTATISTICS

VANESSA C. KOPPE, FERNANDO GAMBIN, JOÃO FELIPE C. L. COSTA,
JAIR C. KOPPE, GARY FALLON and NICK DAVIES
Department of Mining Engineering, Federal University of Rio Grande do Sul
Brazil, RS, POA, Oswaldo Aranha Avenue, 99- 506

Abstract. Modelling mineral deposits requires the use of all possible source of information. Traditionally, core samples from borehole are the most used way to access the ore body, however this method is expensive and provides information restricted to a close neighbourhood within the sample location. Continuity between sampled points needs to be inferred in order to infill values among bore hole locations using interpolation techniques. In contrast, geophysical methods including seismic reflection provide data at much closer intervals, thus approximating continuous sampling along a seismic section. These data are then used to infer spatial continuity, for example the fault of a coal seam in between bore holes. Wave travel time along the seams is recorded by seismic survey at a dense grid. Additionally, sonic wave velocity logged along boreholes can be interpolated at a dense grid. Sonic Logging provides direct and continuous measurements of the sonic wave velocity at all seams down the holes logged. Therefore, this logged sonic velocity can be simulated within a dense grid compatible to the time grid. Multiple velocity grids (equally probable models) are generated within the simulation framework. In combining both grids, i.e. velocity and time, seam depth can be obtained. Consequently risk in depth determination for each seam due to velocity uncertainty can be assessed. Both data types (time and sonic) are subject to various sources of error. Currently, velocity is indirectly determined using processed seismic data, which may breed errors in geologic sections interpretation. The present paper will show results from a Sonic Logging velocity simulation and its uncertainty determination, in order to use the results in calculating seam depth via seismic reflection and additionally provided a measure for error in this parameter. A case study in a major coal deposit illustrates the procedure.

1 Introduction

Modelling mineral deposits is based on a conceptual geological model and on readings derived from samples sparsely taken within this deposit. Usually, the samples are collected by diamond drill holes (core sampling). However, this sampling technique is very expensive (~ US$ 100/m) and provides restricted amount of information, imposing all sort of difficulties in reducing the uncertainties associated with the estimation of geological attributes within the mineral deposit.

Geophysical methods like seismic are a direct method of more coarsely, but cheaply sampling, which can be very accurate in favourable conditions. Seismic methods measure the mechanical wave propagation time from a source to a receiver within rocks. This time can be related to the geometry of seam layers or bed rocks. The depth of the beds is obtained multiplying the time the wave took to travel by the wave's velocity along that rock. Accurate velocities can be obtained from sonic logging sampling. This method can collect velocities at various points along borehole walls.

This study aims to estimate seismic wave velocity in a 3D grid, based on sonic logging data (note that seismic waves velocities can be correlated to sonic logging data). Sonic logging measures slowness, which is an attribute that is defined as the inverse of compressional velocity. For sake of this study slowness will be called and treated as velocity samples. The estimates will be generated using geostatistical methods. The development of an appropriate modelling methodology could facilitate a better depth conversion from seismic data. If the sonic velocity data were converted to a rock mass parameter then a more accurate geotechnical model will also be constructed.

Geostatistics comprises a collection of tools used to estimate values of any attribute of a mineral deposit at unknown locations, supported by its spatial continuity model. The geostatistical tools used on this study include ordinary kriging (Matheron, 1963) and sequential Gaussian simulation (Isaaks, 1990). Sequential Gaussian simulation provides a method of assessing the uncertainty associated with the estimated velocity, which can also be approximate via ordinary kriging variance. However the later must be used with caution given certain limitations (Goovaerts, 1997).

The methodology presented is illustrated by a case study where the uncertainty related to the velocity is determined as well as the corresponding uncertainty of the coal seam depth. Sampling errors generated in seismic and sonic logging also contribute to depth uncertainty, however these errors were not considered in this study. The target coal seam is approximately 210m below the surface. The seam has an average thickness obtained from the 60 core samples) of 2.1m and is extracted by longwall retreat. The overlying stratigraphy is a siliclastic sequence containing at least 9 thinner but of variable thickness coal seams.

2 Case Study

2.1 DATA SET

A coal deposit was used and velocity samples were collected from 60 logged boreholes. Each borehole logged was sampled at 5 cm intervals along 300 m (average hole length). The dataset comprises 228851 sonic wave velocity samples (unit $\mu s/ft$). These samples were obtained by geophysical logging along 60 core and non-core drill holes (Figure 1).

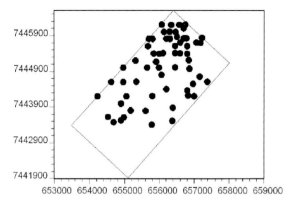

Figure 1- Location map for 60 borehole collars and 3D seismic survey boundary (black line).

2.2 VARIOGRAPHY

The vertical experimental variogram for sonic wave velocity samples showed a fast increase at the first meters due to short scale high variability of this attribute. The horizontal omnidirectional experimental variograms for sonic wave velocity samples showed a high degree of continuity as expected since all the readings tend to belong to the same strata along the horizontal plan.

2.3 KRIGING

The sonic wave velocity attribute was interpolated using ordinary kriging (Matheron, 1963). Figure 2 shows vertical sections sliced from the kriged block model. The smoothing effect is evident in the kriged 3D block model. In a shallow dipping sedimentary environment one might expect near horizontal layering to be evident. While stratigraphic units are correlated with horizontal distance there are changes in their physical property distribution.

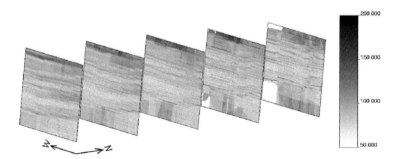

Figure 2- Vertical sections (longitudinal views along XZ plan) at various North (Y) coordinates extracted from the kriged block model. Gray scale represents kriged sonic velocity (μs/ft).

Figure 3 shows sections along XZ plan sliced from the 3D block model representing the kriging standard deviation of each block. Kriging standard deviation calculated at blocks near drill holes provide low values as is observed in Figure 3. Light grey vertical lines (lowest standard deviation values) identify borehole locations.

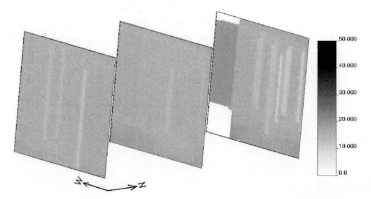

Figure 3- Vertical sections (longitudinal views along XZ plan) at various North (Y) coordinates plotting the standard deviation at every block resulting from kriging. Gray scale represents sonic velocity kriging standard deviation ($\mu s/ft$) at each block.

2.4 SIMULATION

Sequential Gaussian simulation (SGS) (Isaaks, 1990) was selected to be used in this case study. SGS provides realizations (maps) (Deutsch and Journel, 1998) of sonic velocity, where each realization is a possible representation for the attribute being studied. Simulations were conditioned to the 222648 samples collected along the 60 logged boreholes. The number of realisations (20) was considered enough for uncertainty assessing as at this number the ergodic fluctuations on the global mean reached a steady state, i.e. the variance of the mean stabilized. Figure 4 illustrates the same sections as in Figure 2 and shows the vertical sections sliced along XZ plans extracted from a 3D block model obtained by simulation. A granular texture on the grey scale maps is clearly noticed.

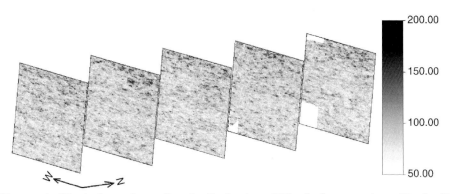

Figure 4- Vertical sections (longitudinal view XZ plan) at various North (Y) coordinates for the sonic velocity calculated at each block of one simulation. Gray scale represents simulated sonic velocity (µs/ft).

Conditional standard deviation was calculated at each simulated node using the values obtained at this node resulting from different realizations.

2.5 ERROR DETERMINATION

The uncertainty about kriged and simulated values can be quantified as the error. Assuming the error follows a normal distribution, the error interval can be calculated using kriging standard deviation or conditional standard deviation for the simulated values as follows (Christman, 1978):

$$\text{Error} = \pm\, t_{\frac{\alpha}{2}, n-1} \frac{\sigma}{\sqrt{n}}$$

where σ is the standard deviation of the values; n is the number of values; $t_{\alpha/2,\, n-1}$ is parameter obtained from t-student distribution which depends on the confidence interval (1-α) and on degrees of freedom (n-1).

The error at each location derived from the kriged or simulated models were determined using confidence interval with 95% probability. It means there is 95% probability for the estimated value to be included in the confidence interval. Figure 5 shows the histograms for the error at all nodes generated by kriging and simulation. Visually the errors obtained via kriging have less variance (smoother) than the errors calculated via simulation. Statistically the error mean and median for the simulation method are lower. Practically, for a 2.1m thick seam at approximately 210m depth this difference is greater than the seam thickness. This difference is due to the smooth effect associated with interpolation methods. For this reason, simulation measure of uncertainty is larger than the one provided via kriging. The ability to improve the depth estimation by greater than a seam thickness is considerable for providing an accurate geological model to mine from.

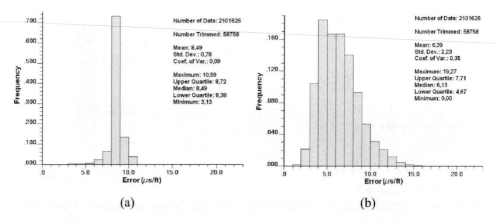

Figure 5- (a) Histogram of error measure at kriged nodes. (b) Histogram of error measure at all simulated nodes.

3 Conclusions

This study was aimed at estimating seismic wave velocity and its associated uncertainty at every node of a 3D grid, based on sonic logging data. The 3D model for velocity was estimated using ordinary kriging and sequential Gaussian simulation.

Ordinary kriging produced the best estimate at a price of smoothing the interpolated values and consequently the error forecasted. Sequential Gaussian simulation (20 realizations) produced 20 estimates at each grid node (one estimate for each realization). These simulated models globally resemble the real mineral deposit as they reproduce ergodically its spatial continuity.

Assuming a normal distribution for the error, a value at each grid node resulting from interpolating using ordinary kriging and sequential Gaussian simulation was calculated. Simulation produced a larger error interval (due to a larger space of uncertainty) but an overall lower mean of errors hence SGS is recommended as a process to derive a sonic velocity model.

References

Christman, R.U., *Estatística Aplicada*, Edgard Blücher Ltda, 1978.
Deutsch, C.V. and Journel, A.G., *GSLIB: Geostatistical Software Library and User's Guide*, 2nd Edition, Oxford University Press, 1998.
Goovaerts, P., *Geostatistics for Natural Resources Evaluation*, Oxford University Press, 1997.
Isaaks, E.H., *The Application of Monte Carlo Methods to The Analysis of Spatially Correlated Data*, Ph.D. Thesis, Leland Stanford Junior University, 1990.
Matheron, G., Principles of Geostatistics, *Economic Geology*, no. 58, 1963, p. 1246-1266.

IMPLEMENTATION ASPECTS OF SEQUENTIAL SIMULATION

STEFAN ZANON and OY LEUANGTHONG
*Department of Civil & Environmental Engineering, 220 CEB,
University of Alberta, Canada, T6G 2G7*

Abstract. Sequential simulation is used throughout the natural resources industry to construct multiple equiprobable numerical models. The sequential methodology is straightforward, but some implementation details require further explanation. This paper explores some of the implementation issues associated with choice of: simulation path, search strategies, number of conditioning data, and the affect of ergodic fluctuations under the Gaussian assumption.

1 Introduction

Sequential simulation (SS) (Johnson, 1987; Journel, 1993) is a stochastic modelling approach that yields multiple realizations based on the same input data. This data could be either continuous or categorical. Depending on the data type, sequential indicator simulation (SIS) (Gomez-Hernandez and Srivastava, 1993), sequential Gaussian simulation (SGS) (Isaaks, 1990), or direct sequential simulation (DSS) (Xu and Journel, 1994; Caers, 2000; Soares, 2001) will be used. This suite of simulation techniques has greatly expanded the tools that are available for building stochastic models, while injecting more variability than their kriging counterparts. The SS work-flow can be described in four basic steps:

1. Choose the stationary domain.
2. Define a path to visit every location.
3. At each location:
   a) search to find nearby data and previously simulated values,
   b) calculate the conditional distribution, and
   c) perform Monte Carlo simulation (MCS) to obtain a single value from the distribution.
4. Repeat step 3 until every location has been visited.

For SIS and SGS, a pre- and post-processing data transformation step is required. In the SIS case, data are transformed into indicator variables; in SGS, data are transformed to be Gaussian via a quantile transform or a Gaussian anamorphosis (Chilès and Delfiner, 1999). The above methodology produces one possible realization, and more realizations can be created by choosing a different random path.

The theory behind each form of SS has been explained numerous times (Isaaks, 1990; Journel, 1993; Goovaerts, 1997; Chilès and Delfiner, 1999; Deutsch, 2002), but it is the details of sequential simulation that warrant further explanation. In the publicly available GSLIB (Deutsch and Journel, 1998) programs like SGSIM or SISIM, for Gaussian and indicator simulation, respectively, the user must specify how key aspects of the simulation will be performed. These decisions can greatly affect the resulting model and the associated CPU requirements. A better understanding of these decisions will help the user to improve their models while balancing efficiency with accuracy.

2 Data and Transformation

Before simulation can be performed the model area must be defined and the input data identified. In general, the data must come from a single underlying distribution. The mean, variance, and higher order statistics are then assumed stationary throughout the area, that is, $E\{Z(\mathbf{u})\} = m$ and $Var\{Z(\mathbf{u})\} = E\{[Z(\mathbf{u}) - m]^2\}$ (Journel and Huijbregts, 1978). If stationarity is violated, the mean and variance will change with location. A trend model can be used to describe these regional changes, and either the trend is removed to create stationary residuals or the trend is used as secondary data in a specialized form of kriging (Deutsch, 2002).

The underlying distribution of the modelling area, as described by the cumulative distribution function (cdf), should be reproduced in every simulation. This cdf is typically determined from the input data, but the data collection process is rarely performed to fairly sample the underlying distribution. To correct this, declustering (Isaaks and Srivastava, 1989; Goovaerts, 1997) can help to remove the affects of non-representative sampling or a reference distribution, based on some secondary data or expert knowledge, can be used as a target distribution.

SGS requires the data to be standard Gaussian with zero mean and unit variance. To achieve this, the input data cdf is transformed through the quantiles to any other cdf. This one-to-one quantile transformation is reversible and allows the mean, variance and shape of the distribution to be changed while preserving the rank of the data (Journel and Huijbregts, 1978). Spikes in the cdf prevent the one-to-one quantile transform and despiking (Verly, 1984) will be required.

In SGS, the original distribution is reproduced by reversing the above transformation. This back transformation requires the data to follow the standard normal distribution; however, statistical fluctuations are inherent in simulation. Fluctuations in the mean and variance should be reasonable and unbiased. Small deviations in normal space can be magnified after back transformation, particularly if the original data follow a skewed distribution.

For example, consider a lognormal distribution, with mean and standard deviation of 8.0, and its corresponding normal score distribution after transformation, $N(0,1)$. The effect of deviations from the standard normal distribution can be assessed by generating *near* standard normal distributions and reversing the above transformation. For this exercise four scenarios will be considered using standard deviations of 0.8 and 1.2, and means of -0.1 and 0.1. When both the mean and standard deviation are low in normal space, $N(-0.1, 0.8)$, the mean and standard deviation in original units are 6.64 and 5.96, respectively. Using the same mean with the higher standard deviation, $N(-0.1, 1.2)$, the original space mean and standard deviation are 8.48 and 9.37. The remaining two scenarios for the high mean, $N(0.1, 0.8)$, and $N(0.1, 1.2)$, result in an original space mean of 7.92 and 9.76, and standard deviations of 6.64 and 10.06, respectively. This example shows how sensitive the summary statistics in original space are to the ergodic fluctuations inherent in stochastic simulation in normal space for a skewed distribution. One proposed solution to mitigate these effects is to apply a standard transform to the simulated values (Journel and Xu, 1994).

3 Simulation Path

At every unsampled location, SS should use all available input data and previously simulated values as conditioning data. No assumption is made about the order in which these locations are visited, but the order will influence the final model. To minimize this influence, the starting location and path should be random (Isaaks, 1990; Tran, 1994). Over multiple realizations the structure in the model will be based on the data and not an artifact of the path.

Alternatives such as the regular path and spiral path (McLennan, 2002) have been considered, but any perceived benefits in CPU efficiency or input data propagation come at the expense of variogram reproduction and accuracy of the local distribution. These paths, along with the random path, can suffer from poor long range variogram reproduction, since the nearby data will preferentially be used as conditioning data. To avoid this problem, a multiple grid search (Tran, 1994) can be incorporated into the random path to improve variogram reproduction.

4 Searching for Local data

Before kriging can be implemented, a search is performed to identify surrounding conditioning data. The user limits this search by specifying a search radius in each principle direction, where the radii should equal or exceed the variogram range to ensure adequate variogram reproduction. Data beyond this range will provide limited information to the kriging estimate.

It is common practice to assign the input data to the grid nodes. This will *exactly* reproduce the input data in the final model and allow the covariance to be quickly calculated using a covariance look-up table. The disadvantage is that only a single data is retained in each grid cell and the remaining data are only used to establish the reference distribution. Also, input data cannot be preferentially used over previously simulated values. The spiral search (Deutsch and Journel, 1998) uses the covariance look-up table to develop a search path based on the decreasing correlation of the surrounding nodes.

When the input data are not assigned to the grid nodes, the spiral search can only locate previously simulated values. The super block search (Journel and

Huijbregts, 1978; Deutsch and Journel, 1998) must then be used to locate the input data. This second search superimposes a coarse grid over the model area, thus creating *super blocks*, and the data inside each block is identified and indexed to that block. The specified search radii are used to construct a template that is centred on the super block containing the point to be estimated. This makes it quick to identify the data inside of the search area. The local data is then exhaustively searched to identify the conditioning data.

The above search routines are only concerned with identifying the most correlated data and ignores their direction. The direction of the data can be taken into account by using the octant search. The octant search divides the surrounding 3D area into eight equal regions. When searching for data, only a maximum number are allowed from each octant. This forces the data to come from different directions at the expense of ignoring closer, but more redundant, data.

5 Kriging

The theory behind SS is based on using every previously simulated value and input data throughout the simulation process (Isaaks, 1990). In practice, only the closest conditioning data are used, up to a maximum number, to keep CPU time reasonable. This assumes the closest data screen the data further away and the additional information from this screened data is deemed small enough to ignore (Isaaks, 1990). The choice of the maximum number is linked to two issues: the speed required to generate a realization, and the accuracy of the kriged estimate and variance.

The impact of the number of data used in kriging on CPU time is controlled by (1) locating the conditioning data, and (2) calculating the kriging weights. For n data, the search is proportional to n, regardless of the search type, and the kriging system calculations are proportional to n^3. So as n increases, the kriging calculations will dominate the CPU time. For example, a 100 x 100 grid was simulated using 300 spatially random data to track the CPU time as n varied from 5 to 300 (Figure 1a). Initially, the change in CPU time is small, but as n increases, the change in the CPU requirement approaches a slope of 3 on a log-log scale.

The uncertainty in kriging is expressed by the kriging variance that is a minimum by construction. Reduction of this variance is only achieved through the addition of more data. Gandin (1963) showed that the change in variance can be bounded when the least informative datum is removed; however, modern computers make the direct calculation of the change quicker than Gandin's method (Zanon, 2004).

For example, 100 conditioning data were randomly chosen and kriging was performed at three arbitrarily chosen test locations (Figure 1b). As n varied from 1 to 100, the kriging estimate and variance were tracked (Figure 2). The best results are achieved when $n = 100$ as indicated by the dotted lines. It is seen that a lower limit of 8 to 10 conditioning data will provide results close to the dotted line, with diminishing returns for $n > 10$.

IMPLEMENTATION ASPECTS OF SEQUENTIAL SIMULATION

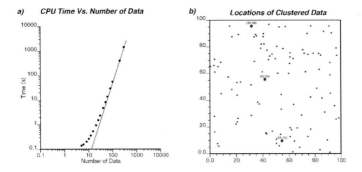

Figure 1. (a) The change in CPU time versus the number of conditioning data. (b) The location of the input data and three test locations (large dots).

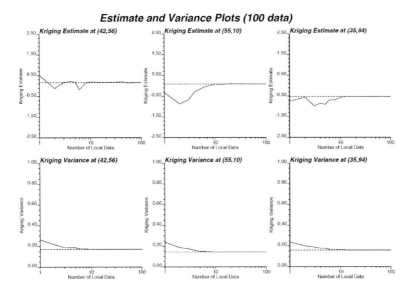

Figure 2. The change in the kriging mean (top) and variance (bottom) for three locations in the area of interest.

6 Final Remarks

Once the modelling process has been completed, the following checks should be performed: reproduction of (1) data values at data locations, (2) the target histogram, (3) the target summary statistics, and (4) the input covariance model. In the multivariate context, this list should also include reproduction of the multivariate distribution and the corresponding summary statistics (Leuangthong, McLennan, and Deutsch, 2004). A visual inspection of the geology can help determine if the model adheres to the expected underlying geological structure.

Failure to satisfy these tests requires some checks and/or changes to the input parameters, this depends on the options available on the software being used. Variance inflation is one cause of poor histogram reproduction. Increasing the number of conditioning data and the search radius, along with the octant search, may help to correct the variance at the cost of increased CPU time. The most common form of poor variogram reproduction is in the long range structure. Using the multiple grid search, along with increasing the number of conditioning data and search radius, can help improve the long range variogram. One general check is to look at the histogram and variogram reproduction of an unconditional simulation. This may help to identify problems caused by the input data and not the program. The assumption of stationarity may be violated and, data permitting, the model should be divided into smaller, more stationary areas.

References

Caers, J., *Adding Local Accuracy to Direct Sequential Simulation*, Mathematical Geology, vol. 32, no. 7, 2000, p. 815-850.

Chilès, J-P., and Delfiner, P., *Geostatistics: Modeling Spatial Uncertainty*, John Wiley & Sons, New York, 1999.

Deutsch, C.V., *Geostatistical Reservoir Modeling*, Oxford University Press, New York, 2002.

Deutsch, C.V. and Journel, A.G., *GSLIB: Geostatistical Software Library and User's Guide*, 2nd Edition, Oxford University Press, New York, 1998.

Gandin, L.S., *Objective Analysis of Meteorological Fields*, Translated from Russian: Israel Program for Scientific Translations (1965), Jerusalem, Israel, 1963.

Gomez-Hernandez, J.J. and Srivastava, R.M., *ISIM3D: an Ansi-C 3 dimentional multiple indicator simulation*, Computer & Geosciences, vol. 16, no. 4, 1993, p. 395-440.

Goovaerts, P., *Geostatistics for Natural Resources Evaluation*, Oxford University Press, New York, 1997.

Isaaks, E.H. and Srivastava, R.M., *An Introduction to Applied Geostatistics*, Oxford University Press, New York, 1989.

Isaaks, E.H., *The Application of Monte Carlo Methods to the Analysis of Spatially Correlated Data*, PhD Thesis, Stanford University, Stanford, CA, 1990.

Johnson, M.E., *Multivariate Statistical Simulation*, John Wiley & Sons, New York, 1987.

Journel, A.G. and Huijbregts, Ch.J., *Mining Geostatistics*, Academic Press, New York, 1978.

Journel, A.G., *Modeling Uncertainty: Some Conceptual Thoughts*, Geostatistics For the Next Century, Kluwer Academic Publications,1993, p. 30-43.

Journel, A.G. and Xu, W., *Posterior identification of histograms conditional to local data*, Mathematical Geology, vol. 26, no. 3, 1994, p. 323-359.

Leuangthong, O., McLennan, J.A., and Deutsch, C.V., *Minimum Acceptance Criteria for Geostatistical Realizations*, Natural Resources Research, accepted April 2004.

McLennan, J.A., *The Effect of the Simulation Path in Sequential Gaussian Simulation*, Centre for Computational Geostatistics Report Four, University of Alberta, Edmonton, Alberta, 2002.

Soares, A., *Direct Sequential Simulation and Cosimulation*, Mathematical Geology, vol. 33, no. 8, 2001, p. 911-926.

Tran, T.T., *Improving Variogram Reproduction on Dense Simulation Grids*, Computers and Geosciences, vol. 20, no. 7/8, 1994, p. 1161-1168.

Verly, G., *The Block Distribution Given a Point Multivariate Normal Distribution*, Geostatistics for Natural Resources Characterization, Part 1, 1984, p. 495-515.

Xu, W., Tran, T.T., Srivastava, R.M. and Journel, A.G., *Integrating Seismic Data in Reservoir Modeling: The Collocated Cokriging Alternative*, Society of Petroleum Engineers, 1992, SPE 24742.

Xu, W. and Journel, A.G., *DSSIM: A General Sequential Simulation Algorithm*, Stanford Center for Reservoir Forecasting Stanford University, 1994.

Zanon, S., *Advanced Aspects of Sequential Gaussian Simulation*, MSc Thesis, University of Alberta, Edmonton, Alberta, 2004.

Quantitative Geology and Geostatistics

1. F.M. Gradstein, F.P. Agterberg, J.C. Brower and W.S. Schwarzacher: *Quantitative Stratigraphy*. 1985   ISBN 90-277-2116-5
2. G. Matheron and M. Armstrong (eds.): *Geostatistical Case Studies*. 1987
    ISBN 1-55608-019-0
3. *Cancelled*
4. M. Armstrong (ed.): *Geostatistics*. Proceedings of the 3rd International Geostatistics Congress, held in Avignon, France (1988), 2 volumes. 1989
    Set ISBN 0-7923-0204-4
5. A. Soares (ed.): *Geostatistics Tróia '92*, 2 volumes. 1993   Set ISBN 0-7923-2157-X
6. R. Dimitrakopoulos (ed.): *Geostatistics for the Next Century*. 1994
    ISBN 0-7923-2650-4
7. M. Armstrong and P.A. Dowd (eds.): *Geostatistical Simulations*. 1994
    ISBN 0-7923-2732-2
8. E.Y. Baafi and N.A. Schofield (eds.): *Geostatistics Wollongong '96*, 2 volumes. 1997
    Set ISBN 0-7923-4496-0
9. A. Soares, J. Gómez-Hernandez and R. Froidevaux (eds.): *geoENV I - Geostatistics for Environmental Applications*. 1997   ISBN 0-7923-4590-8
10. J. Gómez-Hernandez, A. Soares and R. Froidevaux (eds.): *geoENV II - Geostatistics for Environmental Applications*. 1999   ISBN 0-7923-5783-3
11. P. Monestiez, D. Allard and R. Froidevaux (eds.): *geoENV III - Geostatistics for Environmental Applications*. 2001   ISBN 0-7923-7106-2; Pb 0-7923-7107-0
12. M. Armstrong, C. Bettini, N. Champigny, A. Galli and A. Remacre (eds.): *Geostatistics Rio 2000*. 2002   ISBN 1-4020-0470-2
13. X. Sanchez-Vila, J. Carrera and J.J. Gómez-Hernández (eds.): *geoENV IV - Geostatistics for Environmental Applications*. 2004
    ISBN 1-4020-2007-4; Pb 1-4020-2114-3
14. O. Leuangthong and C.V. Deutsch (eds.): *Geostatistics Banff 2004*, 2 volumes. 2005
    Set ISBN 1-4020-3515-2